Chemistry: Concepts and Applications

Chemistry: Concepts and Applications

Editor: Molly Lambert

NY RESEARCH PRESS
New York

Published by NY Research Press
118-35 Queens Blvd., Suite 400,
Forest Hills, NY 11375, USA
www.nyresearchpress.com

Chemistry: Concepts and Applications
Edited by Molly Lambert

International Standard Book Number: 978-1-63238-680-9 (Hardback)

Cataloging-in-Publication Data

Chemistry : concepts and applications / edited by Molly Lambert.
p. cm.
Includes bibliographical references and index.
ISBN 978-1-63238-680-9
1. Chemistry. I. Lambert, Molly.
QD31.3 .C44 2019
540--dc23

Contents

Preface

This book was inspired by the evolution of our times; to answer the curiosity of inquisitive minds. Many developments have occurred across the globe in the recent past which has transformed the progress in the field.

Chemistry is the study of the structure, behavior, properties and changes undergone by chemical compounds during a reaction with other compounds. It is focused on the creation of such compounds by understanding the interactions between atoms and molecules through chemical bonds. Chemistry is sub-divided into various branches such as materials chemistry, inorganic chemistry, nuclear chemistry, analytical chemistry, organic chemistry, theoretical chemistry, etc. The study of phases, energy, bonding, chemical reactions, equilibrium, ions and salts, and acidity and basicity are fundamental to the study of chemistry. This field facilitates the understanding of other basic and applied sciences such as botany, geology, astrophysics, forensics and pharmacology, besides many others. There has been rapid progress in this field and its applications are finding their way across multiple industries. This book attempts to understand the multiple branches that fall under the discipline of chemistry and how such concepts have practical applications. Scientists and students actively engaged in this field will find this book full of crucial and unexplored concepts.

This book was developed from a mere concept to drafts to chapters and finally compiled together as a complete text to benefit the readers across all nations. To ensure the quality of the content we instilled two significant steps in our procedure. The first was to appoint an editorial team that would verify the data and statistics provided in the book and also select the most appropriate and valuable contributions from the plentiful contributions we received from authors worldwide. The next step was to appoint an expert of the topic as the Editor-in-Chief, who would head the project and finally make the necessary amendments and modifications to make the text reader-friendly. I was then commissioned to examine all the material to present the topics in the most comprehensible and productive format.

I would like to take this opportunity to thank all the contributing authors who were supportive enough to contribute their time and knowledge to this project. I also wish to convey my regards to my family who have been extremely supportive during the entire project.

Editor

1

Assessing the Quality of Various Preparations of *Calendula officinalis* using High Performance Thin Layer Chromatography

Snezana Agatonovic-Kustrin[1]*, Anindita Chakrabarti[1], David W Morton[1] and Pauzi A Yusof[2]

[1]*School of Pharmacy and Applied Science, La Trobe Institute of Molecular Sciences, La Trobe University, Edwards Rd, Bendigo, 3550, Australia*

[2]*Physiology Department, Medical School, Universiti Teknologi Mara, Selangor, Malaysia*

Abstract

The primary goal of this study was to develop a simple and reliable High-Performance-Thin-Layer Chromatography method to quantitate active ingredients in commercial topical formulations containing *Calendula officinalis* extract and to investigate the affect of different extraction solvents on the overall quality of the formulation.

The developed method was validated for linearity, precision, accuracy, limit of detection, and limit of quantification. It was found that commercially available formulations containing extracts of *Calendula officinalis* have significant variations in both composition and amounts of active pharmaceutical ingredients, due to different extraction procedures employed and the standardization requirements for extracts used in formulation.

Keywords: Marigold; *Calendula officinalis*; High performance thin layer chromatography; Chlorogenic acid

Introduction

Calendula officinalis of the family Asteraceae, native to Eastern Europe and the Mediterranean, has long been used in both traditional and clinical medicine in wound healing and to help relieve skin inflammations and irritations [1-4]. It is commonly known as marigold, gold bloom and holligold. The anti-inflammatory and anti-oedematous properties of *Calendula officinalis* have been linked to the pentacyclic mono-, di- and trihydroxy triterpenoid fatty acid esters, especially the faradiol esters, faradiol 3-O-laurate, faradiol 3-O-palmitate and faradiol 3-O-myristate (Figure 1) [5-10]. The unesterified faradiol produced by hydrolysis, has been found to have the same effect as an equimolar dose of indomethacin which is a Non-Steroidal Anti-Inflammatory Drug (NSAID) [11]. However, the claimed benefits of these herbal formulations cannot be guaranteed in commercially available preparations unless standardised methods of regulation and testing are introduced. Thus, it has been suggested that the concentrations of the triterpenoid fatty acid esters in *Calendula officinalis* formulations may be an effective method to assess and monitor the quality of products on the market [9].

Since the dried ray and disk florets of the *Calendula officinalis* flower contain the greatest quantities of the faradiol esters [6], European Pharmacopoeia recommends the use of the ligulate ray florets in the production of herbal medicinal products [6,12]. Also, growing and harvesting conditions can affect the chemical composition of *Calendula officinalis*. For example in a study conducted in Brazil [13] the total yield of essential oils obtained from Brazilian flowers was significantly lower than that obtained from another study conducted in France [14]. This was attributed to the difficulties of acclimatization of the plant from the lower temperatures in native areas of France when compared to the higher temperatures in Brazil.

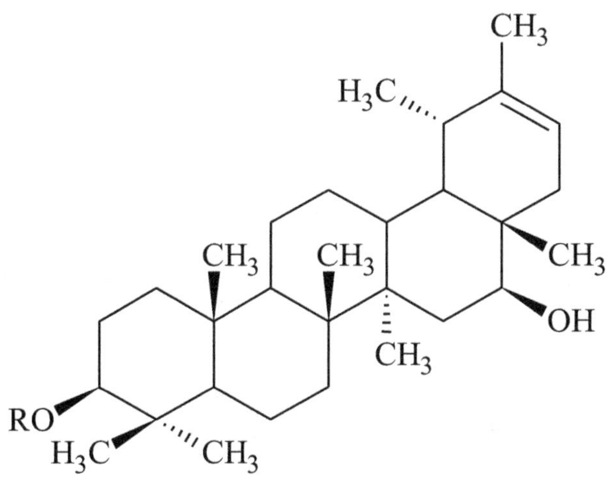

Figure 1: Structure of faradiol 3-O- monoesters; R= laurate, palmitate or myristate.

The main aim of this study was to separate and compare different lipophilic extracts from *Calendula officinalis* flowers using chlorogenic acid as a phytochemical marker. It has been reported that the lipophilic ethyl acetate soluble fraction of the methanol extract of *Calendula officinalis* flowers exhibits the most potent inhibition (84% w/v) of induced inflammation when compared with indomethacin as a reference drug. Alternatively, an aqueous-ethanol extract showed only 20% w/v inhibition of inflammation [8]. However, a number of different extraction methods of *Calendula officinalis*, such as alcoholic and freeze-dried extractions, are used in the manufacture of *Calendula officinalis* formulations. Faradiol esters [9,10,15], the active ingredients, are more soluble in less polar solvents so the method of extraction employed may influence the overall anti-inflammatory activity of *Calendula officinalis* in commercially available products. Another aim of this study was to quantitatively analyze active ingredients in commercially available topical creams and ointments, and to determine how different extraction methods of *Calendula officinalis* flowers affect the overall quality of the product.

Several analytical techniques have been established for the identification and quantification of natural products in herbal

***Corresponding author:** Snezana Agatonovic-Kustrin, School of Pharmacy and Applied Science, La Trobe Institute of Molecular Sciences, La Trobe University, Edwards Rd, Bendigo, 3550, Australia, E-mail: s.kustrin@latrobe.edu.au preparations. However, some of these procedures (e.g. high performance liquid chromatography) are time-consuming, have high operational costs, and also require high technical skills together with sophisticated equipment. Quantitative Thin Layer Chromatography (TLC) has been used widely for analysis of herbal medicinal extracts due to its simplicity of operation, speed, versatility and reproducibility, and relatively low cost, as a number of samples can be analyzed simultaneously on a single plate using only a small amount of solvent as the mobile phase [16-20] (Figure 1).

Experimental

Materials

The standardized *Calendula officinalis* CO_2 extract was obtained from The Herbarie (Batch: 0809, The Herbarie at Stoney Hill Farm, Inc, Prosperity, NC, USA) which contains 6.0-7.0% faradiol esters, 0.25% carotenoids and 0.27% essential oils in fractionated coconut oil. Four different topical formulations, two creams (C1 and C2) and two ointments (O1 and O2) were obtained from manufacturers in Australia and Italy. The samples included the following manufacturers claims; *Calendula officinalis* cream C1 containing Marigold flowers freeze-dried extract 1% (standardized to 1.2% total flavonoids expressed as hyperoside) (Italy), *Calendula officinalis* cream C2 containing 10 mL in 100 g of *Calendula officinalis* extract (1:5) in aqueous cream (Australia), Homeopathic *Calendula officinalis* 10% Ointment O1 (Italy) and *Calendula officinalis* ointment O2 containing *Calendula officinalis* extract equivalent to 71 mg dry flower per 1 g (Australia). Solvents used in the HPTLC mobile phase were ethyl acetate, anhydrous 99.8% (Sigma Aldrich, Germany), n-hexane 95.0% (BDH, England), and/or glacial acetic acid 99.8% (Sigma Aldrich, Germany). Extraction solvents used were dichloromethane 99.5% (Chem-supply, Australia) or ethanol 95.0% (Sigma Aldrich, Germany). Chlorogenic acid or 5-caffeoylquinic acid, minimum 95% (Sigma Aldrich, Germany), was used as an external standard.

Methods

Standard calibration line: A Calendula standard solution was prepared by transferring 5.00 mL of standardized *Calendula officinalis* CO_2 extract of the flower to a 50.0 mL volumetric flask and making it up to volume with dichloromethane. This solution was spotted onto a HPTLC plate in increasing volumes from 0.50-5.0 μL (Figure 2). The plate was developed in a mobile phase of n-hexane: ethyl acetate: acetic acid in a ratio of 20:10:1 (v/v/v). A chlorogenic acid standard curve was prepared by spotting 1.0-10.0 μL of a 1.0 mg/mL chlorogenic acid in ethanol. The plate was developed in a mobile phase of n-hexane: ethyl acetate: acetic acid in a ratio of 20:10:1 (v/v/v).

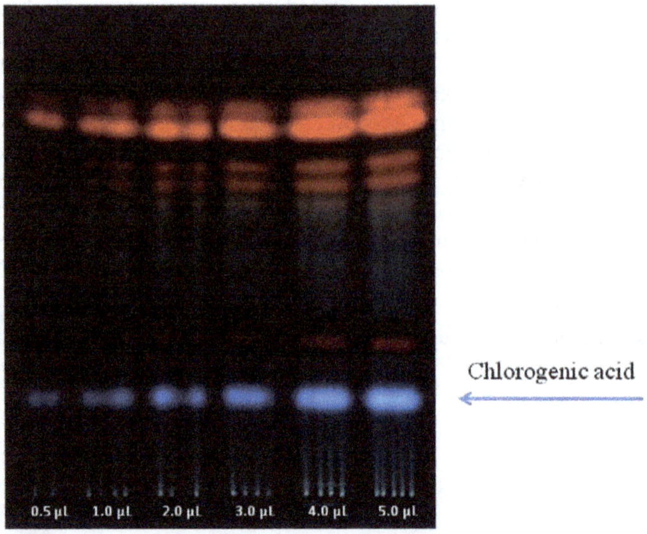

Figure 2: HPTLC chromatogram of standardized *Calendula officinalis* CO_2 extract in dichloromethane solvent. Mobile phase; n-hexane: ethyl acetate: acetic acid in a ratio of 20:10:1 (v/v/v). Spotted from left to right with 0.5 μL, 1 μL, 2 μL, 3 μL, 4 μL, and 5 μL of solution.

Extraction protocols: Samples of organic dried *Calendula officinalis* flowers were obtained from an Australian source (Batch: OCOFOO3EH, All Rare Herbs, Queensland). These flower samples were manually separated to obtain yellow petal (ray floret), green stem (disc floret) and whole flower samples. Ground flower samples (1 g) were extracted either using dichloromethane or ethanol and then filtered using 7.0 cm round filter paper (Whatman, England) and made up to 50 mL in volumetric flasks. Solvent extraction involved heating at approximately 60°C while stirring and then leaving the sample for 24 hours. Freeze dried samples were also prepared using dried flowers that were placed in a freeze drier for 48 hours and then extracted in a similar method using either dichloromethane or ethanol. The extracts stored at 4°C were very stable, having the same qualitative and quantitative composition after 1 month of storage. This is to be expected as flavonoids are substances which generally have a high chemical stability [21].

Alcoholic or dichloromethane extracts from creams and ointments were prepared by mixing 2 g of the formulation in either 15 mL of ethanol or dichloromethane and heating it at 60°C, stirring until dissolved. The solution was then filtered using 7.0 cm round filter paper (No. 1, Whatman, England) into a 25 mL volumetric flask and made up to volume. 10 μL of each extract was directly applied onto HPTLC plates in triplicate and the chromatograms developed and analyzed.

Quantitative HPTLC analysis: Samples were directly applied onto HPTLC Silica gel 60 F254 plates (Merck, Switzerland) (10×10 cm) using a Nanomat 4 applicator (Serial: 1611, Camag®, Muttenz, Switzerland) or using a 10 μL guided plunger syringe (Serial: 10R-GP, SGE, Australia). The distance between the sample bands was set at 10 mm; the distance from the lower plate edge was 8 mm, with 8 bands spotted per plate. Since faradiol esters are not commercially available, a quantitative comparison was carried out according to the external standard method. Chlorogenic acid was used as a reference standard to compare the phytochemical composition in the extracts of flowers and extracts from the commercial formulations.

Plates were developed using saturated vertical twin trough developing chambers (10×10 cm, 10×20 cm) and also using a horizontal (10×10 cm) developing chamber (Camag®, Muttenz, Switzerland) with n-hexane:ethyl acetate:acetic acid (20:10:1, v/v/v) as the mobile phase. The solvent front was allowed to develop until it had travelled a distance of 80 mm which took approximately 25 minutes. After drying the plates with a stream of cold air for 1 minute, the spots were examined using a TLC-Visualiser (Camag®, Muttenz, Switzerland) under transmitted white light at wavelengths of UV 254 nm and UV 366 nm. Images from developed HPTLC plates were digitally captured using a high-resolution 12 bit CCD. Fixed capture parameters (focal length, focus, and aperture) ensured high reproducibility of images from plate to plate [22]. The Just TLC (Sweday®, Sweden) analysis program was used to quantify and compare the area of each spot from the captured chromatogram images.

Results and Discussion

Analytical characteristics of the HPTLC Method

The experimental purpose of this work was to develop a simple and reliable HPTLC procedure for the quantification of active constituents in lipid extracts of *Calendula officinalis*. The first step in chromatographic method development was to select an appropriate mobile phase so that Rf values for the components are between 0.3-0.5. Initial trials with a number of different mobile phases were guided by previous work [5,6,11,23], which included varying ratios of n-hexane, ethanol, ethyl acetate, dichloromethane and acetonitrile. Good separation was previously achieved using a solvent mix of n-hexane: ethyl acetate in a ratio of 8:2 (v/v) [6]. However, the bands tended to merge into each other. Several mobile phases with small amounts of acid modifier to improve separation were investigated. An increase in acidity of the mobile phase by the addition of acetic acid improved the overall separation of compounds. The optimal mobile phase to separate the compounds in *Calendula officinalis* was found to be n-hexane: ethyl acetate: acetic acid in a ratio of 20:10:1 (v/v/v) (Figure 2).

Using the optimal mobile phase mixture, qualitative TLC analysis of the extracts generally revealed seven bands upon development, with the light blue fluorescent zone at an R_f=0.30, corresponding to chlorogenic acid. Note: the chlorogenic acid band is the lowest visible band in Figure 2. Captured color images were visually evaluated and Just TLC software was then used to calculate the area (in pixels) of the bands. The concentration of chlorogenic acid in each extract was determined using a standard curve prepared with chlorogenic acid and was expressed as the weight of chlorogenic acid in 100 mL. It was found that 100 mL of the standardized *Calendula officinalis* CO_2 extract contains 16.750 g of chlorogenic acid or 16.75 % (w/v).

Evaluation of the extraction procedures

Ethanol and dichloromethane were used to extract active components from standard CO_2 extracts of *Calendula officinalis* in order to determine if there is a difference in the amount of active ingredients present in a formulation as a result of the type of solvent used for the extraction. The results of TLC analyses show that dichloromethane extracts dissolved a greater amount of active ingredients from the raw sample than ethanol extracts, as evident in larger band area of chlorogenic acid (2717 pixels with dichloromethane compared to 1827 pixels with ethanol used as extraction solvent). Dichloromethane extracts contained a greater amount and variety of chemical compounds as identifiable, intensely colored bands on the HPTLC plates, in comparison to ethanol extracts. This is supported by previous work that suggests that the active anti-inflammatory faradiol esters in *Calendula officinalis* are more soluble in less polar organic solvents such as dichloromethane [7]. Extractions using individual parts of fresh flowers (i.e. yellow petals and green stems) of *C. officinalis* were performed in order to determine if there are significant differences in the chemical composition of these parts. After extraction (dichloromethane and ethanol), extracts were investigated by HPTLC. Compounds with higher Rf (≥ 0.6) were present in higher amounts in the extracts obtained from the yellow petals, while green stem extracts contained higher amounts of chlorogenic acid (R_f=0.30).

The influence of freeze-drying on sample preparation was analyzed by comparing freeze-dried to non-freeze-dried samples. The non-freeze-dried samples contained an identifiable band for chlorogenic acid, which was absent from the freeze-dried samples. The samples that had not been freeze-dried also displayed bands of greater intensity. This indicates that extraction using dichloromethane and ethanol solvents may be more effective using dry flowers that have not been freeze-dried.

Method validation

The method was validated by testing results for linearity, accuracy, precision, reproducibility, ruggedness and specificity [24,25]. The linearity of the method was established by calculating the areas of the bands for chlorogenic acid from plates spotted with varying concentrations of standardized *Calendula officinalis* CO_2 extract. Two bands on the plate, with Rf values of 0.3 (chlorogenic acid) and 0.7 respectively were selected. The calibration curves were constructed by plotting spot area against applied concentration. The linearity of the calibration curves and the adherence of the method to Beer's law are validated by the high value of the correlation coefficient and the value of intercept on ordinate which is close to zero. The calibration lines demonstrated a linearity (coefficient of correlation) of 0.991 (R_f=0.30) and 0.962 (R_f=0.7), respectively, in the investigated volume range 0.05-0.5 µL of standardized extract.

The area of the bands versus applied amount of standardized extract were plotted to obtain the calibration graph. Both bands obeyed the Beer's law within the investigated concentration range. Rather than to test the significance of the intercept of the line, we wanted to investigate if the equations provide an adequate fit to the experimental data. In order to test the overall fit of the linear regression model and the significance of the regression parameters, analysis of variance (ANOVA) was performed. ANOVA analysis was used to test how variations in the response (concentration of dichloromethane standard sample applied to the plates) affect the variation in responses (band area). High correlation coefficients indicate the good fit of the model to a data. However, they are not the best measure of the effectiveness of the model. Thus, we have tested the effectiveness of the areas of the bands in explaining the model and fitting to the concentration by the Fisher variance ratio (F). The F-test is the ratio of two scaled sums of squares reflecting different sources of variability, so that the calculated F value should be smaller than the critical value, if the hypothesis that the model has no predictive capability is true (null hypothesis). The null hypothesis is rejected if the F ratio is large. The values of F for calibration obtained using band 1 and 2 were 221.63 and 49.18, respectively. These F values were significantly greater than the critical tabular value at a given level of confidence (Fcrit=7.71) indicating that there is not a significant amount of variation in the measured areas that is not explained by the model. Thus, there is not a significant lack of fit between the model and the data.

Band 1 corresponding to chlorogenic acid was selected in order to validate the method in accordance with International Conference on Harmonization (ICH) [26,27] guidelines for method validation. The Limit of Detection (LOD) and Limit of Quantification (LOQ) were calculated according to ICH guidelines [27]. The average LOD, the lowest concentration (amount) of analyte in a sample that can be detected but not necessarily quantified, was 0.02156 µL of the standard solution which corresponds to 2.15 nL of the standardized extract. The LOD is frequently confused with the sensitivity of the method. The sensitivity of an analytical method is the capability of the method to discriminate small differences in the concentration of the test analyte. In practical terms, sensitivity is the slope of the calibration curve that is obtained by plotting the response against the analyte concentration. The LOQ or the minimum concentration that produces quantitative response with acceptable precision was 6.533 nL of the standardized extract.

The International Conference on Harmonization (ICH) guideline for method precision [27] requires precision from at least 9 replications covering the complete investigated concentration range. In this study,

Volume (μL)	Sample 1 (Pixels)	Sample 2 (Pixels)	Sample 3 (Pixels)	SD[a]	LOD[b]	LOQ[c]	RSD (%)[d]
1	251	243	251	4.62	0.032	0.098	1.86
2	375	371	375	2.31	0.016	0.049	0.62
3	523	523	527	2.31	0.016	0.049	0.44
			Average	3.08	0.022	0.065	0.97

aStandard deviation
bLimit of detection
cLimit of quantification
dRelative standard deviation (%).

Table 1: Method validation for the determination of chlorogenic acid concentration from plates spotted with varying volumes of standardized *Calendula officinalis* CO_2 extract solution (CH_2Cl_2 solvent).

Cream/ointment extract	Extraction solvent	Overall pixel count for the bands observed on the HPTLC plate	Amount of chlorogenic acid (mg/100g)
C1d	CH_2Cl_2*	1991	1.4
C1e	ethanol	1695	1.23
C2d	CH_2Cl_2	1455	1.03
C2e	ethanol	1129	0.75
O1d	CH_2Cl_2	500	0.23
O1e	ethanol	485	0.21

*CH_2Cl_2=dichloromethane

Table 2: Overall pixel count (n=3) for bands observed for each cream extract on the HPTLC plate.

the results were obtained at 3 different concentrations with 3 replicates at each concentration (Table 1). Instrumental precision was checked by replicate scanning of the same bands of a selected concentration of *Calendula officinalis* standardized extract and expressed as percentage Relative Standard Deviation (RSD, %) of areas; the average RSD (n=9) was 0.97%. The variability of the method was studied by analyzing selected diluted extracts on the same day (intraday precision), and on different days after an interval of three days (interday precision); results were then again expressed as RSD (%). The intraday precision was found to be 0.65% and the interday precision was found to be 1.69%, highlighting the unstable nature of the *Calendula officinalis* samples. This variability was reduced by spotting samples on the day of preparation and storing the images of HPTLC plates digitally (Table 1).

Analysis of topical formulations

The proposed method was successfully applied to 4 topical formulations, two creams (C1 and C2) and two ointments (O1 and O2). A number of the components in each extract were separated using HPTLC and the concentration of chlorogenic acid (mg in 100 g cream/ointment) determined. As indicated in Table 2, sample extracts from C1 (Italian cream) obtained with dichloromethane (C1d) and with ethanol (C1e) have significantly higher concentrations of chlorogenic acid compared to sample extracts C2d and C2e (Australian creams). The data in Table 2 also indicates that ethanol was less effective compared to dichloromethane in extracting chlorogenic acid from the cream and ointment samples. For the ointments, chlorogenic acid was only quantifiable in the homeopathic Italian ointment extracts (O1d and O1e) in contrast to the Australian ointment extracts where identifiable spots for chlorogenic acid were not visible. Note that the Italian ointment extracts contained much lower concentrations of chlorogenic acid (around 6 times less) compared to the creams investigated (Table 2).

In general, the results of the HPTLC analyses, indicates that the Italian cream has a greater number of high intensity identifiable bands when compared to the Australian cream regardless of the extraction solvent used.

Conclusion

This study has demonstrated the need for standardized testing and regulation of herbal products available to consumers. The commercially available formulations of *Calendula officinalis* have been shown to significantly differ in composition. This work demonstrates that the variations between extraction methods and pharmaceutical dosage forms have considerable impact on the concentration of active ingredients in that product. The literature to support the efficacy of *Calendula officinalis* as an anti-inflammatory cream is strongly supported by evidence [9]; however, the benefits of these available products cannot be guaranteed if appropriate quality control measures and standards are not implemented. Scope for future study includes the development of methods to purify and obtain samples of faradiol esters present in *Calendula officinalis* in order to quantify the amounts of these esters present in *Calendula officinalis* cream and ointment formulations. This study has developed a simple and reliable method for the HPTLC separation of compounds in *Calendula officinalis* formulations. The method is both rapid and sensitive which makes it ideal for routine analysis of formulations containing *Calendula officinalis* extracts.

References

1. Basch E, Bent S, Foppa I, Haskmi S, Kroll D, et al. (2006) Marigold (*Calendula officinalis* L.): an evidence-based systematic review by the Natural Standard Research Collaboration. J Herb Pharmacother 6: 135-159.
2. Muley BP, Khadabadi SS, Banarase NB (2009) Phytochemical Constituents and Pharmacological Activities of *Calendula officinalis* Linn (Asteraceae): A Review. Trop J Pharm Res 8: 455-465.
3. Fuchs SM, Schliemann-Willers S, Fischer TW, Elsner P (2005) Protective effects of different marigold (*Calendula officinalis* L.) and rosemary cream preparations against sodium-lauryl-sulfate-induced irritant contact dermatitis. Skin Pharmacol Physiol 18: 195-200.
4. Fronza M, Heinzmann B, Hamburger M, Laufer S, Merfort I (2009) Determination of the wound healing effect of Calendula extracts using the scratch assay with 3T3 fibroblasts. J Ethnopharmacol 126: 463-467.
5. Hamburger M, Adler S, Baumann D, Förg A, Weinreich B (2003) Preparative purification of the major anti-inflammatory triterpenoid esters from Marigold (*Calendula officinalis*). Fitoterapia 74: 328-338.
6. Zitterl-Eglseer K, Reznicek G, Jurenitsch J, Novak J, Zitterl W, et al. (2001) Morphogenetic variability of faradiol monoesters in marigold *Calendula officinalis* L. Phytochem Anal 12: 199-201.
7. Baumann D, Adler S, Griiner S, Otto F, Weinreich B, et al. (2004) Supercritical carbon dioxide extraction of marigold at high pressures: comparison of analytical and pilot-scale extraction. Phytochem Anal 15: 226-230.

8. Ukiya M, Akihisa T, Yasukawa K, Tokuda H, Suzuki T, et al. (2006) Anti-inflammatory, anti-tumor-promoting, and cytotoxic activities of constituents of marigold (*Calendula officinalis*) flowers. J Nat Prod 69: 1692-1696.

9. Della Loggia R, Tubaro A, Sosa S, Becker H, Saar S, et al. (1994) The role of triterpenoids in the topical anti-inflammatory activity of *Calendula officinalis* flowers. Planta Med 60: 516-520.

10. Neukirch H, D'Ambrosio M, Sosa S, Altinier G, Della Loggia R, et al. (2005) Improved anti-inflammatory activity of three new terpenoids derived, by systematic chemical modifications, from the abundant triterpenes of the flowery plant *Calendula officinalis*. Chem Biodivers 2: 657-671.

11. Zitterl-Eglseer K, Sosa S, Jurenitsch J, Schubert-Zsilavecz M, Della Loggia R, et al. (1997) Anti-oedematous activities of the main triterpendiol esters of marigold (*Calendula officinalis* L.). J Ethnopharmacol 57: 139-144.

12. (2007) European Pharmacopoeia, sixth edition, Directorate for the Quality of Medicines & HealthCare of the Council of Europe (EDQM), Strasbourg.

13. Gazim ZC, Rezende CM, Fraga SR, Filho BPD, Nakamura CV, et al. (2008) Analysis of the essential oils from *Calendula officinalis* growing in Brazil using three different extraction procedures. Rev Bras Cienc Farm 44: 391-395.

14. Chalchat JC, Garry RPH, Michet A (1991) Chemical composition of essential oil of *Calendula officinalis* L. (pot marigold). Flavour and Fragrance Journal 6: 189-192.

15. Loggia RD, Becker H, Issac O, Tubaro A (1990) Topical anti-inflammatory activity of *Calendula officinalis* extracts. Planta Med 56: 658-659.

16. Kivçak B, Akay S (2005) Quantitative determination of alpha-tocopherol in *Pistacia lentiscus, Pistacia lentiscus* var. chia, and *Pistacia terebinthus* by TLC-densitometry and colorimetry. Fitoterapia 76: 62-66.

17. Kivçak B, Mert T (2001) Quantitative determination of alpha-tocopherol in *Arbutus unedo* by TLC-densitometry and colorimetry. Fitoterapia 72: 656-661.

18. Apers S, Naessens T, Pieters L, Vlietinck A (2005) Densitometric thin-layer chromatographic determination of aescin in a herbal medicinal product containing Aesculus and Vitis dry extracts. J Chromatogr A 1112: 165-170.

19. Gunther M, Schmidt PC (2005) Comparison between HPLC and HPTLC-densitometry for the determination of harpagoside from *Harpagophytum procumbens* CO(2)-extracts. J Pharm Biomed Anal 37: 817-821.

20. Fang C, Wan X, Jiang C, Cao H (2005) Comparison of HPTLC and HPLC for determination of isoflavonoids in several kudzu samples. JPC - Journal of Planar Chromatography - Modern TLC 18: 73-77.

21. Crawford DJ, Giannasi DE (1982) Plant chemosystematics. Bioscience 32: 114.

22. (2010) CAMAG Laboratory, Parameters of Planar Chromatography.

23. Cetkovic GS, Djilas SM, Canadanovic-Brunet JM, Tumbas VT (2003) Thin-layer chromatography analysis and scavenging activity of marigold (*Calendula officinalis* L.) extracts. APTEFF 34: 93-102.

24. Reich E, Schibli A, DeBatt A (2008) Validation of high-performance thin-layer chromatographic methods for the identification of botanicals in a cGMP environment. J AOAC Int 91: 13-20.

25. Kumar V, Mukherjee K, Kumar S, Mal M, Mukherjee PK (2008) Validation of HPTLC method for the analysis of taraxerol in *Clitoria ternatea*. Phytochem Anal 19: 244-250.

26. Ferenczi-Fodor K, Végh Z, Nagy-Turák A, Renger B, Zeller M (2001) Validation and quality assurance of planar chromatographic procedures in pharmaceutical analysis. J AOAC Int 84: 1265-1276.

27. Green JM (1996) A practical guide to analytical method validation, International Conference on Harmonization (ICH) of Technical Requirements for the Registration of Pharmaceuticals for Human Use. Anal Chem, Geneva, pp. 305A-309A.

2

Cartilage Oligomeric Matrix Protein as New Marker in Diagnosis of Rheumatoid Arthritis

Mohamed I Aref[1] and Hamdy Ahmed[2*]

[1]*Clinical pathology Department, Al Azhar University, Cairo, Egypt*
[2]*Biochemistry of National Research Center (NRC), Al Azhar University, Egypt*

Abstract

Rheumatoid arthritis (RA) is the most common systemic inflammatory autoimmune disease of unknown etiology. The early in diagnosis of RA is crucial. To facilitate diagnosis during the early stages of RA, when often not all clinical symptoms are manifest, a good serological marker is needed. Among serological markers are rheumatoid factor (RF), erythrocyte sedimentation rate (ESR) C-reactive protein (CRP), anti-cyclic citrullinated peptites (anti-CCP) and cartilage oligomeric matrix protein (COMP). A comparison between those markers in respect to the accuracy was the aim of this study.

Patient and methods

Sixty patients with RA and auto-immune non-RA were selected for this study compared with 20 normal healthy persons. The results showed both COMP and anti-CCP can be help for diagnostic value than other selected parameter.

Keywords: Rheumatic arthritis; Serological markers; Cartilage; Oligomeric matrix protein

Introduction

Rheumatoid arthritis (RA) is a chronic systemic autoimmune inflammatory disorder characterized by inflammation and destruction of articular structures in disorder in association with extra-articular manifestation [1] such as: nodules [2], muscle weakness, nervous system [3], vasculitis, hematological abnormalities [4], skin disease e.g neutrophil dermatitis, ocular [5], lung [6], cardiac [7] and other organs could be involved . Joint destruction in RA results from the invasion of the cartilage and subchondral bone by the hyperplastic synovium, with synovial fibroblasts and inflammatory cells such as macrophages and T cells key rules in this process [8]. Proliferation of synovial membrane following infiltration by immune cells is thought to results in degradation of articular cartilage and bone, causing irreversible damage [9].

Diagnosis of RA depends on a constellation of signs and symptoms that can be supported by serology and radiographs, where involvement of small joints of the hands and feet is often the key of diagnosis [10]. But there's a difficulty in making an early diagnosis for RA, as inflammatory arthritis is a common manifestation of many conditions. Moreover, the classical clinical pattern of RA tend to emerge over time, or incomplete pattern often present in the first few months or even years. Additional, symptoms and signs may be masked by empirical treatment with anti-inflammatory drugs or corticosteroids [11]. Moreover, the damage may be progress in spite of decreased inflammatory activity and erosion may develop in patients without clinical signs of inflammation [12].

Laboratory tests such as ESR and CRP provide useful information for disease activity but are not specific to joint inflammation and correlate poorly with cartilage damage [13]. The presence of RF was used before as a diagnostic marker but now RF titer used as diagnostic and prognostic value in the evaluation of RA. The positive RF test can occur with other diseases such as systemic lupus erythematosus (SLE), Sjogren's syndrome, cryglobulinemia, polymyositis, psoriatic arthritis, scleroderma, polymyagia rheumatic, viral infections, active tuberculosis, tumor, Lyme disease, autoimmune thyroid disease [14].

During the last years a variety of circulating non-RA antibodies have been discovered and reported to be potential diagnostic value. Most of them neither could nor demonstrate to have adequate sensitivity and specificity to form a basic for clinical and therapeutic decisions [15].

Based on the knowledge that mature filaggrine is the target of the AFP and AKA antibodies, synthetic citrulline-containig peptides were developed and tested for their reactivity with RA sera [15]. Citrulline is a nonstandared amino acid, as it is not incorporated into proteins during protein translation. It can be generated by post-translational modification of arginine residues by peptidyl-arginine deiminase enzymes. Antibodies against citrulline-containing peptide which was derived from filaggrin sequences can be detected in up 48% of RA sera with 98 specificity [15,16].

RA sera showed a remarkable variety in the reactivity pattern towered different citrulline-containing peptides, indicating that the amino acids flanking the citrulline residue are important for the antigenicity of the epitope and that anti-citrullinated protein activities such as AFP, AKA and anti-CCP are strongly polyclonal responses. It has been established that these antibodies are produced locally in the synovium of RA patients. However, anti-CCP is now a golden test for diagnosis of RA [17].

A valuable approach to monitor RA would be measuring biological markers of cartilage degradation and repair to reflect variations in joint remodeling. One such potential biological marker of arthritis is cartilage oligomeric matrix protein (COMP). This marker is released into the synovial fluid and other body fluids such as blood. In various studies, COMP has shown promise as a diagnostic and prognostic indicator as a marker of disease severity and the effect of treatment. The present study aimed to evaluate a laboratory marker, COMP in diagnosis of RA and comparing it with other laboratory markers.

***Corresponding author:** Hamdy A Ahmed, Biochemistry of National Research Center (NRC), Al Azhar University, Cairo, Egypt
E-mail: drhamdyahmed@yahoo.com

Material and Methods

This study was conducted on 80 patients divided into three groups:

The 1st group included 40 patients (33 females and 7 males) with RA, fulfilling the American College of Rheumatology diagnostic criteria [18], as a test group.

Inclusion criteria: Patients reserved non-steroidal anti-inflammatory drugs.

Other diseases must be considered which appear similar to RA include: [19],

- Spondyloarthropathies: Ankylosing spondylitis, enteric infections, inflammatory bowel disease, psoriatic arthritis, Reiter's arthritis, Whipple's disease.
- Infections cause acute rheumatic fever, bacterial endocarditis, gonococcal arthritis, Lyme disease. Viral infections (HIV, HBC)
- Metabolic and endorine causes: arthritis of thyroid disease, Gout, hemochromatosis, Pseudogout
- Connective tissue diseases: dermatomycositis, polymyalgia rheumatic, polymyositis , sclerodermia, Still's disease, systemic lupus erythematosus
- Other diseases that can mimic RA: amyloidosis, angioimmunoblastic lymphadenopathy, arthritis associated with oral contraceptives, malignancy, sarcoidosis.

Exclusion criteria: Patient under treatment of cortisone or biological treatment.

The 2nd group included 20 patients (13 females and 7 males) with other rheumatic diseases as SLE, vacuities, dermatomyositis, systemic sclerosis, mixed connective tissue disease and reactive arthritis, as a pathological diseases; a control diseases group with other autoimmune diseases.

The 3rd group included 20 apparently healthy subjects (11 female and 9 male) as a normal control group.

All groups were matched as regard age and sex.

Morning blood samples were collected. CPC, ESR, liver and kidney function tests, CRP and RF were done in the same day of blood collection while the rest of blood left to clotte and centrifuge at 3000. rpm for 15 minutes. Serum was separated and kept in refrigerator at -C20 for other parameters (anti-CCP and COMP).

Complete blood count was done by Sysmes XT2000i series. Liver and kidney function tests (ALT, AST, and Urea creatinine were done by ADVIA 1800 chemistry system.

ESR was done by Wester's method using Westergrent's tube. CRP assay is designed for the quantitative measurement by nephelometry using Hs-CRP reagent on BN-ProSpec Nephelometer.

Qualitative determination of rheumatoid factor (RF) was determined by latex agglutination test provided by SPINREACT, Spain, for research and diagnostic products. Measurement of anti-CCP 2 IgG using ELISA kit provided by the Binding Site Ltd, for research and diagnostic products.

Variable measurement of cartilage oligomeric matrix protein (COMP) using ELISA kit provided Bio Vender-laboratories medicine for research and diagnostic products. Radiological investigations were done on both hands and feet and assessed for erosions and for joint space narrowing in both hand and feet joints.

Results were tabulated and statistical analysis was performed with statistical package for social science (SPSS version 13). All data are expressed as mean ± SD.

The Tests used were :- X mean, SD, Student's T test for testing statistical significant difference between means of two samples. Specificity, sensitivity, positive and negative predictive value accuracy and Pearson correlation test were used. Significant result is considered if $P<0.05$. Highly significant result is considered if $P<0.01$.

Results

The comparison between the RA group and the other autoimmune diseases group for CRP and ESR showed no statistical significant difference while those 2 groups showed a significant difference when compared to the control healthy group (Tables 1-3).

The comparison between the RA group and the other autoimmune diseases group for RF, Anti-CCP and COMP showed a statistical significant difference moreover those 2 groups showed significant differences when compared to the control healthy group (Tables 4-6).

There was high significant positive correlation in group I (RA patient) between anti-CCP versus COMP, CRP, and ESR. While anti-

Variable	ESR group I	ESR group II	CRP group1	CRP group II
Mean ± SD	45.7 ± 23.4	57.4 ± 31.5	49.7 ± 44.8	45.5 ± 43.6
T test	1.632		0.437	
P value	>0.5		>0.5	

Table 1: The comparison between group 1 and group II for ESR and CRP.

Variable	ESR group I	ESR group III	CRP group1	CRP group III
Mean ± SD	45.7 ± 23.4	7.4 ± 4.3	49.7 ± 44.8	2.7 ± 1.4
T test	4.2201		8.3231	
P value	<0.1		<0.1	

Table 2: The comparison between group 1 and group III for both ESR and CRP.

Variable	ESR group II	ESR group III	CRP group1l	CRP group III
Mean ± SD	57.4 ± 31.5	7.4 ± 4.3	45.5 ± 43.6	2.7 ± 1.4
T test	8.333		4.2663	
P value	<0.1		<0.1	

Table 3: The comparison between group II and group III.

CCP had no correlation with other variable in the other autoimmune diseases group II.

There was a high significant positive correlation between COMP versus ESR in RA patients, while there were no significant correlations between COMP and all variables in the other autoimmune disease group (Tables 7-9).

Discussion

Diagnosis of RA depends on a constellation of signs and symptoms that can be supported by serology and radiographs, where involvement of small joints of the hands and feet is often the key of the diagnosis [10]. But there's a difficulty in making an early diagnosis for RA, as inflammatory arthritis is a common manifestation of many conditions. Moreover, the classical clinical pattern of RA tends to emerge over time, or incomplete pattern often present in the first few months or even years. Additionally, symptoms and signs may be masked by empirical treatment with anti-inflammatory drugs or corticosteroids [11].

Recently COMP test was introduced as new marker for diagnosis and prognosis of RA. A comparison between COMP and the well-known markers such as ESR, CRP, RF and anti-CCP was the aim. The present work showed a significant increase of ESR and CRP in RA group compared to control healthy group while there is no significant difference between group I and group II , that is to say those markers cannot used for diagnosis .

As regard to RF, the finding concluded a high positive percentage between RA group compared to either non-RA group or healthy control group. On the other hand a lack of accuracy make RF is out of choice as a marker for diagnosis.

A comparison between RF and COMP were concordance with a cross-sectional study by Andrade et al. and Heidari et al. [20,21]. The average levels of COMP and anti-CCP was superior than RF. Skoumal et al. [22] suggested that this marker can used for prediction in diagnosis of RA in addition to joint destruction.

A comparison between COMP and anti-CCP showed that COMP is more or less in accuracy with anti-CCP. All available data indicate variation in sensitivity and specificity of COMP and anti-CCP across different studies [23,24].

Variable	RF group I	RF group II	Anti-CCP group I	Anti-CCP group II	COMP group I	COMP group II
Mean ± SD	60%	20%	874 ± 741.9	8.2 ± 6.9	1110.1 ± 536.4	44.8 ± 233.7
T test			91.5411		98.41341	
P value			<0.1		<0.1	

Table 4: The comparison between group 1 and group II for RF, anti CCP and COMP.

Variable	RF group I	RF group III	Anti-CCP group I	Anti-CCP group III	COMP group I	COMP group III
Mean ± SD	60%	0%	874 ± 741.9	6.7 ± 5.9	1110.1 ± 536.4	100.3 ± 1.4
T test			31.1461		43.3684	
P value			<0.1		<0.1	

Table 5: The comparison between groups 1 and group III for RF, CCP and COMP.

Variable	RF group II	RF group III	Anti-CCP group II	Anti-CCP group III	COMP group II	COMP group III
Mean ± SD	20%	0%	8.2 ± 6.9	6.7 ± 5.9	44.8 ± 233.7	100.3 ± 1.4
T test						
P value			<0.5		<0.5	

Table 6: The comparison between group II and group III for RF, anti-CCP and COMP.

Variable	Anti-CCP	COMP	RF
Sensitivity	92.5%	90%	60
Specificity	95%	92.5%	90
PVP	94.9%	92.3%	85.7
NPV	92.7%	90.2%	69.2
Accuracy	93.8%	91.3%	75

The results showed that the accuracy was towered anti-CCP and COMP.

Table 7: A summary for accuracy between anti-CCP and COMP.

Correlation	Group I (RA) Coefficient correlation r	Group II (other autoimmune disease) Coefficient correlation r
Anti-CCP with ESR	0.44**	0.11
Anti-CCP with CRP	0.41**	0.33
Anti-CCP with COMP	0.8**	-0.01

*p<0.05 **p<0.01

Table 8: Pearson correlation test between anti-CCP and other studied variables in group I (RA) and group II (other autoimmune diseases).

Correlation	Group I (RA) Coefficient correlation r	Group II (other autoimmune disease) Coefficient correlation r
COMP with ESR	0.4**	0.31
COMP with CRP	0.17	0.14

*p<0.05 **p<0.01

Table 9: Pearson correlation test between COMP and other studied variables in group I (RA) and group II (other autoimmune diseases).

Among several factors that could explain the discrepancy in accuracy between diverse studies including the present study is the presence of high proportion of false positive non-RA among controls. However, other factors such as genetic background may be also responsible for these variations. Also, the difference in scale size of various studies may contribute in this discrepancy.

The correlation analysis between both anti-CCCP and COMP with the clinical signs of RA was significant while not significant with other autoimmune diseases. A prospective study by Lindqvist et al. [25], radiographic changes in hands and feet at 5 and 10 years after inclusion were evaluated and compared with several laboratory markers. The markers analyzed were: ESR, CRP, COMP, RF and anti-CCP. Multiple linear regressions with backward elimination were used to determine the prognostic value of the variables. After 5 years, the presence of IgA RF, serum COMP and anti-CCP were significant associated with more severe damage. Baseline COMP and anti-CCP predicted radiographic outcome after 10 years. A stronger prediction was obtained by combining the prognostic factors. A combination of these measures reflecting different aspects of disease process should be useful for evaluating prognosis in individual patients with early RA.

Feyertag et al. [26] evaluated the changes in a local biomarker, the COMP was better correlated to changes in different clinical measurements in RA than those biomarkers other autoimmune diseases. So, COMP is better in assessment of joint status than other markers which may be masked by the treatment. Indeed, the previous conclusion was also supported by Vilim et al. [27] and Skoumal et al. [28]. Furthermore, Tseng et al. [29] described COMP to be specific marker for the cartilage degradation in RA and not related to the nonspecific inflammatory process.

Conclusion

In this study concluded that anti-CCP is not now the sole specific marker for RA patients. The addition of COPM can enhance the diagnosis especially in very early disease.

References

1. Turesson C, O'Fallon WM, Crowson CS, Gabriel SE, Matteson EL (2003) Extra-articular disease manifestations in rheumatoid arthritis: incidence trends and risk factors over 46 years. Ann Rheum Dis 62: 722-727.

2. Ginsberg MH, Genant HK, Yü TF, McCarty DJ (1975) Rheumatoid nodulosis: an unusual variant of rheumatoid disease. Arthritis Rheum 18: 49-58.

3. Matsuki Y, Suzuki K, Tanaka N, Hirose T, Hosoai K, et al. (1994) Amyloidosis secondary to rheumatoid arthritis associated with plexiform change in bilateral temporal lobes. Intern Med 33: 764-767.

4. Voulgari PV, Kolios G, Papadopoulos GK, Katsaraki A, Seferiadis K, et al. (1999) Role of cytokines in the pathogenesis of anemia of chronic disease in rheumatoid arthritis. Clin Immunol 92: 153-160.

5. Ağildere AM, Tutar NU, Yücel E, Coşkun M, Benli S, et al. (1999) Pachymeningitis and optic neuritis in rheumatoid arthritis: MRI findings. Br J Radiol 72: 404-407.

6. CAPLAN A (1953) Certain unusual radiological appearances in the chest of coal-miners suffering from rheumatoid arthritis. Thorax 8: 29-37.

7. Wisłowska M, Sypuła S, Kowalik I (1999) Echocardiographic findings and 24-h electrocardiographic Holter monitoring in patients with nodular and non-nodular rheumatoid arthritis. Rheumatol Int 18: 163-169.

8. Distler JH, Jüngel A, Huber LC, Seemayer CA, Reich CF 3rd, et al. (2005) The induction of matrix metalloproteinase and cytokine expression in synovial fibroblasts stimulated with immune cell microparticles. Proc Natl Acad Sci U S A 102: 2892-2897.

9. Peake NJ, Khawaja K, Myers A, Jones D, Cawston TE (2005) Levels of matrix metalloproteinase-1 in paird and synovial fluids of juvenile idiopathic arthritis patients, MMP-3 and tissue inhibitor of metalloproteinase-1 in longitudinal study. Rheumatology 44: 1383-1389.

10. Firestein GS (2005) Rheumatoid arthritis: etiology and pathogenesis of RA. In: Harris D, Budd C, Firestein S, Genovese C, Sergent S, et al. (eds.)Text book of Rheumatology (7thedn.) Philadelphia, Pennsylvania, USA, pp. 996-1024.

11. Roberts LJ, Cleland LG, Thomas R, Proudman SM (2006) Early combination disease modifying antirheumatic drug treatment for rheumatoid arthritis. Med J Aust 184: 122-125.

12. van den Berg WB (2001) Uncoupling of inflammatory and destructive mechanisms in arthritis. Semin Arthritis Rheum 30: 7-16.

13. Kushner I, Rzewnicki D, Samols D (2006) What does minor elevation of C-reactive protein signify? Am J Med 119: 166.

14. Caspi D, Anouk M, Golan I, Paran D, Kaufman I, et al. (2006) Synovial fluid levels of anti-cyclic citrullinated peptide antibodies and IgA rheumatoid factor in rheumatoid arthritis, psoriatic arthritis, and osteoarthritis. Arthritis Rheum 55: 53-56.

15. Schellekens GA, de Jong BA, van den Hoogen FH, van de Putte LB, van Venrooij WJ (1998) Citrulline is an essential constituent of antigenic determinants recognized by rheumatoid arthritis-specific autoantibodies. J Clin Invest 101: 273-281.

16. Girbal-Neuhauser E, Durieux JJ, Arnaud M, Dalbon P, Sebbag M, et al. (1999) The epitopes targeted by the rheumatoid arthritis-associated antifilaggrin autoantibodies are post-translationally generated on various sites of (pro) filaggrin by deamination of arginine residues. J Immunol 162: 585-594.

17. Rantapää-Dahlqvist S, de Jong BA, Berglin E, Hallmans G, Wadell G, et al. (2003) Antibodies against cyclic citrullinated peptide and IgA rheumatoid factor predict the development of rheumatoid arthritis. Arthritis Rheum 48: 2741-2749.

18. Aletaha D, Neogi T, Silman AJ, Funovits J, Felson DT, et al. (2010) 2010 rheumatoid arthritis classification criteria: an American College of Rheumatology/European League Against Rheumatism collaborative initiative. Ann Rheum Dis 69: 1580-1588.

19. Hoffman GS (1978) Polyarthritis: the differential diagnosis of rheumatoid arthritis. Semin Arthritis Rheum 8: 115-141.

20. Andrade FD, Bender AL, da Silveira IG, Stein H, von Mühlen CA, et al. (2009) Cartilage oligomeric matrix protein/thrombospondin-5 (COMP/TSP-5) levels do not correlate to functional class in patients with rheumatoid arthritis. Clin Rheumatol 28: 1441-1442.

21. Heidari B, Firouzjahi A, Heidari P, Hajian K (2009) The prevalence and diagnostic performance of anti-cyclic citrullinated peptide antibody in rheumatoid arthritis: the predictive and discriminative ability of serum antibody level in recognizing rheumatoid arthritis. Ann Saudi Med 29: 467-470.

22. Skoumal M, Haberhauer G, Feyertag J, Kittl EM, Bauer K, et al. (2004) Serum levels of cartilage oligomeric matrix protein are elevated in rheumatoid arthritis, but not in inflammatory rheumatic diseases such as psoriatic arthritis, reactive arthritis, Raynaud's syndrome, scleroderma, systemic lupus erythematosus, vasculitis and Sjögren's syndrome. Arthritis Res Ther 6: 73-74.

23. Söderlin MK, Kastbom A, Kautiainen H, Leirisalo-Repo M, Strandberg G, et al. (2004) Antibodies against cyclic citrullinated peptide (CCP) and levels of cartilage oligomeric matrix protein (COMP) in very early arthritis: relation to diagnosis and disease activity. Scand J Rheumatol 33: 185-188.

24. Sharif SK, Gharibdoost F, Kbarian MA, Shahram F, Nadji A, et al. (2007) Comparative study of anti-CCP and RF for the diagnosis of rheumatoid arthritis. Aplar J Rheumatol 10: 121-124.

25. Lindqvist E, Eberhardt K, Bendtzen K, Heinegård D, Saxne T (2005) Prognostic laboratory markers of joint damage in rheumatoid arthritis. Ann Rheum Dis 64: 196-201.

26. Feyertag J, Haberhouer GF, Kittl EM, Bauer K, Skoumal M, et al. (2003) Changes in clinical scorings are correlated to changes in cartilage oligomeric matrix protein (COMP) levels, but not to systemic inflammatory markers in patients with rheumatoid arthritis (RA). Ann Rheum Dis 61 (suppl 1): 75.

27. Vilím V, Olejárová M, Machácek S, Gatterová J, Kraus VB, et al. (2002) Serum levels of cartilage oligomeric matrix protein (COMP) correlate with radiographic progression of knee osteoarthritis. Osteoarthritis Cartilage 10: 707-713.

28. Skoumal M, Kolarz G, Klingler A (2003) Serum levels of cartilage oligomeric matrix protein. A predicting factor and a valuable parameter for disease management in rheumatoid arthritis. Scand J Rheumatol 32: 156-161.

29. Tseng S, Reddi AH, Di Cesare PE (2009) Cartilage Oligomeric Matrix Protein (COMP): A Biomarker of Arthritis. Biomark Insights 4: 33-44.

Biosorption of Pb^{2+} from Natural Water using Date Pits: A Green Chemistry Approach

Salem E Samra[1], Bakir Jeragh[2], Ahmed M EL-Nokrashy[3] and Ahmed A El-Asmy[1,2*]

[1]*Chemistry Department, Faculty of Science, Mansoura University, Mansoura, Egypt*
[2]*Chemistry Department, Faculty of Science, Kuwait University, Kuwait*
[3]*Central Laboratory of drinking water, Dakahliya Comp. for Water, Mansoura, Egypt*

Abstract

Removal of Pb^{2+} ions from aqueous solutions by adsorption onto Date Pits (DP) has been investigated. The date pits (sorbent) may represent an environmental problem. The characteristic parameters (solution pH, initial concentration of Pb^{2+}, sorbent dose, shaking time and temperature) influencing the adsorption process have been examined. pH=7 is found the best one having high floatability of 6 g l^{-1} dose of DP. The Freundlich and Langmuir were applied. The change in Gibbs free energy change ($\Delta G°$), enthalpy ($\Delta H°$) and entropy ($\Delta S°$) were also calculated. Under the optimum experimental conditions employed, the removal of ~95% of Pb^{2+} was attained. The procedure was successfully applied to remove Pb^{2+} from natural water samples. The SEM image of DP before and after lead adsorption shows complete adsorption.

Keywords: Heavy metals; Biosorption; Adsorption

Introduction

Heavy metal toxicity has become a major concern today due to its deleterious effects on health and environment [1,2]. Heavy metals are among the chief pollutants of surface and groundwater [3]. The contamination of natural waters with toxic metals has become one of the major concerns of environmental researchers in recent years due to water importance to environment and mankind [4]. Lead being the most toxic metal, ranks second in the list of prioritized hazardous substances issued by the US Agency for toxic substances and Disease Registry [5]. Lead is attracting wide attention of environmentalists as one of the most toxic heavy metals [6]. Lead has been a major focus in wastewater treatment because it is associated with many health hazards [7]. Lead is an important compound used as an intermediate in several industrial such as plating, paint and dyes, chemicals and allied products, lead acid storage batteries, ceramic and glass industries printing, ammunition, lead smelting and mine tailings, automobile industry, agricultural runoff, chemical spills and municipal wastewaters [2,8,9]. Through the food chain system of soil–plant–animal–human, Pb^{2+} is transferred into animals and human beings [8]. The major biochemical effect of Pb^{2+} is its interference with heme synthesis, which leads to hematological damage [7]. Lead poisoning in human caused severe damage to the kidneys, liver, brain, nervous and reproductive systems. Long term exposure may induce sterility, abortion, and neonatal death [10]. Due to the toxic effects of lead ions, the removal of them from waters and wastewaters is important in terms of protection of public health and environment [11]. Adsorption process was studied and emerged as one of the promising technique due to its low initial cost, simplicity of design, ease of operation and insensitivity to toxic substances [2]. It is necessary to have a low-cost material to treat large volumes of waste water. The use of low-cost sorbents has been thoroughly investigated instead of other more expensive materials, for example, natural and waste materials coming from industrial, agriculture and forestry activities have high capacity for removing metal ions [12]. The removal efficiency of new and inexpensive adsorbents can be tested first in model aqueous solutions (distilled water) and then in spiked ground and drinking water. One cheap and easily available material having possibilities as a suitable sorbent for Pb^{2+} ions is date pits. The date palm, *Phoenix dactylifera*, is the oldest tree known to be cultivated by man. Since ancient times, the date palm has been a significant source of food for both human and livestock [13]. The date tree (*Phoenix dactylifera* L.) is an important staple food and a strategic plant in many arid regions of the world. Date fruits are an important food item, with plenty vitamins and minerals. They are eaten fresh or are dried and stored for later consumption [14]. Individual date fruits contain a pit which, depending on the variety, accounts for about 10% to 18% weight of the fruits. The pits are generally used as complementary feed materials or as a conventional soil fertilizer also used for extracting oil for cosmetic and pharmaceutical purposes. Date pits contained 7.1-10.3% moisture, 5.0-6.3% protein; 9.9-13.5% fat; 65-69% neutral detergent fiber; and 1.0-1.8% ash. Total carbohydrate content of date pits is 71.9-73.4% and 3.8-5.8% total sugars [15,16]. We reported here the use of date pits as an adsorbent for removal of Pb^{2+} from aqueous solutions.

Experimental

Sorbent samples and solutions

The date pits samples used in this study were obtained from some date's factories located in Mansoura city, Egypt. The samples were collected, washed with water and dried for 2 h in large trays in an oven at 125°C, allowed to cool, crushed and sieved with size (25-63 μm). The samples were packed into stoppered bottles and stored in a desiccator for future use. All the solutions were prepared from certified reference materials. Aqueous solutions were prepared in deionized water. The Pb^{2+} stock solution 1002 ± 2 mg/l was prepared from [$Pb(NO_3)_2$] in HNO_3 0.5 mol/l (Merck, Germany). The working solutions were made by diluting with deionized water. Further dilutions were prepared daily as required.

Apparatus

A VWR model 3500 digital shaker was used for shaking solutions. Sartotius digital balance was used for all weights, the infrared spectroscopy was undertaken via a Mattson 5000 FT-IR

***Corresponding author:** Ahmed A El-Asmy, Chemistry Department, Faculty of Science, Kuwait University, Kuwait, E-mail: aelasmy@yahoo.com spectrophotometer using the KBr disc method. The measurements were carried out using Atomic Absorption Spectrophotometer AA240FS (Varian, Australia). Also, the stirring of solutions was performed with a magnetic stirrer, Jenway 1000. The pH was measured using pH meter (symphony, USA) provided with a glass electrode.

Procedure

Unless stated otherwise, all batch sorption experiments were done at room temperature (25 ± 2°C). Known volumes of Pb^{2+} stock solution were pipetted into quick-fit glass bottles containing 0.3 g of DP sorbent in 50 ml aqueous solution to give concentrations ranging from 1 mg l^{-1} to 30 mg l^{-1}. Since the pH of any resulting solution was 3.0, no further controlling of pH was necessary since it was suitable for most adsorption experiments. The resulting solution was then shaken at 250 rpm and samples were taken at fixed time periods (1, 3, 5, 10, 15, 30, 60, 120 min) in order to study the kinetics of the adsorption process. Preliminary experiments showed that 60 min was sufficient for adsorption of Pb^{2+} onto DP. The samples were subsequently filtered off and the residual Pb(II) concentration in the filtrate was determined by atomic absorption spectrometry (at wavelength of 217 nm, lamp current 5 mA, slit width 1 nm, acetylene as fuel, air as support). The percentage adsorption of Pb^{2+} from the solution was calculated from the relationship:

$$\% \text{ Adsorption} = (C_i - C_r)/C_i \times 100$$

Where C_i corresponds to the initial concentration of Pb^{2+} ions and C_r is the residual concentration in the filtrate after shaking for a definite time period. The metal uptake q (mg/g) was calculated as:

$$q = [(C_i - C_r)/m].V$$

Where m is the quantity of sorbent (g) and V is the volume of the suspension (L). To assess the applicability of the procedure, another series of experiments was conducted on 50 ml of clear and pre-filtered natural water samples with an initial pH adjusted to 3.0. These suspensions contained 5 mg l^{-1} Pb^{2+}, ES 0.3 g and were shaken for the optimum time (60 min) at 250 rpm.

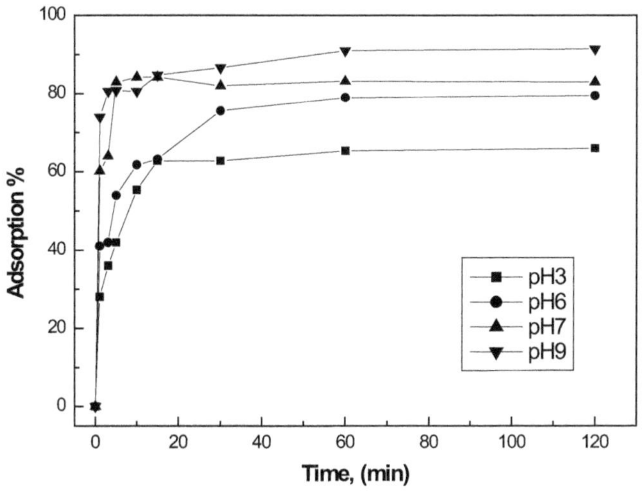

Figure 1: Adsorption % of 5 mg l^{-1} Pb^{2+} by 0.3 g DP vs. shaking time at different pH.

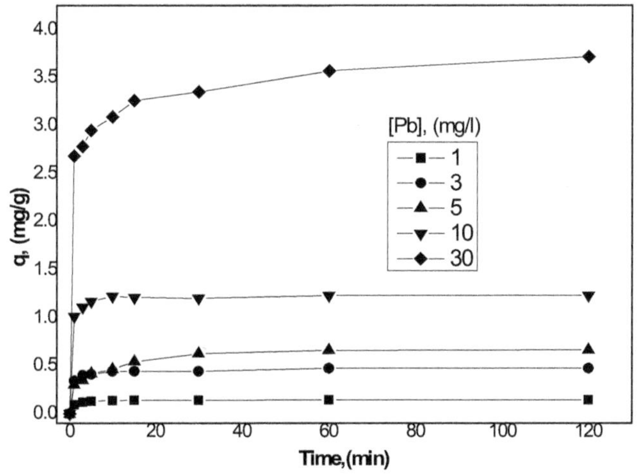

Figure 2: Amount of Pb^{2+} concentration (mg l^{-1}) adsorbed on DP at different shaking times.

Results and Discussion

The IR spectrum of date pits showed bands at 3340-3440 cm^{-1} indicating the stretching of inter and intramolecular hydrogen bonding in cellulose and lignin. The band at 2924 cm^{-1} indicates symmetric or asymmetric ν(C–H) of aliphatic acids. The band at 1728 cm^{-1} due to ν(C=O) is attributed to the non-ionic carboxyl groups ν(COOH or –$COOCH_3$), and may be assigned to the carboxylic acids or their esters. Asymmetric stretching vibrations of ionic carboxylic groups (–COO–), appeared at 1632, and that at 1084 cm^{-1} is assigned to ν(C–OH) of alcoholic groups and carboxylic acids [17,18].

Effect of pH

One of the most important parameters in adsorption process is the pH of the medium. Moreover, the sorption of Pb^{2+} by DP is also influenced by the surface properties of the sorbent and lead species present in aqueous solution. Variation in pH can affect the surface charge of the adsorbent and the degree of ionization and speciation of the metal adsorbate [19]. Pb^{2+} uptakes on DP may involve coordination, ion exchange and adsorption [20]. The pH dependence of the binding showed that ion exchange, electrostatic interactions, hydrogen bond and other phenomena are involved in the binding mechanism [20]. Figure 1 illustrates the effect of pH on Pb^{2+} adsorption by DP. In the ion-exchange mechanism, metal ions bind to anionic sites by displacing protons from acidic groups. In the complexation mechanism, metal ions sequestration is viewed as the coordination of metal ions to surface functional groups [20]. It was confirmed that DP is dominated by negatively charged sites that are largely hydroxyl groups. Pb^{2+} may form complexes with surface functional groups of DP such as cellulose-OH and phenolic-OH through ion-exchange reactions.

Effect of Pb^{2+} concentration and sorbent dose

The effect of initial Pb^{2+} concentration (1-30 mg l^{-1}) on its adsorption by DP (0.3 g) after shaking from 1 to 120 min was shown in Figure 2. It was found that the increase of initial Pb^{2+} concentration, the q [metal uptake (mg/g)] increases. The increase of q with increasing C_0 [initial metal ion concentration] was expected due to the increase of the sorbed ion concentration per unit weight of DP. Moreover, increasing the Pb^{2+} concentration increases the diffusion of Pb^{2+} in the boundary layer resulting in higher sorption by DP. Varying the amounts of DP from 0.05 to 1 g, on the adsorption of Pb^{2+} [5 mg l^{-1}] of pH 6 was depicted in Figure 3. The data show that the adsorption increases as the amount of DP increases. This agrees well with the data shown in

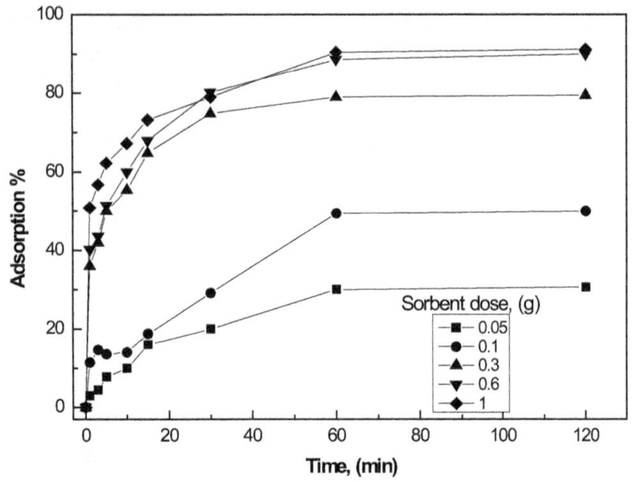

Figure 3: Effect of DP amount on the adsorption % of Pb^{2+} at different shaking times.

Figure 2. Moreover, 6 g l^{-1} dose of DP was an optimum one for further experiments.

Kinetics of the adsorption process

The adsorption of Pb^{2+} is quite rapid at the first stage suggesting that the adsorption occurs mainly at the surface of the sorbent and to some extent by the internal macro-pores, transitional pores and micro-pores. The uptake increases with increasing the initial Pb^{2+} concentration attributing to higher collision between the Pb^{2+} and sorbent. One hour of shaking was found suitable for maximum adsorption. When the data in Figure 2 were re-plotted against the square root of the shaking time ($t^{1/2}$), the obtained linear correlation (Figure 4) may verify the Morris-Weber equation:

$$q = K_d (t)^{1/2}$$

where q is the amount of Pb^{2+} adsorbed (mg/g) and K_d is the initial rate of intraparticle diffusion (mg $g^{-1}min^{-0.5}$). This indicates that an intra-pore diffusion mechanism was involved in the adsorption of Pb^{2+} by DP. Figure 4 shows two distinct regions: an initial linear portion due to the boundary layer effect [21,22] and a second due to intra-particle diffusion effect [23]. However, the fact that the line depicted in Figure 4 is nonlinear indicating that intra-pore diffusion is not the controlling step in sorption of Pb^{2+} by DP [24,25]. The data agree with those of Juang et al. [26]. The value of the rate constant K_d is 0.037 mg $g^{-1}min^{-0.5}$, which gives indication about the mobility of the Pb^{2+} toward the DP surface. The kinetic data was examined by Bangham equation [27]:

$$\text{Log log } [C_i / (C_i - q_m)] = \log (K_o m / 2.303V) + \alpha \text{ Log } t$$

Plot of Log log $[C_i/(C_i-q_m)]$ vs. log t gives a straight line (Figure 5). The results show that the diffusion of Pb^{2+} into DP pores played a role in the adsorption process and is similar to those described elsewhere [28]. The calculated K_o and α constants are 0.0037 and 0.148, respectively. The kinetic data obtained in Figure 5 for Pb^{2+} adsorption by DP were tested by Lagergren equation, as cited by Gupta and Shukla [29]:

$$\log (q_e - q_t) = \log q_e - K_{ads} t / 2.303$$

where q_t is the amount of Pb^{2+} adsorbed (mg g^{-1}) at any given time t (min), qe is the amount of metal ion adsorbed (mg g^{-1}) at equilibrium and K_{ads} is the pseudo-first order reaction rate constant for adsorption (g $mg^{-1}min^{-1}$). The linear plot of log (q_e – q) versus t (Figure 6) shows the first-order nature of the process. The K_{ads} is 0.0925 g $mg^{-1}min^{-1}$. The pseudo second order kinetic model may be expressed by the equation:

$$t/q_t = 1/k_2 q_e^2 + 1\ t/q_e$$

where k_2 (g mg^{-1} min^{-1}) is the equilibrium rate constant for the pseudo-second-order adsorption and can be obtained from the plot of t/qt against t (Figure 7) and it is calculated to be 0.493 (g mg^{-1} min^{-1}).

Also, the kinetics of the adsorption was examined by linear form of Elovich model which describes a number of reaction mechanisms including bulk and surface diffusion and the activation and deactivation of catalytic surfaces. It gives a straight line by plotting (q_t) vs. (ln t). It is represented by [10]:

$$qt = 1/\beta \ln(\alpha\beta) + 1/\beta \ln t$$

Where α (mg $g^{-1}min^{-1}$) and β (mg $g^{-1}min^{-1}$) are the constants of the adsorption and are determined from a plot depicted in Figure 8. Elovich

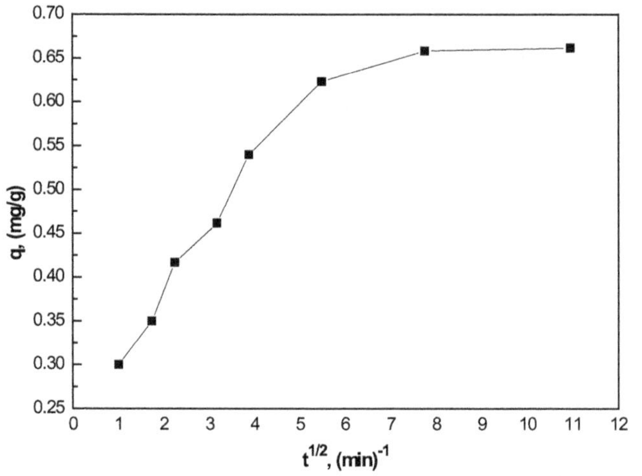

Figure 4: Plot of the amount of Pb^{2+} adsorbed onto 0.3 g DP vs. square root of t at pH 6.

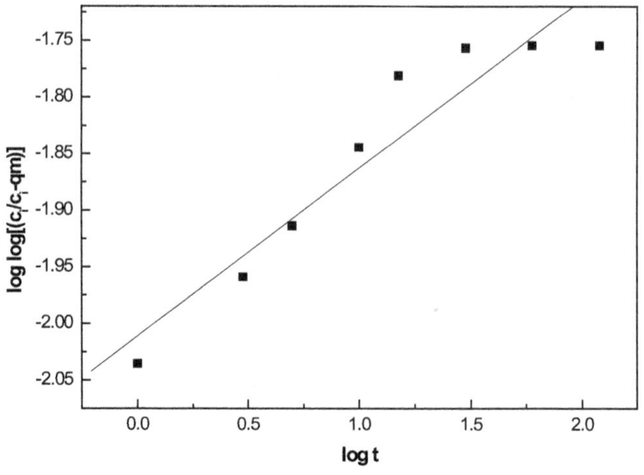

Figure 5: Plot of log $[C_i/(C_i - qm)]$ vs. log t for the adsorption of Pb^{2+} onto 0.3 g DP at pH 6.

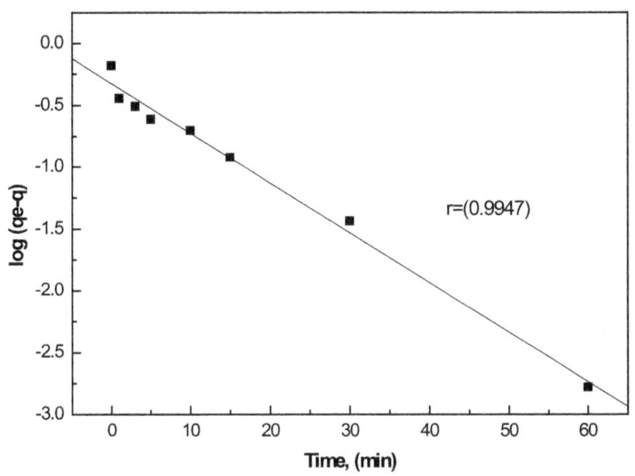

Figure 6: Plot of log (q_e - q) *vs.* stirring time (t) for the adsorption of 5 mg l^{-1} Pb^{2+} by 0.3 g DP at pH 6.

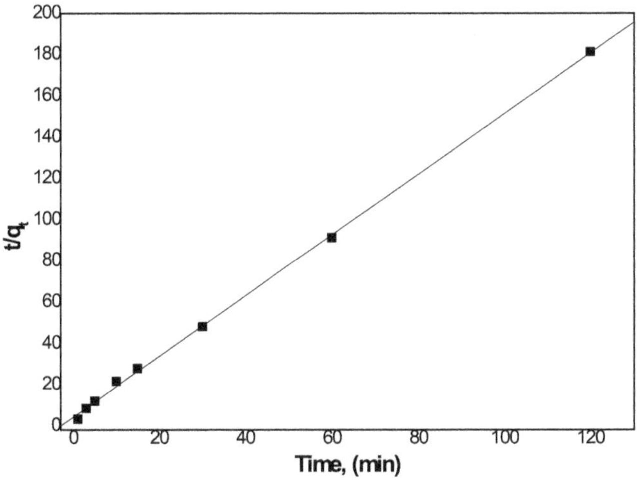

Figure 7: Plot of (t/q_t) vs. stirring time (t) for the adsorption of Pb^{2+} by 0.3 g DP at pH 6.

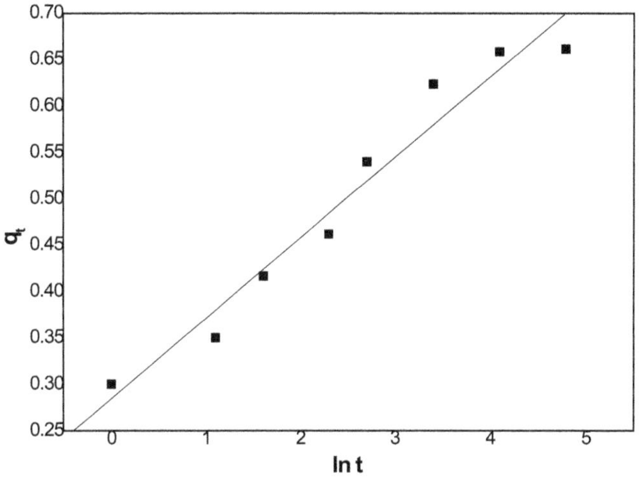

Figure 8: Plot of (q_t) vs. (ln t) for the adsorption of Pb^{2+} by 0.3 g DP at pH 6.

Kinetic model	Parameter	Value
Pseudo-first-order	K_{ads} (g mg^{-1} min^{-1})	0.0925
	$q_{e\,(exp)}$ (mg g^{-1})	0.660
	$q_{e\,(theo)}$ (mg g^{-1})	0.473
	r^2	0.9947
Pseudo-second-order	K_2 (g mg^{-1} min^{-1})	0.493
	$q_{e\,(exp)}$ (mg g^{-1})	0.660
	$q_{e\,(theo)}$ (mg g^{-1})	0.679
	r^2	0.9997
Elovich model	α (mg g^{-1} min^{-1})	0.8335
	β (mg g^{-1} min^{-1})	11.521
	r^2	0.9788
Intraparticle diffusion model	K_d (g/mg min)	0.039
	r^2	0.90
Bangham's equation	K_o	0.0037
	A	0.148
	r^2	0.947

Table 1: Kinetic model parameters for adsorption of Pb^{2+} onto DP.

model is based on a kinetic principle assuming that, the adsorption sites increase exponentially with adsorption which implies a multilayer adsorption. The Elovich coefficients related to initial adsorption rate (a) and surface coverage (b) were calculated to be 0.834 mg $g^{-1}min^{-1}$ and 11.521 mg $g^{-1}min^{-1}$ respectively. All kinetic data for the adsorption of Pb^{2+} onto DP, calculated from the related plots, are summarized in Table 1.

Adsorption isotherms

Adsorption isotherms can be generated based on numerous theoretical models where Langmuir and Freundlich models are the most commonly used. The Langmuir model assumes that the uptake of Pb^{2+} occurs on a homogenous surface by monolayer adsorption without any interaction between the adsorbed ions. The linear form of the Langmuir equation applied to the Pb^{2+} adsorption data in Figure 2 was:

$$1/q_e = 1/q_{max} + (1/\, q_{max}\, k_L)\ 1/\, C_e$$

Where q_e is the amount of Pb^{2+} (mg g^{-1}) adsorbed at equilibrium, C_e is the final equilibrium concentration (mg L^{-1}), k_L is the Langmuir

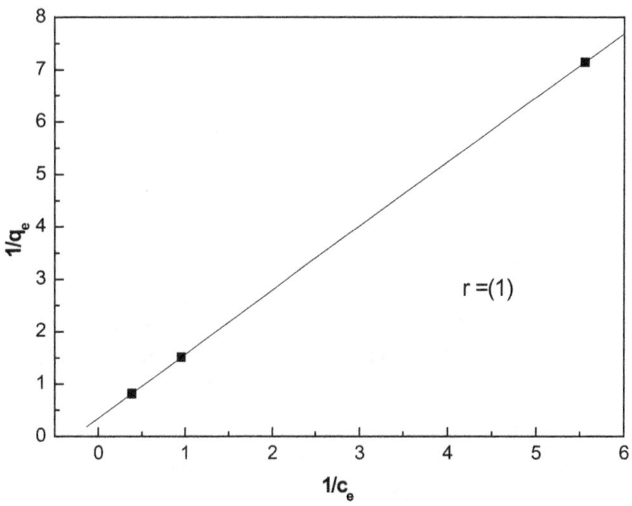

Figure 9: Langmuir plot for Pb^{2+} adsorption onto 0.3 g DP.

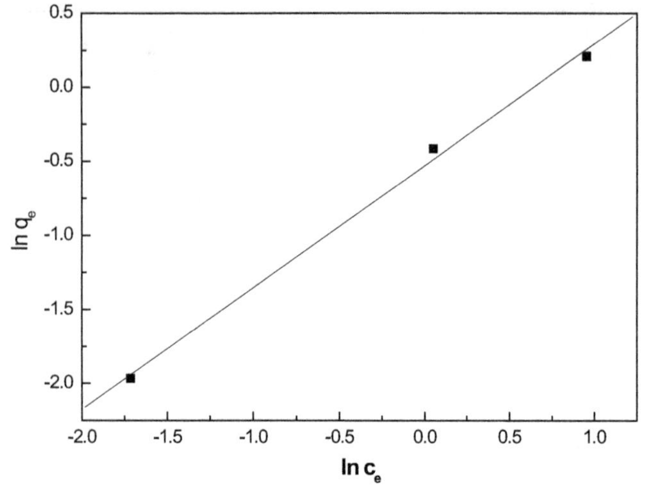

Figure 10: Freundlich plot for Pb^{2+} adsorption onto 0.3 g DP.

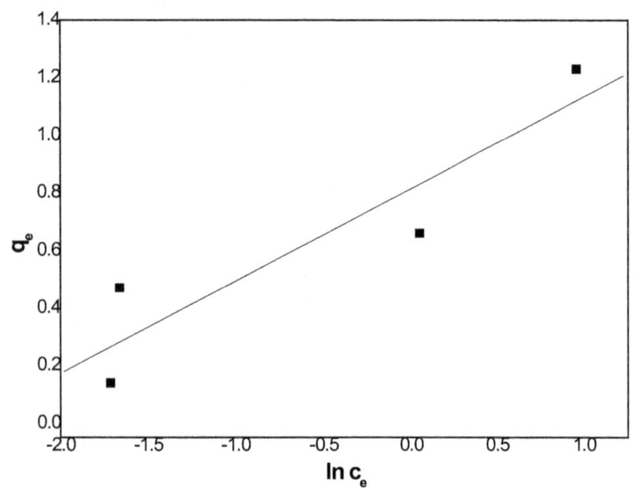

Figure 11: Plot of q_e vs. ln c_e for the adsorption of Pb^{2+} by 0.3 g DP at pH 6.

equilibrium constant (L mg^{-1}) and q_{max} is the maximum adsorption capacity (mg g^{-1}). Figure 9 shows that the plot of $1/q_e$ *vs.* $1/C_e$ gave a straight line suggesting the applicability of the Langmuir model. The values of maximum adsorption capacity (q_{max}), Langmuir cons. (k_L) and the correlation coefficient (r) were 2.891 mg g^{-1}, 0.283 L mg^{-1} and 1, respectively. Also, the equilibrium parameter R_L was found to be 0.0334 indicating that the adsorption process is favorable.

The essential feature of the Langmuir isotherm can be expressed in terms of dimensionless separation cons or equilibrium parameter (R_L) to predict the type of adsorption process, which is defined as: $R_L = 1/(1+K_L C_0)$. The values of R_L indicate the type of isotherm to be irreversible ($R_L=0$), favorable ($0<R_L<1$), linear ($R_L= 1$) or unfavorable ($R_L>1$) [30]. The R_L value was found to be 0.0334, less than 1 and greater than 0.

On the other hand, the Freundlich equation is an empirical equation based on adsorption on a heterogeneous surface. The equation is commonly represented by:

$$\ln q_e = \ln k_F + 1/n \ln C_e$$

Where K_F (l g^{-1}) and 1/n are the Freundlich isotherm constants characteristic of the system, which indicate the adsorption capacity and the adsorption intensity (n), respectively. Figure 10 shows the applicability of this equation on the adsorption of Pb^{2+} on DP. The parameters K_F and n for Pb^{2+} adsorption onto DP were calculated from intercept and the slope of the figure giving values of 0.589 l g^{-1} and 1.216, respectively with a correlation coefficient r=0.998. It was known that favorable adsorption occurs when 1/n>1 [31].

Temkin isotherm is represented by the linear equation as follows:

$$q_e = B \ln K_t + B \ln C_e$$

where K_t (L g^{-1}) is the equilibrium binding constant corresponding to the maximum binding energy and constant B=RT/b represents the heat of adsorption, while R is the universal gas constant, T is the absolute temperature in Kelvin and 1/b indicates the adsorption potential of the adsorbent. Both K_t and B can be determined from a plot q_e versus C_e (Figure 11) [10]. The equilibrium binding cons. (K_t) and the heat of adsorption (b) were calculated from the intercept and slop giving the values of 12.767 L g^{-1} and 7.742 kJ mol^{-1} respectively. The isotherm parameters for the adsorption of Pb^{2+} onto DP are given in Table 2. The sorption data was modeled by Dubinin-Radushkevich (D–R) isotherm

Equilibrium model	Parameter	Value
Langmuir isotherm	q_{max} (mg g^{-1})	2.891
	k_L (l mg^{-1})	0.283
	R_L	0.0334
	r^2	1
Freundlich isotherm	K_F (mg g^{-1})	0.589
	N	1.216
	r^2	0.9983
Temkin isotherm	K_T	12.767
	B (l g^{-1})	0.3204
	b (kJ mol^{-1})	7.742
	r^2	0.9239
Dubinin-Radushkevich model	q_{max} (mol g^{-1})	2.4×10^{-6}
	β (mol^2 k J^{-2})	0.199
	E (K J mol^{-1})	1.585
	r^2	0.9542

Table 2: Equilibrium model parameters for adsorption of Pb^{2+} onto D.P.

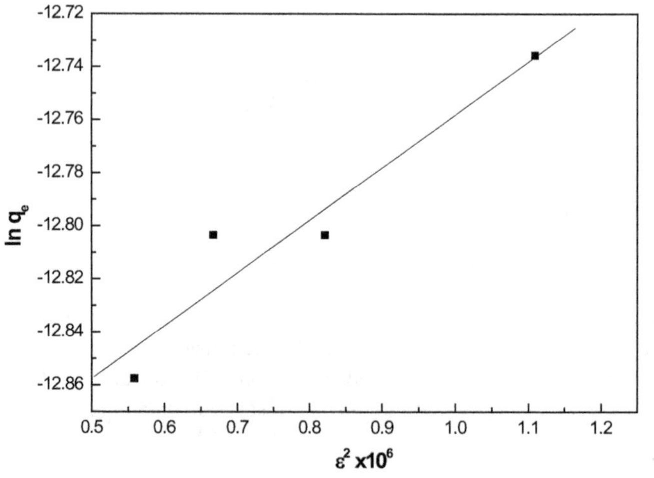

Figure 12: D-R isotherm for Pb^{2+} adsorption onto DP.

equation (Figure 12) to determine the adsorption type (physical or chemical). The linear form of this model is expressed by [27,32]:

$$\ln q_e = \ln q_m - \beta\varepsilon^2$$

Where q_e is the amount of the metal adsorbed per unit dosage of DP (mol l^{-1}), q_m the monolayer capacity (mol g^{-1}), and β is the activity-coefficient related to mean sorption energy ($mol^2\ J^{-2}$) and ε is the Polanyi potential described as: $\varepsilon = RT \ln(1 + 1/C_e)$. The mean sorption energy, E (kJ mol^{-1}), can be calculated by the equation [24,32]:

$$E = (2\beta)^{-1/2}$$

As seen in Figure 12, the slope of the D–R plot gives β constant and was evaluated as 0. 199 ($mol^2\ kJ^{-2}$). The sorption energy (E) was found to be 1.585 kJ mol^{-1}. It is generally assumed that if the sorption energy is below 8 kJ mol^{-1}, the sorption can be affected by physical forces such as Vander Der Walls forces, while if E is between 8 and 16 kJ mol^{-1}, the sorption is governed mainly by ion exchange. Sorption may be governed by particle diffusion if $E > 16$ kJ mol^{-1} [33]. Therefore, the E value calculated for the adsorption of Pb^{2+} onto DP showed that the sorption may be physical in nature.

Effect of temperature and thermodynamic parameters

The temperature has two major effects on the sorption process. Increasing the temperature is known to increase the rate of diffusion of the sorbate; changing the temperature will change the equilibrium capacity of the sorbent for a particular sorbate. In this study, a series of experiments were conducted on the adsorption of 5 mg l^{-1} Pb^{2+} onto 0.3 g of DP at 278, 283, 298, 313 and 333 K to investigate the effect of temperature on the sorption dynamics at different stirring times. The results depicted in (Figure 13) showed that the sorption increases as the temperature increases confirming that the process is endothermic in nature. Such results may either be attributed to the creation of some new active sites on the sorbent or to the acceleration of some originally slow adsorption steps. Moreover, the enhancement of mobility of Pb^{2+} from the bulk of solution towards the adsorbent surface should also be taken into consideration. This agrees well with the literature data [34,35]. Moreover, there was a decrease in the equilibration time to reach to a 100% for lead adsorption.

In order to investigate the thermodynamic parameters for the adsorption of Pb^{2+} by DP, the distribution coefficient K_d (l g^{-1}) was calculated at 278, 283, 313, 333 K according to the following equation [31,36]:

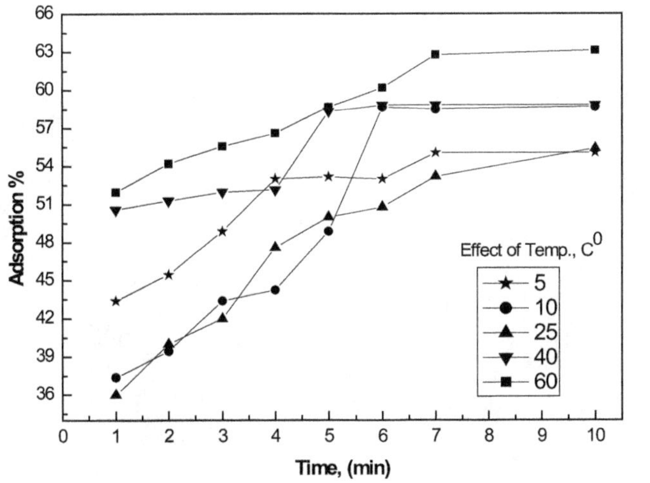

Figure 13: Effect of temperature on the sorption % of Pb^{2+} onto DP at different shaking times.

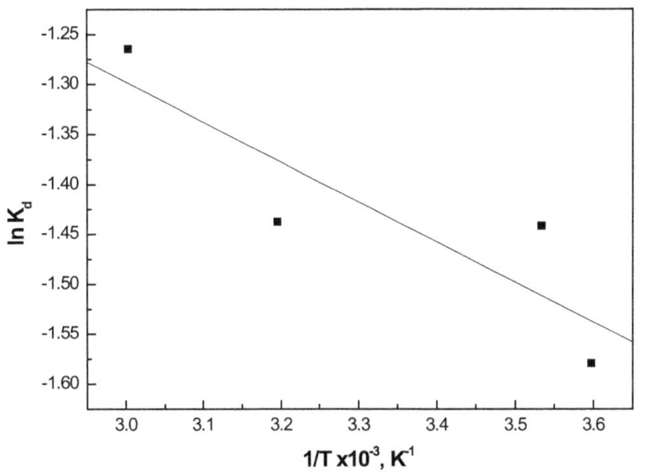

Figure 14: Thermodynamic distribution coefficient (K_d) calculated for the adsorption of Pb^{2+} on DP as a function of temperature.

Sample (location)	Added (mg l^{-1})	Adsorbed (mg l^{-1})	Re %
Distilled water	5	3.95	79.0
Tap water (our laboratory)	5	4.52	90.5
Nile water (Mansoura City)	5	4.49	89.8
Underground water (Mansoura City)	5	4.38	87.6
Sea water (Sharm El-Sheikh)	5	4.32	86.4

Table 3: Recovery of Pb^{2+} added to some water samples using 0.3 g of DP sorbent.

$$K_d = q_e / C_e$$

The K_d values at 278, 283, 313, 333 K are 0.206, 0.236, 0.238, and 2.82 L g^{-1} respectively. These results show that the K_d increases with temperature and revealing that the sorption of Pb^{2+} by DP may be endothermic. The enthalpy change (ΔH°) and entropy change (ΔS°) were calculated from the slope and intercept of the plot of ln K_d against 1/T, respectively, as depicted in (Figure 14) and according to the following equation [31,37,38]:

$$\ln K_d = \Delta S^\circ/R - \Delta H^\circ/RT$$

The Gibbs free energy change (ΔG°) was calculated by: $\Delta G^\circ = -RT \ln K_d$, where R is the universal gas constant (8.314 J mol^{-1} K^{-1}) and T is the absolute temperature (K).

The calculated enthalpy change ΔH° was found to be 3.74 J mol^{-1}. The positive value of ΔH° clarified that the sorption process is endothermic. The entropy change (ΔS°) was found to be 0.105 J mol^{-1} K^{-1}. According to Sari et al. [28], this result showed that Pb^{2+} in bulk phase (aqueous solution) is in a much chaotic distribution compared with the relatively ordered state of solid phase (sorbent surface). Moreover, the Gibbs free energy change (ΔG°) was -0.239, -0.243, -0.269, and -0.287 kJ mol^{-1} for the adsorption of Pb^{2+} at 278, 283, 313 and 333 K, respectively. The negative ΔG° values indicate that the adsorption of Pb^{2+} on DP is spontaneous thermodynamically.

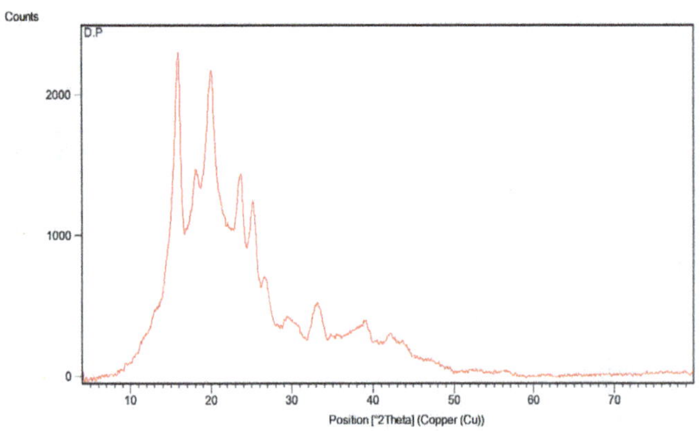

Figure 15 (a): XRD of DP before treatment.

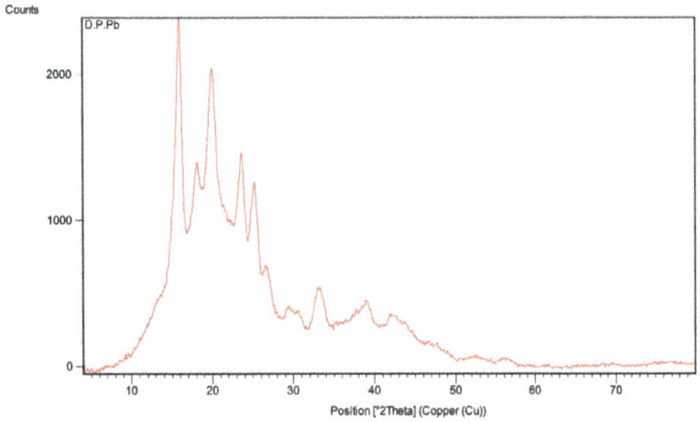

Figure 15 (b): XRD of DP after treatment with Pb^{2+}.

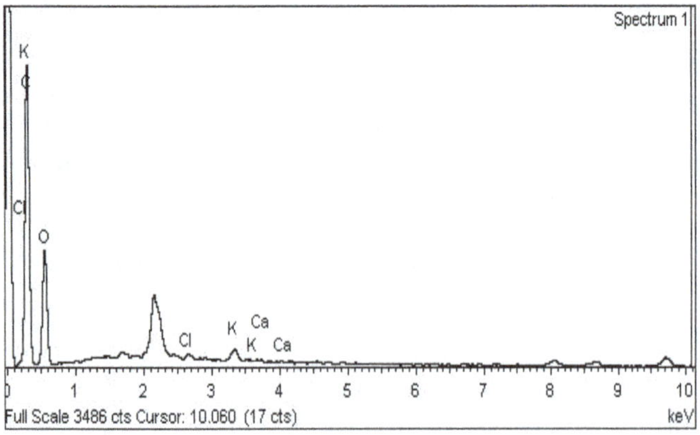

Figure 16 (a): EDX of DP before treatment.

Application

To investigate the applicability of the recommended procedure, a series of experiments were performed to recover 5 mg l^{-1} of Pb^{2+} added to aqueous and some natural water samples. The adsorption experiments were carried out using 50 ml clear and filtered. The results obtained are listed in Table 3.

XRD, EDX and SEM studies

Powered XRD studies help in understanding the changes occurred on the structure of DP sorption. XRD data of DP before and after adsorption are shown in Figure 15a and b and it provided evidence of modification in the surface morphologies. The EDX graph was shown in Figure 16a and b which also show the presence of the Pb^{2+} after its adsorption on DP (Figure 16b). To explain the morphological changes in the surface of the adsorbents in the coverage of pores of the DP due to the adsorption of Pb(II), SEM image of DP before and after lead adsorption is shown in Figure 17 a and b.

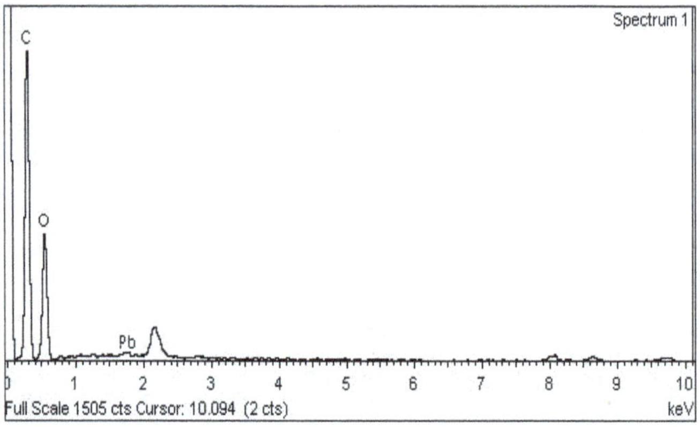

Figure 16 (b): EDX of DP after treatment with Pb^{2+}.

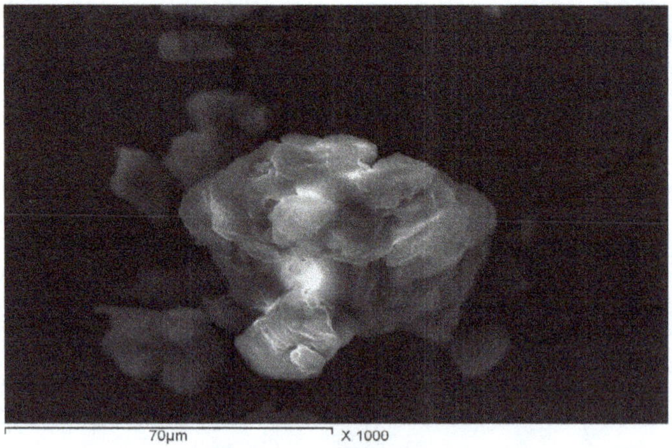

Figure 17 (a): SEM of DP before treatment.

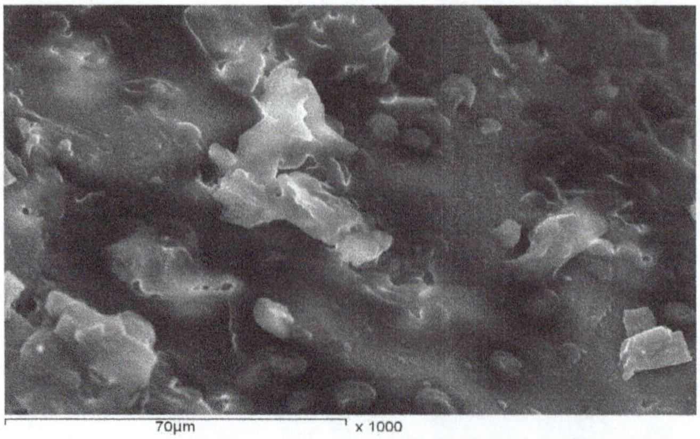

Figure 17 (b): SEM of DP after treatment with Pb^{2+}.

Conclusion

Date pits were used to remove Pb^{2+} ions from aqueous solutions. The parameters (solution pH, initial concentration of Pb^{2+}, sorbent dose, shaking time and temperature) have been examined. pH=6 is found the best one having high floatability of 6 g l^{-1} dose of DP. The change in ΔG^o, ΔH^o and ΔS^o were calculated. The removal of ~95% of Pb^{2+} was attained. The procedure was successfully applied to remove Pb^{2+} from natural water samples.

References

1. Li X, Zheng W, Wang D, Yang Q, Cao J, et al. (2010) Removal of Pb (II) from aqueous solutions by adsorption onto modified areca waste: Kinetic and thermodynamic studies. Desalination 258: 148-153.
2. Wang XS, Lu ZP, H. Miao HH, He W, et al. (2011) Kinetics of Pb (II) adsorption on black carbon derived from wheat residue. Chemical Engineering Journal 166: 986-993.
3. Ucurum M (2009) A study of removal of Pb heavy metal ions from aqueous solution using lignite and a new cheap adsorbent (lignite washing plant tailings). Fuel 88: 1460-1465.
4. Prado AGS, Moura AO, Holanda MS, Carvalho TO, Andrade RDA, et al. (2010) Thermodynamic aspects of the Pb adsorption using Brazilian sawdust samples: Removal of metal ions from battery industry wastewater. Chemical Engineering Journal 160: 549-555.
5. Reddy DHK, Harinath Y, Seshaiah K, Reddy AVR (2010) Biosorption of Pb(II) from aqueous solutions using chemically modified *Moringa oleifera* tree leaves. Chemical Engineering Journal 162: 626-634.
6. Lu D, Cao Q, Cao X, Luo F (2009) Removal of Pb(II) using the modified lawny grass: mechanism, kinetics, equilibrium and thermodynamic studies. J Hazard Mater 166: 239-247.
7. Martín-Lara MA, Rodríguez IL, Blázquez G, Calero M (2011) Factorial experimental design for optimizating the removal conditions of lead ions from aqueous solutions by three wastes of the olive-oil production. Desalination 278: 132-140.
8. Liu Y, Liu Z, Gao J, Dai J, Han J, et al. (2011) Selective adsorption behavior of Pb(II) by mesoporous silica SBA-15-supported Pb(II)-imprinted polymer based on surface molecularly imprinting technique. J Hazard Mater 186: 197-205.
9. Adie Gilbert U, Emmanuel IU, Adeyemo Adebanjo A, Olalere G (2011) Biosorptive removal of Pb^{2+} and Cd^{2+} onto novel biosorbent: Defatted Carica papaya seeds. Biomass and Bioenergy 35: 2517-2525.
10. Momcilovic M, Purenovic M, Bojic A, Zarubica A, Randelovic M (2011) Removal of lead(II) ions from aqueous solutions by adsorption onto pine cone activated carbon. Desalination 276: 53-59.
11. Unlü N, Ersoz M (2006) Adsorption characteristics of heavy metal ions onto a low cost biopolymeric sorbent from aqueous solutions. J Hazard Mater 136: 272-280.
12. Palma G, Freer J, Baeza J (2003) Removal of metal ions by modified *Pinus radiata* bark and tannins from water solutions. Water Res 37: 4974-4980.
13. Aldhaheri A, Alhadrami G, Aboalnaga N, Wasfi I, Elridi M (2004) Chemical composition of date pits and reproductive hormonal status of rats fed date pits. Food Chemistry 86: 93-97.
14. Waezi-Zadeh M, Ghazanfari A, Noorbakhsh S (2010) Finite element analysis and modeling of water absorption by date pits during a soaking process. J Zhejiang Univ Sci B 11: 482-488.
15. Saad EM, Mansour RA, El-Asmy A, El-Shahawi MS (2008) Sorption profile and chromatographic separation of uranium (VI) ions from aqueous solutions onto date pits solid sorbent. Talanta 76: 1041-1046.
16. Hamada JS, Hashim IB, Sharif FA (2002) Preliminary analysis and potential uses of date pits in foods. Food Chemistry 76: 135-137.
17. Al-Ghouti MA, Li J, Salamh Y, Al-Laqtah N, Walker G, et al. (2010) Adsorption mechanisms of removing heavy metals and dyes from aqueous solution using date pits solid adsorbent. J Hazard Mater 176: 510-520.
18. Iqbal M, Saeed A, Zafar SI (2009) FTIR spectrophotometry, kinetics and adsorption isotherms modeling, ion exchange, and EDX analysis for understanding the mechanism of $Cd(^{2+})$ and $Pb(^{2+})$ removal by mango peel waste. J Hazard Mater 164: 161-171.
19. Pehlivan E, Yanik BH, Ahmetli G, Pehlivan M (2008) Equilibrium isotherm studies for the uptake of cadmium and lead ions onto sugar beet pulp. Bioresour Technol 99: 3520-3527.
20. Al-Ghouti MA, Li J, Salamh Y, Al-Laqtah N, Walker G, et al. (2010) Adsorption mechanisms of removing heavy metals and dyes from aqueous solution using date pits solid adsorbent. J Hazard Mater 176: 510-520.
21. Liao D, Zheng W, Li X, Yang Q, Yue X, et al. (2010) Removal of lead(II) from aqueous solutions using carbonate hydroxyapatite extracted from eggshell waste. J Hazard Mater 177: 126-130.
22. Crank J (1965) The Mathematics of Diffusion. Carlendon Press, Oxford, London.
23. McKay G, Otterbern MS, Sweeney AG (1980) The removal of colour from effluent using various adsorbents—III. Silica: Rate processes. Water Research 14: 15-20.
24. Ghazy SE, Ragab AH (2007) Removal of copper from water samples by sorption onto powdered limestone. Indian Journal of Chemical Technology 14: 507-514.
25. Weber WJ, Morris JC (1963) Intraparticle diffusion during the sorption of surfactants onto activated carbon. J Sanit Eng Div Am Soc Civ Eng 89: 53-61.
26. Juang RS, Wu FC, Tseng RL (2000) Mechanism of Adsorption of Dyes and Phenols from Water Using Activated Carbons Prepared from Plum Kernels. J Colloid Interface Sci 227: 437-444.
27. Ghazy SE, Ragab AH (2007) Removal of Lead from Water Samples by Sorption onto Powdered Limestone. Separation Science and Technology 42: 653- 667.
28. Qadeer R, Hanif J (1994) Kinetics of Uranium (VI) Ions Adsorption on Activated Charcoal from Aqueous Solutions. Radiochimca Acta 65: 259-264.
29. Gupta GS, Shukla SP (1996) An inexpensive adsorption technique for the treatment of carpet effluents by low cost materials. Adsorp Sci Tech 13: 15-26.
30. Krika F, Azzouz N, Ncibi MC (2011) Adsorptive removal of cadmium from aqueous solution by cork biomass: Equilibrium, dynamic and thermodynamic studies. Arabian Journal of Chemistry.
31. Erdem M, Altundogan HS, Tumen F (2004) Removal of hexavalent chromium by using heat-activated bauxite. Minerals Engineering 17: 1045-1052.
32. Sari A, Tuzen M, Soylak M (2007) Adsorption of Pb(II) and Cr(III) from aqueous solution on Celtek clay. J Hazard Mater 144: 41-46.
33. Plante B, Benzaazoua M, Bussière B, Biesinger MC, Pratt AR (2010) Study of Ni sorption onto Tio mine waste rock surfaces. Applied Geochemistry 25: 1830-1844.
34. Al-Asheh S, Banat F (2001) Adsorption of Zinc and Copper Ions by the Solid Waste of the Olive Oil Industry. Adsorption Science & Technology 19: 117-129.
35. Yubin T, Fangyan C, Honglin Z (1998) Adsorption of Pb^{2+}, Cu^{2+} and Zn^{2+} Ions on to Waste Fluidized Catalytic Cracking (FCC) Catalyst. Adsorption Science & Technology 16: 595-606.
36. Mane VS, Deo Mall I, Chandra Srivastava V (2007) Kinetic and equilibrium isotherm studies for the adsorptive removal of Brilliant Green dye from aqueous solution by rice husk ash. J Environ Manage 84: 390-400.
37. Donat R, Akdogan A, Erdem E, Cetisli H (2005) Thermodynamics of Pb^{2+} and Ni^{2+} adsorption onto natural bentonite from aqueous solutions. J Colloid Interface Sci 286: 43-52.
38. Gupta SS, Bhattacharyya KG (2005) Interaction of metal ions with clays: I. A case study with Pb(II). Applied Clay Science 30: 199-208.

4

An Overview on the Thermodynamic Techniques used in Food Chemistry

Prabal Giri[1]* and Churala Pal[2]

[1]Department of Chemistry, Guskara Mahavidyalaya, Burdwan 713128, West Bengal, India
[2]Department of Chemistry, Basanti Devi College, Kolkata 700029, West Bengal, India

Abstract

The thermal behaviour of food strongly depends on its composition. The goals of food processing are to inactivate spoilage and pathogenic microorganisms and to maintain this status in storage. Using calorimetric techniques, many physicochemical effects can be observed in the temperature range between -50°C and 300°C. Biophysical techniques namely isothermal titration (ITC) and Differential Scanning Calorimetry (DSC) are used to characterize the structure and properties of food materials before and after processing to develop a fundamental understanding of the impact of processing and storage conditions. The data resulting from such studies can be used to predict the physical properties of foods under optimized condition.

Keywords: Isothermal calorimetry; Differential Scanning Calorimetry; Proteins; Food processing; β–glucosidase

Introduction

Diverse biophysical techniques including thermal and non-thermal methods are utilized in processing and preservation of food materials and in manufacture of value added products. The food processing aims the inactivation of spoilage and pathogenic microorganisms during storage. Alteration takes place in food components, including carbohydrates, lipids, vitamins, and most importantly, proteins during processing. Such changes lead to structural and functional changes in foods at the micro - and macromolecular levels that affect the physical, organoleptic, and nutritional properties of the food [1].

Food materials are complex biological systems containing a heterogeneous, heterophase mixture of high and low molecular weight components and their aggregates and complexes. The structure of foods may incorporate a broad range having three states of matter including dilute to concentrated liquids, solids, and mixtures of multiliquid, liquid-solid, liquid-gas, and solid-gas structures [2]. Several biophysical techniques are used to address the broad variety of structures and compositions developing a basic understanding of the impact of processing and storage conditions. The data resulting from such studies can be used to predict the physical properties of foods so that food processing and storage conditions are optimized.

Calorimetry presents itself as particularly well suited for analysis of food materials. Specifically, because many food processing methods involve thermal treatment (heating, cooling, freezing) of the materials, thermal characterization of food systems and their components leads to data that can be related directly to the processing protocols [3-5]. Determination of thermal properties of food materials, such as specific heat as a function of temperature, is essential for heat transfer and energy balance calculations [6]. Generation of a reliable database to develop equations predicting thermal properties of food materials for optimization of food processes can be accomplished by using calorimetry. Moreover, food materials and their components go through conformational and phase transitions. In the investigation of foods using thermal analysis and calorimetric techniques, many physicochemical effects can be observed in the temperature range between -50°C and 300°C [7]. These thermal phenomena may be either endothermic or exothermic as presented in Table 1. Calorimetry data can be analyzed to evaluate the thermal and thermodynamic stability of various phases for a rational design of food product formulations and process conditions. Response to some perturbation caused from heat effects can be measured by Calorimetry. Two distinct ultrasensitive calorimetric techniques namely Isothermal Titration Calorimetry (ITC) and Differential Scanning Calorimetry (DSC) cover the application of thermodynamic techniques in food science. In DSC perturbation is a change in temperature of the sample whereas in ITC, the perturbation is the introduction of new material into the sample [8-11].

While the calorimetric technique is powerful, the validity and utility of the data depend strongly on the careful use of the equipment and correct interpretation of data. Due to the ultrasensitive instrumental set up, calorimetric data depends strongly on the conditions maintained during the experiment [12]. In choosing the calorimetry parameters, one should follow the following guidelines as provided by Haines [13]:

a. **Time scale:** Especially in dynamic measurement systems, for events to be detected the experimental time scale should match the time scale of the observed event.

b. **Magnitude of the heat flow:** If the energy associated with the transition is small, it can lead to ambiguities in its detection. Increasing the scanning rate enhances the signal; however, it may cause deviation from equilibrium conditions, which requires models beyond the standard equilibrium thermodynamics treatment of calorimetric data.

c. **Moisture loss during experiment:** Biological samples in general are high moisture content materials. If the sample cell is not sealed well, the moisture content of the sample will change due to evaporation during the course of experiment. This may lead to overestimation of the transition temperature as well as the transition enthalpy change.

d. **Interpretation of overlapping peaks:** Biological samples may contain multiple components that undergo thermally induced transitions at similar temperatures.

The advantages of using calorimetry for study of food components can be summarized as follows [14-16]:

***Corresponding author:** Prabal Giri, Department of Chemistry, Guskara Mahavidyalaya, Burdwan 713128, West Bengal, India
E-mail: prabalgiri@yahoo.co.in

Types of food	Yeast	Bacteria	Carbohydrate	Protein	Fat	Starch	Enzyme
outcome (exothermic)	fermentation	growth, metabolism, fermentation	crystallization, decomposition	aggregation, crystallization	crystallization, oxidation	retrogradation, oxidation	aggregration, enzymatic reaction
outcome (endothermic)	-	-	melting, glass transition	denaturation	denaturation	gelatanization, glass transtion	denaturation

Table 1: Exothermic and endothermic outcomes on food materials.

a) Calorimetric methodology does not impose any prerequisite on the physicochemical properties of the experimental sample unlike spectroscopic method to have in pure state or presence of chromophore.

b) Model independent and direct determination of enthalpy change (ΔH) and change in heat capacity at constant pressure (ΔC_p) is feasible.

c) Materials do not have to be uniform or have to be a homogeneous mixture. In fact, in addition to pure materials, the technique can be used to evaluate the interactions among the components in a complex system and how the interactions are altered by the processing.

d) Sample preparation for calorimetric analysis is very simple and can be handled without trouble by the researchers.

Today, the instruments are highly developed for accurate measurement of thermal events. The theory behind the technique is well developed, which facilitates interpretation of the data [17,18]. In this paper, the special emphasis has given on the working principle, methods and application of the thermodynamic techniques used namely DSC and ITC used frequently in food chemistry.

Differential Scanning Calorimetry

Differential Scanning Calorimetry, which measures heat capacity as a function of temperature, is an well-established thermal analysis technique that detects and monitors thermally induced conformational transitions and phase transitions as a function of temperature [19,20]. During temperature scanning, depending on the complexity of the material, many peaks or inflection points (one to several) reflecting the thermally induced transitions can be observed. The direction of the peak corresponds to the nature of the transition, being heat absorbing (endotherms) or heat releasing (exotherms). While melting of solids and denaturation of proteins display endotherms, crystallization of carbohydrates and aggregation of proteins manifest themselves as exotherms.

DSC is a thermoanalytical technique in which the difference in the amount of heat required to increase the temperature of a sample and reference are measured as a function of temperature. Both the sample and reference are maintained at nearly the same temperature throughout the experiment. Generally, the temperature program for a DSC analysis is designed such that the sample holder temperature increases linearly as a function of time. The reference sample should have a well-defined heat capacity over the range of temperatures to be scanned. The basic principle underlying this technique is that, when the sample undergoes a physical transformation such as phase transitions, more (or less) heat will need to flow to it than the reference to maintain both at the same temperature. Whether more or less heat must flow to the sample depends on whether the process is exothermic or endothermic. By observing the difference in heat flow between the sample and reference, differential scanning calorimeters are able to measure the amount of heat absorbed or released during such transitions. Thus the result of a DSC experiment is a curve of heat flux versus temperature. DSC can be used to measure a number of characteristic properties like crystallization temperature of solids, oxidative stability of samples, used in the pharmaceutical and polymer industries etc. The temperatures for the endothermic and exothermic transitions and the heat involved in such transitions are measured using a calorimeter. Inflection points are indicative of glass transitions; that is, transitions from a glassy to rubbery state. The transition temperatures (T_{peak}/ T_g /T_m) reflect the thermal stability of the phase or state going through the transition. One can extract from calorimetry data values for the thermal and thermodynamic changes in free energy (ΔG), enthalpy (ΔH), entropy (ΔS), and heat capacity (ΔC_p) of the various transitions in addition to determination of the bulk heat capacity of the material [20].

In this context, it would be quiet unjustified if the older technique Differential Thermal Analysis (DTA) is not mentioned under discussion. In this technique it is the heat flow to the sample and reference that remains the same rather than the temperature. When the sample and reference are heated identically, phase changes and other thermal processes cause a difference in temperature between the sample and reference. Both DSC and DTA provide similar information. DSC measures the energy required to keep both the reference and the sample at the same temperature whereas DTA measures the difference in temperature between the sample and the reference when they are both put under the same heat. DTA is an older technique than DSC. So DSC is more sophisticated and improved than DTA. DTA instrument can be used at very high temperatures and in aggressive environments where DSC instrument may not work. In DSC, influence of sample properties on the area of the peak is comparatively lower than in DTA [21,22].

The basis for thermodynamic study of food materials is that the relevant initial and final states (pre-processing and post processing states) can be defined and the energetic and structural differences between these states can be measured using calorimetric instrumentation. To this end, calorimetry can be used to evaluate the effect of other physical and chemical variables by comparing the thermo grams of the materials before and after exposure to the variable outside the calorimetry [23]. However, in most food processing food ingredients are mixed or diluted with a liquid (water, milk) or with a powder (sugar, salt, yeast). For simulation of such transformations and interactions, the limited volume and the lack of in situ mixing constitute the major drawbacks of the DSC technique.

Brief instrumental procedure of DSC

Experimental set up of a differential scanning calorimeter is represented in Figure 1(A). In a series of DSC scans, both the sample and reference cells are loaded with buffer solution, equilibrated at 10°C for 15 min and scanned from 10 to 120°C at a scan rate of 50°C/hour [20]. The buffer scans were repeated till reproducible and on cooling, the sample cell is rinsed and loaded with sample. The DSC thermograms of excess heat capacity versus temperature plots are analyzed using a variety of softwares like Origin 7.1, CALISTO, Pyris™, Proteus 7.0 etc. available in the market. The area under the experimental heat capacity (C_p) curve is used to determine the calorimetric transition enthalpy (ΔH_{cal}) given by the equation

$$\Delta H_{cal} = \int C_p dT$$

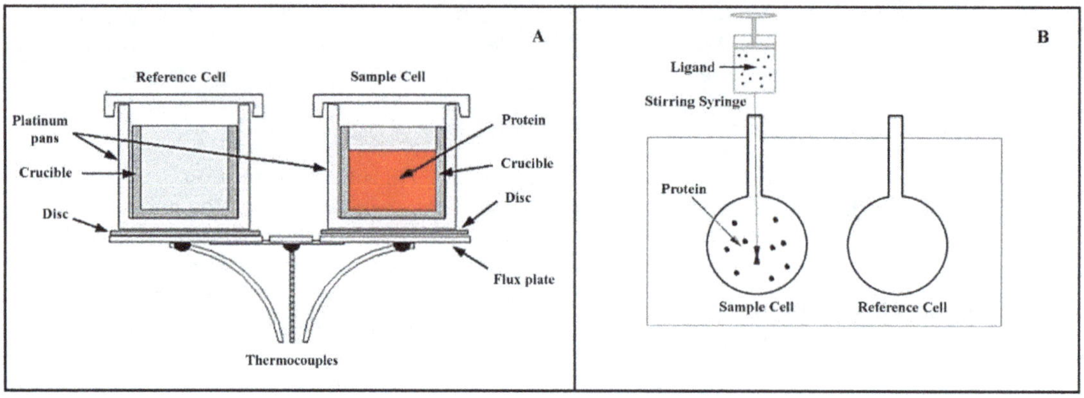

Figure 1: Experimental set up of (A) Differential scanning and (B) Isothermal titration calorimeter.

Where T is the absolute scale temperature in Kelvin. This calorimetrically determined enthalpy is model-independent and is thus unrelated to the nature of the transition. The temperature at which excess heat capacity is at a maximum defines the transition temperature (T_m). The model-dependent van't Hoff enthalpy (ΔH_v) is obtained by shape analysis of the calorimetric data and the cooperativity factor is obtained from the ratio ΔH_{cal} /ΔH_v [24]. Comparison of the model independent calorimetrically determined ΔH_{cal} value with the ΔH_v value assesses the validity of the assumptions employed in the derivation of the van't Hoff relation. Specifically, it is assumed that the transition from the ordered, low temperature form to the disordered, high temperature form passes through no thermodynamically significant intermediate states (two-state assumption); that is, there is no partial unfolding of the protein in the denaturation pathway. The ΔH_v reports the enthalpy change associated with disruption of a single cooperative unit, the fraction of the protein that acts as a single thermodynamic unit.

Isothermal Titration Calorimetry

Most biological processes involve one or more binding events. The types of binding reactions are varied and include, but are not limited to, assembly of protein subunits into functional enzyme complexes, formation of enzyme-inhibitor complexes, formation of protein-nucleic acid complexes, enzyme-substrate binding, and enzyme-cofactor binding. These binding processes can be described in terms of the standard thermodynamic parameters. A predictive understanding of the binding process can be achieved by measurement of the binding thermodynamic parameters. All of the binding processes enumerated above are amenable to analysis by some variation of ITC experiment [25]. Because the binding sites for small molecules to proteins tend to be well defined and small in number, and because most of the binding reactions involve a nonzero enthalpy change, protein -small molecule interactions frequently are particularly well suited to examination by isothermal titration calorimetry.

Here, we consider a simple association (without reaction) defined by the equilibrium $nL + M \leftrightarrow ML_n$ of a small molecule L, with a protein or other macromolecule M, with n identical, non-interacting binding sites. Other types of complex equilibria are beyond the scope of this paper. An Isothermal titration calorimetry instrument consists of two identical cells composed of a highly efficient thermal conducting material (Hasteloy) surrounded by an adiabatic jacket. Sensitive thermopile/thermocouple circuits detect temperature differences between the two cells and the cells and the jacket. Heaters located on both cells and the jacket is activated when necessary to maintain identical temperatures between all components. In an ITC experiment, the macromolecule solution is generally placed in the sample cell. The reference cell contains buffer or water minus the macromolecule. Prior to the injection of the titrant, a constant power (<1 mW) is applied to the reference cell. This signal directs the feedback circuit to activate the heater located on the sample cell. This represents the base line signal. The direct observable measured in an ITC experiment is the time -dependent input of power required to maintain equal temperatures in the sample and reference cell. During the injection of the titrant into the sample cell, heat is taken up or evolved depending on whether macromolecular association reaction is endothermic or exothermic. For an exothermic reaction, the temperature in the sample cell will increase and the feedback power will be deactivated to maintain equal temperatures between the two cells for endothermic reactions, the reverse will occur meaning the feedback circuit will increase power to the sample cell to maintain the temperature. The heat absorbed or evolved during a calorimetric titration is proportional to the fraction of bound ligand. For the initial injections, all or most of the added ligand is bound to the macromolecule, resulting in large exothermic or endothermic signals depending on the nature of association. As the ligand concentration increases the macromolecules becomes saturated and subsequently less heat is evolved or absorbed on further addition of titrant as presented in Figure 1. The amount of heat evolved on addition of ligand can be represented by the equation

$$Q = V_o \Delta H_b [M]_t K_a [L] / (1 + K_a [L])$$

Where V_o is the volume of the cell, ΔH_b is the enthalpy of binding per mol of ligand, $[M]_t$ is the total macromolecule concentration including bound and free fractions, K_a is the binding affinity and [L] is the free ligand concentration [26].

Brief Instrumental Procedure of ITC

ITC experiments are performed on a Microcal VP-ITC microcalorimeter (MicroCal, Northampton, MA, USA) at 20°C. Origin 7.0 software, supplied by the manufacturer, is used for data acquisition and manipulation, as presented in Figure 1(B). All the solutions used for ITC experiments are degassed prior to use under vacuum (140 mbar, 8 min). In case of ligand-protein and complexation, degassed ligand solutions are injected from the rotating syringe (290 rpm) into the isothermal sample chamber containing 1.42 mL of protein. Corresponding control experiments to determine the heat of dilution of ligand into buffer and food component into buffer are also performed. Each injection generated a heat burst curve (micro calories

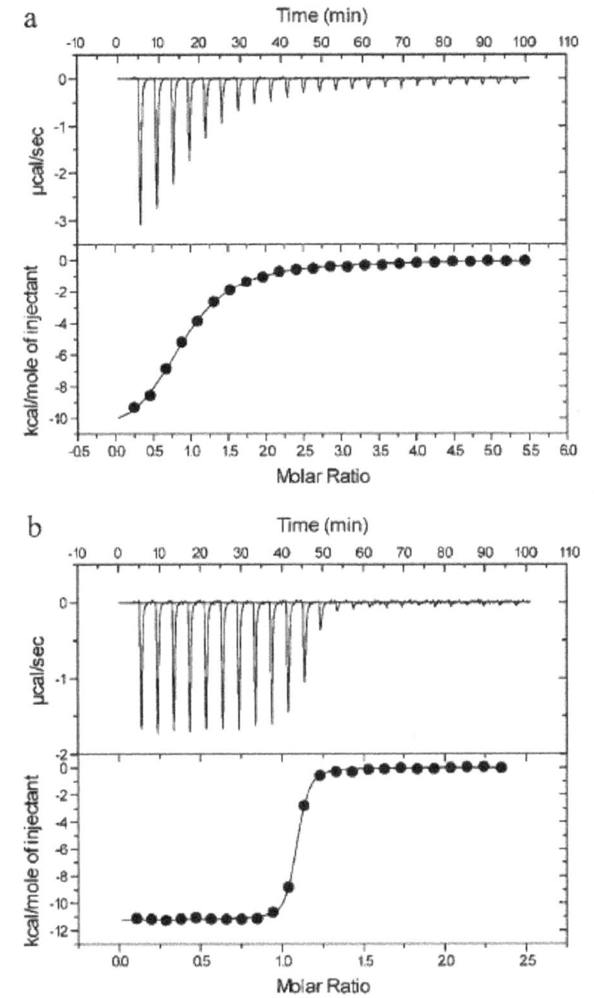

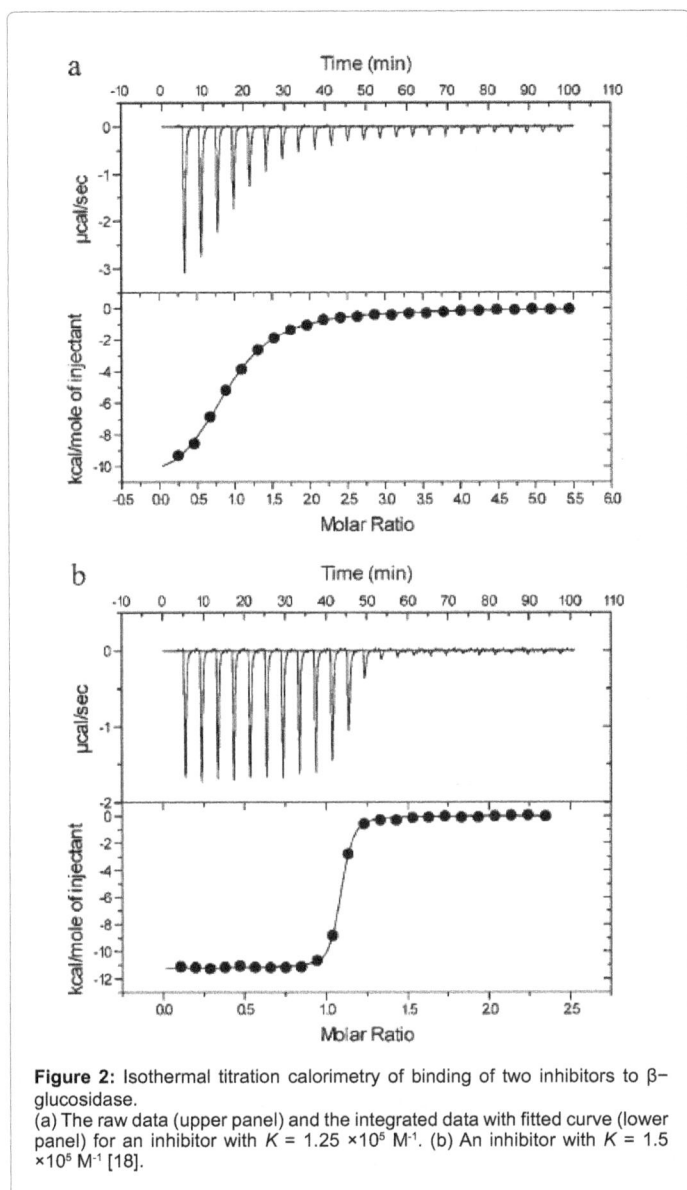

Figure 2: Isothermal titration calorimetry of binding of two inhibitors to β-glucosidase.
(a) The raw data (upper panel) and the integrated data with fitted curve (lower panel) for an inhibitor with K = 1.25 ×10^5 M^{-1}. (b) An inhibitor with K = 1.5 ×10^5 M^{-1} [18].

per second) the area under which is determined by integration (using market available software) that gives the measure of the heat of reaction associated with the injection. The heat associated with each control is subtracted from the corresponding heat associated with the ligand-food component injection to give the heat of ligand binding for that injection. The resulting data were analyzed using Origin software to estimate the binding affinity (K_a), the binding stoichiometry (N), and the enthalpy of binding (ΔH_o). The free energies (ΔG_o) are calculated using the standard relationship [11],

$$\Delta G_o = -RT \ln K_a$$

Where R is 1.987cal mol^{-1} K^{-1} and T is represented in Kelvin for the appropriate temperature.

The binding free energy coupled with the binding enthalpy derived from the ITC data allowed the calculation of the entropic contribution to the binding ($T\Delta S_o$), where ΔS_o is the calculated binding entropy using the standard relationship,

$$T\Delta S_o = \Delta H_o - \Delta G_o$$

Each experiment is repeated at least three times and the error values that reflect the standard deviations among the different runs were always less than 10%, which indicates the quality of data. The ITC profile of binding of two inhibitors to β-glucosidase is shown in Figure 2 [18].

Calorimetry of Proteins

As the food industry begins to take advantage of recent developments in protein chemistry by introducing enzymes and structural proteins into modern food materials and their processing, detailed understanding of protein chemical and physical properties becomes increasingly important. Development of a predictive understanding of the energetics-structure-function relationships will be required to fully exploit the possibilities presented to engineer proteins with novel substrate specificity or enhanced physical properties, including thermal stability, pH, and ionic strength optima.

To fully exploit the structural and catalytic properties of proteins it is critical to develop a predictive understanding of their functions and stability as a function of temperature and solution conditions. Monitoring the unfolding of a macromolecule induced by exposure to elevated temperature is a classical method for evaluating stability. DSC is particularly well suited to characterization of protein stability. The thermodynamic characterization of the protein unfolding process derived from DSC data can be used to predict the stability of the protein at any temperature. Binding equilibria between a protein and a small molecule effector, such as a cofactor or drug, or a second protein subunit, can alter the protein's denaturation temperature. If the second molecule binds more tightly to the folded form of the protein than to the unfolded form, the denaturation will shift to higher temperature. Conversely, preferential binding to the unfolded form shifts the denaturation to lower temperature. DSC thermograms are particularly well suited to measure very tight binding based on the observed binding - induced changes in the heat capacity versus temperature profiles. The simulated DSC thermogram of a small globular protein is shown in Figure 3 [27].

Isothermal calorimetry is an extremely general technique as it measures heat production rate, which accompanies nearly all physical, chemical and biological processes. The applications of ITC in food sciences include studies on wound respiration, quantification of cell

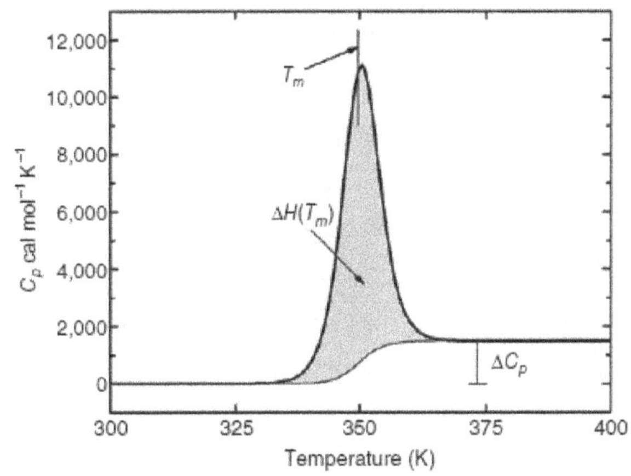

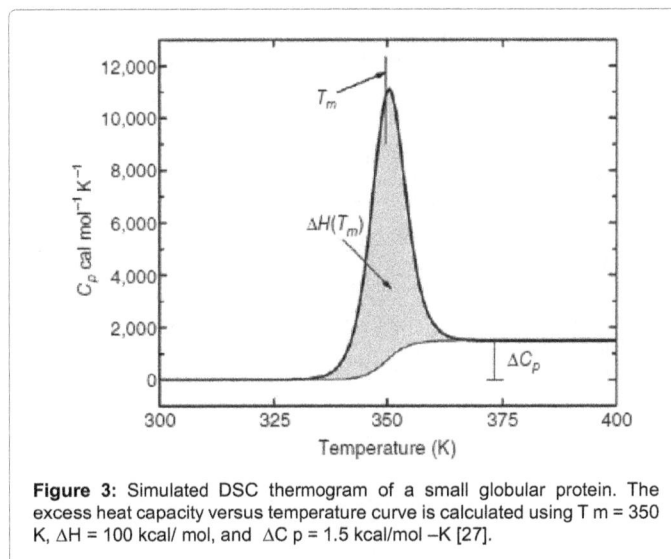

Figure 3: Simulated DSC thermogram of a small globular protein. The excess heat capacity versus temperature curve is calculated using T m = 350 K, ΔH = 100 kcal/ mol, and ΔC p = 1.5 kcal/mol –K [27].

damage during blanching and microbiological fermentation of milk, beer and pre-biotic foods [28-30].

Conclusion

Modern ultrasensitive calorimetry provides powerful tools for understanding the stability of proteins in solution, the forces that maintain their folded structures, and their interactions with other macromolecules and small molecules. Differential Scanning Calorimetry quantifies the thermal (T_m) and thermodynamic (ΔG) stabilities of the protein. The thermodynamic origins of the stability (ΔH, ΔS, and ΔC_p) can be interpreted to dissect the forces maintaining the folded structure and how they depend on the structure of the protein and the solution conditions. The model-independent values derived directly from the DSC thermogram can be compared to model-dependent values derived from the shape of the DSC curves to gain insight into the unfolding mechanism and aggregation. Isothermal titration calorimetry provides a means to directly measure the heat of interaction (ΔH) of a protein with another macromolecule or with small molecules.

Application of models for the association reaction provides detailed thermodynamic characterization (K_a, ΔG, ΔS and ΔC_p) of the binding process. Taken together, the techniques of modern solution calorimetry provide a predictive understanding of the stability of a protein and its interactions with other molecules as a function of temperature and solution conditions.

References

1. Lewis MJ (1990) Physical properties of foods and food processing systems. Woodhead Publishing Ltd., Abington Hall, Abington, Cambridge CB1 6AH, England.
2. Wilkinson C, Dijksterhuis GB, Minekus M (2000) From food structure to texture. Trends Food Sci Tech 11: 442-450.
3. Biliaderis CG (1983) Differential scanning calorimetry in food research-A review. Food Chem 10: 239-265.
4. Roos YH (2003) Thermal analysis, state transitions and food quality. J Therm Anal Calorim 71: 197-203.
5. Ghai R, Falconer RJ, Collins BM (2012) Applications of isothermal titration calorimetry in pure and applied research--survey of the literature from 2010. J Mol Recognit 25: 32-52.
6. Kaletunc G (2007) Prediction of heat capacity of cereal flours: a quantitative empirical correlation. J Food Eng 82: 589-594.
7. Aguilera JM (2005) Why food microstructure? J Food Eng 67: 3-11.
8. Sturtevant JM (1987) Biochemical Applications of Differential Scanning Calorimetry. Annu Rev Phy Chem 38: 463-488.
9. Freire E (1995) Differential scanning calorimetry. Methods Mol Biol 40: 191-218.
10. Jelesarov I, Bosshard HR (1999) Isothermal titration calorimetry and differential scanning calorimetry as complementary tools to investigate the energetics of biomolecular recognition. J Mol Recognit 12: 3-18.
11. Lewis EA, Murphy KP (2005) Isothermal titration calorimetry. Methods Mol Biol 305: 1-16.
12. Hohne GWH, Hemminger W, Flammersheim HJ (2003) Differential Scanning Calorimetry: an Introduction for Practitioners. 2nd Edition. Springer – Verlag, Berlin; New York, USA.
13. Haines PJ (1995) Thermal Methods of Analysis, Principles, Applications and Problems. Blackie Academic & Professional, Glasgow, USA.
14. Tan CP, CheMan YB (2002) Recent developments in differential scaning calorimetry for assessing oxidative deterioration of vegetable oils. Trends Food Sci Tech 13: 312-318.
15. Gabbott P (2008) Principles and applications of thermal analysis. John Wiley & Sons Ltd., The Atrium, Southern Gate, Chichester, West Sussex, PO 19 8SQ, England.
16. Singh J, Kaur L, McCarthy OJ (2007) Factors influencing the physic-chemical, morphological, thermal and rheological properties of some chemically modified starches for food applications- a review. Food Hydrocolloid 21: 1-22.
17. Sturtevant JM (1977) Heat capacity and entropy changes in processes involving proteins. Proc Natl Acad Sci U S A 74: 2236-2240.
18. Zechel DL, Boraston AB, Gloster T, Boraston CM, Macdonald JM, et al. (2003) Iminosugar glycosidase inhibitors: structural and thermodynamic dissection of the binding of isofagomine and 1-deoxynojirimycin to beta-glucosidases. J Am Chem Soc 125: 14313-14323.
19. Ladbury JE, Doyle ML (2004) Biocalorimetry 2 Applications of calorimetry in the biological sciences. John Wiley & Sons Ltd., The Atrium, Southern Gate, Chichester, West Sussex, PO 19 8SQ, England.
20. Giri P (2009) Biophysical studies on the interaction of plant alkaloids with polyadenylic acid structures. PhD Thesis, Jadavpur University, Kolkata, India.
21. Vold MJ (1949) Differential thermal analysis. Anal Chem 21: 683-688.
22. McElhaney RN (1982) The use of differential scanning calorimetry and differential thermal analysis in studies of model and biological membranes. Chem Phys Lipids 30: 229-259.
23. Farkas J, Mohacsi-Farkas C (1996) Application of DSC in food research and food quality assurance. J Thermal Anal 47: 1787-1803.
24. Hinz HJ, Schwarz FP (2001) Measurement and analysis of results obtained on biological substances with differential scanning calorimetry. The Journal of Chemical Thermodynamics 73: 745-759.
25. Hemminger W, Hohne G (1984) Calorimetry: Fundamentals and Practice, Verlag Chemie, Weinhein, Germany.
26. Pierce MM, Raman CS, Nall BT (1999) Isothermal titration calorimetry of protein-protein interactions. Methods 19: 213-221.
27. Prabhu NV, Sharp KA (2005) Heat capacity in proteins. Annu Rev Phys Chem 56: 521-548.
28. Smith BN, Hansen LD, Breidenbach RW, Criddle RS, Rank DR, et al. (2000) Metabolic heat rate and respiratory substrate changes in aging potato slices. Thermochim Acta 349: 121-124.
29. Gomez Galindo F, Roculli P, Wadso L, Sjoholm I (2005) The potential of isothermal calorimetry in monitoring and predicting quality changes during processing and storage of minimally processes fruits and vegetables. Trends Food Sci Technol 16: 325-331.
30. Wadso L, Gomez Galindo F (2009) Isothermal calorimetry for biological applications in food science and technology. Food Control 20: 956-961.

Removal of Methylene Blue (Mb) Dye from Aqueous Solution by Bioadsorption onto Untreated *Parthenium hystrophorous* Weed

Million Mulugeta* and Belisti Lelisa

Department of Chemistry, Arba Minch University, Arba Minch, Ethiopia

Abstract

Nowadays, the application and search of alternative cheap and ecofriendly adsorbents to replace activated carbon was made. It has been a major focus for the removal of dyes from waste water. In this study untreated *Parthenium hystrophorous* weed (PHW) was used to remove a textile dye (Methylene Blue (MB)) from an aqueous solution by adsorption technique. The factors influencing the adsorption were also investigated. The MB dye removal by the PHW was significantly dependent on contact time, pH, dye concentration, adsorbent dose and pH. The optimum equilibrium conditions for removal of MB dye by PHW were; contact time of 2 hrs, at pH 8 and an adsorbent dose of 0.8 g. The adsorption data better fits Langmuir isotherm model well and the maximum adsorption capacity of the PHW was found to be 23.8 mg g^{-1}. The results obtained in this study indicated that PHW will be an attractive candidate for removing cationic dyes from the dye wastewater.

Keywords: *Parthenium hystrophorous;* Weed; Adsorption; Methylene blue

Introduction

Water pollution by dyes is a worldwide problem particularly in textile industry where large quantities of dye effluents are discharged from the dyeing process. Considering both volume and composition, effluent from the textile industry was declared as one of the major sources of wastewater in the world. Dyes are also widely used in many industries such as rubber, paper, plastic, cosmetic etc. There are more than 10,000 commercially available dyes with over 7×10^5 tons of dyestuff being produced annually across the world [1]. The total dye consumption of the textile industry worldwide is more than 10^7 kg/year, and about 90% ending up on fabrics. Dye producers and consumers are interested in the stability and fastness of dyes and consequently, are producing dyestuffs which are more difficult to degrade after being used. It is estimated that 10-15% of the dye is lost during the dyeing process and released with the effluent [2].

So, it is vital to treat these polluted waste waters efficiently as well as effectively in terms of cost by using Bio-sorbents. Biomaterials that are available in large quantities may have a potential to be used as low cost adsorbents, because they represent unused resources that are widely available and environmentally friendly [3]. Today, many industries commonly used activated carbon as adsorbent agent for dye removal. Commercially available Activated Carbons (AC) are usually derived from natural materials such as wood, coconut shell, lignite or coal, but almost any carbonaceous material may be used as precursor for the preparation of carbon adsorbents [4,5].

A wide variety of carbons have been prepared from agricultural and wood wastes, such as bagasse [6], coir pith [7], banana pith [8], date pits [9], corn cob [10], maize cob [10], straw, rice husk [11,12], rice hulls, fruit stones and nutshells [13], pinewood, sawdust [12] and etc.

Moreover, the overlying cost of activated carbon and associated problems of regeneration has force a new research in order to find other alternative low cost adsorbents agent. Since preparation of activated carbon from bio-resources requires a great amount of energy and as well as regeneration of pollutants from the activated carbon poses a great deal of bulky process. In addition to biomaterials, microorganisms have also been used as metal sorbents. Bacteria, fungi, yeast and algae have been reported to remove heavy metals from aqueous solutions [14].

Nowadays, there are numerous numbers of low cost, commercially available adsorbents which had been used for dye removal. However, as the adsorption capacities of the above adsorbents are not very large, the new adsorbents which are more economical, easily available, environmentally friendly and highly effective are still needed [15,16]. A variety of materials are used as adsorbents for dyes removal, and various studies have been published on its adsorption on activated carbon, starch xanthate, alumina, low-grade manganese ore, crushed coconut shell, fly ash, sawdust, rice husk carbon, wood charcoal, bituminous coal, and lignite.

Past recent research was still searching of an effective biosorbent that can be used to remove dyes from aqueous solution. However, study on *Parthenium hystrophorous* weed adsorbent is only few in this field. Hence there is a need to investigate dry *Parthenium hystrophorous* ability to MB from industrial waste waters. Therefore, the aim of this study was to investigate the potential of untreated *Parthenium hystrophorous* weed, as a non-conventional adsorbent in the removal of a Methylene Blue dye from aqueous solutions. Using these alien weeds as biosorbent can be an alternative way to monitor their spreading in the environment to some extent.

Materials and Methods

Apparatus and instruments

All glassware (conical flasks, measuring cylinders, beakers, pipettes etc.) were manufactured by Borosil / Rankem. Whatman No-1(125 mm) filter paper, Sieve to get 1mm particle size and all the instruments used in the experiment are listed below: (Table 1)

Chemicals and reagents

Analytical grade reagents; Methylene blue (C.I. 52015, S.D. Fine Chemicals) were used to prepare standard solutions of the adsorption studied. HCl and NaOH (Blulux Laboratories (p) Ltd- 121001) used to

***Corresponding author:** Million Mulugeta, Department of Chemistry, Arba Minch University, Arba Minch, P.O. Box: 21, Ethiopia
E-mail: million.mulugeta@amu.edu.et

Instrument	Manufacturer
Electronic Balance	OHAUS, Switzerland
pH meter	Jenway/(MP 220)
Oven	Shivaki(Contherm 260M)
Shaker	Orbital shaker SO1, UK
Spectrophotometer	Sanyo (Uv/Vis-65)
Electrical mill	(IKA WERKE), UK

Table 1: List of all instruments used

adjust pH and buffer solutions (E. Merck) used to calibrate pH meter.

Experimental site

The adsorbent (*Parthenium* weed) was collected from the Arba Minch City, Ethiopia in randomly sampling technique and the collected samples were mixed to prepare adsorbent composite sample.

Biosorption study

Adsorbent collection and preparation: Adsorbent is the material upon whose surface the adsorption takes place is called an adsorbent. For this study *Parthenium hystherophoresis* weed was used as a low cost biosorbent. The collected weed adsorbent was then washed with distilled water for several times to remove all the dirt particles. The washed materials were cut into small pieces (1-3 cm) and dried in a hot air oven at 60°C for 48 hr. Then ground and finally screened to obtain a particle size range of 1 mm size sieve and then stored in plastic bottles for further use.

Adsorbate: The basic dye, methylene blue (C.I. 52015, S.D. Fine Chemicals, 85% dye content, chemical formula C16H18N3SCl, FW 319.86, nature basic blue, and λ_{max} 665 nm) has been used in this study. The MB was chosen in this study because of its known strong adsorption onto solids. An accurately weighed quantity of the dye was dissolved in double distilled water to prepare the stock solution (1000 mg/L). Experimental solutions of desired concentration were obtained by successive dilution (Figure 1).

Batch adsorption studies

Batch mode adsorption studies for individual parameters were carried out using 250 ml Erlenmeyer flask. The effects of different parameters such as adsorbate concentration, adsorbent dose, contact time and pH were studied. The Erlenmeyer flasks were pretreated with the respective adsorbate for 24 hours to avoid adsorption of the adsorbate on the container walls. Standard solutions of the MB dye were mixed with the *Parthenium hystherophoresis* adsorbent and agitated at different agitation rate on a mechanical shaker. This was carried out by varying the MB dye concentration. Finally, the resulting suspension of each of the dye was filtered using a Whatman No.1 filter paper and the filtrate was analyzed for the corresponding MB dye concentration. Removal efficiency was finally calculated by using the relationship.

$$\text{Adsorption}\ (\%) = (C_o - C_e)/C_o) \times 100$$

Where C_0 = the initial concentration (mg/L) and C_e = final concentration (mg/L) of the MB dye being studied. The adsorption capacity of the *Parthenium hystherophoresis* is the concentration of the MB dye on the adsorbent mass and was calculated based on the mass balance principle.

$$q_e = (C_o - C_e) \times V/m$$

Where q_e = adsorption capacity of *Parthenium hystherophoresis* (mg/g)

V = the volume of reaction mixture (L)

m = the mass of adsorbent used (g)

C_0 = the initial concentration (mg/L) of the MB dye and

C_e= final concentration (mg/L) of the MB dye

Effect of contact time: Contact time is one of the most important parameters for the assessment of practical application of adsorption process.

50 ml of the working solution which is 50 mg/L concentration was put in each different conical flask. An adsorbent dose of 0.2 g/50 ml and an initial pH of 8 were used. All the flasks were put in the shaker at 150 rpm and 25°C for a predetermined time period ranging from 20 to 140 minutes, on a 20 minutes interval. Other parameters were kept constant. Then, flasks were withdrawn from the shaker, solution was filtered and absorbance of the solutions was measured. A graph was plotted with % Q vs. contact time.

Effect of adsorbent dose: 50 ml of the working solution was put in each different conical flask. Then, different adsorbent dose from (0.1, 0.2, 0.4, 0.6, 0.8 and 1 gm) was added in each flask other parameters are constant. And all the flasks were kept inside the shaker at 150 rpm and 25°C for 100 min.

After 100 minutes, the flasks were withdrawn from the shaker and the dye solutions were separated from adsorbents by using filtration. The absorbance of all the solutions was then measured. A graph was plotted with percent removal (%Q) vs. adsorbent dose. %Q is expressed as,

Where,

%Q = percentage of dye adsorbed

C_o = initial dye concentration (mg/lit)

C_e = equilibrium time solution concentration (mg/lit)

The amount of dye adsorbed onto *Parthenium hystrophorous*, q_e (mg/g), was calculated as follows:

$$q_e = \frac{(C_o - C_e)\ V}{W}$$

Where, C_o and C_e are the initial and equilibrium time solution concentrations of the dye (mg/L), respectively, V the volume of the solution (L) and W the weight of the dry adsorbent used (g).

Effect of initial pH: 50 ml of the 50 ppm methylene blue dye solution was put in each different conical flask. The optimum adsorbent dose as obtained from the above study (0.8 gm) was put in each flask. The pH of each flask was adjusted in the range of 2-10 with dilute HCl (0.1 M) and NaOH (0.1 M) solution by the use of pH meter. Then, all the flasks were kept inside the shaker at 150 rpm and 25°C for 100 minutes. After that, flasks were withdrawn, solution was filtered and absorbance of the solutions was measured. A graph was plotted with % Q vs. initial pH.

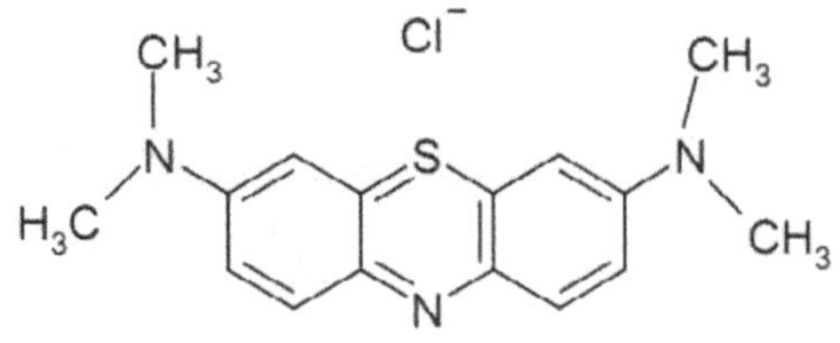

Figure 1: Structure of methylene blue dye

Effect of adsorbate concentration: 50 ml of the methylene blue dye solution with concentration ranging from 10 mg/L, 50 mg/L, 100 mg/L, 200 mg/L, 400 mg/L was put in each different conical flask. The optimum contact time, adsorbent dose and pH as obtained from the above studies was put in each flask. Then, all the flasks were kept inside the shaker at 150 rpm and 25°C. After that, flasks were withdrawn, solution was filtered and absorbance of the solutions was measured. A graph was plotted with % Q vs. adsorbate concentration.

Each experiment result was an average of three independent adsorption tests. Blank runs with only the adsorbents in 50 mL of double distilled water and 50 mL of dye solution without any adsorbent, were conducted simultaneously at similar conditions to account for any color leached by the adsorbent and adsorbed by the glass containers. Each experimental point is the average of three independent runs and all the results were reproducible with ±3 error limit.

Data analysis

The data generated were analyzed by using Microsoft Excel to compute the mean, standard deviation and linear regression values.

Results and Discussion

The effect of contact time on the removal of MB dye

The effect of contact time on the removal of unmodified *Parthenium hystrophorous* at different equilibration time is given in Figure 2. It indicates that the rate of color removal increased depending on the contact time. For the first 60 minutes, the percentage removal for MB dye by the adsorbent was rapid and thereafter it proceeds at a slower rate and finally attains saturation at different contact time for different initial concentration of the dye.

The rate of removal of the adsorbate is higher in the beginning due to the large surface area of the adsorbent available for the adsorption of dye ions [17]. After a certain period, only a very low increase in the dye uptake was observed because there are few active sites on the surface of sorbent. From the contact time studied, it was revealed that 120 minutes of agitation time is sufficient to reach equilibrium when 50 mg/L of dyes concentration was employed. Therefore, equilibrium time of 2 hours was selected for the adsorption of both dyes for further studies. A similar observation was reported for the adsorption of MG on oil palm trunk fiber [18].

Effect of adsorbent dosage

The percentage removal of dyes was studied by varying the adsorbent dose between 0.1 g and 1.00 g at a dye concentration of 50 mg/L. An attempt to enhance MB dye removal was evaluated by examining the effect of adsorbent dosage. As shown in Figure 3, it is apparent that the removal percentage of MB dye increases as the adsorbent amount increases and then becomes constant. The removal increased with increased amount of adsorbent dose up to a maximum efficiency (>90%), after which an increase in adsorbent dosage does not further improve the dye removal, implying that a complete dye removal could not be achieved even though using large amount of the adsorbents. An adsorbent dose of 0.8 g was selected for subsequent studies equilibrium studies. When too much adsorbent was added into the dye solution, the transportation of dye ions to the active adsorption sites will be limited as well, hence reduced the adsorption efficiency [15].

Influence of initial pH

The effect of initial pH on bio-adsorption percentage of MB dye was examined over a range of pH values from 2 to 10 and the results are presented in Figure 4. As elucidated in Figure 4, the dye removal was minimum at the initial pH 2. The dye adsorbed increased as the pH was increased from pH 2 to 7. Incremental dye removal was not significant beyond pH 8. For this reason, pH 8 was selected for future equilibrium studies.

After adsorption experiments, it was found that at low pH, the dyes become protonated, the electrostatic repulsion between the protonated dyes and positively charged adsorbent sites results in decreased adsorption. Higher adsorption at increased pH may be due to increased protonation by the neutralization of the negative charges at the surface of the adsorbent; which facilitates the diffusion process and provides more active sites for the adsorbent. These findings obtained in this study are in line with the results obtained by the following scholars [15].

Effect of adsorbate concentration

The adsorption of the dyes on carbon was studied by varying the carbon concentration (10-400 mg/L). The percentage of adsorption increased up to dye concentration of 50 ppm as the carbon concentration increased shown in Figure 5. This is attributed to increased weed's surface area and availability of more adsorption sites. It is apparent that the percent removal of MB dye increases rapidly with increase in the concentration of the MB dye weed due to the greater availability of the exchangeable sites or surface area at higher concentration of the sorbent.

But it gradually decreases as result of less available sites for adsorption as depicted in Figure 5. According to Anjaneya et al. [19], lower decolorization percentage at high dye concentration was reported and it was expected to happen because the inhibitory effects of high dye concentrations which have been observed in this study. Initial concentration provides a significant driving force to overcome all mass transfer resistances of the dye between the aqueous and solid phases and thus, this suggests that initial dye concentration affected dye decolorization percentages.

Adsorption isotherm

The Langmuir and Freundlich equation were employed to study the adsorption isotherms of dye.

The linearized form of the Langmuir equation [20] is as follows

$$C_e / q_e = 1/(aQ_m) + C_e / Q_m$$

where C_e (mg/L) is the concentration of the dye solution at

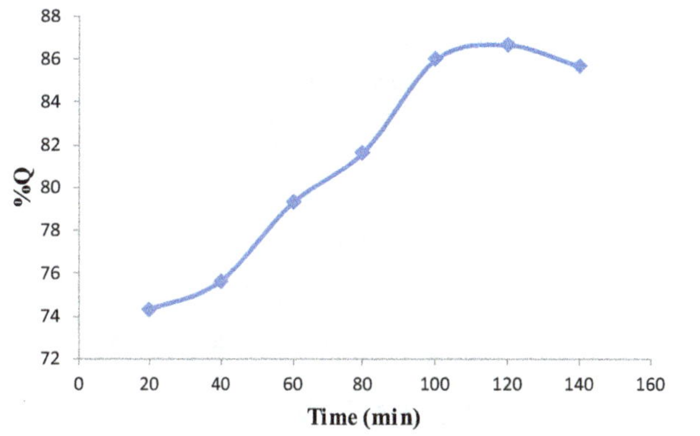

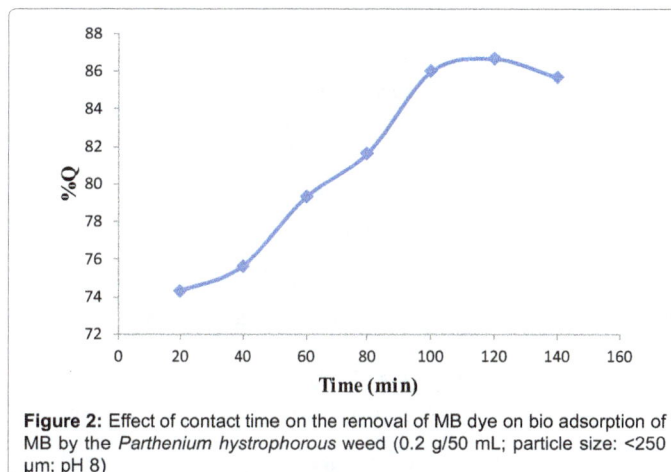

Figure 2: Effect of contact time on the removal of MB dye on bio adsorption of MB by the *Parthenium hystrophorous* weed (0.2 g/50 mL; particle size: <250 μm; pH 8)

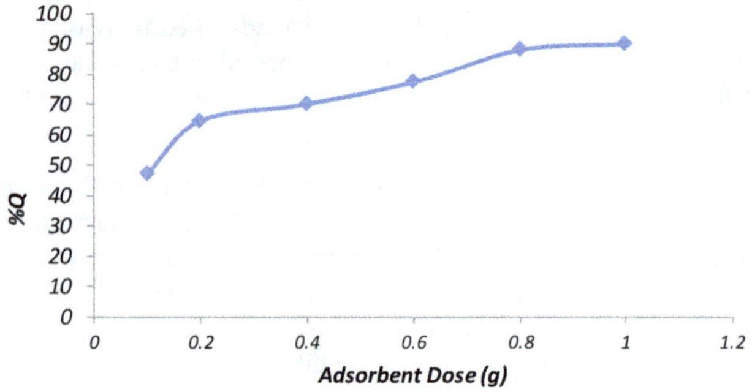

Figure 3: Effect of adsorbent dosage on bio adsorption of MB by *Parthenium hystrophorous* weed (pH 8: particle size: <250 m; contact time: 2 h, 50 mg/L)

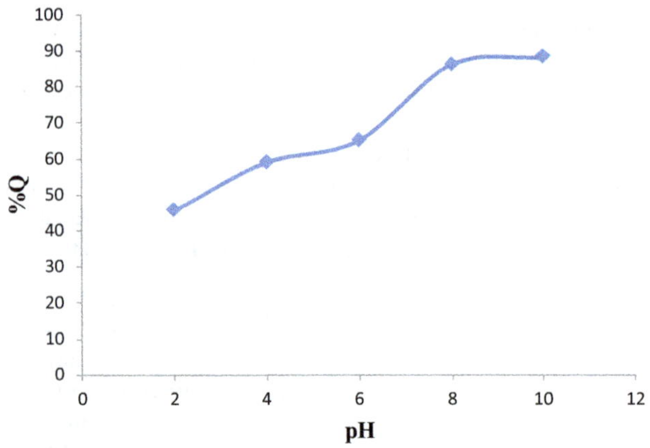

Figure 4: Influence of initial pH on bio adsorption of MB by *Parthenium hystrophorous* weed (adsorbent dose: 0.8 g/50 mL; particle size: <250 m; contact time: 2 h)

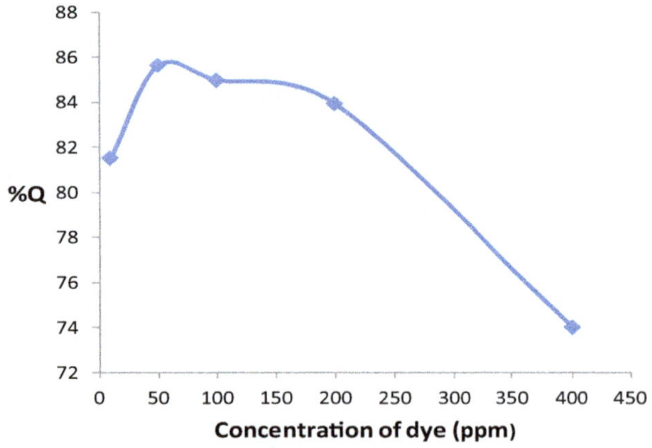

Figure 5: Effect of contact time on the removal of MB Dye on bio adsorption of MB by *Parthenium hystrophorous* weed (0.8 g/50 mL; particle size: <250 m; contact time: 2 h: pH 8)

equilibrium, q_e (mg/g) is the amount of dye adsorbed at equilibrium, Q_m is the maximum adsorption capacity and represents a practical limiting adsorption capacity when the adsorbent surface is fully covered with monolayer adsorbent molecule and a is Langmuir constant. The Q_m and a values are calculated from the slopes ($1/Q_m$) and intercepts ($1/aQ_m$) of linear plots of $1/C_e$ versus $1/q_e$ as shown in Figure 6.

The linearized form of the Freundlich equation [21] is as follows:

$$lnQ_e = ln\,K + (1/n)\,lnC_e$$

Where Q_e is the amount of dye adsorbed at equilibrium, C_e is the concentration of the dye solution at equilibrium and $1/n$ is empirical constant and indicate adsorption capacity and intensity, respectively. Their values were obtained from the intercepts (ln K) and slope ($1/n$) of linear plots of log Q_e versus log C_e as shown in Figure 7 [22].

The Q_m and a values in the Langmuir equation, the K and 1/n values in the Freundlich equation are given in the Table 2 and Figures 6 and 7. From the results in Table 2, it could be concluded that the adsorption of MB followed the Langmuir model.

Conclusion

This study confirmed that the bioadsorbent prepared from *parthenium hystrophorous*, a low cost agricultural weed, could selectively remove MB from an aqueous solution. The amount of MB dye adsorbed was found to be dependent on solution pH, adsorbent concentration, initial dye concentration, and contact time. The basic MB dye adsorption decreased at low pH values in accordance with a presupposed ion-exchange mechanism of the adsorption. The optimal pH for favorable adsorption of dye was 8. The change adsorbent dose had an effect on the bioadsorption of MB dye. The adsorption equilibrium was reached in approximately after 2 hours. From this study the adsorption equilibrium data fitted the Langmuir isotherm equation. Even though the removal efficiency of PHW is equivalent to conventional bio-adsorbents and so it is cheaply available. By using this invasive weed as economically and cheap friendly adsorbent

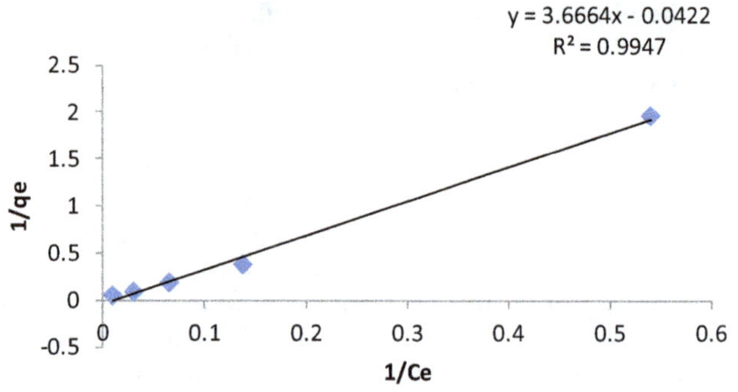

Figure 6: Langmuir adsorption isotherms of MB dye using *Parthenium hystrophorous*

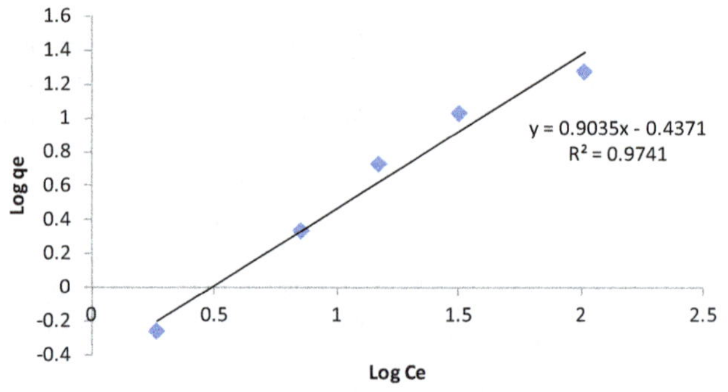

Figure 7: Langmuir Adsorption isotherms of MB dye using *Parthenium hystrophorous*

Langmuir		Freundlich	
Qm (mg/g)	a	K	1/n
23.8	3.66	2.28	0.903

Table 2: The Qm, a values in the Langmuir equation, the K and 1/n values in Freundlich equation

considerable dye removal from industrial effluents can be achieved. So it can be substituting other expensive bio-adsorbents. With the experimental data obtained in this study, it is possible to design and optimize an economical treatment process for the dye removal from industrial effluents by PHW. So as it poses killing two birds with one stone which is removing the invasive weed and dye simultaneously will be a novel adsorbent.

References

1. Robinson T, McMullan G, Marchant R, Nigam P (2001) Remediation of dyes in textile effluent: a critical review on current treatment technologies with a proposed alternative. Bioresour Technol 77: 247-255.
2. Mohammed MA, Shitu A, Ibrahim A (2014) Removal of Methylene Blue Using Low Cost Adsorbent: A Review. Research Journal of Chemical Sciences 4: 91-102.
3. Deans JR, Dixon BG (1992) Uptake of Pb^{2+} and Cu^{2+} by novel biopolymers. Water Research 26: 469-472.
4. Rozada F, Calvo LF, García AI, Martín-Villacorta J, Otero M (2003) Dye adsorption by sewage sludge-based activated carbons in batch and fixed-bed systems. Bioresour Technol 87: 221-230.
5. Rodriguez-Reinoso F (1997) Activated carbon: structure, characterization, preparation and applications. Introduction to Carbon Technologies. Marsh H, Heintz EA, Rodriguez-Reinoso F (Editors), Universidad de Alicante, Secretariado de Publicaciones.
6. Valix M, Cheung WH, McKay G (2004) Preparation of activated carbon using low temperature carbonisation and physical activation of high ash raw bagasse for acid dye adsorption. Chemosphere 56: 493-501.
7. Namasivayam C, Kavitha D (2002) Removal of Congo red from water by adsorption onto activated carbon prepared from coir pith, an agricultural solid waste. Dyes Pigments 54: 47-58.
8. Kadirvelu K, Palanivel M, Kalpana R, Rajeswari S (2000) Activated carbon prepared from an agricultural by-product for the treatment of dyeing industry wastewater. Bioresour Technol 74: 263-265.
9. Banat F, Al-Asheh S, Al-Makhadmeh L (2003) Evaluation of the use of raw and activated date pits as potential adsorbents for dye containing waters. Process Biochem 39: 193-202.
10. Juang RS, Wu FC, Tseng RL (2002) Characterization and use of activated carbons prepared from bagasses for liquid-phase adsorption. Colloid Surf A: Physicochem Eng Aspect 201: 191-199.
11. Mohamed MM (2004) Acid dye removal: comparison of surfactant-modified mesoporous FSM-16 with activated carbon derived from rice husk. J Colloid Interface Sci 272: 28-34.
12. Malik PK (2003) Use of activated carbons prepared from sawdust and rice-husk for adsorption of acid dyes: a case study of acid yellow 36. Dyes Pigments 56: 239-249.
13. Aygün A, Yenisoy-Karakas S, Duman I (2003) Production of granular activated carbon from fruit stones and nutshells and evaluation of their physical, chemical and adsorption properties. Micropor Mesopor Mater 66: 189-195.
14. Wang J, Chen C (2009) Biosorbents for heavy metals removal and their future. Biotechnol Adv 27: 195-226.
15. AbdurRahman FB, Akter M, Zainal Abedin M (2013) Dyes Removal from Textile Wastewater Using Orange Peels. International Journal of Scientific & Technology Research 2: 47-50.
16. Fahmi MR, Abidin CZA, Rahmat NR (2011) Characteristics of colour and COD removal of Azo dyes by advanced oxidation process and biological treatment. International Conference on Biotechnology and Environmental Management 18: 13-18.
17. Hameed BH, Ahmad AA (2009) Batch adsorption of methylene blue from aqueous solution by garlic peel, an agricultural waste biomass. J Hazard Mater 164: 870-875.
18. Hameed BH, El-Khaiary M (2008) Batch removal of malachite green from aqueous solutions by adsorption on oil palm trunk fibre: Equilibrium isotherms and kinetic studies. Journal of Hazardous Materials 154: 237-244.
19. Anjaneya O, Souche SY, Santoshkumar M, Karegoudar TB (2011) Decolorization of sulfonated azo dye Metanil Yellow by newly isolated bacterial strains: *Bacillus* sp. strain AK1 and *Lysinibacillus* sp. strain AK2. J Hazard Mater 190: 351-358.
20. Langmuir I (1918) The Adsorption of Gases on Plane Surfaces of Glass, Mica and Platinum. J Am Chem Soc 40: 1361-1403.
21. Teng H, Hsieh CT (1998) Influence of Surface Characteristics on Liquid-Phase Adsorption of Phenol by Activated Carbons Prepared from Bituminous Coal. Ind Eng Chem Res 37: 3618-3624.
22. Pekkuz H, Uzun I, Güzel F (2008) Kinetics and thermodynamics of the adsorption of some dyestuffs from aqueous solution by poplar sawdust. Bioresour Technol 99: 2009-2017.

6

Structural Investigation of HSP70-HSP90 and HSP90-TDF Interactions

Urmi Roy, Alisa G Woods, Izabela Sokolowska and Costel C Darie*

Department of Chemistry and Biomolecular Science, Clarkson University, USA

Abstract

Tumor Differentiation Factor (TDF) is a pituitary protein, which is secreted into the blood stream and targets breast and prostate. The end effect of TDF on these tissues is differentiation of breast and prostate cells. However, it is not yet clear how TDF induces cell differentiation. Studies in our laboratory determined that the potential TDF receptor candidates are: HSPA8, a member of the 70 kDa heat shock protein family and HSP90 protein. Our previous studies also indicated that TDF may have an inducible receptor, composed of both HSP70 and HSP90, and that TDF signaling depends on the interaction of these proteins. Here we provide additional insights about the proposed interaction between HSP70 and HSP90 and about the HSP90-TDF interaction.

Keywords: Tumor differentiation factor; Protein-protein interactions; Molecular modeling

Introduction

Tumor differentiation factor (TDF) is a recently identified protein, produced in the brain. Work in our lab identified TDF in the brain, specifically in the pituitary, but also in other regions. TDF immunostaining specifically co-localized with markers specific for neurons, but not with markers specific to astrocytes [1].

The TDF protein produced by the pituitary is likely secreted in the blood stream and targets breast and prostate tissue [2]. *In vitro* work using proteomics suggested that the TDF receptor candidates are members of the heat shock family, specifically heat shock protein 70 (HSP70) and heat shock protein 90 (HSP90) [3-5].

Proteomics allows identification and characterization of proteins in large scale [6-10] and TDF receptor candidates (and TDF ligand) were identified by affinity chromatography followed by mass spectrometry [3-5,11]. Our earlier studies also indicated that TDF may have an inducible receptor, composed of HSP70 and HSP90 [3-5]. Furthermore, we also suggested that TDF signaling depends on the interaction of HSP70 and HSP90 proteins with TDF [12]. However, the function of TDF is still not understood and many of these studies still need additional theoretical and experimental confirmation. In addition, the TDF crystal structure has not been established, thus making us rely mostly on structural biology-based work [13]. Therefore, a better understanding of TDF requires additional studies. To further understand the function of TDF, we employed structural biology to investigate the interaction of HSP70, HSP90 and TDF. Here we provide additional insights about the possible interaction between TDF and its potential HSP70 and HSP90 receptors.

Methods

For protein interaction and docking experiments, we have taken the homology model of 3C7NB, a member of the HSP70 as a ligand protein [4] and HSP90-beta as receptor protein. The 3C7NB is the open, weakly ADP bound form [14]. For HSP90, we have selected the homology model based on gi20149594 and template crystal structure 2IOQ Chain B [15]. This model receptor protein is developed using Swiss model [16,17]. 2IOQ is an open form of HSP90.

Protein-protein docking was carried out using GRAMM-X Docking Web Server v.1.2.0 [18,19] and verified by Patch dock and Fire dock servers [20-23]. Descriptions of these docking experiments are described elsewhere [3-5]. For the TDF-HSP90 interaction, we used the same homology model of HSP90 (2IOQ.pdb) as receptor protein [15], but used the model TDF structure as a ligand protein [11,12]. This model TDF protein was developed using I-Tasser server [24,25]. In our previous experiments, we have used TDF-P1 as a ligand [12].

To further substantiate we have performed another set of experiment using HSP90 model receptor based on template structure 2CG9B.PDB (Chain B) [26]. That part could be found in the supplemental materials (Figures S1-S6).

Results and Discussion

HSP70-HSP90 interaction

HSP70 proteins may interact with the proteins from the HSP90 family. Both proteins were also experimentally determined that interact with TDF. To investigate the details on TDF-HSP70-HSP90 interaction, we conducted docking experiments involving HSP70 and HSP90 proteins, and TDF-HSP90 proteins. HSP90 is composed of N Terminal Domain (NTD), Middle Domain (MD) and C Terminal Domain (CTD). HSP90 forms a dimer at the C-terminal domain [27]. Figure 1A describes the structure of the model receptor protein (HSP90, based on 2IOQ.pdb) colored from N-terminal to C-terminal. Three receptor cavities for potential ligand-binding are also labeled by numbers in this figure. Figure 1B displayed the hydrophobic surface of the model receptor protein. From Figure 1B it is clear that the ligand binding pockets are somewhat hydrophobic in nature. Figure 1C described the structural aspects of model ligand protein. The structure of the model ligand protein is based on HSP70 (HSPA8, 3C7NB.pdb) (Figure 1C). The structure of the model HSP70 is composed of three parts N-terminal nucleotide binding domain (NBD), substrate binding domain (SBD), and a linker that connects the SBD and NBD.

Figures 2 and 3 display the two tentatively identified docking sites (cavity 3 and cavity 1 of Figure 1A) of HSP70 ligand on the HSP90 model receptor model. Interfacial residues are also displayed in Figures 2C-2D and 3C-3D. These docking sites are identified using GRAMM-X docking server. One additional ligand protein binding pocket (cavity

***Corresponding author:** Costel C Darie, Biochemistry & Proteomics Group, Structural Biology & Molecular Modeling Unit, Department of Chemistry & Biomolecular Science, Clarkson University, 8 Clarkson Avenue, Potsdam, NY, 13699-5810, USA, E-mail: cdarie@clarkson.edu

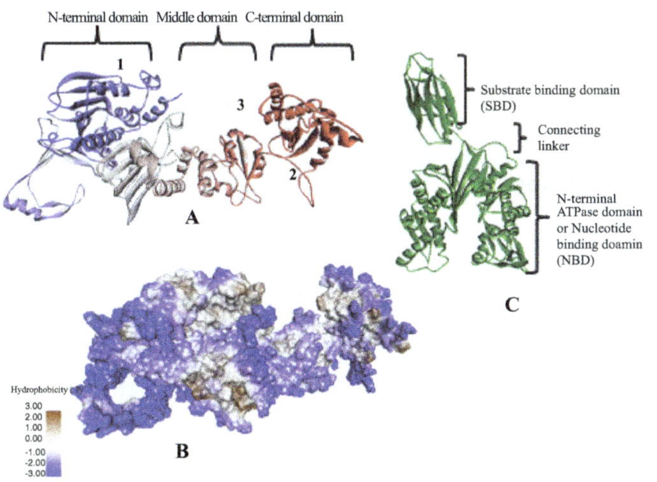

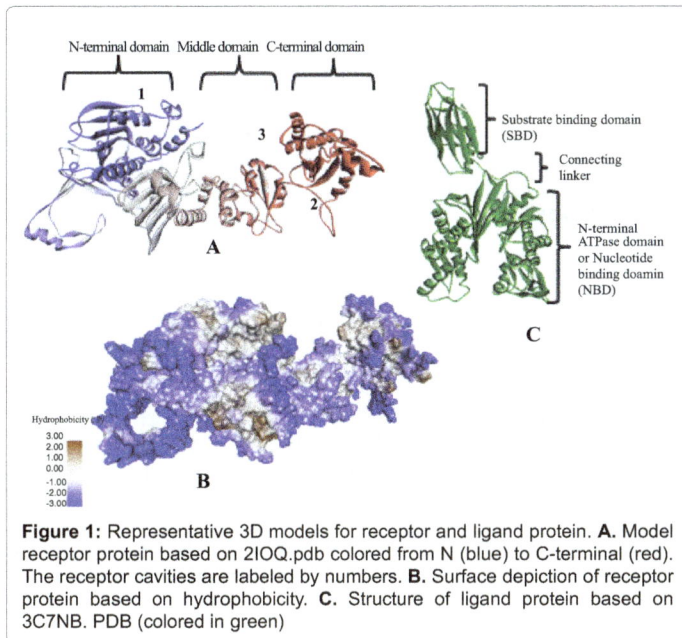

Figure 1: Representative 3D models for receptor and ligand protein. **A.** Model receptor protein based on 2IOQ.pdb colored from N (blue) to C-terminal (red). The receptor cavities are labeled by numbers. **B.** Surface depiction of receptor protein based on hydrophobicity. **C.** Structure of ligand protein based on 3C7NB. PDB (colored in green)

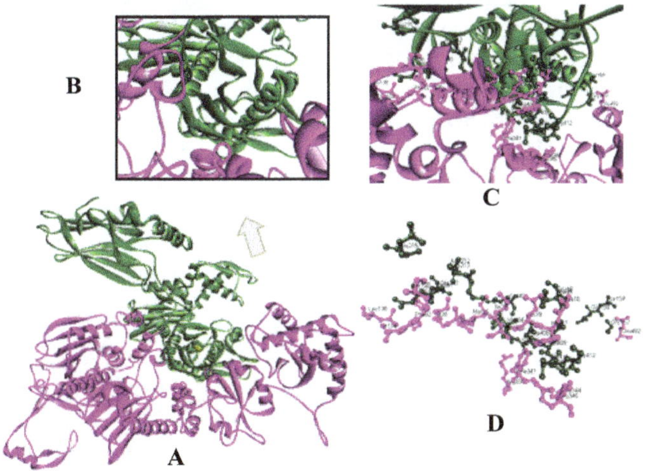

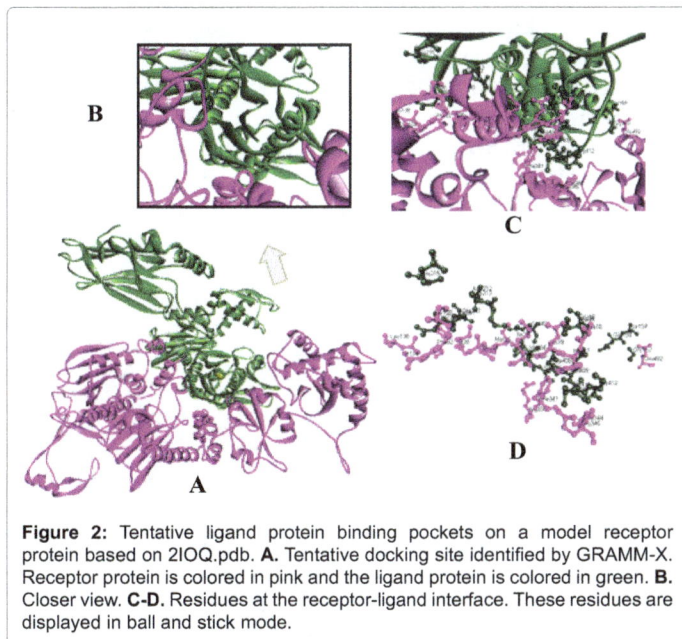

Figure 2: Tentative ligand protein binding pockets on a model receptor protein based on 2IOQ.pdb. **A.** Tentative docking site identified by GRAMM-X. Receptor protein is colored in pink and the ligand protein is colored in green. **B.** Closer view. **C-D.** Residues at the receptor-ligand interface. These residues are displayed in ball and stick mode.

2 of Figure 1A) as displayed in Figure 4 is identified using Patch dock and Fire dock server. Figure 4C and 4D are the residues of the receptor and ligand protein at the interfacial region. In Figure 2, the NBD of the ligand protein is docked onto the receptor cavity 3. In Figure 3, the SBD of the ligand protein is docked onto the receptor cavity 1. In Figure 4, the SBD is docked onto the receptor protein cavity 2. From the above computational study we can see that the SBD of HSP70 is most frequently bound to HSP90. An additional feature is observed while identifying the tentative docking site using Patch dock and Fire dock. In this particular instance, the tentative docking site of the ligand protein is traversed between the receptor protein cavity 2 and 3 of Figure 1A. When docking by GRAMM-X, the ligand protein has a higher tendency to dock in cavity 3 than cavity 1 of model receptor based on 2IOQ. In case of patch dock and fire dock web-server however, the ligand protein is mostly docked at cavity 2 and 3 than cavity 1. All these figures are made using Accelrys Discovery Studio 3.5 [28]. Overall, we have found three tentative docking sites for interaction between HSP70 and HSP90 and that From Figure 1B, we can assume that these pockets are somewhat hydrophobic in nature.

Figure 5 displays the predicted protein-protein interaction network for HSP70 kDa protein 8 (HSPA8). These analyses are performed using the The Biological General Repository for Interaction Datasets (BioGRID) [29,30] and the graphic views are generated using the web based BioGRID graphical network viewer [http://thebiogrid.org/]. The numbers of total interactions are 414 (Figure 5A). If "interactor interactions" are included then the total numbers of interactions for HSPA8 are 7160 (Figure 5B).

HSP90-TDF interaction

We also investigated the interaction between TDF ligand protein [11,12] and HSP90 receptor based on template structure 2IOQ.pdb [15]. In this case, when docking by GRAMM-X, the ligand TDF

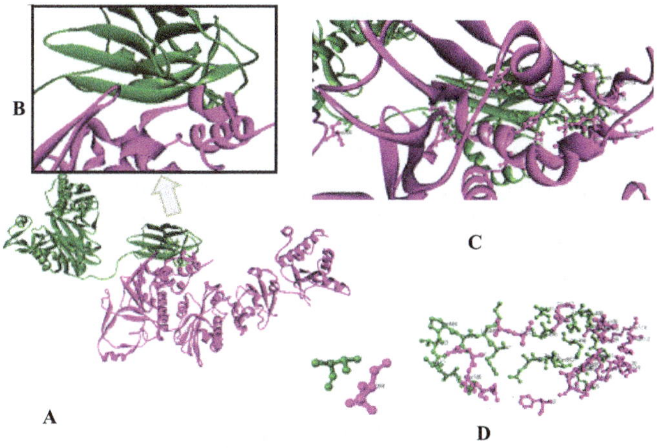

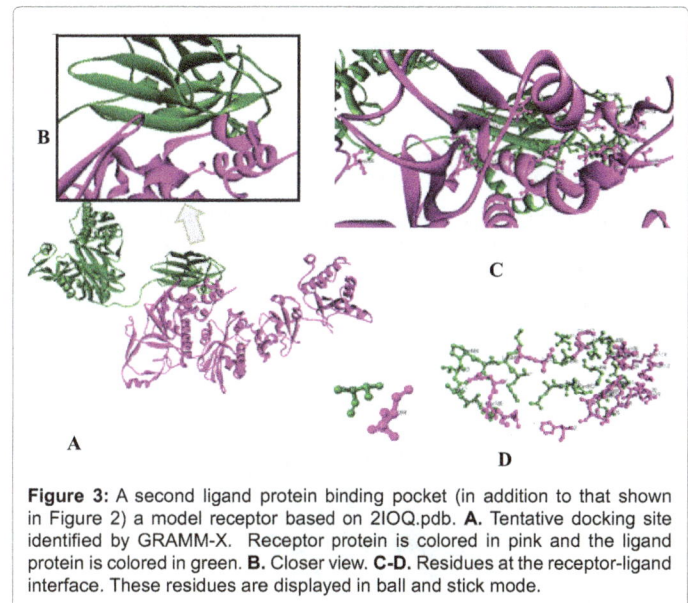

Figure 3: A second ligand protein binding pocket (in addition to that shown in Figure 2) a model receptor based on 2IOQ.pdb. **A.** Tentative docking site identified by GRAMM-X. Receptor protein is colored in pink and the ligand protein is colored in green. **B.** Closer view. **C-D.** Residues at the receptor-ligand interface. These residues are displayed in ball and stick mode.

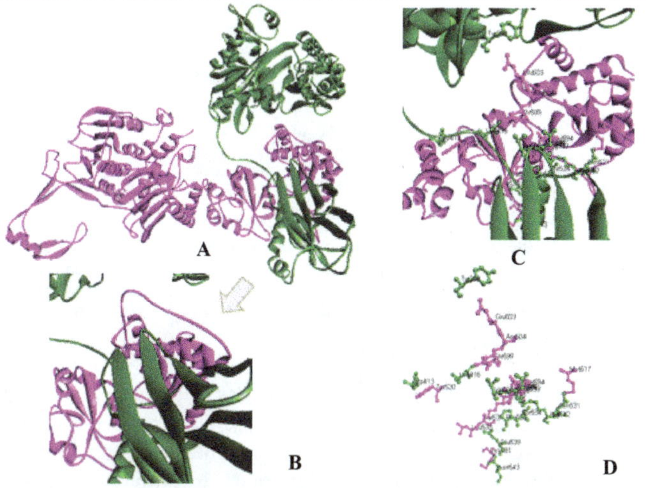

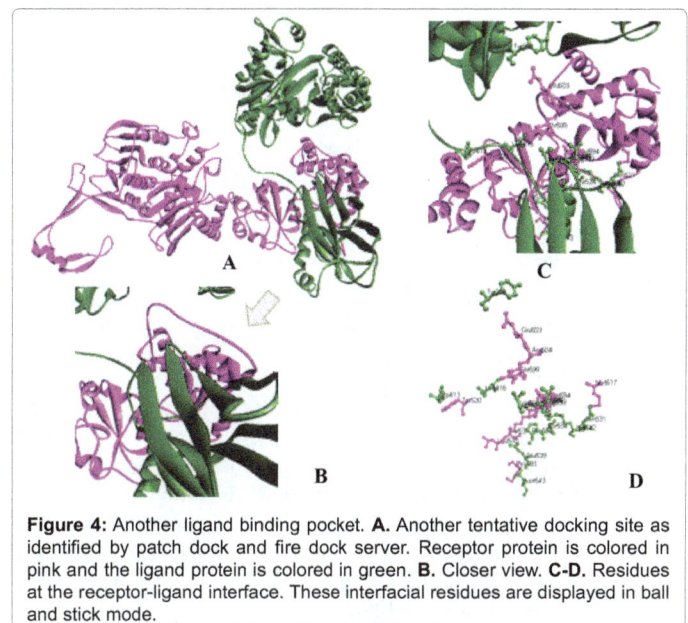

Figure 4: Another ligand binding pocket. **A.** Another tentative docking site as identified by patch dock and fire dock server. Receptor protein is colored in pink and the ligand protein is colored in green. **B.** Closer view. **C-D.** Residues at the receptor-ligand interface. These interfacial residues are displayed in ball and stick mode.

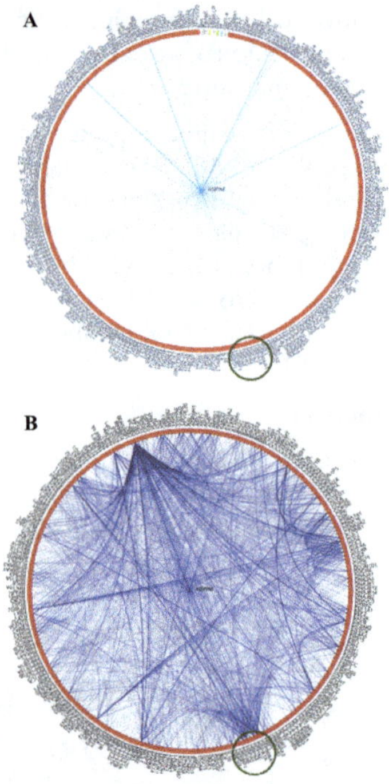

Figure 5: Prediction of interactions between HSP70 (Heat shock 70kDa protein 8; HSPA8) and HSP90. Other heat shock proteins including hsp90 are in close up. B. Graphic view "interactor interactions" of Heat shock 70kDa protein 8 (HSPA8). These figures are generated using the web-based graphical network viewer BioGRID 3.2.102.

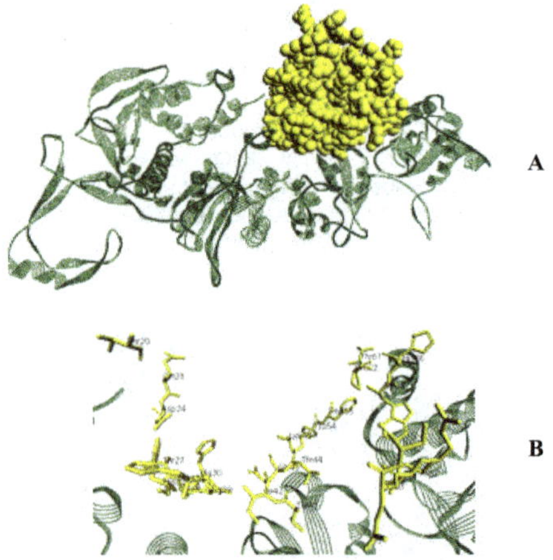

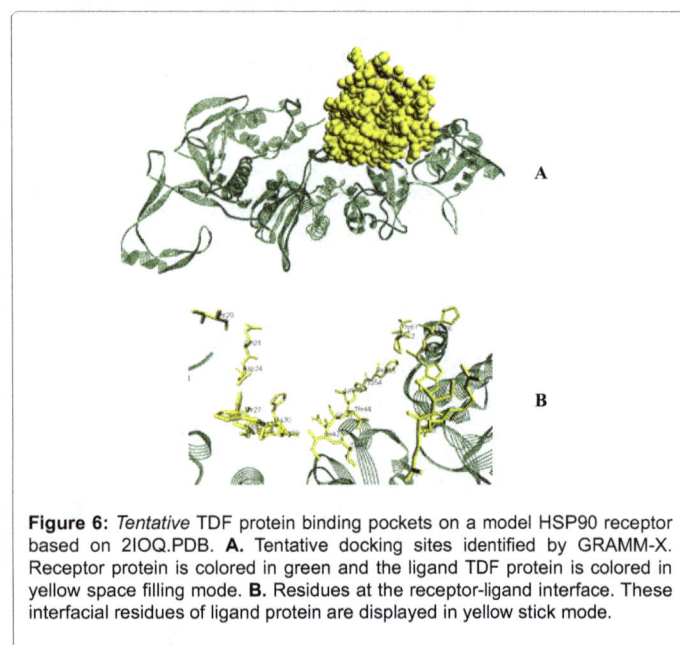

Figure 6: *Tentative* TDF protein binding pockets on a model HSP90 receptor based on 2IOQ.PDB. **A.** Tentative docking sites identified by GRAMM-X. Receptor protein is colored in green and the ligand TDF protein is colored in yellow space filling mode. **B.** Residues at the receptor-ligand interface. These interfacial residues of ligand protein are displayed in yellow stick mode.

protein has a higher tendency to dock in cavity 3 than cavity 1 of model receptor. And in this case the TDF protein does not seem to dock strongly at cavity 2 of the model receptor. These docking site using GRAMM-X webserver are displayed in Figure 6A. Interfacial residues of TDF structure are displayed in Figures 6B. In case of patch dock and fire dock web-server the ligand TDF protein however mostly docked at cavity 3 than cavity 1 and 2. Figures 6A and 6B were made using Accelrys Discovery Studio 3.5 [28].

The protein-protein interaction network (physical and genetic) for HSP90 protein is shown in Figure 7. The numbers of total predicted interactions are 906 (Figure 7A). These calculations are based on interspecies interactions and self-interaction. If "interactor interactions" are included (Figure 7B) then the total numbers of interactions are 8456 [29,30].

Conclusions

Overall, the current data suggest that TDF may indeed interact, in addition to HSP70, with HSP90. As we demonstrated previously, TDF may activate a pathway that is specific to breast and prostate cancer cells but not sensitive to other cancer or normal fibroblast or fibroblast-like cells. Our previous report also suggested that TDF-R may be a multi-subunit inducible receptor, composed of HSP70 and HSP90, and that TDF signaling depends on the interaction of these proteins. Thus TDF interacts with its receptors and induces cell differentiation through a unique, non-steroid mechanism [3-5,12].

From the above structural study we can speculate that TDF protein has a higher tendency to bind to the MD of HSP90 receptor protein. The client proteins and co-chaperones are supposed to bind in the MD of HSP90 [31-35]. The NTD of HSP90 contains a lid structure that is closed in ATP bound conformation but remains open in the ADP bound and apo forms [36]. In the case of HSP70, we find that the ligand typically binds to the SBD of HSP70, which is consistent with earlier reported findings [37,38]. Nevertheless, some binding of the ligand protein is found in the linker region of HSP70, and this type of binding has also been reported in the literature [39,40]. For HSP70, the α helical lid in the SBD plays a role in substrate binding. Here, the substrate is encompassed by the lid in a cavity comprised of β sheets. Structural reordering of helical lid is possible before and after substrate binding [41]. Ongoing experimental investigations in our laboratory

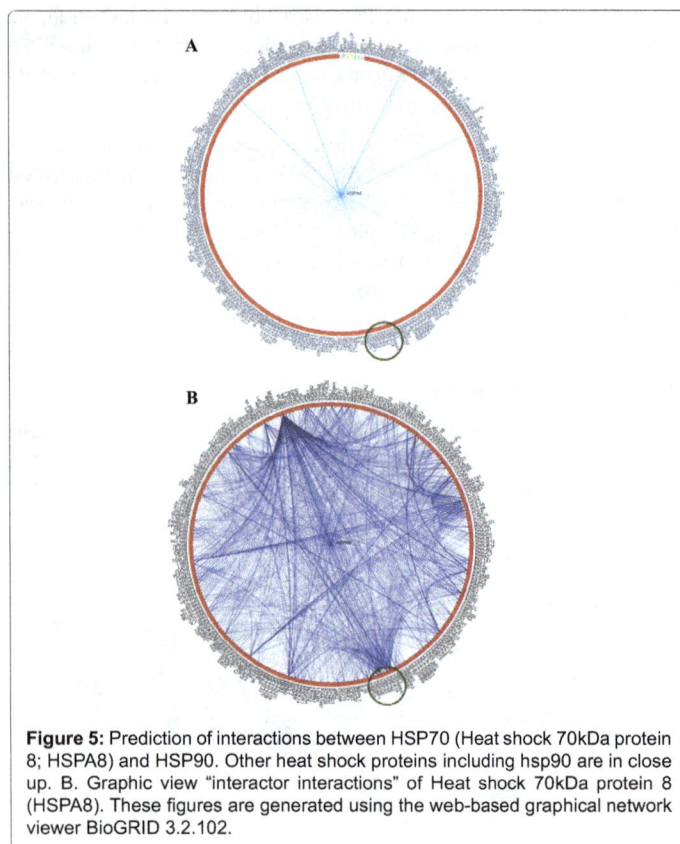

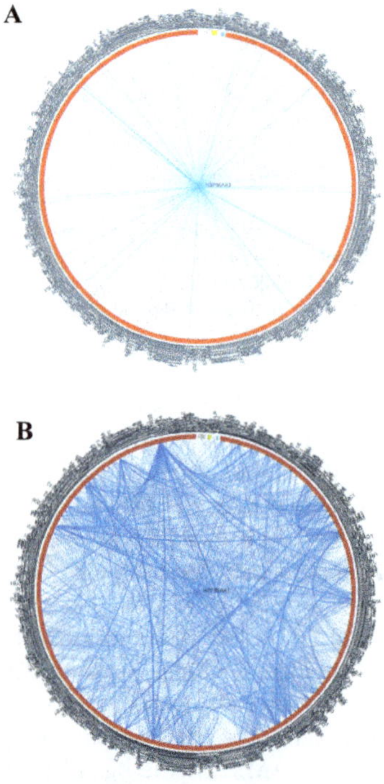

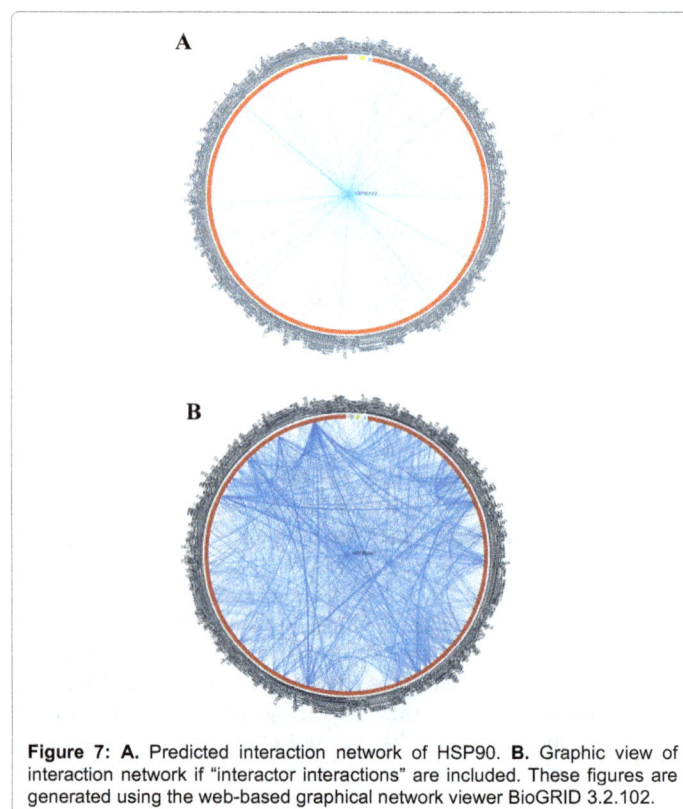

Figure 7: **A.** Predicted interaction network of HSP90. **B.** Graphic view of interaction network if "interactor interactions" are included. These figures are generated using the web-based graphical network viewer BioGRID 3.2.102.

will hopefully shed some light in the current interactions identified by structural biology.

Although HSP90 system is largely studied, there are many unanswered questions [42-45]. To our knowledge, complex formation between Hsp90 and Hsp70 are mediated through co-chaperones such as Hop/p60 or Sti1 in yeast [46-48]; no strong experimental evidence or crystal structure is available at this time to verify direct interactions of Hsp70's with hsp90's. The present work proposes the idea of a complex formation between HSP70 and HSP90 (as a TDF receptor), and a recent study on multiple myeloma also suggests the possibility of such direct interaction between Hsp70 and Hsp90 [49]. Nevertheless, since no experimental evidence is available to verify this interaction, this proposal remains an open topic for further investigation.

Acknowledgement

This work was supported in part by the Keep A Breast Foundation (KEABF-375-35054) and by support from the U.S. Army research office through the Defense University Research Instrumentation Program (DURIP grant #W911NF-11-1-0304). This work was also supported in part by Volverine Packing Co (Jim & Marsha Bonahoom) and Bonhomie Wine imports (Charlie & Claire Woods). CCD was supported during the summer of 2013 by the David A. Walsh '67 fellowship.

References

1. Woods AG, Sokolowska I, Deinhardt K, Sandu C, Darie CC (2013) Identification of tumor differentiation factor (TDF) in select CNS neurons. Brain Struct Funct.
2. Platica M, Ivan E, Holland JF, Ionescu A, Chen S, et al. (2004) A pituitary gene encodes a protein that produces differentiation of breast and prostate cancer cells. Proc Natl Acad Sci U S A 101: 1560-1565.
3. Sokolowska I, Woods AG, Gawinowicz MA, Roy U, Darie CC (2013) Characterization of tumor differentiation factor (TDF) and its receptor (TDF-R). Cell Mol Life Sci 70: 2835-2848.
4. Sokolowska I, Woods AG, Gawinowicz MA, Roy U, Darie CC (2012) Identification of a potential tumor differentiation factor receptor candidate in prostate cancer cells. FEBS J 279: 2579-2594.
5. Sokolowska I, Woods AG, Gawinowicz MA, Roy U, Darie CC (2012) Identification of potential tumor differentiation factor (TDF) receptor from steroid-responsive and steroid-resistant breast cancer cells. J Biol Chem 287: 1719-1733.
6. NgounouWetie AG, Sokolowska I, Woods AG, Wormwood KL, Dao S, et al. (2013) Automated mass spectrometry-based functional assay for the routine analysis of the secretome. J Lab Autom 18: 19-29.
7. Sokolowska I, NgounouWetie AG, Woods AG, Darie CC (2012) Automatic determination of disulfide bridges in proteins. J Lab Autom 17: 408-416.
8. Sokolowska I, Woods AG, Wagner J, Dorler J, Wormwood K, et al. (2011) Mass spectrometry for proteomics-based investigation of oxidative stress and heat shock proteins. Oxidative Stress: Diagnostics, Prevention, and Therapy 1083: 369-411.
9. Woods AG, Sokolowska I, Yakubu R, Butkiewicz M, LaFleur M, et al. (2011) Blue native page and mass spectrometry as an approach for the investigation of stable and transient protein-protein interactions. Oxidative Stress: Diagnostics, Prevention, and Therapy 1083: 341-367.
10. Darie CC,Shetty V, Spellman DS, Zhang G, Xu C, et al. (2008) Blue Native Page and Mass Spectrometry Analysis of Ephrin Stimulation-Dependent Protein-Protein Interactions in Ng108-Ephb2 Cells. Applications of Mass Spectrometry in Life Safety 3-22.
11. Roy U, Sokolowska I, Woods AG, Darie CC (2012) Structural investigation of tumor differentiation factor. Biotechnol Appl Biochem 59: 445-450.
12. Roy U, Sokolowska I, Woods AG, Darie CC (2013) Tumor Differentiation Factor (TDF) and its receptor (TDF-R): Is TDF-R an inducible complex with multiple docking sites? Mod Chem appl 1: 108.
13. Roy U, Woods AG, Sokolowska I, Darie CC (2013) Structural evaluation and analyses of tumor differentiation factor. Protein J 32: 512-518.
14. Schuermann JP, Jiang J, Cuellar J, Llorca O, Wang L, et al. (2008) Structure of the Hsp110: Hsc70 nucleotide exchange machine. Mol Cell 31: 232-243.
15. Shiau AK, Harris SF, Southworth DR, Agard DA (2006) Structural Analysis of E. coli hsp90 reveals dramatic nucleotide-dependent conformational rearrangements. Cell 127: 329-340.
16. Arnold K, Bordoli L, Kopp J, Schwede T (2006) The SWISS-MODEL workspace: a web-based environment for protein structure homology modelling. Bioinformatics 22: 195-201.
17. Kiefer F, Arnold K, Künzli M, Bordoli L, Schwede T (2009) The SWISS-MODEL Repository and associated resources. Nucleic Acids Res 37: D387-392.
18. Tovchigrechko A, Vakser IA (2005) Development and testing of an automated approach to protein docking. Proteins 60: 296-301.
19. Tovchigrechko A, Vakser IA (2006) GRAMM-X public web server for protein-protein docking. Nucleic Acids Res 34: W310-314.
20. Duhovny D, Nussinov R, Wolfson HJ (2002) Efficient Unbound Docking of Rigid Molecules. Algorithms in Bioinformatics,Lecture Notes in Computer Science 2452: 185-200.
21. Schneidman-Duhovny D, Inbar Y, Nussinov R, Wolfson HJ (2005) Patch Dock and Symm Dock: servers for rigid and symmetric docking. Nucleic Acids Res 33: W363-367.
22. Andrusier N, Nussinov R, Wolfson HJ (2007) FireDock: fast interaction refinement in molecular docking. Proteins 69: 139-159.
23. Mashiach E, Schneidman-Duhovny D, Andrusier N, Nussinov R, Wolfson HJ (2008) FireDock: a web server for fast interaction refinement in molecular docking. Nucleic Acids Res 36: W229-232.
24. Roy A, Kucukural A, Zhang Y (2010) I-TASSER: a unified platform for automated protein structure and function prediction. Nat Protoc 5: 725-738.
25. Zhang Y (2008) I-TASSER server for protein 3D structure prediction. BMC Bioinformatics 9: 40.
26. Ali MM, Roe SM, Vaughan CK, Meyer P, Panaretou B, et al. (2006) Crystal structure of an Hsp90-nucleotide-p23/Sba1 closed chaperone complex. Nature 440: 1013-1017.
27. Pearl LH, Prodromou C (2000) Structure and in vivo function of Hsp90. CurrOpinStruct Biol 10: 46-51.
28. http://accelrys.com/products/discovery-studio/
29. Stark C, Breitkreutz BJ, Reguly T, Boucher L, Breitkreutz A, et al. (2006) BioGRID: a general repository for interaction datasets. Nucleic Acids Res 34: D535-539.
30. http://thebiogrid.org/
31. Li J, Soroka J, Buchner J (2012) The Hsp90 chaperone machinery: conformational dynamics and regulation by co-chaperones. BiochimBiophys Acta 1823: 624-635.
32. Meyer P, Prodromou C, Liao C, Hu B, Roe SM, et al. (2004) Structural basis for recruitment of the ATPase activator Aha1 to the Hsp90 chaperone machinery. EMBO J 23: 1402-1410.
33. Sato S, Fujita N, Tsuruo T (2000) Modulation of Akt kinase activity by binding to Hsp90. Proc Natl Acad Sci U S A 97: 10832-10837.
34. Street TO, Lavery LA, Agard DA (2011) Substrate binding drives large-scale conformational changes in the Hsp90 molecular chaperone. Mol Cell 42: 96-105.
35. Fontana J, Fulton D, Chen Y, Fairchild TA, McCabe TJ, et al. (2002) Domain mapping studies reveal that the M domain of hsp90 serves as a molecular scaffold to regulate Akt-dependent phosphorylation of endothelial nitric oxide synthase and NO release. Circ Res 90: 866-873.
36. Blacklock K, Verkhivker GM (2014) Allosteric regulation of the Hsp90 dynamics and stability by client recruiter cochaperones: protein structure network modeling. PLoS One 9: e86547.
37. Zhu X, Zhao X, Burkholder WF, Gragerov A, Ogata CM, et al. (1996) Structural analysis of substrate binding by the molecular chaperone DnaK. Science 272: 1606-1614.
38. Mayer MP, Bukau B (2005) Hsp70 chaperones: cellular functions and molecular mechanism. Cell Mol Life Sci 62: 670-684.
39. NgounouWetie AG, Sokolowska I, Woods AG, Roy U, Deinhardt K, et al. (2014) Protein-protein interactions: switch from classical methods to proteomics and bioinformatics-based approaches. Cell Mol Life Sci 71: 205-228.

40. Yu H, Li S, Yang C, Wei M, Song C, et al. (2011) Homology model and potential virus-capsid binding site of a putative HEV receptor Grp78. J Mol Model 17: 987-995.

41. Schlecht R, Erbse AH, Bukau B, Mayer MP (2011) Mechanics of Hsp70 chaperones enables differential interaction with client proteins. Nat Struct Mol Biol 18: 345-351.

42. Wang AM, Miyata Y, Klinedinst S, Peng HM, Chua JP, et al. (2013) Activation of Hsp70 reduces neurotoxicity by promoting polyglutamine protein degradation. Nat Chem Biol 9: 112-118.

43. Mahalingam D, Swords R, Carew JS, Nawrocki ST, Bhalla K, et al. (2009) Targeting HSP90 for cancer therapy. Br J Cancer 100: 1523-1529.

44. Leach MD, Klipp E, Cowen LE, Brown AJ (2012) Fungal Hsp90: a biological transistor that tunes cellular outputs to thermal inputs. Nat Rev Microbiol 10: 693-704.

45. Li J, Buchner J (2013) Structure, function and regulation of the hsp90 machinery. Biomed J 36: 106-117.

46. Zhao R, Houry WA (2007) Molecular interaction network of the Hsp90 chaperone system. Adv Exp Med Biol 594: 27-36.

47. Pratt WB, Morishima Y, Osawa Y (2008) The Hsp90 chaperone machinery regulates signaling by modulating ligand binding clefts. J Biol Chem 283: 22885-22889.

48. Johnson BD, Schumacher RJ, Ross ED, Toft DO (1998) Hop modulates Hsp70/Hsp90 interactions in protein folding. J Biol Chem 273: 3679-3686.

49. Chatterjee M, Andrulis M, Stühmer T, Müller E, Hofmann C, et al. (2013) The PI3K/Akt signaling pathway regulates the expression of Hsp70, which critically contributes to Hsp90-chaperone function and tumor cell survival in multiple myeloma. Haematologica 98: 1132-1141.

7

The Use of Probabilistic Neural Network and UV Reflectance Spectroscopy as an Objective Cultured Pearl Quality Grading Method

Snezana Agatonovic-Kustrin* and David W Morton

School of Pharmacy and Applied Science, La Trobe Institute of Molecular Sciences, La Trobe University, Australia

Abstract

Pearl quality and value are determined as a combination of different features, with mollusk species, nacre thickness, luster, surface, shape, color and pearl size, being the most important. A pearl grader has to quantify visual observations and to assign a grading level to a pearl. The aim of this work was to reduce subjectivity in the assessment of some aspects of pearl quality by using artificial neural networks to predict pearl quality parameters from UV reflectance spectra. Given the good predictability of our previous model that used multilayer perceptron ANN modeling of UV-Visible spectra to predict the grade of pearls, we wanted to simplify and improve the model by reducing the spectral input to UV only and by using classifier neural network modelling. It is hypothesized that as UV light is of higher energy than visible light, it may penetrate further into the surface of the pearl, and hence the corresponding UV diffuse reflectance spectrum may provide more information that can be used to assess pearl quality.

The developed models were successful in predicting mollusk pearl growing species, pearl and donor color, luster, and surface complexity. The simplified models have been built resulting in more accurate prediction of selected pearl quality parameters when compared with the previous reported model.

Keywords: Probabilistic neural network; Diffuse reflectance ultraviolet spectroscopy; Pearl grading; Pearl quality

Introduction

A pearl grader has to quantify visual observations and to assign a grading level to a pearl. However, there is no international standard method for overall pearl grading [1] and identical pearls may be graded differently by different pearl graders. Usually pearls are classified according to their origin (mussel species) and then graded by assessing the size, nacre thickness, shape, color, luster, and surface (Table 1).

When pearl grading, the appearance of the surface of a pearl is one of the most important characteristics in determining its overall desirability and value. Ideally, the pearl's surface should be smooth, clean and shiny. The presence of imperfections and blemishes on the pearl surface can significantly decrease the value of a pearl, with only 30% of the cultured pearls harvested categorized as high quality. The surface of a cultured pearl is examined in terms of the number, size, type and location of the imperfection (Table 2).

Blemishes may range from small spots to big chips, or cracks or chalky (calcareous) bumps on the pearl surface [2]. However, in grading South Sea pearls, big ridges forming rings (usually more than three rings) in a pearl is categorized as a circled pearl. Small spots (non-calcareous) on the surface of a pearl are usually removed by polishing the pearl after harvesting.

1	Brilliant, very bright pearl, like a mirror. Light appears to reflect from within the pearl (inner glow).
2	Excellent. A bright pearl but has a slightly blurred reflection
3	High luster. A pearl with minimal inner reelection but blurred
4	Modest luster. A pearl that appears slightly opaque, the reflection is not clear
5	Poor luster. Opaque to the point appearing milky. Commercially not for sale.

Table 1: Lustre category.

A1	No blemishes or one very small blemish that can be removed by drilling.
B1	One to three very small blemishes in close proximity with the majority of the pearl surface being clear
B2	Three or more blemishes but with at least one clean face visible on the pearl
C1	Minor blemishes all over the pearl surface (no clean face) or one to two large blemishes that affect 70% of the pearl surface. Wrinkled or scratched pearls fall into this category.
C2	Blemishes on entire surface, spots are calcified
D1	A commercially reject pear. A pear that does not fall into above categories.

Table 2: Surface complexity of the pearls.

Given the good agreement of the previous study that employed the multilayer perceptron (MLP) artificial neural network (ANN) [3] (one of the most commonly used networks), to correlate diffuse reflectance UV-Visible spectra with pearl quality parameters, we wanted to investigate if reducing the number and type of spectral inputs (to UV only) and using the more specific probabilistic neural network (PNN) for correlation we could further improve model predictability. It is well established that the presence of redundant inputs that are not highly related to the outputs, increases the size of a neural network and may result in the network taking longer to train and may provide misleading results. It is hypothesized that as UV light is of higher energy than visible light, it may penetrate further into the surface of the pearl, and hence the corresponding UV diffuse reflectance spectrum may provide more information that can be used to assess pearl quality. A PNN is predominantly a classifier that combines attributes of statistical pattern recognition and feed forward artificial neural networks [4]. In the past decade the ANN modeling technique has found applications for recognition and classification of spectra from a variety of spectroscopic methods [5-9]. Different network architectures, including a MLP, radial basis function (RBF), self-organizing map (SOM), and probabilistic neural network (pnn), have been proposed for classification purposes. However, the PNN has proven to provide a better general solution to pattern classification problems by following Bayesian estimation theory, a statistically developed approach [10,11].

***Corresponding author:** Snezana Agatonovic-Kustrin, School of Pharmacy and Applied Science, La Trobe Institute of Molecular Sciences, La Trobe University, Edwards Rd, Bendigo, 3550, Australia, E-mail: s.kustrin@latrobe.edu.au

Experimental

Data collection

This study was carried out on twenty eight of naturally-colored cultured pearls obtained from commercial pearl farms, 11 freshwater pearls from Zhuji (Zhejiang, China), 4 Akoya pearls from Japan, 5 Tahitian pearls from the South pacific and 8 pearls from a farm in Bali, Indonesia. Pearls were graded according to the South Sea Pearl Grading System issued by Atlas Pacific, Ltd. [12] and had not been subjected to any color or luster enhancing treatments (Table 3). Pearls were classified according to mollusk species, pearl quality (shape, color, luster, and surface complexity), and donor color.

Apparatus

Since pearls are opaque in nature, spectroscopy measurements were performed using diffuse reflectance spectroscopy. The diffuse reflectance spectra were collected using a Cary 50 UV-Vis spectrophotometer (Varian, Inc.) with a remote diffuse reflectance accessory (DRA) probe (Barrelino TM, Harrick Scientific). The scan rate was 9600 nm/min. The spectra were measured using appropriate baseline correction at 100% transmittance. Prior to scanning, the white level was calibrated with a wavelength reflectance standard (Labsphere®), in which approximately 100% reflectance across the entire spectrum is designated as a white reference standard. The spectra were acquired at two different locations on each pearl sample in order to assess surface homogeneity, so that spectral data were independent on the shape of the pearl.

Statistica Neural Networks 9.0 (StatSoft, Tulsa, OK, US) was used to model the spectral data (inputs) and correlate it to pearl quality graded values (outputs). The recorded spectra were post-processed to smooth the noise according to the central moving average method by calculating 102 average spectral values, each from twenty consecutive wavelength [13]. These averaged spectral values were used as inputs to the ANN model. From measured spectral data, training (70% or a 20 pearl data set), testing (15% or a 4 pearl data set) and validation sets (15% or a 4 pearl data set) were randomly selected before each training run. Spectral data were used as inputs and pearl quality assessment parameters (mussel species, pearl color and shape complexity, donor color, and donor condition) were used as categorical outputs to train, test, and validate 6 different ANNs.

Artificial neural networks (ANNs)

An artificial neural network is an information processing model that is inspired by the way biological nervous systems process information and learns from examples. Neural networks are composed of a large number of highly interconnected processing elements or artificial neurons that are organized in successive layers. Each neuron that receives information, process the information and produces an output. Neurons that receive information from outside the network (i.e. spectral data) are called input neurons. Neurons that receive information from other neurons are hidden neurons and neurons whose outputs are used as target values (pearls quality descriptors) are output neurons. Connection weights and the number of hidden neurons in an ANN are adjustable parameters that are optimized during the learning phase.

Pearl property	Model topology*	Pearl property	Model topology*
Mollusc species	Linear 93-4	Lustre	PNN 94-20-5
Pearl colour	Linear 99-12	Surface complexity	PNN 98-20-6
Donor colour	PNN 97-20-4		

*Number of inputs-hidden neurons-outputs.

Table 3: Developed ANN models.

There are different ways in which information can be processed by a neuron and different ways of connecting the neuron to one another. Different neural network structures can be constructed by using different processing elements and by the specific way in which they are connected.

After running 30 tests, each evaluating 250 different ANN topologies, linear (LNN) and probabilistic neural networks (PNN) were selected due to their superiority in network performance. A general principal is that a simple model should always be chosen in preference to a complex model if the complex model does not fit the data better. In terms of function approximation, LNN model without hidden layers is the simplest. The network simply multiplies the input by the weights and produces the output.

The architecture of PNN is feed forward, but differs in the way that learning occurs. A PNN is a supervised learning algorithm but includes no weights in its hidden layer. Instead the training data set is acting as the weights to the hidden node. Thus, weights are not adjusted at all. Every training case is copied to the hidden layer of the network, which applies a Gaussian function to the data. The output layer represents each of the possible classes for which the input data can be classified. As no training is required, classifying an input vector is fast, as is simply depends on the number of classes and input data that are present.

Results and Discussion

The UV diffuse reflectance spectrum is a unique property of a pearl and different pearls will generally have different spectra due to differences in nacre composition (Figures 1-3) [3].

All of the investigated pearls show a decrease in diffuse reflectance at around 260 nm followed by a peak at around 280 nm and then a decrease at around 340 nm, regardless of their color (Figures 1-3). Seven Hybrid ANN models with different topology were trained, tested and validated to correlate UV spectra with mollusk pearl growing species, pearl and donor color, luster, and surface complexity (Table 3).

Nondestructive methods for determining the parentage of pearls have practical importance since commercial value of pearls depends on the mother mollusk. The model developed to classify mollusk species was a linear network with 93 spectral inputs and 4 different categories as outputs for four different mollusk species (*P. maxima, P. margaritifera, P. fucata martensi* and freshwater mussel (species unknown for the

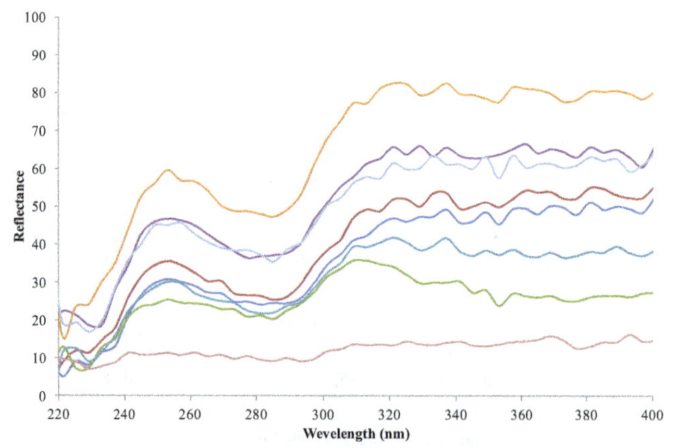

Figure 1: UV Spectral data of eight South sea pearls from *Pinctada maxima.*

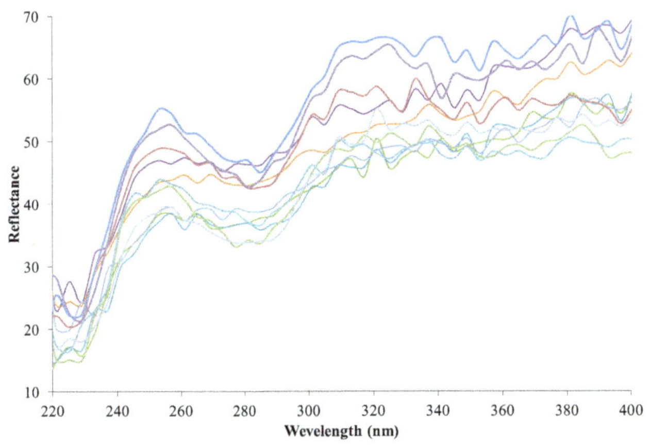

Figure 2: UV spectral data of eleven freshwater pearls from unknown species of freshwater mussels.

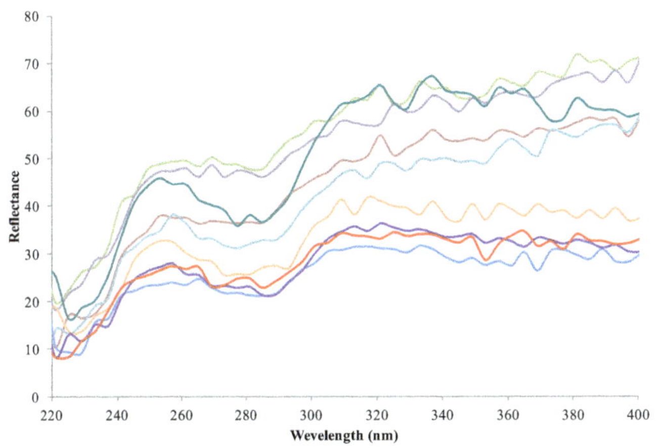

Figure 3: UV spectral data of four Akoya pearls from *Pinctada fucata martensi* and five Tahitian pearls from *Pinctada margaritifera*.

freshwater pearls used)). It clearly distinguished between freshwater and saltwater cultured pearls. However for the testing data set, it made mistakes by predicting *P. margaritifera* instead *P. maxima* (sample 2) and *P. fucata martensi* (Japan's Akoya oyster) instead of *P. margaritifera* (sample 27), while for the validation set it predicted *P. maxima* instead of *P. margaritifera* (sample 26) (Table 4).

The natural colors observed in a pearl are determined by the type of oyster or freshwater mussel that produces the pearl, the water conditions, and the nature of the mantle cells that were used to supply the graft implant. The donor tissue is chosen from oysters with attractive colors in the nacre lining their shells, which gives an indication as to the color of the resulting pearl [14,15]. Pearl color is a more subjective indicator of a pearl value, with white South Sea pearls, especially those with a pink overtone, having the highest commercial value [16]. A pearl's color is defined by the nature and the relative proportions of the organic pigments in the nacre [2] (body color), together with reflection and refraction of light [17] (overtone of secondary color).

Using the developed LNN model (99-12), donor color was successfully correlated with the UV spectral data with only one mistake observed (predicting yellow instead unknown donor color (sample 27). However in the case of the pearl color, the developed PNN 97-20-4 model made a mistake by predicting pink (pearl 2), lavender (pearl 26) and cream with overtone (pearl 27) instead of white (pearl 4) and various colors (pearls 26 and 27) (Table 4).

All natural colors of cultured pearls originate from a mixture of organic pigments, carotenoids or polyenes [18,19]. The *P. fucata* oyster produces white or cream pearls and doesn't naturally produce black pearls. Cultured pearls from *P. margaritifera* and *P. maxima* are distinguished by their color [1] with *P. maxima* producing pearls with a base color of gold or silver, while *P. margaritifera* produce predominately black color based pearls. *P. margaritifera* naturally secretes a black pigment, which, depending on the quantity, gives predominantly black pearls, with a basic color ranging from black to gray. *P. maxima* are a large oyster, also called the gold or silver-lip oyster depending on the color of its shell lip. The oyster from Northern Australia is the silver-lip oyster which tends to produce mostly white pearls with a silver overtone. The gold-lip oyster in the sea around Indonesia, Thailand and the Philippines, produces yellow or creamy South Sea pearls [20]. Freshwater cultured pearls occur in four main hues, white, gray, orange and purple. However, due to various combinations of tone and saturation, a broader range of color appearances is observed.

A pearl's color is described by its main body color, overtone or glow, and overtone of secondary color. Usually body colors are enhanced by at least one overtone color. An overtone color is a translucent color that may sometimes appear on a pearl together with its main body color [1] and slightly alters the body color [2]. However, a pearl's overtone color(s) is distinct from its basic color, and this result in pearls in the same color category having very different looks and hues. The phenomenon of overtone of secondary color shows glittering of various colors in a pearl due to the interference and diffraction of light interacting with the specific structure on the pearl's surface [1,14].

Both, yellow and gold cultured pearls exhibit broad absorption from 330 to 460 nm [21], with two absorption maxima, one between 350 and 365 nm and the second weaker maximum from 420 to 435 nm. The strength of these absorption maxima increases with increasing saturation of the yellow color. The absence of the UV absorption indicates treated color, regardless of the treatment method used. The relative intensity of the absorption band at 356 nm is positively correlated to the concentration of the yellow organic pigments present in the nacre [22] (Table 4).

For pearl luster, model (PNN 94-20-5) predicted luster predicted luster 3, luster 1 and luster 2 instead luster 2, luster 2 and luster 1 for the samples 10, 15 and 26. The pearl's surface luster is critical in evaluating pearl quality. It depends upon the reflection and refraction of light from the translucent layers of the nacre (Table 1). The luster of a pearl may be closely related to the homogeneity, light transmittance and quality of the nacre. A pearl's luster is a measure of its brilliance and the reflectivity of a pearl. High-quality pearls are bright and shiny; with high reflectivity (mirror-like reflectivity) while lower-quality pearls have a chalky or dull appearance. The error in the prediction of the surface complexity was minor. However, there was an issue with South Sea pearl sample coded 2, where the model predicted B2 (one to three very small blemishes) instead of C1 (minor blemishes all over the pearl or one or two large blemishes) (Table 2). It is important to note that no predictive model will be 100% accurate. We also need to recognize that pearls graded by humans are also not 100% accurately graded. For example, there can be differences in peoples' perception of luster. Given that our developed model is dependent on actual graded pearls, any error made in the grading of these pearls will affect the accuracy of the

	Mollusc Species		**Pearl colour**		**Donor colour**		**Pearl lustre**		**Surface Complexity**	
	Species	**Linear 93-4**	**graded**	**PNN99-12:1**	**graded**	**PNN 97-20-4**	**graded**	**PNN 94-20-5**	**graded**	**PNN98-20-6**
	*P.maxima*	*P. maxima*	white	white	silver	silver	3	3	C2	C2
	*P. maxima*	*P. margaritifera*	white	white/ pinkish	silver	silver	3	3	C1	B1
	*P. maxima*	*P. maxima*	gold	gold	yellow	yellow	2	2	A1	A1
	*P. maxima*	*P. maxima*	white	pink	yellow	yellow	3	3	C1	B1
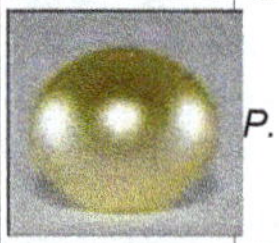	*P. maxima*	*P. maxima*	cream with various overtone	cream with various overtone	yellow	yellow	2	2	B2	B2
	*P. maxima*	*P. maxima*	cream with cream overtone	cream with cream overtone	yellow	yellow	1	1	B1	B1
	*P. maxima*	*P. maxima*	white with pink overtone	white with pink overtone	yellow	yellow	2	2	B2	B2
	*P. maxima*	*P. maxima*	reject	reject	silver	silver	reject	reject	reject	reject
	Freshwater mussel*	Freshwater mussel	white	white	white	white	1	1	B1	B1
	Freshwater mussel*	Freshwater mussel	white	white	white	white	2	3	B1	B1
	Freshwater mussel*	Freshwater mussel	white	white	white	white	3	3	B1	B1
	Freshwater mussel*	Freshwater mussel	white	white	white	white	3	3	B1	B1
	Freshwater mussel*	Freshwater mussel	white	white	white	white	4	4	B1	B1

	Freshwater mussel*	Freshwater mussel	lavender	lavender	white	white	1	1	B1	B1
	Freshwater mussel*	Freshwater mussel	lavender	lavender	white	white	2	1	B1	B1
	Freshwater mussel*	Freshwater mussel	lavender	lavender	white	white	3	3	B1	B1
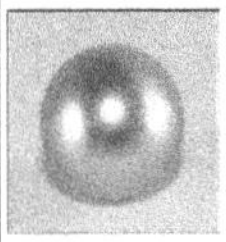	Freshwater mussel*	Freshwater mussel	pink	pink	white	white	1	1	B1	B1
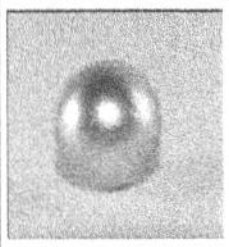	Freshwater mussel*	Freshwater mussel	pink	pink	white	white	2	2	B1	B1
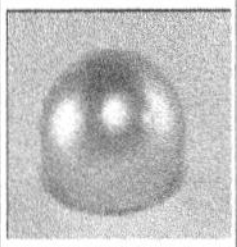	Freshwater mussel*	Freshwater mussel	pink	pink	white	white	3	3	B1	B1
	*P. fucata martensi*	*P. fucata martensi*	white	white	white	white	1	1	B1	B1
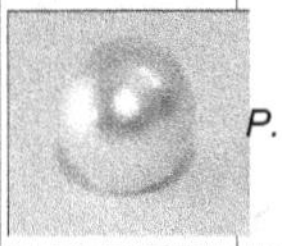	*P. fucata martensi*	*P. fucata martensi*	white	white	white	white	2	2	B1	B1
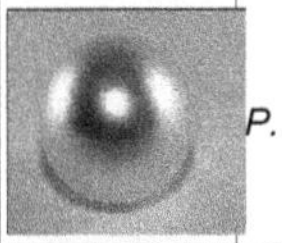	*P. fucata martensi*	*P. fucata martensi*	white	white	white	white	3	3	B1	B1
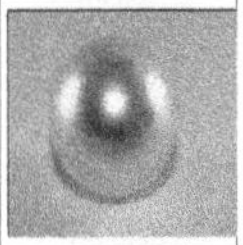	*P. fucata martensi*	*P. fucata martensi*	white/ pinkish	white/pinkish	white	white	4	4	B1	B1
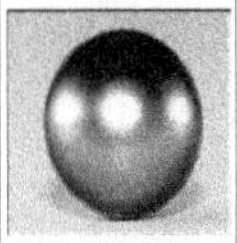	*P margaritifera*	*P. margaritifera*	silver with green overtone	silver with green overtone	unknown	unknown	1	1	B2	B2
	*P. margaritifera*	*P. margaritifera*	silver with various overtone	silver with various overtone	unknown	unknown	1	1	B2	B2
	*P. margaritifera*	*P. maxima*	various colour	lavender	unknown	unknown	1	2	B2	C1
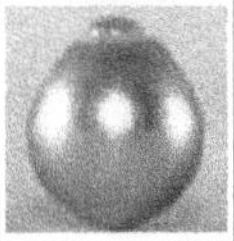	*P. margaritifera*	*P. fucata martensi*	various colour	cream with cream overtone	unknown	yellow	1	1	B2	B2

	*P. margaritifera*	*P. margaritifera*	various colour	various colour	unknown	unknown	2	2	C1	C1

*Species unknown.

Table 4: Mollusc species graded and predicted pearl and donor colour, lustre and surface complexity using the optimised ANN models.

model. It is important to note that the advantage of using a predictive model of this type is that it is not subject to the errors associated with the use of human graders provided an accurate set of graded pearls is used in the modelling process and a predictive model of high accuracy is produced. This approach will minimize the difference between pearls graded using a predictive model and graders.

Conclusions

Diffuse reflectance UV spectroscopy combined with PNN data modeling, successfully classified and graded 28 different pearls, resulting in more accurate prediction of selected pearl quality parameters (mollusk species, pearl and color donor, luster, and surface complexity) when compared with the previous model [3]. 25 out of 28 mollusk species were correctly classified (90%) and similarly 90% of pearl color was correctly predicted with developed models. In the case of donor color only 1 prediction was wrong and for the luster and surface complexity relative error was 7% (2 out of 28). The simplified LNN and PNN models have been developed without any loss of accuracy in prediction. There are still few prediction errors observed, most likely to be attributable to the relatively small number of pearl samples. The proposed method may thus provide non-destructive and objective evaluation of pearl quality within minutes and rapid scanning of large number of pearls using simple and cost effective apparatus.

References

1. Wegst UGK, Ashby MF (2004) The mechanical efficiency of natural materials. Philosophical Magazine 84: 2167-2181.
2. Olmos P, Diaz JC, Perez JM, Gomez P (1991) A new approach to automatic radiation spectrum analysis. IEEE Transactions on Nuclear Science 38: 971-975.
3. Agatonovic-Kustrin S, Morton DW (2012) The use of UV-visible reflectance spectroscopy as an objective tool to evaluate pearl quality. Mar Drugs 10: 1459-1475.
4. Ancona F, Colla A, Rovetta S, Zunino R (1997) Implementing probabilistic neural networks. Neural Comput & Applic 7: 37-51.
5. Taylor JJU, Strack E (2008) Pearl production, Elsevier, Amsterdam.
6. Agatonovic-Kustrin S, Tucker IG, Schmierer D (1999) Solid state assay of ranitidine HCl as a bulk drug and as active ingredient in tablets using DRIFT spectroscopy with artificial neural networks. Pharm Res 16: 1477-1482.
7. Sarikaya M, Aksay IA (1995) Biomimetics: Design and Processing of Materials. AIP Press, Woodbury, NY, USA.
8. Amendolia SR, Doppiu A, Ganadu ML, Lubinu G (1998) Classification and quantitation of 1H NMR spectra of alditols binary mixtures using artificial neural networks. Anal Chem 70: 1249-1254.
9. Bos M, Weber HT (1991) Comparison of the training of neural networks for quantitative x-ray fluorescence spectrometry by a genetic algorithm and backward error propagation. Anal Chim Acta 247: 97-105.
10. Romero RD, Touretzky DS, Thibadeau RH (1997) Optical Chinese character recognition using probabilistic neural networks. Pattern Recognition 30: 1279-1292.
11. Specht DF (1990) Probabilistic Neural Networks. Neural Networks 3: 109-118.
12. Shadmehr R, Angell D, Chou PB, Oehrlein GS, Jaffe RS (1992) Principal component analysis of optical emission spectroscopy and mass spectrometry: Application to reactive ion etch process parameter estimation using neural networks. J Electrochem Soc 139: 907-914.
13. Milosevic M, Berets SL (2006) Accessories and Sample Handling for Mid-Infrared Diffuse Reflection Spectroscopy. In: Chalmers JM, Griffiths P (eds.) Handbook of Vibrational Spectroscopy. John Wiley & Sons, Ltd, USA.
14. Acosta-Salmón H, Martinez-Fernández E, Southgate PC (2004) A new approach to pearl oyster broodstock selection: can saibo donors be used as future broodstock? Aquaculture 231: 205-214.
15. Mamangkey NG, Agatonovic S, Southgate PC (2010) Assessing pearl quality using reflectance UV-Vis spectroscopy: does the same donor produce consistent pearl quality? Mar Drugs 8: 2517-2525.
16. Gilbert PU, Metzler RA, Zhou D, Scholl A, Doran A, et al. (2008) Gradual ordering in red abalone nacre. J Am Chem Soc 130: 17519-17527.
17. Raman CV, Krishnamurti D (1954) On the chromatic diffusion halo and other optical effects exhibited by pearls. Proceedings of the Indian Academy of Sciences 39A: 265-271.
18. de Oliveira VE, Castro HV, Edwards HGM, de Oliveira LFC (2009) Carotenes and carotenoids in natural biological samples: a Raman spectroscopic analysis. Journal of Raman Spectroscopy 41: 642-650.
19. Karampelas S (2007) Pigments in natural-color corals. Gems & Gemmology 43: 95-96.
20. Edge R, McGarvey DJ, Truscott TG (1997) The carotenoids as anti-oxidants--a review. J Photochem Photobiol B 41: 189-200.
21. Elen S (2001) Spectral reflectance and fluorescence characteristics of natural-color and heat-treated "golden" south sea cultured pearls. Gems & Gemology 37: 114-123.
22. Li-jian Q, Yi-lan H, Zeng CG (2008) Colouration Attributes and UV-NIS Reflection Spectra of Various Golden Seawater Cultured Pearls. Gems & Gemnology 10: 1-8.

Classic Radionuclide $^{188}W/^{188}Re$ Generator (Experiments, Design and Construction)

8

Marcin Konior* and Edward Iller

National Centre for Nuclear Research, Radioisotope Centre Polatom, 05-400 Otwock, Andrzej Sołtan 7, Poland

Abstract

Rhenium-188 belongs to the group of beta-gamma emitters. A radiometric characteristic of radiation emitted by Rhenium-188 creates advantageous conditions for medical applications of this radionuclide. The Radioisotope Center Polatom Poland has developed and implemented the technology for routine production of carrier free ^{188}Re. A production line for the preparation of sterile, isotonic solution carrier-free ^{188}Re sodium perrhenate (VII) has been constructed. On the basis of collected experiences, the manufacturing of $^{188}W/^{188}Re$ generators, in which chromatographic column is loaded with alumina has been established. At the present time the generators with activity of 3.7-37 GBq are available.

Keywords: 188W/188Re generator; Rhenium-188; Radioisotope generator

Introduction

Rhenium-188 is an important beta-gamma emitter, which emits beta particles with an average energy of 784 keV and a maximal energy of 2.11 keV as well as gamma photons with energy of 155 keV (15%) and a half-life time of 17 hours. There are important properties which make this radioisotope of interest for clinical use. The beta particles have a penetration range in soft tissues of about 8-10 mm which is convenient for the destruction of tumor tissue. In addition the gamma emission permits quantitative gamma camera imaging for evaluation of biokinetics and dosimetric estimations.

Rhenium -188 exhibits chemical reactivity similar to technecium-99m and can be converted to chemical forms required for preparation of various therapeutic radiopharmaceuticals.

A number of important therapeutic applications of rhenium-188, which have been developed over the two decades, have demonstrated uses of rhenium-188 as a cost effective alternative to more expensive and/or less readily available therapeutic radioisotopes. Clinical trials include the use of ^{188}Re-HEDP and -DMSA for the treatment of metastatic bone pain and various ^{188}Re-labeled HDD/Lipiodol and DEDC/Lipiodol agents for radioembolytic therapy of Hepatocellular Carcinoma (HCC) [1-12].

More recently, the use of ^{188}Re colloid for radionuclide synovectomy has been found effective for treatment of refractory disease [13-15].

Rhenium-188 is produced from the decays of tungsten-188. The ^{188}W parent is obtained by double neutron capture of ^{186}W in a nuclear reactor following the reaction scheme:

$$^{186}W \xrightarrow{n,\gamma(\sigma=37,9b)} {}^{187}W \xrightarrow{n,\gamma(\sigma=64\ b)} {}^{188}W \xrightarrow{n,\gamma(\sigma=12\ b)} {}^{189}W$$

Because this reaction proceeds in two steps and the intermediary product ^{188}W is short lived ($T_{1/2}$ = 23,8 h), the reaction requires a thermal neutron flux $>1^{*}10^{15}$ n/cm^2. Such conditions are available at the SM in Dimitrovgrad the Russian Federation and HFIR in Oak Rige, Tennessee USA. In Figure 1 Dependence of ^{188}W specific activity versus time of irradiation and the flux of thermal neutrons is presented.

The radionuclide generator is an efficient and convenient device for obtaining ^{188}Re solution. Several types of $^{188}W/^{188}Re$ generators have been described in the literature in which different material, Dowex, zirconium oxides, alumina, tungstate of Zr, Ti, Co, Mo were used [13,16]. Some other sorbents with high content of tungsten such as gel metal oxide composites, hydroxyapatites and Polymeric Zirconium Compounds (PZC), as well as new nanomaterials have been employed in the preparation of $^{188}W/^{188}Re$ generators also a simple electrochemical method for the separation of ^{188}Re from ^{188}W and obtained ^{188}Re suitable for radiolabeling biomolecules has been presented [17-24]. However the best elution efficiency was obtained when the aluminium oxide was applied for loading of the generator column [25-27].

Material and Methods

Laboratory works

In our experiments the adsorption of tungstate solution on two types of alumina (50-200 μm and 63-200 μm) were investigated for a wide range of pH from 9.7 to 2.5 and temperature values from 0 to 34°C. After loading of a generator column the alumina was activated using 0.9% NaCl in 0,001m HCl to obtain a final pH 3. Tungsten-188 in the form of tungstenic acid was slowly fed onto the column. After tungsten deposition, the alumina column was washed with 0.9% NaCl. Next the ^{188}Re was eluted with 0.9% NaCl. The obtained sodium perrhenate solution was purified and concentrated in the chromatographic system. The system consists of two columns with ion-exchange resins. The solution first passed through the column with cation exchanger AG-50W-X4 (Bio-Rad) resin, 200-400 mesh hydrogen form. The Ag^+ cations deposited on the bed react with Cl^- anions to form unsoluble AgCl, whereas the Na^+ cations replace H^+. This column essentially "traps" all of the chloride anions. The second column is filled with the anion exchanger Sem-Pack Plus QMA Light resin. The $^{188}ReO_4^-$ ions pass through the column and are trapped. The column is the eluted with 1-2ml of saline and concentrated solution of sodium perrhenate is obtained [26].

Production line to provide carrier-free rhenium-188

Based on the previously developed procedures for loading and eluting of the generator, and also taking into account the requirements

***Corresponding author:** Marcin Konior, National Centre for Nuclear Research, Radioisotope Centre Polatom, 05-400 Otwock, Andrzej Sołtan 7, Poland
E-mail: Marcin.Konior@polatom.pl

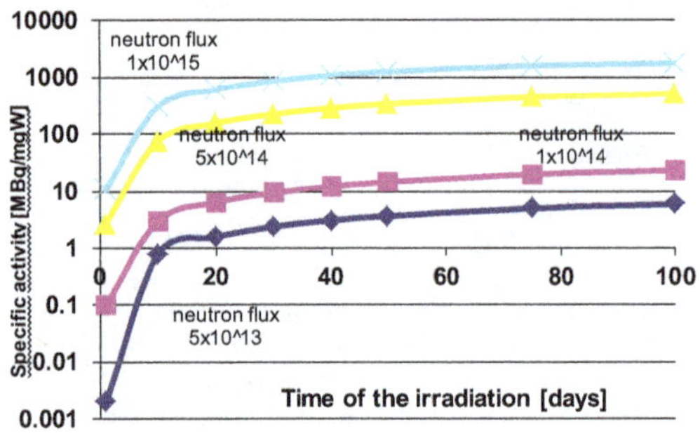

Figure 1: Dependence of the W-188 specific activity rests on the time of the irradiation and the flux of thermal neutrons

formulated for the final product, a production line for processing of sodium tungstate was constructed.

The line consists of five lead-shielded chambers in which the following operations are performed:

1. Unloading of active material

2. Preparation of ^{188}W solution and loading of ^{188}W solution onto the alumina column

3. Elution of ^{188}Re in the form of sodium perrhenate

4. Concentration of the elute and dispensing ^{188}Re solution in to vials.

5. Sterilization and removal of vials from production line.

The ^{188}Re solution is recommended for medical applications as a precursor for radiopharmaceuticals production, therefore the air inside the production chambers must meet GMP purity requirements.

The air in chamber 1 and 5 should be classified as Class C. Also the air inside the room where the production line is installed should fulfil the D-class requirements of purity [28].

Quality control

An important aspect of the production is Quality Control of the final product, especially when the product is intended for medical purposes.

Radiochemical purity of eluated ^{188}Re perrhenate solution was evaluated by means of paper chromatography using 0.9% NaCl as a developing solution. Chemical purity was determined using ICP-Optical Emission Spectrometer (Optima 33000XL, Perkin-Elmer). Radionuclide purification of the eluates was checked by γ-spectrometry with HPGe detector, which included the overall assessment of radionuclide impurities related to the ^{188}Re activity and ^{188}W breakthrough. The quantitative specifications of sodium perrhenate (VII) solution are summarized in Table 1.

Radionuclide 188W/188Re generator

On the basis of our development study and availability of necessary facilities the production of $^{188}W/^{188}Re$ generators has been undertaken at the Radioisotope Centre POLATOM Poland. Aluminium oxide (Alumina A, ICN, MP Biomedicals) was stirred in deionized water and decanted several times until the supernatant was clear. The decanted alumina was then dried for 4 h at 80°C. Dry alumina (2 g) was loaded onto a column, with diameter of 7.5 mm and length of 40 mm, which was fixed on the top with a bed of glass wool and closed with rubber seals and aluminium caps. The column was autoclaved for 0.5 h at 121°C.

The tungsten oxide irradiated in SM nuclear reactor at Dimitrovgrad (Russian Federation) with specific activity of between 27 and 133 GBq per gram of tungsten is dissolved in NaOH and converted to sodium tungstate. Directly before the adsorption of tungstate on the generator alumina column, the solution of sodium tungstate is heated and sodium hypochlorite (0.5 ml NaOCl 0.5 M per gram of tungsten), 80% acetic acid (1 ml CH_3COOH per gram of tungsten) and 32% hydrochloric acid is added in order to lower the solution pH to 2.5. In Figure 2 the pH changes of sodium tungstate solution during acidification are illustrated.

The purpose of this operation is to transform the sodium tungstate into tungstenic acid solution at pH 2-3. The acid solution is slowly passed (flow 0.1 ml/min) onto the generator column placed in a lead shield. Under such conditions the adsorption of tungstenic acid solution on the alumina column achieved a value up to 99%. The column was washed with 20-50 ml of 0.9% NaCl solution (pH 5.0-5.5) at flow rate of 0.1 ml/min in order to remove the unbound ^{188}W. Two methods were evaluated for the elution of the generator. The first involves use of a peristaltic pump to pass saline through the generator column under positive pressure. Application of this solution enables as to join a few generators and increase the total activity of rhenium solution up to 74 GBq. In the second method the vacuum technique was applied. The saline is removed at the generator column outlet under negative pressure using evacuated vials as in the $^{99}Mo/^{99m}Tc$ generator which is a simple and reliable method. The example of the elution profile of a generator is shown in Figure 3.

Results and Discussion

As observed from the data of Figure 3, 99% of ^{188}Re activity is collected in 2-3 ml of eluted solution. Such results allow obtaining of high specific activity ^{188}Re and indicate the possibility of avoiding of concentrated eluate post elution. Standard elutions of the $^{188}W/^{188}Re$ generator were performed using 8 ml of saline solution (2 × 4 ml). Evaluation of the prototype generator has been conducted over six months.

As shown in Figure 4, maximum activity of ^{188}Re we can be obtained after 72 hours from the previous elution, but an elution yield of 60% is available after 24 hours. In the case of consistence in eluate of considerable amount of ^{188}W (breakthrough) the obtained solution could be purified by adsorption of tungstenic anions on additional alumina columns. In our experiments the ^{188}W concentration in

Tests	Requirement	Method
Identification Characters Identity Gamma-ray spectrum	colourless, clear solution Presence of specifics γ lines: Eγ=155.06 Eγ=477.96 Eγ=633.00	Visual inspection Gamma-ray spectrometry
pH	5.5-7.5	Potentiometry
Radionuclide purity	>99.9%	Gamma-ray spectrometry
Radiochemical purity	≥ 98%	Paper chromatography
Chemical purity	Not more than: 5 ppm Pb, Al, Ba, Ni 10 ppm B, Zn 15 ppm W 20 ppm Si, Mg, Ca	ICP-OES
Sterility	Sterile	Direct inoculation
Bacterial endotoxins	<0.125 EU/ml	LAL test

Table 1: Specifications of 188Re sodium perrhenate (VII) solution

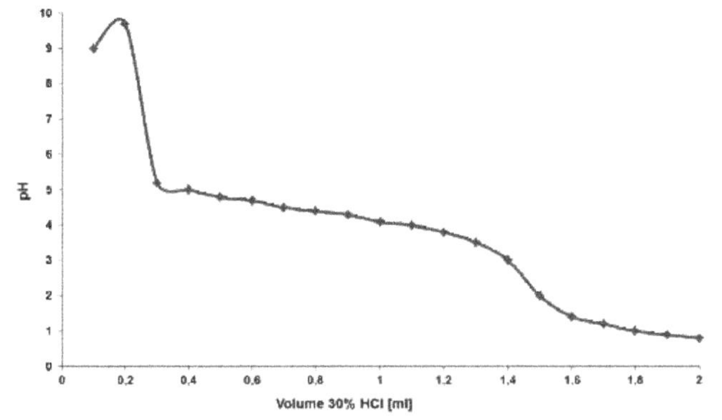

Figure 2: Graph of pH changes of the sodium tungstate solution while adding hydrochloric acid

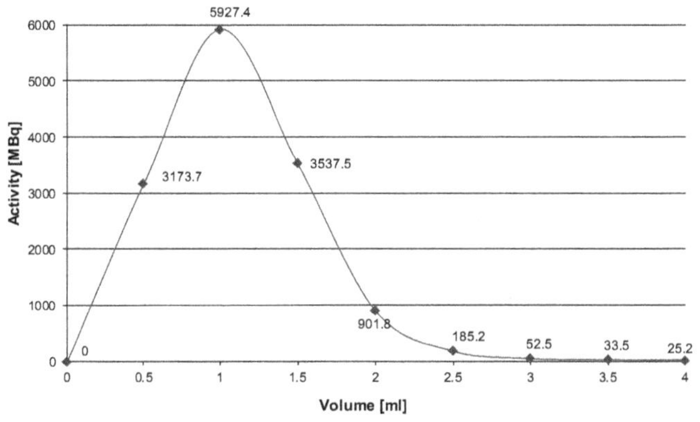

Figure 3: Elution profile of generator

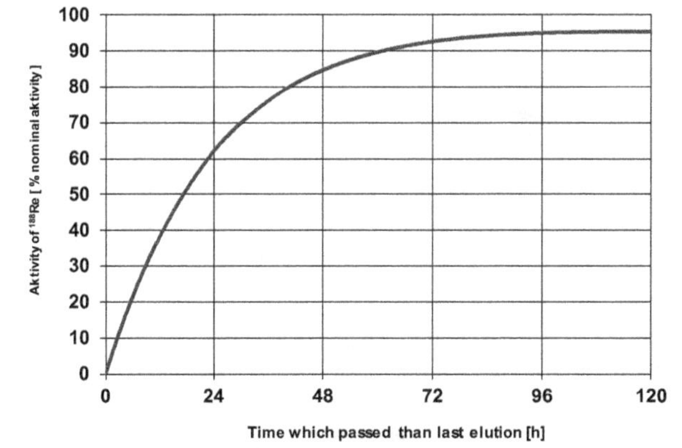

Figure 4: Increase activity of ^{188}Re on the generator

sodium perrhenate did not exceed 10.0 ppm; therefore the application of a purification column was unnecessary.

Table 2 compares the requirement parameters of eluates and obtained data during examinations of Polish $^{188}W/^{188}Re$ generators are listed. In the Figure 5 the cross section of $^{188}W/^{188}Re$ generator is presented. Table 3 presents the technical parameters of the generator. All plastic parts which are used for generator construction were examined for resistance to long-time irradiation of beta particles. For generators which allow us to obtain ^{188}Re solution with activity of 18.5 GBq, the radiation dose on lead shield is less than 1 mSv/h.

Conclusion

Radioisotope Centre Polatom, Poland to develop the technology for manufacturing of $^{188}W/^{188}Re$ generators. The production line for preparation of carrier-free ^{188}Re in form of sterile, isotonic solution of sodium perrhenate (VII) has been constructed.

The ^{188}Re sodium perrhenate solution is obtained from a single stationary generator or from a series of generators attached in tandem. The total activity of ^{188}Re eluate achieved value 74 GBq with radiochemical purity of 99.9%.

On the basis of collected experiences, during operations of the production line, the manufacturing of $^{188}W/^{188}Re$ generators, in using alumina has been established. At the present time, 37 GBq level generators are available, from RC Polatom. Maximum activity of ^{188}Re we can be obtained after 72 hours from the previous elution, but an elution yield of 60% is available after 24 hours. In the case of consistence in eluate of considerable amount of ^{188}W (breakthrough) the obtained solution could be purified by adsorption of tungstenic anions on additional alumina columns. The radiopharmaceuticals and radiochemicals containing rhenium-188 are used in clinical trials of cancer radioimmunotherapy, palliation of skeletal bone pain, endovascular brachytherapy as well as in the development of novel radiopharmaceuticals in pre-clinical stage. Its high energy beta radiation with soft tissues penetration range of about 8-11 mm is suitable for irradiation of medium or large tumors, while the low energy and abundance of gamma photons make the isotope suitable for quantitative gamma camera imaging for evaluation of biokinetics and dosimetric estimations. References

Parameter		Requirement	Result	Method
Radionuclide purity	188W	< 0.5 %	< 0.01 %	ICP-OES
	others	< 0.1 %	< 0.1 %	
Radiochemical purity		>98%	99,8%	
Chemical purity	Al	<10.0 ppm	5.5 ppm	
	W	<10.0 ppm	<10.0 ppm	
pH of the elute		4.5-6.0	5.0	potentiometry
Elution yield	in first 4 ml	>90%	98.0%	ionization chamber
	in first 8 ml	>95%	99.7%	

Table 2: Monitoring parameters of generator elute

Chemical form	**sodium perrhenate $Na^{188}ReO_4$ in 0.9% NaCl solution**
Nominal activity of ^{188}Re	3.7 GBq-37 GBq (100-1000 mCi) 2 generators can be coupled serially to be eluted in a single procedure
Elution yield	> 90 % of nominal activity in first 4 ml fraction of elute > 95 % of nominal activity in first and second 4 ml fractions of elute
Radiochemical purity	> 98%
Chemical purity	Al <10.0 ppm W <10.0 ppm
Radionuclide purity	^{188}W <0.5 % others <0.1 %
pH of the eluate	4.5-6.0
Weight of the generator	16 kg
Expiration time	at least 6 months from the day of calibration
Calibration	maximal 7 days after production

Table 3: Technical parameters of generator $^{188}W/^{188}Re$

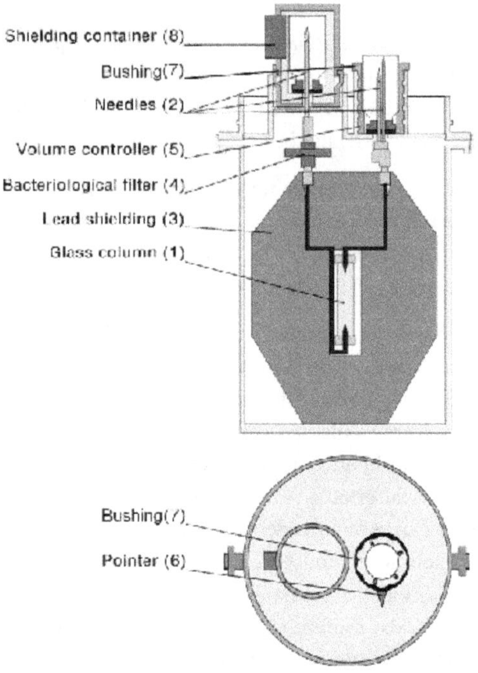

Figure 5: Cross section of $^{188}W/^{188}Re$ generator

References

1. Liepe K, Hliscs R, Kropp J, Runge R, Knapp FF Jr, et al. (2003) Dosimetry of 188Re-hydroxyethylidene diphosphonate in human prostate cancer skeletal metastases. J Nucl Med 44: 953-960.
2. Zhang H, Tian M, Li S, Liu J, Tanada S, et al. (2003) Rhenium-188-HEDP therapy for the palliation of pain due to osseous metastases in lung cancer patients. Cancer Biother Radiopharm 18: 719-726.
3. Kothari K, Pillai MR, Unni PR, Shimpi HH, Noronha OP, et al. (1999) Preparation of [186Re]Re-DMSA and its bio-distribution studies. Appl Radiat Isot 51: 43-49.
4. Kothari K, Pillai MR, Unni PR, Shimpi HH, Noronha OP, et al. (1999) Preparation, stability studies and pharmacological behavior of [186Re]Re-HEDP. Appl Radiat Isot 51: 51-58.
5. Bernal P, Ozorio M, Guiterrez C, Esguerra R, Cerquerra AM, et al. (2003) Evaluation of rhenium-188 lipiodol in the treatment of liver cancer: Experience in Columbia between 2001-2003. J Nucl Med 44: 176.
6. Chaudakshetrin P, Osorio M, Somanesan S, Sundram FX, Padhy AK, et al. (2003) Rhenium-188 lipiodol therapy of liver cancer: optimization of conjugate-view imaging of Re-188 for patient-specific dosimetry. J Nucl Med 44: 324.
7. Paeng JC, Jeong JM, Yoon CJ, Lee YS, Suh YG, et al. (2003) Lipiodol solution of 188Re-HDD as a new therapeutic agent for transhepatic arterial embolization in liver cancer: preclinical study in a rabbit liver cancer model. J Nucl Med 44: 2033-2038.
8. Sundra MFX, Yu S, Somanesan S, Jeong JM, Bernal P, et al. (2002) Phase I study of transarterial rhenium-188-HDD lipiodol in treatment of inoperable primary hepatocellular carcinoma-a multicentre evaluation. World J Nucl Med 1: 5-11.
9. Jeong JM, Kim YJ, Lee YS, Ko JI, Son M, et al. (2001) Lipiodol solution of a lipophilic agent, (188)Re-TDD, for the treatment of liver cancer. Nucl Med Biol 28: 197-204.
10. Boschi A, Uccelli L, Duatti A, Colamussi P, Cittanti C, et al. (2004) A kit formulation for the preparation of 188Re-lipiodol: preclinical studies and preliminary therapeutic evaluation in patients with unresectable hepatocellular carcinoma. Nucl Med Commun 25: 691-699.
11. Lambert B, Bacher K, De Keukeleire K, Smeets P, Colle I, et al. (2005) 188Re-HDD/lipiodol for treatment of hepatocellular carcinoma: a feasibility study in patients with advanced cirrhosis. J Nucl Med 46: 1326-1332.
12. Liepe K, Kropp J, Runge R, Kotzerke J (2003) Therapeutic efficiency of rhenium-188-HEDP in human prostate cancer skeletal metastases. Br J Cancer 89: 625-629.
13. Le Van So (2003) Preparation of titanium-tungstate gel based 188W/188Re generator. Proceedings of the Fifth National Conference on Nuclear Physics and Techniques, Ho Chi Min City, Vietnam. 22-28 April, 250-264.
14. Shin CY, Son M, Ko JI, Jung MY, Lee IK, et al. (2003) DA-7911, 188Rhenium-tin colloid, as a new therapeutic agent of rheumatoid arthritis. Arch Pharm Res 26: 168-172.
15. Wang SJ, Lin WY, Hsieh BT, Shen LH, Tsai ZT, et al. (1995) Rhenium-188 sulphur colloid as a radiation synovectomy agent. Eur J Nucl Med 22: 505-507.
16. Dadachov MS, Le VS, Lambrecht RM, Dadachova E (2002) Development of a titanium tungstate-based 188W/188Re gel generator using tungsten of natural isotopic abundance. Appl Radiat Isot 57: 641-646.
17. Monroy-Guzman F, Badillo Almaraz VE, Rivero Gutirrez T, Galico Cohen L (2009) Development of inorganic adsorbents as matrices of generators for therapeutic radionuclides. IAEA Technical Reports Series No. 470, Chapter 16, 161-173.
18. Dadachov MS, Lambrecht RM, Hetherington E (1994) An improved tungsten-188/rhenium-188 gel generator based on zirconium tungstate. J Radioanal Nucl Chem Lett 188: 267-278.
19. Van So Le, Nguyen CD, Bui VC, Vo CH (2009) Preparation of inorganic polymer sorbents and their application in radionuclide generator technology. IAEA Technical Reports Series No. 470, Chapter 20, 217-229.
20. Chakravarty R, Dash A, Kothari K, Pillai MRA, Venkatesh M (2009) A novel 188W/188Re electrochemical generator with potential for medical application. Radiochimica Acta 97: 300-317.
21. Chakravarty R, Dash A, Venkatesh M (2009) Separation of clinical grade ^{188}Re from ^{188}W using polymer embedded nanocrystalline titania. Chromatographia 69: 1363-1372.
22. Chakravarty R, Shukla R, Ram R, Venkatesh M, Tyagi AK, et al. (2011) Exploitation of nano alumina for the chromatographic separation of clinical grade 188Re from 188W: a renaissance of the 188W/188Re generator technology. Anal Chem 83: 6342-6348.
23. Iller E, Polkowska-Motrenko H, Lada W, Wawszczak D, Sypula M, et al. (2009) Studies of gel metal-oxide composite samples as filling materials for 188W/188Re generator column. J Radioanal Nucl Chem 281: 83-86.
24. Iller E, Wawszczak D, Konior M, Polkowska-Motrenko H, Milczarek JJ, et al. (2013) Synthesis and structural investigations of gel metal oxide composites WO_3-ZrO_2, WO_3-TiO_2, WO_3-ZrO_2-SiO_2, and their evaluation as materials for the preparation of $^{188}W/^{188}Re$ generator. Appl Radiat Isot 75: 115-127.
25. Mikolajczak R, Zuchlinska M, Korsak A, Iller E, Pawlak D, et al. (2009) Development of a 188W/188Re generator. IAEA Technical Report Series No. 470, Chapter 17, 175-184.
26. Knapp FF, Turner JH, Jeong J-M, Padhy AK (2004) Issues associated with the use of Tungsten-188/Rhenium-188 generator and concentrator system and preparation of Re-188 HDD, A report. Word J Nucl Med 3: 137-143.
27. Lee JS, Lee JS, Park UJ, Son KJ, Han HS (2009) Development of a high performance (188)W/(188)Re generator by using a synthetic alumina. Appl Radiat Isot 67: 1162-1166.
28. Iller E, Zelek Z, Konior M, Sawlewicz K, Staniszewska J, et al. (2007) Technological line for production of carrier-free 188Re in the form of sterile, isotonic solution of sodium perrhenate (VII). Trends in Radiopharmaceuticals (ISTR-2005) 1: 323-331.

9

Incidence of *Listeria monocytogenes* in Meat Product Samples by Real-Time PCR

Miroslava Kačániová[1*], Maciej Kluz[2], Jana Petrová[1], Martin Mellen[3], Simona Kunová[1], Peter Haščík[1] and Ľubomír Lopašovský[1]

[1]*Faculty of Biotechnology and Food Sciences, Slovak University of Agriculture in Nitra, Nitra, Slovak Republic*

[2]*Department of Biotechnology and Microbiology, University of Rzeszow, Rzeszow, Poland*

[3]*Hydina Slovakia, s.r.o., Nová Ľubovňa 505, Nová Ľubovňa 065 11, Slovakia*

Abstract

The aim of this study was to trace a contamination of meat products with *Listeria monocytogenes*. Step One real time Polymerase Chain Reaction (PCR) was used. We used the PrepSEQ Rapid Spin Sample Preparation Kit for isolation of DNA and MicroSEQ® *Listeria monocytogenes* Detection Kit for the real-time PCR performance. We found out the strains of *Listeria monocytogenes* in hundred samples of meat products with no incubation. There was Internal Positive Control (IPC) in 40 samples. Our results showed that the real time PCR assay tested in this study might detect *Listeria monocytogenes* in meat product samples with no incubation sensitively.

Keywords: Meat product samples; Real time PCR, *Listeria monocytogenes*

Introduction

It is known that the incidence of food-borne illnesses in both industrialized as well as in non-industrialized countries [1,2] has increased. This could be the result of some major changes in food production, preservation, storage and consumption as well as in globalization and liberation of food trade and importation of foods.

Listeria monocytogenes is a human pathogen which is widely distributed in the environment [3,4]. Meat products are one of the major sources of *L. monocytogenes* [5-7]. The International Commission on Microbiological Specification for Foods concluded that 100 CFU of *L. monocytogenes* per gram of food at the time of consumption is acceptable for consumers. This statement is in regard to the fact that the clinical cases of listeriosis are usually associated with high loads of *L. monocytogenes,* as it is difficult to eradicate listeriae from the environment of the food processing plants [8].

Several studies of real-time PCR-based detection of *L. monocytogenes* in food have been published [9,10] and several validated real-time PCR based kits are commercially available (iQ-Check *Listeria*, Bio-Rad, TaqMan *Listeria*, Applied Biosystems, BAX system PCR Assay *Listeria monocytogenes*, DuPont Qualicon). The real-time PCR detection is preceded by a single-step or a two-step enrichment using media of different selectivity in these complete methods. These methods are used by different enrichment and DNA preparation approaches to reach the increasing numbers of live cells to a detectable level and dilution of dead *L. monocytogenes* cells as well as food-borne PCR inhibitors [11].

The conventional testing methods for the detection of *L. monocytogenes* in food involves growth in pre enrichment medium, followed by growth on selective medium and a battery of confirmatory biochemical and serological tests [12]. These methods are labor-intensive and time-consuming and they often take up to 10 days. The rapid alternative method is real-time (RTi)-PCR which allows an accurate and unambiguous identification and a precise quantification of nucleic acid sequences [4,13]. Furthermore the lack of post-PCR steps reduces the risk of cross-contamination and allows high throughput and automation.

O'Grady et al. [14,15] described a qPCR assay for the specific detection of *L. monocytogenes* in food samples. Positive deviation was observed in ten analyzed samples. There was a possible explanation for these discordant results are that DNA from dead or viable but non culturable *L. monocytogenes* cells detected by the alternative method in the food matrix. Alternatively, these positive deviations might results which indicated this new method is more sensitive than the *Listeria* precis method for detecting *L. monocytogenes* in RTE pork products. Furthermore, negative deviation was not detected which demonstrates the robustness of the alternative method as food components such as organic compounds, calcium ions, glycogen and lipids have been demonstrated to inhibit PCR [16].

Therefore, the aim of this study was to determine the level of contamination in different meat products samples that are ready for human consumption in Slovakia by isolating and identifying of *Listeria monocytogenes* by Step One real time PCR.

Material and Methods

Sample design

Altogether 100 samples collected in Slovakia from 2009 to 2013. Samples included ready-to-eat meat product samples (100 g each) were collected from supermarkets in the Slovakian localities-mentioned above. Among the samples were Soft Meat Products (SMP): Ratatouille sausage (n=12), ham sausage (n=8), ipeľská sausage (n=12), fine sausage (n=7), ludová salami (n=11), durable heat-treated (DHT): inovecká salami (n=15), cingovská salami (n=18), touristic salami (n=17). After sampling products were placed in clean sterile plastic bags, kept on ice and transported immediately to the laboratory for testing.

DNA extraction

DNA extraction was performed with PrepSEQ Rapid Spin Sample Preparation Kit (Applied Biosystems, USA) as a pre-preparation step for the Step One real-time PCR. A sample of 750 μL was centrifuged

***Corresponding author:** Miroslava Kačániová, Department of Microbiology, Faculty of Biotechnology and Food Sciences, Slovak University of Agriculture, Tr. Andreja Hlinku 2, Nitra 949 76, Slovak Republic
E-mail: miroslava.kacaniova@gmail.com

for 3 minutes at maximum speed (12,000). Supernatant was discarded and 50 μL of Lysis Buffer was added to the pellet. Every sample was incubated for 10 minutes at 95°C [17].

MicroSEQ® *Listeria monocytogenes* detection kit

Real-time Polymerase Chain Reaction (PCR) was used for detection of *Listeria monocytogenes* and TaqMan® probe was applied for detection of the amplified sequence. Eight-tube strips which contained assay beads compatible with Step One™ Systems. Samples of 30 μL were loaded to the lyophilized beads. MicroAmp®48-Well Base and the MicroAmp®Cap Installing Tool to the tubes were used. MicroAmp® Fast 48-Well Tray on the sample block of the Step One System was performed. Real-time PCR procedure for the detection of strain is described in Figure 1.

Real-time PCR

We used PCR pathogen diagnostic AmpliSens kits, kits for the isolation of bacterial DNA - PrepSEQ Rapid Spin Sample Preparation Kit, MicroSEQ® *Listeria monocytogenes* Detection Kit and SensiFast SYBR Hi-ROX kit for testing.

TaqMan® probes labeled with both a fluorophore and aquencher dye were used in real-time PCR assays to detect amplification of specific DNA targets. We used three fluorophore detection chemistries that include FAM™ and VIC® dye-labeled TaqMan® MGB probe-based assays, VIC®and TAMRA™ dye-labeled probe-based assays and ROX™ as passive reference dye. FAM™, which has an emission of 520 nm, has become the most commonly used fluorophore for single plex qPCR reactions. TAMRA™ will efficiently quench the fluorescence of FAM™ until the probe hybridizes to the target and is cleaved by the 5'exonuclease activity of the polymerase. Thermal cycling conditions were: 2 minutes of incubation at 95°C, followed by 40 cycles of 1 sec. denaturation at 95°C and 20 seconds annealing and elongation at 60°C. Required data were collected during each elongation step. PCR products were detected by monitoring the increase of fluorescence on the reporter dye at each PCR cycle. Applied Biosystems software plotted the normalized reporter signal, ΔRn, (reporter signal minus background) against the number of amplification cycles and also determined the threshold cycle (Ct) value; i.e. the PCR cycle number at which fluorescence increases above a defined threshold level was used [17]. Samples with an increasing fluorescence signal were considered positive, regardless of the internal control amplification. Samples with no increasing fluorescence signal but with amplified internal control were considered negative. Samples with no fluorescence signal for both specific target and internal control were considered inhibited.

Sample (swabs from the inside of samples)

↓

Isolation of DNA

↓

(PrepSEQTM Rapid Spin Sample Preparation Kit)

↓

Pursuance real time PCR

↓

(MicroSEQ *Listeria monocytogenes* Detection Kit)

Figure 1: Real-time PCR procedure for the detection of the prevalence of *Listeria monocytogenes*.

Results and Discussion

Altogether 40 samples out of 100 tested were positive for *Listeria monocytogenes* with Step One real time PCR in our study. Regarding a type of the product, 17 out of 50 samples of Durable Heat-Treated (DHT) and 23 out of 50 samples of soft meat products (SMP) were positive for *Listeria monocytogenes*.

The results were evaluated based Ct values. The lowest value of 6.10 Ct was measured in a sample of the tourist sausage and the highest Ct value of 35.46 was recorded in the sample ipeľská sausage and the average Ct value was 21.90 as it is shown in Table 1.

This shows us the positive samples of monitored pathogen and in which the cycle curves to ascend. In our case, the standard for pathogens was zero, so we quantified the amount of pathogens in a sample. Internal positive control (IPC) was positive for all our study samples indicating the correct course of the reaction.

We found out 40% samples which were positive of *Listeria monocytogenes*. *Listeria* species hasbeen associated with a wide variety of food sources particularly poultry, red meat and meat products [18]. The present study has recorded a high occurrence of *Listeria* spp. in pies (33%) and chicken stew (28%) (Figure 2). These findings are in line with those of other authors [18-20]. In Egypt, prevalence of *Listeria* (41%) in meat and chicken products was [18] lower than reported in Malaysia (73.9%) in imported frozen beef [21] and Turkey (83.3%) in raw minced meat [20]. We know that foods are complex matrices. There are several publications that report about on filtration based protocols for PCR detection of various pathogenic species [22]; however, they have never been used with quantitative purposes. Most available detection systems require selective enrichment that steps to overcome the problem of potential PCR inhibitors, especially for low pathogen concentrations [23-26]. Remarkably, our method does not require any culture steps, meaning that results can be obtained considerably quicker (Table 2).

Regarding contamination of samples expressed as a percentage of all samples were evaluated ratatouille sausage SMP - 3.00%, fine sausage SMP - 2%, ham sausage SMP - 3.00%, ipeľská sausage SMP - 7.00%, ľudová salami SMP - 6.00%, touristic salami DHT - 4.00%, inovecká salami DHT - 5.00%, cingovská salami DHT - 8.00% as it is shown in Table 2.

The presence of *L. monocytogenes* in a set of 250 samples of RTE meats, which were taken from different stores in Macon and Lee Counties, AL was investigated using microbiological culture by USDA technique and IMS+RT-PCR. There were found no *L. monocytogenes* in the fifty samples that were taken randomly from the 250 samples by means of IMS+RT-PCR. The same result was obtained by the standard methods. Whereas out of the total 250 samples checked first by microbiological methods, *L. monocytogenes* was found in 5 samples (2%), two deli chickens and one Turkey deli meat, one Turkey sausage and one beef frankfurter [27].

A different real-time PCR assay for the detection of *L. monocytogenes* in food has been described [28-35]. The commercial availability of real-time PCR reagents and kits makes it easier for food companies to adapt real-time PCR testing to their laboratories. They also facilitate the development of common testing protocols and standards so that proper collaborative studies can be performed [36].

Conclusion

In terms of practicability, DNA-based method has potential to

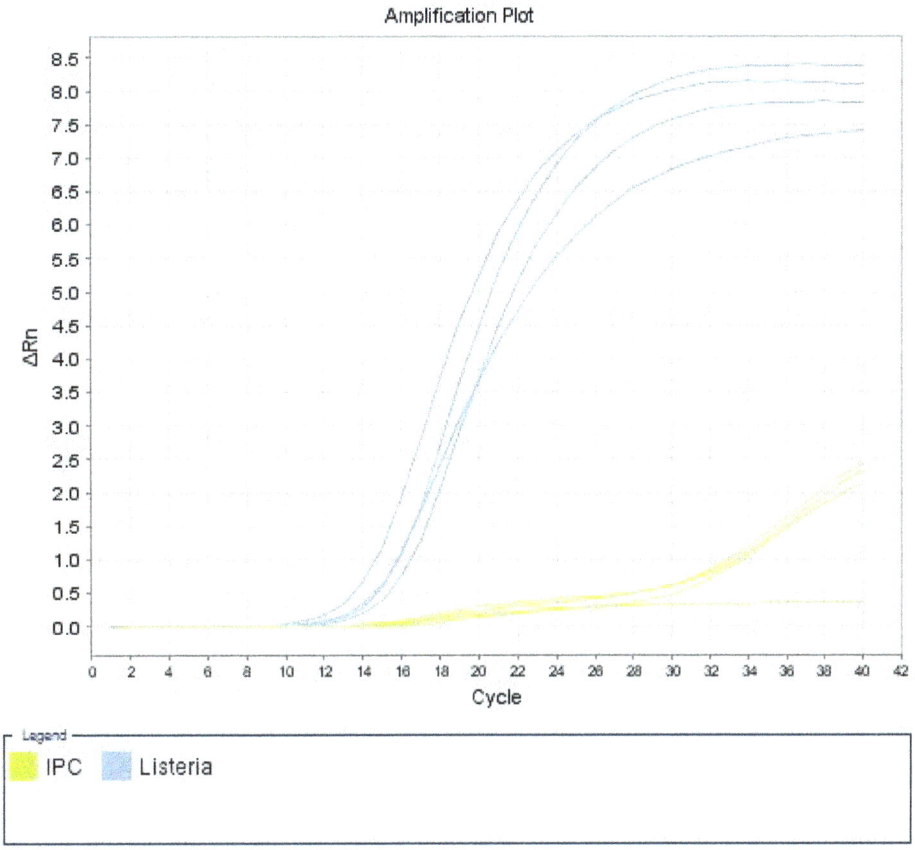

Figure 2: Real-time PCR detection of *Listeria monocytogenes* without incubation in meat product samples with internal positive control (IPC).

		Ct-threshold cycle				Ct-threshold cycle	
Number of sample	**Meat product**	***Listeria monocytogenes***	**IPC**	**Number of samples**	**Meat product**	***Listeria monocytogenes***	**IPC**
1	**Ratatouille sausage SMP**	16.23	29.69	21	**Ľudová salami SMP**	24.82	27.63
2		16.98	25.11	22		29.06	27.53
3		16.59	17.31	23		23.38	26.66
4		15.94	16.11	24	**Inovecká salami DHT**	7.47	18.52
5		14.77	14.76	25		27.46	29.46
6	**Ham salami SMP**	15.89	15.98	26		14.34	17.88
7		19.44	28.60	27		11.23	20.28
8		24.27	29.18	28		32.96	29.07
9	**Ipeľská sausage SMP**	23.06	29.39	29	**Čingovská salami DHT**	9.78	29.74
10		19.49	26.39	30		24.64	29.28
11		19.11	27.29	31		25.00	28.03
12		28.95	30.94	32		24.74	29.05
13		35.46	32.91	33		24.85	27.80
14		31.90	32.14	34		20.34	25.21
15		32.25	31.97	35		24.70	28.11
16	**Fine sausage SMP**	34.39	32.58	36		23.91	28.19
17		32.94	31.90	37	**Touristic salami DHT**	21.64	25.09
18	**Ľudová salami SMP**	30.97	31.45	38		33.84	21.08
19		13.89	20.13	39		11.74	30.63
20		11.58	16.13	40		6.10	29.19

Table 1: Real-time PCR detection of internal positive control of *Listeria monocytogenes* in meat product samples.

	Number of samples			Number of samples (%)		
Meat product	**Negative**	**Positive**	**Total**	**Negative**	**Positive**	**Total**
Ratatouille sausage SMP	7	5	12	7.00	5.00	12.00
Fine sausage SMP	5	2	7	5.00	2.00	7.00
Ham sausage SMP	5	3	8	5.00	3.00	8.00
Ipeľská sausage SMP	5	7	12	5.00	7.00	12.00
Ľudová salami SMP	5	6	11	5.00	6.00	11.00
Meat products SMP	27	23	50	27.00	23.00	50.00
Touristic salami DHT	13	4	17	13.00	4.00	17.00
Inovecká salami DHT	10	5	15	10.00	5.00	15.00
Čingovská salami DHT	10	8	18	10.00	8.00	18.00
Meat product DHT	33	17	50	33.00	17.00	50.00
Total meat product	60	40	100	60.00	40.00	100.00

Table 2: Incidence of *Listeria monocytogenes* in meat product samples.

be used for the positive screening of *L. monocytogenes*, and the major advantage the DNA based method is that it can detect *L. monocytogenes* without enrichments in meat products. However, it does need not more steps to perform and only approximately predicts the level of contamination in meat product. Our results indicate that the Step One real-time PCR assay indicated in this study could sensitively detect *Listeria monocytogenes* in meat product samples. The rapid real-time PCR-based method performed very well in comparison with the conventional method. It is a fast, simple, specific and sensitive way to detect nucleic acids which could be used in clinical diagnostic tests in the future.

Acknowledgement

The study was supported by the project: The research leading to these results has received funding from the European Community under project no 26220220180: Building Research Centre, Agro Biotech.

References

1. Centers for Disease Control and Prevention (CDC) (2000) Multistate outbreak of listeriosis--United States, 2000. MMWR Morb Mortal Wkly Rep 49: 1129-1130.
2. CDC (2002) Public health dispatch: Outbreak of Listeriosis-Northeastern United States, 2002. MMWR 51: 950-951.
3. Vázquez-Boland JA, Kuhn M, Berche P, Chakraborty T, Domínguez-Bernal G, et al. (2001) *Listeria* pathogenesis and molecular virulence determinants. Clin Microbiol Rev 14: 584-640.
4. Klein D (2002) Quantification using real-time PCR technology: applications and limitations. Trends Mol Med 8: 257-260.
5. Rocourt J, Hogue A, Toyofuku H, Jacquet C, Schlundt J (2001) *Listeria* and listeriosis: risk assessment as a new tool to unravel a multifaceted problem. Am J Infect Control 29: 225-227.
6. Wing EJ, Gregory SH (2002) *Listeria monocytogenes*: clinical and experimental update. J Infect Dis 185 Suppl 1: S18-24.
7. Peccio A, Autio T, Korkeala H, Rosmini R, Trevisani M (2003) *Listeria monocytogenes* occurrence and characterization in meat-producing plants. Lett Appl Microbiol 37: 234-238.
8. Rodríguez-Lázaro D, Hernández M, Pla M (2004) Simultaneous quantitative detection of *Listeria* spp. and *Listeria monocytogenes* using a duplex real-time PCR-based assay. FEMS Microbiol Lett 233: 257-267.
9. Navas J, Ortiz S, Lopez P, Jantzen MM, Lopez V, et al. (2006) Evaluation of effects of primary and secondary enrichment for the detection of *Listeria monocytogenes* by real-time PCR in retail ground chicken meat. Foodborne Pathog Dis 3: 347-354.
10. Rossmanith P, Krassnig M, Wagner M, Hein I (2006) Detection of *Listeria monocytogenes* in food using a combined enrichment/real-time PCR method targeting the prfA gene. Res Microbiol 157: 763-771.
11. Oravcová K, Trncíková T, Kaclíková E (2007) Comparison of three real-time PCR-based methods for the detection of *Listeria monocytogenesin* food. J Food Nutr Res 46: 63-67.
12. Rodríguez-Lázaro D, Hernández M, Scortti M, Esteve T, Vázquez-Boland JA, et al. (2004) Quantitative detection of *Listeria monocytogenes* and *Listeria innocua* by real-time PCR: assessment of hly, iap, and lin02483 targets and AmpliFluor technology. Appl Environ Microbiol 70: 1366-1377.
13. Norton DM (2002) Polymerase chain reaction-based methods for detection of *Listeria monocytogenes*: toward real-time screening for food and environmental samples. J AOAC Int 85: 505-515.
14. O' Grady J, Sedano-Balbás S, Maher M, Smith T, Barry T (2008) Rapid real-time PCR detection of *Listeria monocytogenes* in enriched food samples based on the ssrA gene, a novel diagnostic target. Food Microbiol 25: 75-84.
15. O'Grady J, Ruttledge M, Sedano-Balbás S, Smith TJ, Barry T, et al. (2009) Rapid detection of *Listeria monocytogenes* in food using culture enrichment combined with real-time PCR. Food Microbiol 26: 4-7.
16. Rodríguez-Lázaro D, Hernandez M (2006) Molecular methodology in Food Microbiology diagnostics: trends and current challenges. IUFoST World Congress, pp. 1085-1099.
17. Pochop J, Kačániová M, Hleba L, Lopasovský L, Bobková A, et al. (2012) Detection of *Listeria monocytogenes* in ready-to-eat food by Step One real-time polymerase chain reaction. J Environ Sci Health B 47: 212-216.
18. El-Malek AMA, Ali SFH, Hassanein R, Moemen AM, Elsayh KI, et al. (2010) Occurrence of *Listeria* species in meat, chicken products and human stools in Assiut city, Egypt with PCR use for rapid identification of *Listeria monocytogenes*. Vet World 3: 353-359.
19. Molla B, Yilma R, Alemayehu D (2004) *Listeria monocytogenes* and other *Listeria* species in retail meat and milk products in Addis Ababa, Ethiopia. Ethiop J Health Dev 18: 208-212.
20. Yucel N, Citak S, Onder M (2005) Prevalence and antibiotic resistance of *Listeria* species in meat products in Ankara, Turkey. Food Microbiol 22: 241-245.
21. Hassan Z, Purwati E, Radu S, Rahim RA, Rusul G (2001) Prevalence of *Listeria* spp and *Listeria monocytogenes* in meat and fermented fish in Malaysia. Southeast Asian J Trop Med Public Health 32: 402-407.
22. Rijpens NP, Herman LM (2002) Molecular methods for identification and detection of bacterial food pathogens. J AOAC Int 85: 984-995.
23. Al-Soud WA (2000) Optimisation of diagnostic PCR a study of PCR inhibitors in blood and sample pretreatment. Ph.D. thesis. Lund University, Lund, Sweden.

24. Abu Al-Soud W, Rådström P (2000) Effects of amplification facilitators on diagnostic PCR in the presence of blood, feces, and meat. J Clin Microbiol 38: 4463-4470.

25. Al-Soud WA, Rådström P (2001) Purification and characterization of PCR-inhibitory components in blood cells. J Clin Microbiol 39: 485-493.

26. Bhagwat AA (2003) Simultaneous detection of *Escherichia coli* O157:H7, *Listeria monocytogenes* and *Salmonella* strains by real-time PCR. Int J Food Microbiol 84: 217-224.

27. Abdelgadir AMMA, Srivastava KK, Gopal Reddy P (2009) Detection of *Listeria monocytogenes* in Ready-to-Eat Meat Products. Am J Anim Vet Sci 4: 101-107.

28. Nogva HK, Rudi K, Naterstad K, Holck A, Lillehaug D (2000) Application of 5'-nuclease PCR for quantitative detection of *Listeria monocytogenes* in pure cultures, water, skim milk, and unpasteurized whole milk. Appl Environ Microbiol 66: 4266-4271.

29. Hough AJ, Harbison SA, Savill MG, Melton LD, Fletcher G (2002) Rapid enumeration of *Listeria monocytogenes* in artificially contaminated cabbage using real-time polymerase chain reaction. J Food Prot 65: 1329-1332.

30. Koo K, Jaykus LA (2003) Detection of *Listeria monocytogenes* from a model food by fluorescence resonance energy transfer-based PCR with an asymmetric fluorogenic probe set. Appl Environ Microbiol 69: 1082-1088.

31. Rodríguez-Lázaro D, Jofré A, Aymerich T, Hugas M, Pla M (2004) Rapid quantitative detection of *Listeria monocytogenes* in meat products by real-time PCR. Appl Environ Microbiol 70: 6299-6301.

32. Rodríguez-Lázaro D, Jofré A, Aymerich T, Garriga M, Pla M (2005) Rapid quantitative detection of, *Listeria monocytogenes* in salmon products: evaluation of pre-real-time PCR strategies. J Food Prot 68: 1467-1471.

33. Rodríguez-Lázaro D, Pla M, Scortti M, Monzó HJ, Vázquez-Boland JA (2005) A novel real-time PCR for *Listeria monocytogenes* that monitors analytical performance via an internal amplification control. Appl Environ Microbiol 71: 9008-9012.

34. Berrada H, Soriano JM, Picó Y, Mañes J (2006) Quantification of *Listeria monocytogenes* in salads by real time quantitative PCR. Int J Food Microbiol 107: 202-206.

35. Oravcová K, Kaclíková E, Krascsenicsová K, Pangallo D, Brezná B, et al. (2006) Detection and quantification of *Listeria monocytogenes* by 5'-nuclease polymerase chain reaction targeting the actA gene. Lett Appl Microbiol 42: 15-18.

36. Janzten MM, Navas J, Corujo A, Moreno R, López V, et al. (2006) Review. Specific detection of *Listeria monocytogenes* in foods using commercial methods: from chromogenic media to real-time PCR. Spanish J Agric Res 3: 235-247.

10

Synthesis Characterization and Antimicrobial Activities of Azithromycin Metal Complexes

Saeed Arayne M[1]*, Najma Sultana[2], Sana Shamim[2] and Asia Naz[2]

[1]Department of Chemistry, University of Karachi, Pakistan

[2]Research Institute of Pharmaceutical Sciences, Faculty of Pharmacy, University of Karachi, Pakistan

Abstract

Azithromycin is a well-established antimicrobial agent which has been widely prescribed for the treatment of respiratory tract infections owing to its high efficacy and safety. Various essential metal complexes of azithromycin were synthesized and characterized by techniques as UV, FT-IR, NMR, atomic absorption and elemental analysis. Spectroscopic studies of complexes suggested that the $-N(CH_3)_2$ and hydroxyl group of desosamine sugar moiety present in azithromycin has been involved in complexation i.e., azithromycin ligand (L) behaves bidentately for complexation with different metal ions such as Mg (II), Ca (II), Cr (III), Mn (II), Fe (III), Co (II), Ni (II), Cu (II), Zn (II) and Cd (II). These complexes were then subjected to *in-vitro* antibacterial and antifungal studies against several Gram positive, Gram negative bacteria and fungi. ANOVA studies illustrates that all the tested complexes exhibited significantly mild to moderate antibacterial activity against all bacterial strains and highly significant against fungus *C. albican*.

Keywords: Azithromycin; Metal complexes; Spectroscopic techniques; Antibacterial; Antifungal studies; ANOVA

Introduction

The interaction of metal ions with drugs has been recognized internationally as an important area for research [1] and also evident by huge financial support by National Institute of Health (NIH), USA program [2] and two European Union Cost Collaborative programs. Metals due to their variable oxidation states, number and types of coordinated ligands, and coordinative geometry after complexation can provide variety of properties.

On the other side, the ligands can not only control the reactivity of the metal, but also play critical roles in determining the nature of interactions involved in the recognition of biological target sites, such as DNA, enzymes and protein receptors. These variables provide enormous potential diversity for the design of metallodrugs [3]. Synthesized metal complexes might prove to have altered therapeutic activity or may have toxic effects. Therefore it is worth emphasizing point to synthesize these complexes and determine their biological activity. Most widely prescribed drugs in chemotherapy are metal-based drugs (Platinum drugs) and there are ranges of iron, copper, cobalt, gold, molybdenum complexes possessing potent anti-cancerous activity [4,5]. Metal complexes have proved potent antimicrobial agents and are in common day-to-day use in medicine such as silver bandages [6] for treatment of burns, zinc antiseptic creams, and metal clusters as anti-HIV drugs.

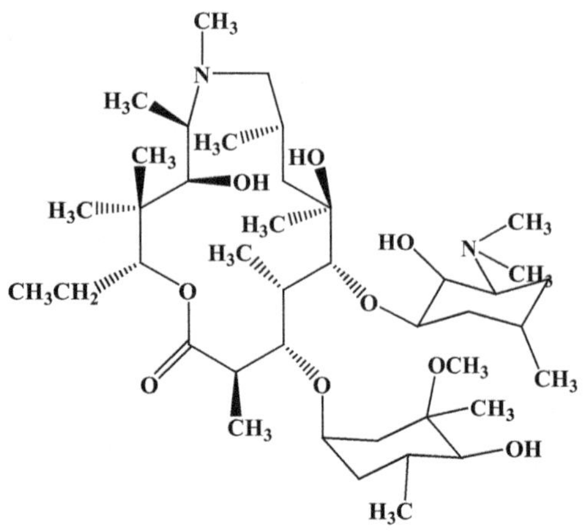

Figure 1: Chemical structure of azithromycin.

Therefore, the potential for further development of metal-based drugs and treatments as antimicrobial agents is enormous and have great importance with the evolution of drug-resistant bacteria and threats from a range of viral diseases. Numerous clinical trials for the usage of metals in therapeutics have been carried out worldwide for assessing metal based drug's efficacy in a wide diversity of human problems, including malaria, upper respiratory tract infections, urinary tract infections, sinusitis infections, vaginal yeast infections, ENT infections, cuts and fungal skin infections and even for sexually transmitted diseases like gonorrhea etc. proving it to be an antibiotic alternative at a convenient dosage [2,3,7,8].

Macrolide antibiotic azithromycin (Figure 1), the first representative of the azalide class, is entrenched antimicrobial agent that has been widely prescribed for the treatment of Respiratory Tract Infections (RTI's) owing to its high efficacy and safety [9,10]. Azithromycin contains methyl substituted nitrogen in the 15-membered macrolide aglycone ring and has greater stability than macrolide antibiotics in the presence of acids, leading to good absorption in the digestive tract [11]. Macrolide antibiotics exert their antibacterial activity by binding to ribosomal 23S RNA and thus blocking the bacterial protein synthesis [12]. The lactone ring in azithromycin, substituted with a number of hydroxyl and amine functional groups is involved in interaction with metal ions [13]. The bioavailability of azithromycin is affected by the coadministration of medications containing multivalent cations,

***Corresponding author:** Saeed Arayne M, Department of Chemistry, University of Karachi, Karachi-75270, Pakistan, E-mail: msarayne@gmail.com aluminum and magnesium containing antacids and decreases the bioavailability by approximately 24%, but has no effect on the area under the plasma concentration-time curve. Oral azithromycin should be administered at least one hour before or two hours after aluminum and magnesium containing antacids [14]. Work done by Sher, et al. on interaction of azithromycin with copper(II) ion [13] and Rjoob, et al. also supports the interaction of azithromycin with iron sulfate and iron perchlorate salts for the characterization of complexes that can be formed after interaction [15].

Therefore, as part of our continuous efforts focused on the *in vitro* activity of cephalosporins in presence of essential and trace elements [16-18], fluoroquinolone interactions with essential and trace elements [19-24] and macrolides including clarithromycin and erythromycin synergism [25,26] and roxithromycin antagonism [27] with essential and trace elements impelled us to study the antibacterial and antifungal activities of newly synthesized essential metal complexes of azithromycin. These complexes were synthesized as later are already present in our body and food or may be co-administered along with azithromycin as part of multivitamin combinations. The stoichiometry's of the complexes were determined by conductometric titrations prior to synthesis. One-way analysis of variance (ANOVA) studies were conducted to check the differences between the zone of inhibitions of all synthesized complexes and reference standard. Post hoc Dunnett's test [31] was applied to the data and differences were considered.

Materials and Methods

Materials and instruments

Azithromycin sample was gifted by Platinum Pharmaceuticals (PVT) Ltd. While, the hydrated metal salts ($MgCl_2.6H_2O$, $CaCl_2.2H_2O$, $CrCl_3.6H_2O$, $MnCl_2.H_2O$, $FeCl_3.6H_2O$, $CoCl_2.6H_2O$, $NiCl_2.6H_2O$, $CuCl_2.2H_2O$, $ZnCl_2$, $CdCl_2.H_2O$), other solvents and reagents of analytical grade were purchased from Merck Marker (PVT) LTD. FT-IR and ^{1}H-NMR spectra were recorded on Prestige-21 Shimadzu FT-IR instrument. The samples were scanned in the form of KBr pellets. ^{1}H-NMR instrument was Bruker AMX 400 MHz. Chemical shifts were reported in ppm using tetramethylsilane (TMS) as an internal standard. CHN analysis was done on elemental analyzer Carlo Erba 1106. Atomic absorption studies were carried out by Perkin-Elmer "AAnalyst 700" atomic absorption spectrometer using "Analyst" software and conductometric titrations were carried on Vernier Lab Pro™. Data acquisition and analysis was carried out by using Logger pro 3.2 software. Chlorine was determined by titration with $Hg(NO_3)_2$.

Conductometric titration

For conductometric titration, metal chloride solutions of 1 mM were prepared and titrated with 1 mM ligand solution at 25°C [17]. Results obtained were then extrapolated to determine the stoichiometric ratio.

Synthesis of azithromycin metal complexes

Metal complexes were synthesized by mixing a hot methanolic solution of ligand (1 mM) with 0.5 mM solution of metallic chlorides in the ratio of 2:1. The reaction mixture was continuously refluxed on a water bath for 3.0-3.5 h at 60°C [19]. The solutions were then filtered and left for crystallization at room temperature for two to three weeks. In each case, a fine solid product was obtained which was washed, dried and their melting points were noted. These were than subjected for spectroscopic and microbial studies after their characterization. Unfortunately X-ray diffraction analysis could not be performed as we were not able to obtain appropriate mono crystals

Antimicrobial activity

The synthesized metal complexes were screened for their antibacterial activity against Gram-positive organisms as *Bacillus subtilis, Micrococcus luteus, Staphylococcus aureus,* and *Streptococcus features* and Gram-negative organisms which include *Salmonella typhi, Klebsiella pneumoniae, Proteus mirabilis, Pseudomonas aeruginosa, Escherichia coli, Citrobacter* and *Shigella flexneri*. A disk diffusion assay was employed for the evaluation of antibacterial activity [28,29], 6mm filter paper discs were impregnated with 5, 10 and 20 ppm dilutions and were allowed to remain at 37°C till complete diluents evaporation and kept under refrigeration before antibacterial activity was assessed [28]. They were grown routinely overnight in a nutrient broth (Merck) at 37°C.

The antibacterial discs of drug and drug based metal complexes were applied over each of the culture plates previously seeded with the 0.5 McFarland turbidity cultures of the test bacteria and then incubated at 37°C for 18-24 h. The release of drugs into the surrounding agar medium shown by growth inhibition of microorganisms was evaluated. The growth inhibitory effect was determined by measuring the zone of growth inhibition around the disk.

Antifungal activity

Same procedure was repeated for antifungal activity (as done for antibacterial activity) against series of fungi (*C.albicans, F.solani, T.rubrub, A.parasitieus, A.effusis and S.cervicis*). Dilutions were made in same manner for soaking discs as before. Sabraoud dextrose agar was then prepared and autoclaved at 121°C for 15 minutes, cooled and then poured in Petri dishes. Streaking was done in same way as done for antibacterial activity and dishes then incubated for 48 hrs at 37°C. Finally the zones of inhibition were carefully measured.

Statistical study

One-way analysis of variance (ANOVA) studies were conducted by using Statistical Package for the Social Sciences (SPSS Inc., Chicago, IL, USA) software with the level of significance chosen at $p \leq 0.05$ (any values lesser than 0.05 were measured significant).

Results and Discussion

Conductometric titration

The stoichiometry of the complexes in methanolic solution was determined by conductometric titration [30] at 298 K. The conductance value was measured during the titration of 0.1 mM metal ion solution against a 0.1 mM azithromycin solution (40 mL). The conductance was corrected for dilution by means of the following equation, assuming that conductivity is a linear function of dilution:

$$\Omega_{corr} = \Omega_{obs} (v_1 + v_2) / v_1$$

Where Ω is the electrolytic conductivity, v_1 is the initial volume and v_2 is the volume of the metal solution added. The drug to metal ratio was determined by plotting a graph of corrected conductivity versus the volume of titrant added and the end point was determined shown in Figure 2. The general trend of the conductograms is a steady increase of the conductance values of the solution after each addition of the metal up to equivalence point where a sudden change in the slope occurs [30]. This behavior is recognized to the formation of ion-pair in solution because of complexation reaction. The electrical conductance of the complexes point out that M:L ratio is 1:2 in all complexes.

From the conductance measurement data an attempt was made to synthesize azithromycin-metal complexes in the ratio of 2:1 (L:M) using

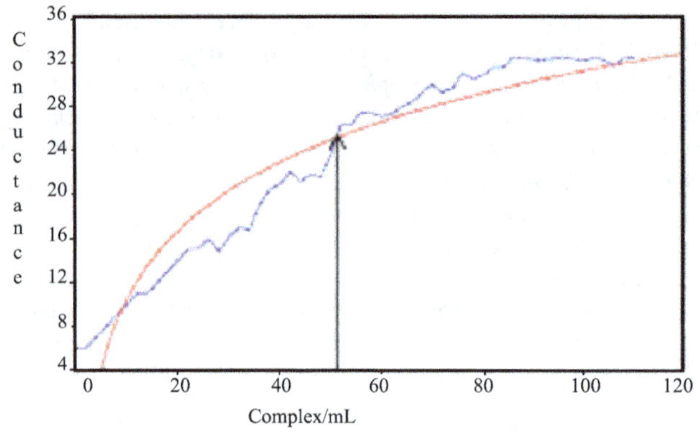

Figure 2: Conductometric titration curve.

Derivatives	M.P (°C)	State	Color	% Yield
Azithromycin	115	crystalline	white	45
$[Mg(Azi)_2(H_2O)_2]Cl_2$	130	crystalline	white	35
$[Ca(Azi)_2(H_2O)_2]Cl_2$	150	crystalline	white	55
$[Cr(Azil)_2\ (H_2O)_2]2Cl.2H_2O$	140	crystalline	light green	45
$[Mn\ (Azi)_2(H_2O)_2].4H_2O$	168	powder	white	42
$[Fe(Azi)_2\ (H_2O)_2]2Cl_2.2H_2O$	166	crystalline	white	51
$[Co\ (Azi)_2(H_2O)_2]$	139	crystalline	white	45
$[Ni\ (Azi)_2(H_2O)_2]Cl_2.2H_2O$	134	powder	white	58
$[Zn\ (Azi)_2(H_2O)_2].2H_2O$	160	crystalline	purple	40
$[Zn\ (Azi)_2(H_2O)_2].2H_2O$	148	powder	white	60
$[Cd\ (Azi)_2(H_2O)_2]\ Cl_2.4H_2O$	170	crystalline	light pink	52

Table 1: Physiochemical parameters of azithromycin and azithromycin– complexes.

methanol as solvent. Their physical parameters as color, melting points and % yields of are given in Table 1. These complexes are insoluble in water, slightly soluble in chloroform but completely soluble in methanol and DMSO. Techniques as IR, NMR and elemental analysis were used to characterize these complexes.

Infrared absorption studies

The assignments of IR bands were made by comparing the spectra of the complexes with azithromycin. In the reported spectra of azithromycin, there were two very strong absorption peaks at 1750 cm^{-1} and 1652 cm^{-1} due to lactone and ketonic carbonyl groups, respectively [9]. The absorption peaks between 1000 and 1250 cm^{-1} are due to the ethers and amine functions. The CH_2 bending is evident by peaks between 1340 and 1460 cm^{-1} and alkane stretching peaks appeared among 2800-2980 cm^{-1}. Hydrogen bonded OH and water molecule appeared as bands between 3350 and 3650 cm^{-1} with peak maxima at 3550 cm^{-1}.

Azithromycin is more likely to form complexes, with metals as it possess number of lone pair rich sites and amine substituted lactone ring [13]. In metal complexes of azithromycin, some very prominent peak shifting has been observed along with change in intensities of several important peaks indicating azithromycin has undergone complexation reaction with metals as shown in Figures 3A and 3B. In azithromycin chromium complex, the aliphatic amine stretch at 1100 cm^{-1} diminished to a significant extent than in azithromycin while the peak of 1200 cm^{-1} was shifted to 1160 cm^{-1}. The intensity of OH band decreased considerably and its maxima shifted from 3550 cm^{-1} to 3500 cm^{-1}. In azithromycin calcium complex same results were observed, i-e the OH absorption bands were shifted to 3400 cm^{-1} as shown in Figure 3A. The aliphatic amine stretch diminished to a significant extent that was present at 1100 cm^{-1} in azithromycin while the peak of 1200 cm^{-1} shifted to 1160 cm^{-1}. The same results were observed in azithromycin-copper, cobalt, cadmium, manganese, magnesium, ferric, zinc, nickel and calcium complex. In the light of these observations, it can be fairly concluded that the N $(CH_3)_2$ group of desosamine and hydroxyl group, have been utilized in the complex formation. The anionic part is out side the sphere and is represented by a dotted line.

NMR studies

The ^{1}HNMR spectra of azithromycin is harmonized with reported spectra which showed the resonances at δ_H 3.352 (3 H, s), 2.316 (3 H, s) and 2.288 (6 H, s) assigned to the 3”-OCH_3 absorption of cladinose, the 9a-NCH_3 group of the 15-membered aglycone ring, and the 3’-$N(CH_3)_2$ group of desosamine, respectively [9]. The same chemical shifts (3.352, 2.316 and 2.287) are also observed in the ^{1}H NMR spectra of metal containing compounds but the signals of 9a-NCH_3 group of the 15-membered aglycone ring were found less intensive. Additionally, the intense signal at 2.927 ppm observed in the spectrum of azithromycin is absent in the spectra of coordinative compounds of Cu(II), Mn(II), Fe(III), Cr(III), Co(II), Ni(II), and Ca(II) and much obscure in case of Mg(II), Zn(II) and Cd(II). Moreover, in the region δ_H 3.00-3.20, a triplet in the spectra of the coordinative compounds is observed instead of a quadruplet seen in the spectrum of azithromycin. It confirmed that the hydroxyl group and 9a-NCH_3 group of the 15-membered aglycone ring of azithromycin are bonded to metal ions. On the basis of the

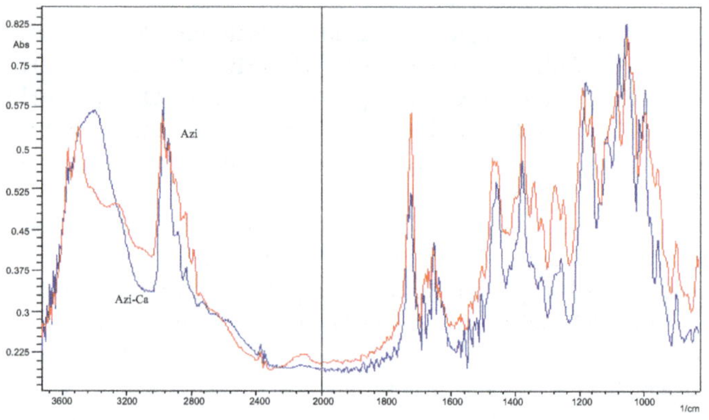

Figure 3a: Infrared spectra of azithromycin and its calcium based complex.

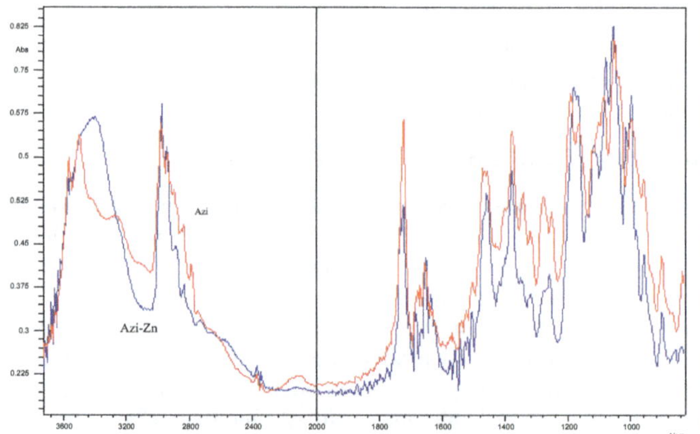

Figure 3b: Infrared spectra of azithromycin and its Zinc based complex.

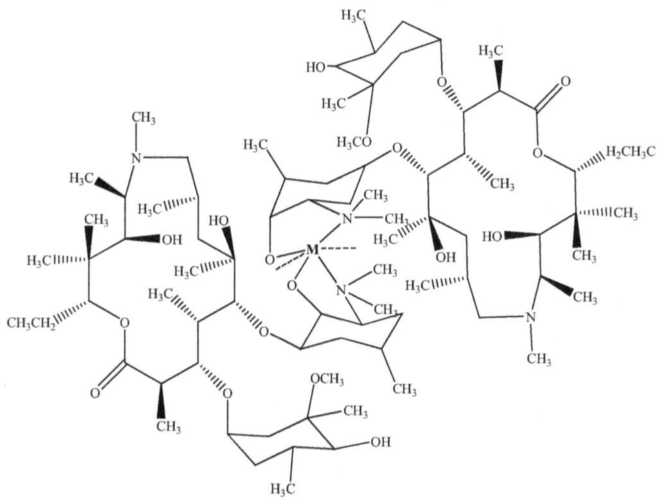

Figure 4: Proposed structures of the azithromycin-metal complexes.

Compound	C% Found(Calc)	H% Found(Calc)	N% Found(Calc)	Metal% Found(Calc)
Azithromycin	61.36 (61.39)	9.73 (9.78)	3.62 (3.67)	-
$[Mg(Azi)_2(H_2O)_2]Cl_2$	56.52 (56.60)	9.30 (9.31)	3.36 (3.38)	1.45 (1.47)
$[Ca(Azi)_2(H_2O)_2]Cl_2$	55.99 (56.06)	9.04 (9.05)	3.30 (3.35)	2.39 (2.40)
$[Cr(Azil)_2 (H2O)_2]2Cl_2.2H_2O$	54.48 (54.50)	9.40 (9.30)	3.22 (3.26)	4.04 (4.12)
$[Mn (Azi)_2(H_2O)_2].4H_2O$	56.71 (56.74)	9.45 (9.42)	3.34 (3.39)	3.30 (3.33)
$[Fe(Azi)_2 (H_2O)_2]2Cl_2.2H_2O$	54.32 (54.38)	9.00 (9.01)	3.20 (3.25)	3.20 (3.24)
$[Co (Azi)_2(H_2O)_2]$	57.80 (57.87)	9.32 (9.34)	3.42 (3.46)	6.41 (6.44)
$[Ni (Azi)_2(H_2O)_2]Cl_2.2H_2O$	54.20 (54.29)	9.00 (8.99)	3.24 (3.25)	3.39 (3.40)
$[Zn (Azi)_2(H_2O)_2].2H_2O$	57.11 (57.07)	9.35 (9.33)	3.42 (3.41)	3.40 (3.41)
$[Zn (Azi)_2(H_2O)_2].2H_2O$	57.42 (57.64)	9.50 (9.30)	3.49 (3.45)	4.01 (4.02)
$[Cd (Azi)_2(H_2O)_2] Cl_2.4H_2O$	52.49 (52.65)	8.70 (8.72)	3.05 (3.15)	6.33 (6.32)

Table 2: Elemental analysis of azithromycin and its metal complexes.

above mentioned data and literature data concerning the structure and stereochemical configuration of azithromycin, the purposed structure of ML_2 is shown in Figure 4.

Elemental analysis

The structure of the complexes suggested form the elemental analyses agree well with their proposed formulae as given in Table 2. The found values of elemental analysis agree well with calculated percentages of CHN data are in a well agreement with each other and prove the molecular formulas of the complexes.

Biological study

The results of Minimal Inhibitory Concentrations (MIC) in µg/mL concentration of Azithromycin metal complexes against various Gram positive and Gram negative microorganisms are given in Table 3. The results of the disk diffusion assay method for each bacterial and fungal strain are shown in Table 4. All data are presented as zone of inhibition, diameter in mm. One way analysis of variance (ANOVA) was carried out to check the differences between the zone of inhibitions of all synthesized complexes and standard. Post hoc Dunnett's test [31] was applied to the data and differences were considered significant at $p \leq 0.05$. ANOVA showed the significance differences between all synthesized complexes with comparison to azithromycin.

Antibacterial study

Post hoc Dunnett's test analyzed that against *P. mirabilis* all complexes have significantly decreased ($p<0.001$) activity except $Co(Azi)_2$ complex which showed significant increase ($p<0.001$) in antibacterial activity. Significant differences also existed between all synthesized complexes with azithromycin against S. *typhi* however only $Ni(Azi)_2$ complex showed significant ($p<0.001$) increased antibacterial activity. Significant decrease ($p<0.001$) was exhibited by all complexes against *E. coli* and *P. aeruginosa* except $Cr(Azi)_2$ which showed significant increase ($p<0.001$) against *P. aeruginosa*. Azithromycin based $Mg(Azi)_2$, $Cr(Azi)_2$, $Fe(Azi)_2$, $Cd(Azi)_2$ and $Zn(Azi)_2$ complexes were shown to have significantly ($p<0.001$) increased antibacterial activity against *K. pneumonia*. These complexes $Ni(Azi)_2$ and $Ca(Azi)_2$ also showed significant increase ($p<0.001$) at 10 and 20 µg concentration against *S. flexneri* while, remaining complexes have significant ($p<0.001$) decrease in activity. Post hoc Dunnett's test analysis also assures that all synthesized metal complexes showed significant decrease ($p<0.001$) in activity at all concentration levels against *Citrobacter*, *S. features* and *M. luteus*. Only $Zn(Azi)_2$, $Fe(Azi)_2$ and $Ca(Azi)_2$ complexes exhibits significant increase ($p<0.001$) in activity against *B. subtilis*. Against *S. aureus* all the complexes showed significant decrease ($p<0.001$) except $Co(Azi)_2$ which showed significant increase ($p<0.001$) in activity. Microbiological screening of all synthesized complexes revealed that against *P. mirabilis* $Co(Azi)_2$ complex and against *S.typhi* $Mn(Azi)_2$, $Fe(Azi)_2$ and $Cu(Azi)_2$ complexes exhibit enhanced activity

Compound	Minimum Inhibitory Concentration (MIC, µg/mL)											
	BS	CT	MRSA	PA	ML	SF	PM	CA	EC	SF!	KP	ST
Azi	2	8	8	10	8	8	9	4	2	4	4	2
Azi – Mg	5	4	8	4	5	4	8	5	6	3	5	3
Azi – Ca	2	2	6	5	8	4	5	6	5	3	6	4
Azi – Cr	2	2	8	4	4	5	4	4	3	3	7	2
Azi – Mn	6	1	4	6	4	8	4	6	5	4	4	8
Azi – Fe	1	6	4	4	8	12	14	13	7	6	9	7
Azi – Co	6	2	4	4	5	4	7	5	7	8	7	2
Azi – Ni	3	4	4	4	4	7	7	7	6	5	7	5
Azi – Cu	2	4	6	4	3	4	7	4	6	5	3	4
Azi – Zn	4	1	6	3	2	5	5	5	6	5	2	7
Azi – Cd	8	3	4	8	5	6	5	4	4	9	2	5

BS: *Bacillus subtilis*; CT: *Citrobacterium;* MRSA: Methicillin-Resistant *Staphylococcus aureus;* PA: *Pseudomonas aeruginosa*; ML: *Micrococcus luteus*; SF: *Streptococcus features*; PM: *Proteus mirabilis*; CA: *Candida albicans*; EC: *Escherichia coli*; SF!: *Shigella flexenri*; KP: *Klebsiella pneumoniae*; ST: *Shigella typhi*

Table 3: The minimal inhibitory concentrations (MIC, µg/mL) of Azithromycin metal complexes against various Gram positive and Gram negative microorganisms

Compound	ST	BS	CT	MRSA	PA	ML	SF	PM	CA	EC	SF!	KP
Azi	22	-	30	30	20	35	20	30	20	22	24	22
Azi – Mg	22	-	30	25	18	-	20	22	20	20	15	20
Azi – Ca	25	30	20	22	24	22	24	21	20	25	20	22
Azi – Cr	25	30	20	23	-	22	20	18	20	-	25	-
Azi – Mn	28	-	28	26	-	-	17	20	15	21	14	20
Azi – Fe	22	25	30	20	-	22	25	29	30	22	25	25
Azi – Co	22	-	30	35	25	25	20	38	23	20	22	19
Azi – Ni	20	30	18	23	21	24	22	20	18	30	22	21
Azi – Cu	22	-	25	-	25	22	20	22	-	16	20	22
Azi – Zn	28	30	20	20	-	22	25	25	20	22	23	18
Azi – Cd	20	-	26	-	19	-	25	28	-	20	22	25

BS: *Bacillus subtilis*; CT: *Citrobacterium;* MRSA: Methicillin-Resistant *Staphylococcus aureus;* PA: *Pseudomonas aeruginosa*; ML: *Micrococcus luteus*; SF: *Streptococcus features*; PM: *Proteus mirabilis*; CA: *Candida albicans*; EC: *Escherichia coli*; SF!: *Shigella flexenri*; KP: *Klebsiella pneumoniae*; ST: *Shigella typhi*; (-): Inactive

Table 4: Antimicrobial activity of Azithromycin metal complexes (20 µg/8 mm disc), as compared to azithromycin.

in comparison to parent compound. While against *E.coli*, $Fe(Azi)_2$ complex and against *P. aureogenosa* $Ca(Azi)_2$, $Cr(Azi)_2$, $Fe(Azi)_2$, $Co(Azi)_2$ and $Ni(Azi)_2$ complexes showed greater activity. Against *K. pneumonea* $Cr(Azi)_2$, $Fe(Azi)_2$ and $Cd(Azi)_2$ while against *S. flexneri* $Ni(Azi)_2$ complexes exhibits increased activity whereas, other metal-Azi complexes have activities equivalent or less than azithromycin (Azi). Against *Citrobacter* all complexes exhibits activities equivalent to azithromycin. Against Gram positives as *M. luteus*, *S. faetures* and *S. aureus* all complexes showed slight increased or equivalent activity in comparison to azithromycin. While against *B. subtilus* $Ca(Azi)_2$, $Cr(Azi)_2$, $Fe(Azi)_2$, $Ni(Azi)_2$ and $Zn(Azi)_2$ have increased activity whereas, azithromycin didn't show any activity.

Antifungal study

Azithromycin was not used as antifungal agent before, to the best of our knowledge; it is tested as antifungal agent for the first time along with its synthesized complexes. Surprisingly, all of the complexes and azithromycin showed excellent activity only against *C. albican* and did not exhibit any activity against *F.solani*, *T. rubrub*, *A. purasiticus*, *A. effuris* and *S. cervicis*. Post hoc ANOVA studies reveals that $Mg(Azi)_2$, $Co(Azi)_2$, $Fe(Azi)_2$, $Zn(Azi)_2$ and $Ca(Azi)_2$ complexes showed significant increase ($p<0.001$) in antifungal activity against *C. albicans* in comparison to azithromycin.

Conclusion

The azithromycin (ligand, L) here, shows a bidentate behavior and was found that $N(CH_3)_2$ group of desosamine and hydroxyl group of azithromycin underwent complexation with selected metals. Among all synthesized drug based metal complexes, $Ca(Azi)_2$, $Fe(Azi)_2$, $Co(Azi)_2$, $Ni(Azi)_2$ and $Zn(Azi)_2$ complexes exhibits increased activity against both Gram negative and Gram positive organisms. All the other tested complexes exhibited mild to moderate antibacterial activity while, $Mg(Azi)_2$ and $Co(Azi)_2$ possess enhanced antifungal activity in comparison to azithromycin. Therefore, it can be concluded that azithromycin can be used as antifungal agent along with synthesized complexes.

Acknowledgment

Ms Sana Shamim wish to thank Higher Education Commission, Pakistan, for providing scholarship under Indigenous 5000 Ph.D Fellowship Program Batch IV.

References

1. Rehman W, Baloch MK, Badshah A (2008) Synthesis, spectral characterization and bio-analysis of some organotin(IV) complexes. Eur J Med Chem 43: 2380-2385.
2. Gordon Research Conferences (2000) Metals in Medicine: Targets, Diagnostics, and Therapeutics. June 28-29, 2000. Natcher Conference Center National Institutes of Health, Bethesda, Maryland, USA.
3. Kostova I (2010) Metal-containing drugs and novel coordination complexes in therapeutic anticancer applications-part II. Anticancer Agents Med Chem 10: 352-353.
4. Kelland L (2007) The resurgence of platinum-based cancer chemotherapy. Nat Rev Cancer 7: 573-584.
5. Hambley TW (2007) Developing new metal-based therapeutics: challenges and opportunities. Dalton Trans: 4929-4937.
6. Caldwell MD (1990) Topical wound therapy-an historical perspective. J Trauma 30: 116-122.
7. Moues CM, Heule F, Legerstee R (2009) Topical Negative Pressure in Wound Care Effectiveness and guidelines for clinical application. Ostomy and wound management 55: 16-32.
8. Rafique S, Idrees M, Nasim A, Akbar H, Athar A (2010) Transition metal complexes as potential therapeutic agents. Biotechnology and Molecular Biology Reviews 5: 38-45.
9. Bukvić Krajacić M, Novak P, Cindrić M, Brajsa K, Dumić M, et al. (2007) Azithromycin-sulfonamide conjugates as inhibitors of resistant Streptococcus pyogene. Eur J Med Chem 42: 138-145.
10. Schonfeld W, Mutak S (2002) Macrolide Antibiotics. 73-95. Birkhauser Verlag Basel, Switzerland.
11. (2006) Physician Desk Reference, Medical economic company Inc., 64th edition, USA.
12. Riedel KD, Wildfeuer A, Laufen H, Zimmermann T (1992) Equivalence of a high-performance liquid chromatographic assay and a bioassay of azithromycin in human serum samples. J Chromatogr 576: 358-362.
13. Sher A, Rau H, Greiner G, Haubold W (1996) Spectroscopic and polarographic investigations of copper(II)-azithromycin interactions under equilibrium conditions. International Journal of Pharmaceutics 133: 237-244.
14. http://www.pharmaselect.cz/soubory/F/USP_DI_2005.pdf
15. El-Rjoob AW, Al-Mustafa J, Taha Z, Abous M (2008) Spectroscopic and conductometric investigation of the interaction of azithromycin with iron (II) ion. Jordan Journal of Chemistry 3: 199-209.
16. Arayne MS, Sultana N, Rafiq K (2001) *In vitro* activity of ceftizoxime and ceftazidime in presence of essential and trace elements. Pak J Pharm Sci 14: 57-64.
17. Arayne MS, Sultana N, Khanum F, Ali MA (2002) Antibacterial studies of cefixime copper, zinc and cadmium complexes. Pak J Pharm Sci 15: 1-8.
18. Sultana N, Arayne MS, Afzal M (2003) Synthesis and antibacterial activity of cephradine metal complexes: part I complexes with magnesium, calcium, chromium and manganese. Pak J Pharm Sci 16: 59-72.
19. Sultana N, Arayne MS, Sharif S (2004) Levofloxacin interactions with essential and trace elements. Pak J Pharm Sci 17: 67-76.
20. Arayne S, Sultana N, Haroon U, Mesaik MA (2009) Synthesis, characterization, antibacterial and anti-inflammatory activities of enoxacin metal complexes. Bioinorganic Chemistry and Applications 914105.
21. Sultana N, Naz A, Arayne MS, Mesaik MA (2010) Synthesis, characterization, antibacterial, antifungal and immunomodulating activities of gatifloxacin–metal complexes. Journal of Molecular Structure 969: 17-24.
22. Sultana N, Arayne MS, Gul S, Shamim S (2010) Sparfloxacin–metal complexes as antifungal agents-Their synthesis, characterization and antimicrobial activities. Journal of Molecular Structure 291: 285-291.
23. Sultana N, Arayne MS, Rizvi SBS, Haroon U, Mesaik MA (2013) Synthesis, spectroscopic and biological evaluation of some levofloxacin metal complexes. Medicinal Chemistry Research 22: 1371-1377.
24. Sultana N, Humza E, Arayne MS, Haroon U (2011) Effect of metal ions on the *in vitro* availability of enoxacin, it's *in vivo* implications, kinetic and antibacterial studies. Quimica Nova 4: 186-189.
25. Sultana N, Arayne MS, Sabri R (2002) Clarithromycin synergism with essential and trace elements. Pak J Pharm Sci 15: 43-54.
26. Sultana N, Arayne MS, Sabri R (2005) Erythromycin synergism with essential and trace elements. Pak J Pharm Sci 18: 35-39.
27. Arayne MS, Sultana N, Sabri R (2004) Roxithromycin antagonism with essential and trace elements. Pak J Pharm Sci 17: 65-75.
28. World Health Organization (2003) Manual for the Laboratory Identification and Antimicrobial Susceptibility Testing of Bacterial Pathogens of Public Health Importance in the Developing World 103-162.
29. Akinyemi KO, Oladapo O, Okwara CE, Ibe CC, Fasure KA (2005) Screening of crude extracts of six medicinal plants used in South-West Nigerian unorthodox medicine for antimethicillin resistant *staphylococcus aureus* activity. BMC Complementary and Alternative Medicine 5: 1472-1483.
30. Lingane JJ (1958) Electro analytical Chemistry, 2nd edition, Interscience Publishers, New York, USA.
31. Mazumder UK, Gupta M, Bhattacharya S, Karki SS, Rathinasamy S, et al. (2004) Antineoplastic and antibacterial activity of some mononuclear Ru(II) complexes. J Enzyme Inhib Med Chem 19: 185-192.

Effect of Thermal Annealing on the $Cd(OH)_2$ and Preparation of Cdo Nanocrystals

Nadana Shanmugam*, Balan Saravanan, Rajaram Reagan, Natesan Kannadasan, Kannadasan Sathishkumar and Shanmugam Cholan

Department of Physics, Annamalai University, Annamalai Nagar, Chidambaram 608 002, Tamilnadu, India

Abstract

Nanosized β-$Cd(OH)_2$ were successfully synthesized via simple chemical precipitation method using cadmium nitrate as a precursor in a solution of sodium hydroxide. The CdO nanoparticles were harvested from β-$Cd(OH)_2$ by thermal decomposition at 400°C. The structural, optical, and magnetic properties of the as prepared and annealed products of β-$Cd(OH)_2$ were studied. The morphology of the CdO nanocrystals annealed at 400°C analyzed by FE-TEM exhibits pseudo spherical morphology with sizes around 60 nm.

Keywords: Nanosized β -$Cd(OH)_2$; Decomposition; Annealed products; Magnetic properties

Introduction

In the past two decades, new terms with prefix nano have captured ample space among the scientific community owing to the unusual physical and chemical properties of nanomaterials. Usually on the basis of size, morphology and structure applications of nanomaterials are justified. Recently, interests in the study of Transparent Conductive Oxide (TCO) nanomaterials have gained special attention due to their important applications in the current technology. Among the transparent conducting metal oxide semiconductor materials, CdO is an important n-type semiconductor material with direct band gap of 2.2-2.7 eV and indirect band gap of 1.36-1.98 eV [1]. CdO finds its potential applications in the field optoelectronics devices such as solar cells, phototransistors, photodiodes, transparent electrodes, catalysts and gas sensors [2-11]. Despite cadmium is toxic, CdO is widely used as a photo catalyst for effluent treatment [4,12,13]. Many researchers have reported the preparation of CdO nanostructures with different methods such as chemical vapour deposition, sol-gel, laser ablation, spray pyrolysis and hydrothermal methods. Nowadays, the usage of the simple chemical precipitation method in comparison with other methods increases among researchers because of its less time consuming and less expensive nature. Lotf Ali, et al. have synthesized $Cd(OH)_2$ and CdO nanocrystals by the solvothermal method. They predicted the conversion of nanosized $Cd(OH)_2$ into CdO at 500°C [14]. Siraj et al. have studied the magnetic properties of Al-doped CdO thin films and reported their para and ferromagnetic behaviors [15].

Herein, we demonstrate a simple chemical precipitation method for the synthesis of nanocrystalline β-$Cd(OH)_2$. Nanocrystals of CdO can be obtained through the thermal decomposition using the as-prepared β-$Cd(OH)_2$ as precursor.

Materials and Methods

Chemicals

Cadmium nitrate hexahydrate [$Cd(NO_3)_2 \cdot 6H_2O$], sodium hydroxide [NaOH] were purchased from Merck and were used as received since they were of analytical reagent grade with 99% purity. Ultra-pure water was used for all procedures of sample preparation and dilution.

Synthesis of CdO nanocrystals

In the preparation of CdO nanocrystals from cadmium nitrate hexahydrate ($Cd(NO_3)_2 \cdot 6H_2O$) and sodium hydroxide (NaOH), 0.5 M of $Cd(NO_3)_2 \cdot 6H_2O$ in 50 ml of deionized water and 2 M of NaOH in 50 ml water were mixed up dropwise. The entire mixture was stirred magnetically until a white precipitate of cadmium hydroxide hexahydrate was formed. The resultant precipitate was filtered and then washed alternately with deionized water and ethanol for 3 times to remove the impurities. Further, the precipitate of cadmium hydroxide hexahydrate ($Cd(OH)_2 \cdot 6H_2O$) was dried in hot air oven at 100°C for 4 h and cadmium hydroxide was harvested in the nanosize. The obtained product was thermally annealed at different temperatures (200, 300, 400, 600 and 800°C) for 2 h. The formation of CdO took place at 400°C upon thermal annealing.

Growth mechanism

Formation of $Cd(OH)_2$ in the presence of NaOH can be explained on the basis of buffer action of cadmium ions. Cadmium ions in the solution become hydrated and transformed to solid cadmium hydroxide through stepwise coordination of hydroxyl ions. However, depending upon the concentration of the base and the synthesis temperature, cadmium hydroxide is transformed into cadmium oxide through dehydration.

Apparatus

The prepared products were characterized by powder X-Ray Diffraction (XRD) on a X'PERT PRO diffractometer with Cu-Ka radiation (k=1.5406 Å). From the line broadening, the size of the particle was estimated by the Scherrer equation. FT-IR analysis was made to characterize the functional groups of the precursor and nanosized cadmium oxide using SHIMADZU-8400 with a resolution of 4 cm^{-1}. The Photoluminescence (PL) emission spectra of the samples were recorded with a Spectroflurometer (Jobin Yvon, FLUOROLOG–FL3-11). Vibrating Sample Magnetometer (VSM) is used to identify the nature of magnetic species in the material. To study the morphology and size of the nanocrystals FE-TEM (Model JSM 2100F JEOL, Japan) analysis was made.

***Corresponding author:** Nadana Shanmugam, Department of Physics, Annamalai University, Annamalai Nagar, Chidambaram 608 002, Tamilnadu, India, E-mail: quantumgosh@rediffmail.com

Results and Discussion

Thermal analysis

To understand the thermal behavior of β-$Cd(OH)_2$, TG-DTA analyses were carried out between room temperature and 1000°C with a heating rate of 20°C/min in nitrogen atmosphere. Figure 1 shows the TG and DTA traces of β-$Cd(OH)_2$. Five prominent weight losses were observed at the end set temperatures 157°C (2.63%), 244°C (9.61%), 398°C (2.79%), 729°C (2.58%), and 1000°C (11.39). The initial weight loss of 2.63% observed between room temperature and 157°C could be ascribed to the removal of water molecules adsorbed on the surface of the $Cd(OH)_2$ nanoparticles. The dehydration process involved in the first stage is given by the following chemical equation

$$Cd(OH)_2{}^{\cdot}X_{ads}H_2O \rightarrow Cd(OH)_2 + X\,H_2O$$

From the initial weight loss of 2.63%, it is possible for us to estimate the amount of water exists in the as prepared $Cd(OH)_2$. From the weight loss, the water absorption in molar fraction has been calculated as 0.0215 M (moles of H_2O per mole of $Cd(OH)_2$). Therefore, the possible dehydration reaction is modified as

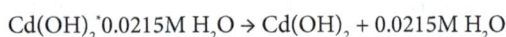

$$Cd(OH)_2{}^{\cdot}0.0215M\,H_2O \rightarrow Cd(OH)_2 + 0.0215M\,H_2O$$

The second stage of weight loss occurring between 157-244°C may be ascribed to the transformation of β-$Cd(OH)_2$ into γ-$Cd(OH)_2$. The third stage of weight loss predicted between 244-398°C is due to the decomposition of γ-$Cd(OH)_2$ into CdO. The fourth stage of minimum weight loss of 2.58% recorded between 398-729°C indicating the improved crystallinity of the CdO. A final and fifth stage of steep weight loss of 11.39% ascribed to the transformation of CdO nanocrystals into CdO_2. The DTA curve of β-$Cd(OH)_2$ shows two endothermic peaks at 204°C and 354°C corresponding to the removal of adsorbed water and decomposition of $Cd(OH)_2$ into CdO respectively. Further, the DTA curve exhibits two exothermic peaks at 272°C and 695°C which are related to the phase transformation of $Cd(OH)_2$ and conversion of CdO into CdO_2 respectively. The obtained DTA results support the results of the TG curve.

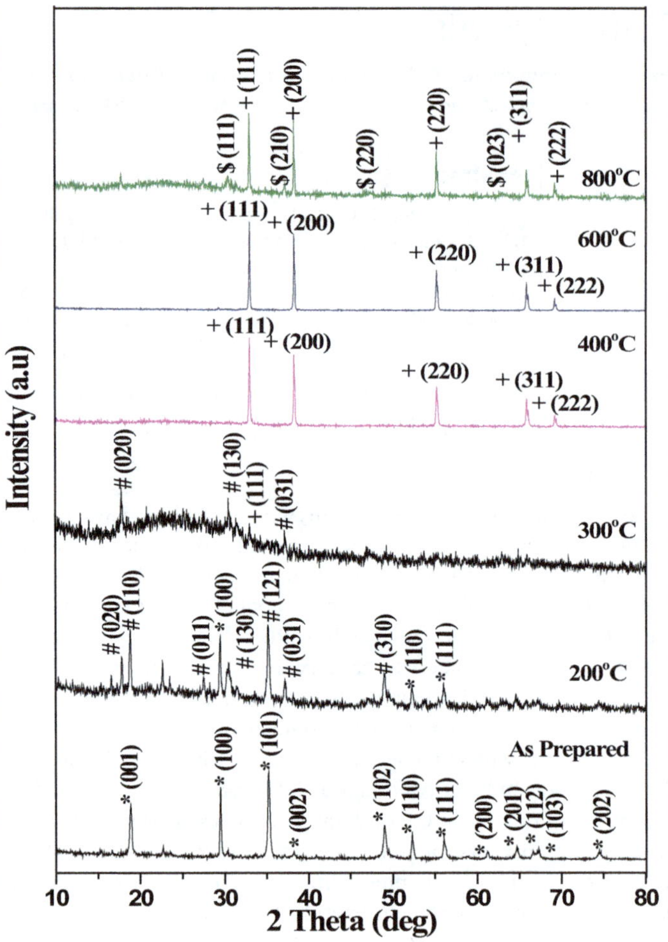

Figure 2: XRD patterns of as prepared and annealed CdO nanocrystals (*Cd(OH)$_2$-Hexogonal; # Cd(OH)$_2$-Monoclinic; + CdO-Cubic; $ CdO$_2$-Cubic).

XRD analysis

The XRD patterns of as-synthesized and annealed products are shown in Figure 2. As synthesized products exhibited twelve diffraction peaks corresponding to the (001), (100), (101), (002), (102), (110), (111), (200), (201), (112), (103) and (202) planes of hexagonal β-$Cd(OH)_2$ nanoparticles. The XRD patterns of the products annealed at 200°C show the presence of hexagonal as well as monoclinic (JCPDS: 20-0179) phases of γ-$Cd(OH)_2$. Further, on annealing at 300°C, in addition to the (020), (130) and (031) planes of monoclinic phase of γ-$Cd(OH)_2$, a dominating cubic (111) peak of CdO is also seen. After annealing at high temperatures of 400 and 600°C, the diffraction peaks could be indexed as (111), (200), (220), (311) and (222) planes of cubic CdO (JCPDS: 05-0640). The XRD patterns of the sample annealed at 400°C show the formation of pure cubic phase of CdO. However, the sample annealed at 600°C shows sharp diffraction peaks with reduced peak width as a result of improved crystallinity. Further annealing of the sample at 800°C leads to the formation of both cubic CdO and CdO_2.

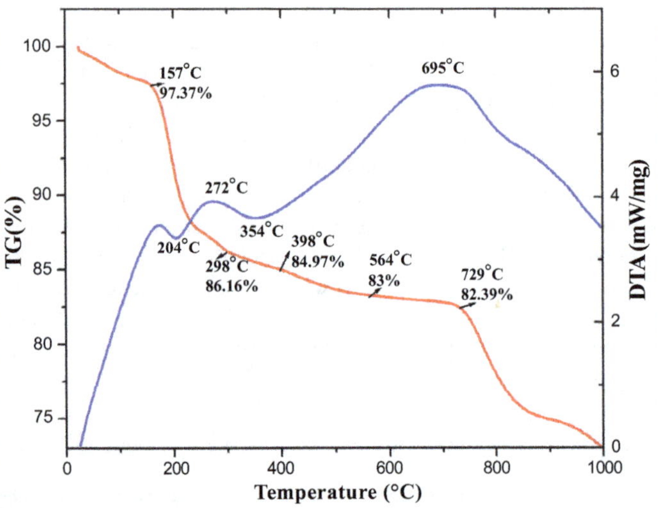

Figure 1: TG-DTA Curves of nanosized $Cd(OH)_2$.

The average grain size was calculated from the XRD patterns using the Debye Scherrer's formula [16] for the as prepared and annealed products. The crystal structure, lattice parameters, and particle size of the as prepared and annealed products are given in Table 1. As can be seen from the table, the grain size increases with annealing temperature and thus the crystallization of the products is improved. The lattice constants calculated for all the products almost identical to the JCPDS values, especially the lattice parameters of the products annealed after 400 and 600°C exactly match the JCPDS values.

FT-IR analysis

FT-IR analysis was used to study the β-$Cd(OH)_2$ powder during heating. The heated samples were ground with KBr and pressed into pellets. IR spectra were recorded on samples after heat treatment at 200, 300, 400, 600 and 800°C. As shown in Figure 3 the IR spectra of the as-prepared sample annealed at 100°C show a sharp and intense band at 3605 cm^{-1} arising from the stretching vibrations of structural OH groups confirming they are β-$Cd(OH)_2$ [11,17]. Usually β-$Cd(OH)_2$ can show a band at around 3605 cm^{-1}, whereas γ-$Cd(OH)_2$ may provide absorption bands at around 3588 and 3531 cm^{-1}. Absorption bands observed around 3447 and 1640 cm^{-1} are respectively assigned to the stretching and bending vibrations of H_2O molecules. Further, on heating at 200°C, the sharpness of the band at 3603 cm^{-1} decreases and a new band at 3522 cm^{-1} emerges. This indicates the conversion of β-$Cd(OH)_2$ into γ-$Cd(OH)_2$. Upon heating at 300°C, the absorption band at 3603 cm^{-1} completely disappeared. Since the absorption bands of γ-$Cd(OH)_2$ are predicted at 3524 and 3584 cm^{-1}, the observation confirms a complete transformation from β-$Cd(OH)_2$ to γ-$Cd(OH)_2$ [18]. According to the literature, the bands in between 800-1400 cm^{-1} belong to the Cd-O vibration [19]. In addition, peaks around 685 and 447 cm^{-1} could be ascribed to the Cd-O stretching mode [14]. After being at 400°C, the formation of CdO is characterized by the sharp bands positioned at 1383, 686 and 447 cm^{-1}. On further annealing (600 and 800°C) the characteristic peaks of Cd-O are broadened as a result of increased particle size.

Temperature (°C)	Product	Crystal structure	Lattice parameters (Å)		Particle size (nm)
			Slandered values	Calculated values	
100	$Cd(OH)_2$	Hexagonal	a=3.494 c=4.710	a=3.4973 c=4.791	41.29
200	$Cd(OH)_2$	Hexagonal		a=3.5002 c=4.8132	38.96
	$Cd(OH)_2$	Monoclinic	a=5.63 b=10.18 c=3.4127	a=5.0231 b=9.9507 c=3.4127	34.52
300	$Cd(OH)_2$	Monoclinic		a=5.0131 b=9.9317 c=3.4012	99.82
	CdO	Cubic	a=4.695	4.7004	19.15
400	CdO	Cubic		4.6963	64.63
600	CdO	Cubic		4.6960	86.46
800	CdO	Cubic	a=4.695	4.6984	98.61
	CdO_2	Cubic	a=5.313	5.0789	17.20

Table 1: Crystal structure, lattice parameters, and particle size of the as prepared and annealed products.

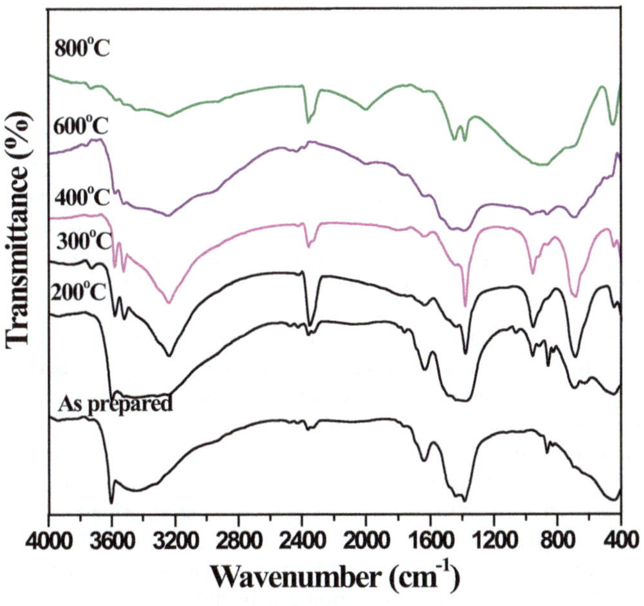

Figure 3: FT-IR spectra of as prepared and annealed CdO nanocrystals.

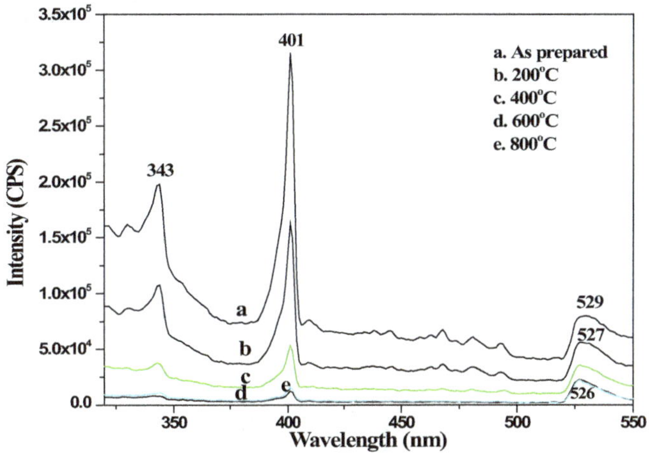

Figure 4: PL emission spectra of as prepared and annealed CdO nanocrystals.

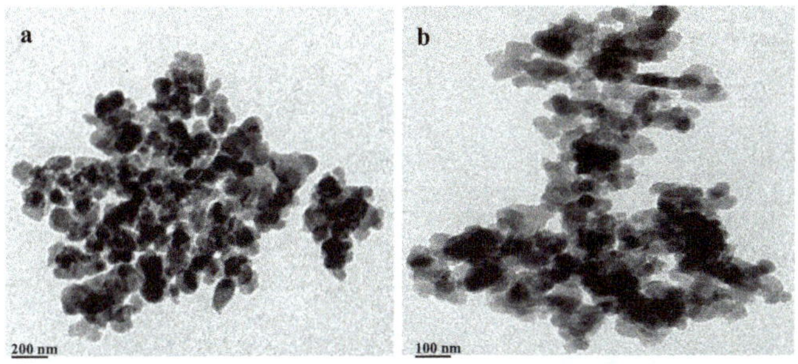

Figure 5: FE-TEM micrographs of CdO nanocrystals annealed at 400°C.

Photoluminescence

The room temperature Photoluminescence (PL) spectra of as prepared and annealed products with 250 nm excitation are shown in Figure 4. All the products show three emission peaks positioned at 343, 401 and 527 nm. The peak appearing at 343 and 401 nm are assigned to the near band edge emission of CdO originating from excitonic transitions between the electrons in the conduction bands and the holes in the valence bands. The emission peak at 527 nm may be ascribed to structural defects such as vacancies and surface traps [20,21]. With such visible emission, the CdO nanocrystals can be utilized in the industry of high-quality monochromatic laser.

As a general behavior, the PL spectra of the CdO nanomaterials showed a relatively broad less intense UV and visible emission bands as the annealing temperature is raised. This is due to the fact that the oxygen vacancy concentration decreases after annealing at high temperatures. However, the position of the emission bands are not majorly changed as the annealing temperature is raised from 200-800°C. This suggests that these emission bands are weakly associated with the band gap properties.

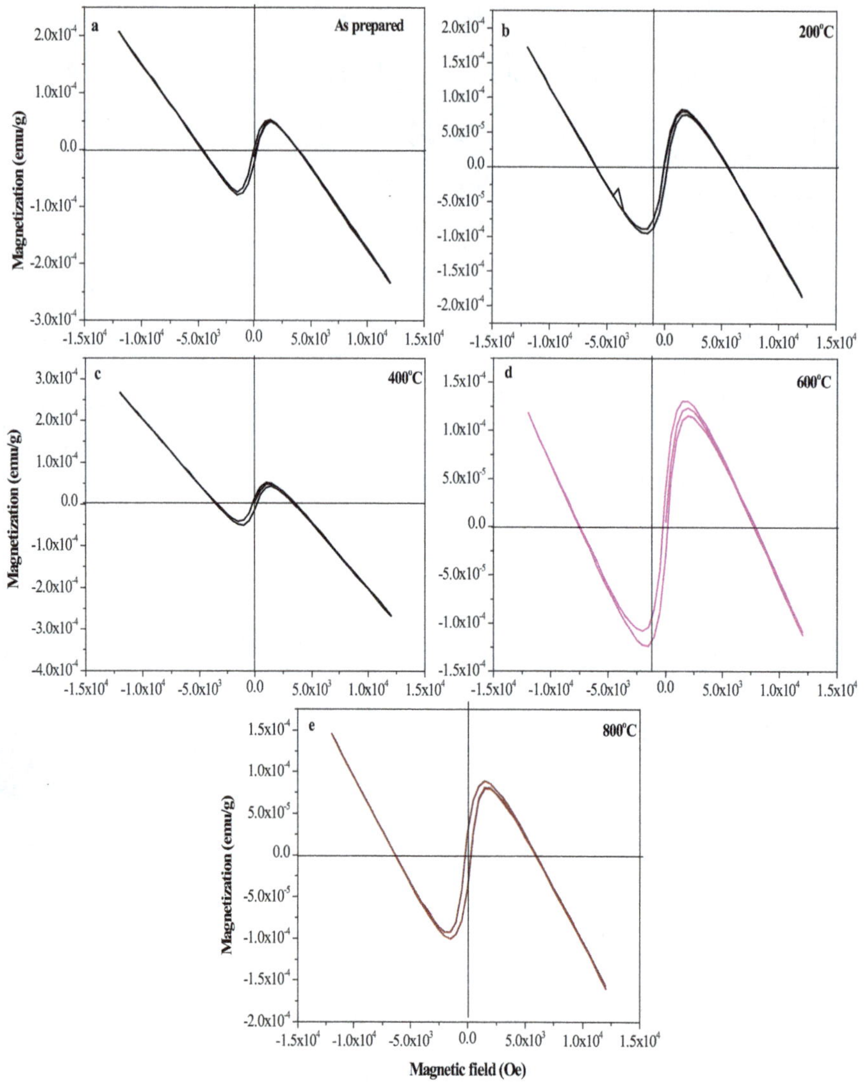

Figure 6: M-H loops of as prepared and annealed CdO nanocrystals.

Field emission-transmitting electron microscope analysis

FE-TEM analysis was used to evaluate the shape and size of the CdO nanoparticles. FE-TEM micrographs of CdO annealed at 400°C are shown in Figure 5a and b. The annealed sample is constituted of pseudo spherical 60 nm sized particles. The entire observed particles have almost same diameter. The value of crystallite size obtained from the FE-TEM analysis is in good agreement with the value obtained from the Scherrer's formula.

Magnetic characterization

The magnetic behavior of CdO nanoparticles has not been much investigated so far. Vibrating sample magnetometer was used to study the magnetic properties of CdO nanocrystals at different growth temperatures. Figure 6 a-e shows the hysteresis loops for as prepared and annealed CdO nanocrystals. It can be observed that both as-prepared and annealed formulations reveal typical paramagnetic behavior. The paramagnetism of the products is clearly shown by coercivity (H_c) saturation magnetization (M_s) and remnant magnetization (M_r) listed in Table 2. The saturation magnetization is the maximum induced magnetic moment that can be obtained in a magnetic field, beyond this field no further increase in magnetization occurs. As shown in Figure 6c and Table 2, the effect of 400°C of annealing resulted in an increase of the saturation magnetization by almost 20%. Coercivity is the reverse magnetic field required to reduce the net magnetization to zero. For magnetic materials, it is necessary to reduce the coercivity as a way to control the energy losses. As shown in second column in Table 2, as prepared and annealed at 400°C CdO nanocrystals have relatively lower coercivity compared with that of other annealed products. Simply, remnant magnetization (M_r) can be defined as the remaining magnetic momentum after realizing the magnetic field. Low remnant magnetization materials are classified as magnetically clean materials.

Temperature (°C)	Coercivity (H_c)	Magnetization $\times 10^{-6}$emu	Retentivity $\times 10^{-6}$
As prepared	1921.4	220.90	12.540
200	2839.7	179.67	16.317
400	1895.3	268.10	13.453
600	4078.5	127.14	31.729
800	3095.1	152.68	32.833

Table 2: The paramagnetism of the products is clearly shown by coercivity (H_c) saturation magnetization (M_s) and remanent magnetization (M_r) listed.

In some distinct fields, low remnant magnetization is highly desirable for instance in data storage applications. As shown in the last column in Table 2, both as prepared $Cd(OH)_2$ and CdO annealed at 400°C have low remnant magnetization.

Conclusion

In conclusion, nanometer sized particles of CdO have been successfully synthesized by thermal decomposition of β-$Cd(OH)_2$ at 400°C. The results of XRD and FT-IR analyses confirmed the formation of CdO phase. Thermal annealing on CdO has a considerable effect of increasing the particle size. The prepared CdO showed visible emission at 527 nm that can be used in the preparation of gas sensors. FE-TEM analysis of CdO shows the pseudo spherical particles with diameter around 60 nm. The CdO nanocrystals annealed at 400°C show low values of coercivity and remnant magnetization suggesting potential usage in data storage applications.

References

1. Kuo TJ, Huang MH (2006) Gold-catalyzed low-temperature growth of cadmium oxide nanowires by vapor transport. J Phys Chem B 110: 13717-13721.
2. Ferro R, Rodriguez JA (2000) Influence of F-doping on the transmittance and electron affinity of CdO thin films suitable for solar cells technology. Solar Energy Materials and Solar Cells 64: 363-370.
3. Guo Z, Li M, Liu J (2008) Highly porous CdO nanowires: preparation based on hydroxy- and carbonate-containing cadmium compound precursor nanowires, gas sensing and optical properties. Nanotechnology 19: 245611.
4. Li J, Ni Y, Liu J, Hong J (2009) Preparation, conversion, and comparison of the photocatalytic property of $Cd(OH)_2$, CdO, CdS and CdSe. Journal of Physics and Chemistry of Solids 70: 1285-1289.
5. Liu Y, Zhang YC, Xu XF (2009) Hydrothermal synthesis and photocatalytic activity of CdO2 nanocrystals. J Hazard Mater 163: 1310-1314.
6. Lu H, Liao L, Li J, Wang D, He H, et al. (2006) High surface-to-volume ratio ZnO microberets: low temperature synthesis, characterization, and photoluminescence. J Phys Chem B 110: 23211-23214.
7. Marinakos SM, Anderson MF, Ryan JA, Martin LD, Feldheim DL, (2001) Encapsulation, permeability, and cellular uptake characteristics of hollow nanometer-sized conductive polymer capsules. J Phys Chem B 105: 8872-8876.
8. Mondal S, Chattopadhyay T, Das S, Maulik SR, Neogi S, et al. (2012) CdO and CdS nanoparticles from pyrolytic method: Preparation, characterization and photocatalytic activity. Indian Journal of Chemistry 51: 807-811.
9. Ortega M, Santana G, Morales-Acevedo A (2000) Optoelectronic properties of CdO/Si photodetectors. Solid-State Electronics 44: 1765.
10. Reyes MED, Delgado GT, Perez RC, Marin JM, Angel OZ (2012) Optimization of the photocatalytic activity of CdO+$CdTiO_3$ coupled oxide thin films obtained by sol-gel technique. Journal of Photochemistry and Photobiology A: Chemistry 228: 22-27.
11. Ristic M, Popovic S, Music S (2004) Formation and properties of $Cd(OH)_2$ and CdO particles. Materials Letters 58: 2494-2499.
12. Abdulkarem AM, Elssfah EM, Yan NN, Demissie G, Yu Y (2013) Photocatalytic activity enhancement of CdS through in doping by simple hydrothermal method. Journal of Physics and Chemistry of Solids 74: 647-652.
13. Saravanan R, Shankar H, Prakash T, Narayanan V, Stephen A (2011) ZnO/CdO composite nanorods for photocatalytic degradation of methylene blue under visible light. Materials Chemistry and Physics 125: 277-280.
14. Saghatforoush LA, Sanati S, Mehdizadeh R, Hasanzadeh (2012) Solvothermal synthesis of $Cd(OH)_2$ and CdO nanocrystals and application as a new electrochemical sensor for simultaneous determination of norfloxacin and lomefloxacin. Superlattices and Microstructures 52: 885-893.
15. Siraj K, Khaleeq-ur-Rahman K, Hussain SI, Rafique MS, Anjum S (2011) Effect of deposition temperature on structural , surface, optical and magnetic properties of pulsed laser deposition Al-dopped CdO thin films. Journal of Alloys and Compounds 509: 6756-6762.
16. Shanmugam N, Cholan S, Kannadasan N, Sathishkumar K, Viruthagiri G (2014) Effect of polyvinylpyrrolidone as capping agent on Ce^{3+} doped flowerlike ZnS nanostructure. Solid State Sciences 28: 55-60.
17. Weckler B, Lutz HD (1996) Near-infrared spectra of M(OH)Cl (M=Ca, Cd, Sr), Zn(OH)F, -$Cd(OH)_2$, $Sr(OH)_2$, and brucite-type hydroxides $M(OH)_2$ (M=Mg, Ca, Mn, Fe, Co, Ni, Cd). Spectrochimica Acta Part A: Molecular and Biomolecular Spectroscopy 52: 1507-1513.
18. Schmidt M, Lutz HD (1991) γ-$Cd(OH)_2$, A common hydroxide or an aquoxy-hydroxide? Materials Research Bulletin 26: 605-612.
19. Nakamoto K (2009) Infrared and Raman Spectra of Inorganic and Coordination Compounds, Part B, 6th edition, John Wiley & Sons, New Jersey, USA.
20. Johnson JC, Yan HQ, Yang PD, Saykally RJ (2003) Optical Cavity Effects in ZnO Nanowire Lasers and Waveguides. J Phys Chem B 107: 8816-8828.
21. Stichtenoth D, Ronning C, Niermann T, Wischmerier L, Voss T, et al. (2007) Optical size effects in ultrathin ZnO nanowires. Nanotechnology 18: 435701-435705.

12

An Impact of Biofield Treatment on Spectroscopic Characterization of Pharmaceutical Compounds

Mahendra Kumar Trivedi[1], Shrikant Patil[1], Harish Shettigar[1], Ragini Singh[2] and Snehasis Jana[2*]

[1]*Trivedi Global Inc., 10624 S Eastern Avenue Suite A-969, Henderson, NV 89052, USA*

[2]*Trivedi Science Research Laboratory Pvt. Ltd., Hall-A, Chinar Mega Mall, Chinar Fortune City, Hoshangabad Rd., Bhopal- 462026, Madhya Pradesh, India*

Abstract

The stability of any pharmaceutical compound is most desired quality that determines its shelf life and effectiveness. The stability can be correlated to structural and bonding properties of compound and any variation arise in these properties can be easily determined by spectroscopic analysis. The present study was aimed to evaluate the impact of biofield treatment on these properties of four pharmaceutical compounds such as urea, thiourea, sodium carbonate, and magnesium sulphate, using spectroscopic analysis. Each compound was divided into two groups, referred as control and treatment. The control groups remained as untreated and treatment group of each compound received Mr. Trivedi's biofield treatment. Control and treated samples of each compound were characterized using Fourier-Transform Infrared (FT-IR) and Ultraviolet-Visible (UV-Vis) spectroscopy. FT-IR spectra of biofield treated urea showed the shifting of C=O stretching peak towards lower frequency (1684→1669 cm^{-1}) and N-H stretching peak towards higher frequency (3428→3435 cm^{-1}) with respect to control. A shift in frequency of C-N-H bending peak was also observed in treated sample as compared to control i.e. (1624→1647 cm^{-1}). FT-IR spectra of thiourea showed upstream shifting of NH_2 stretching peak (3363→3387 cm^{-1}) as compared to control, which may be due to decrease in N-H bond length. Also, the change in frequency of N-C-S bending peak (621→660 cm^{-1}) was observed in treated thiourea that could be due to some changes in bond angle after biofield treatment. Similarly, treated sample of sodium carbonate showed decrease in frequency of C-O bending peak (701→690 cm^{-1}) and magnesium sulphate showed increase in frequency of S-O bending peak (621→647 cm^{-1}) as compared to control, which indicated that bond angle might be altered after biofield treatment on respective samples. UV-Vis spectra of biofield treated urea showed shift in lambda max (λ_{max}) towards higher wavelength (201→220 nm) as compared to control sample, whereas other compounds i.e. thiourea, sodium carbonate, and magnesium sulphate showed the similar λ_{max} to their respective control. These findings conclude that biofield treatment has significant impact on spectral properties of tested pharmaceutical compounds which might be due to some changes happening at atomic level of compounds, and leading to affect the bonding and structural properties of compounds.

Keywords: Urea; Thiourea; Sodium carbonate; Magnesium sulphate; Biofield treatment; Fourier transform infrared spectroscopy; Ultraviolet-visible spectroscopy

Introduction

Pharmaceutical industries are an important component of health care systems which are largely driven by scientific discovery and development of various chemical and biological agents for human and animal health. The pharmaceutical industry is based primarily upon many organic and inorganic chemicals, which are used as raw materials, serve as reactants, reagents, catalysts, counter ions and solvents. However these chemicals exhibit a wide range of pharmacological activity and toxicological properties [1]. Although the pharma industries are dominated by organic compounds and drugs, the inorganic compounds also focus their attention due to their therapeutic potential such as neurological, anticancer, antimicrobial, antiulcer, antiviral, anti-inflammatory, cardio vascular and insulin-mimetic agents. Moreover, inorganic compounds also play an important role as counter ions in drugs, which influence the solubility, stability, and hygroscopicity of active pharmaceutical ingredients [2]. The compounds selected in this study for biofield treatment are urea, thiourea, sodium carbonate and magnesium sulphate, which have wide applications in pharmaceutical industry.

Urea, a white crystalline powder is commonly used in denaturing and solubilising proteins in the biopharmaceutical industry. It serves an important role in the metabolism of nitrogen-containing compounds by animals and is the main nitrogen-containing substance in the urine of mammals. It is small hydrophilic molecule, present in all taxa, and widely used as protein denaturant in *in vitro* unfolding/refolding experiments [3]. It is also used clinically as emollient and keratolytic agent in treatment of skin related diseases [4].

Thiourea is an organosulfur compound which is utilized in organic synthesis of various compounds and pharmaceuticals like sulfathiazoles, thiouracils, tetramizole and cephalosporins. Moreover, it was also used as thyroid depressant during 1940s [5,6].

Sodium carbonate, commonly known as washing soda, is sodium salt of carbonic acid. Naturally it exists in the form of crystalline heptahydrate; however it readily effloresces to form a white powder which is monohydrate [7]. Sodium carbonate (Na_2CO_3) is a food additive and used as carbonating agent, anti-caking agent, raising agent, and stabilizer. Its activities are also reported as an alkalizing agent, used in lotion or bath in the treatment of scaly skin in pharmaceuticals [8].

Magnesium sulphate is commonly known as Epsom salt, and used both externally and internally in pharmaceutical preparation. In addition, Epsom salt is also used as bath salts and for isolation tanks. Oral

***Corresponding author:** Snehasis Jana, Trivedi Science Research Laboratory Pvt. Ltd., Hall-A, Chinar Mega Mall, Chinar Fortune City, Hoshangabad Rd., Bhopal- 462026, Madhya Pradesh, India
E-mail: publication@trivedisrl.com

magnesium sulphate is commonly used as a saline laxative or osmotic purgative. Magnesium sulphate is the main compound for preparation of intravenous magnesium [9,10]. In all these four compounds, stability plays a crucial role in pharmaceutical preparations, which is directly related to its structural and atomic bonding properties. Currently, in pharmaceutical industries stability of these compounds is modulating through altering temperature and pH conditions [11]. Thus, it is important to evaluate an alternate strategy, which could alter the structural and bonding properties and that can affect the stability in these compounds.

Biofield is the scientific term for the biologically produced ultra-fine electromagnetic energy field that can function for regulation and communication within the organism [12]. It is already demonstrated that electrical current exists inside the human body in the form of vibratory energy particles like ions, protons, and electrons and they generate magnetic field in the human body [13,14]. This electromagnetic field of the human body is known as biofield and energy associated with this field is known as biofield energy [15,16]. Thus, a human has the ability to harness the energy from environment or universe and can transmit the energy into any living or non-living object around this Globe. The object(s) always receive the energy and respond into useful way via biofield energy. This process is termed as biofield treatment. Mr Trivedi's biofield treatment (The Trivedi Effect®) is recognized as an alternate approach to alter the several physical and structural properties of metal powder at atomic level [17-21]. The biofield treatment has also transformed the characteristics in several other fields like biotechnology [22,23], microbiology [24,25], and in agricultural science [26,27].

IR spectroscopy which deals with the infrared region and UV-Vis spectroscopy which deals with ultraviolet-visible spectral region of the electromagnetic spectrum are used in analytical chemistry for the quantitative determination of different analytes, such as transition metal ions, highly conjugated organic compounds, and biological macromolecules [28]. They can provide analytically useful information on a large variety of compounds, ranging from small inorganic ions to large organic molecules [29]. Recently, effect of biofield treatment on ceramic oxide nano powders was studied using infrared spectroscopy, which reported that structural and bond properties were altered after treatment [30-33]. Hence based on the outstanding results achieved on different materials and considering the pharmaceutical applications of these four compounds, the present study was undertaken to evaluate the impact of biofield treatment on the spectroscopic characteristics of urea, thiourea, sodium carbonate, and magnesium sulphate.

Materials and Methods

Study design

The samples of urea, thiourea, sodium carbonate and magnesium sulphate were procured from Sigma-Aldrich, MA, USA; and each compound was divided into two parts i.e. as control and treatment group. The control samples remained as untreated, and treatment samples were handed over in sealed pack to Mr. Trivedi for biofield treatment under laboratory condition. Mr. Trivedi provided this treatment through his energy transmission process to the treated groups without touching the samples. After that, the control and treated samples of each compound were analysed using Fourier Transform Infrared (FT-IR) spectroscopy and Ultraviolet-Visible (UV-Vis) spectroscopy. Infrared and UV-Vis spectroscopy are particularly useful techniques in identifying organic as well as inorganic structures [34].

FT-IR spectroscopic characterization

The samples were crushed into fine powder for analysis. The powdered sample was mixed in spectroscopic grade KBr (1:20) in an agate mortar and pressed into 3 mm thick pellets with a hydraulic press. FT-IR spectra were recorded on Shimadzu's Fourier transform infrared spectrometer (Japan) with frequency range of 4000-500 cm^{-1} at room temperature. The FT-IR spectroscopic analysis of urea, thiourea, sodium carbonate and magnesium sulphate (control and treated) were carried out to evaluate the impact of biofield treatment at atomic and molecular level like bond strength, stability, rigidity of structure etc. The FTIR spectroscopy applied to determine any change in structural and bonding properties due to its ability to characterize the functional group and fingerprint region of very small quantities of samples.

UV-Vis spectroscopic analysis

The UV-Vis spectral analysis was measured using Shimadzu UV-2400 PC series spectrophotometer over a wavelength range of 200-400 nm with 1 cm quartz cell and a slit width of 2.0 nm. This analysis was performed to evaluate the effect of biofield treatment on structural property of different pharmaceutical compounds such as urea, thiourea, sodium carbonate and magnesium sulphate. With UV-Vis spectroscopy it is also possible to investigate electron transfers between orbitals or bands of atoms, ions and molecules existing in the gaseous, liquid and solid phase.

Results and Discussion

FT-IR spectroscopic analysis

Infrared (IR) spectroscopy is based on the vibrations of the atoms in a molecule. When a molecule absorbs infrared radiation, its chemical bonds vibrate and can stretch, contract or bend [35]. FT-IR spectra of control and treated samples of urea are shown in Figure 1. IR spectra of control urea sample showed in plane and out of plane N-H stretching at 3428 cm^{-1}. Other peaks showed C=O stretching at 1684

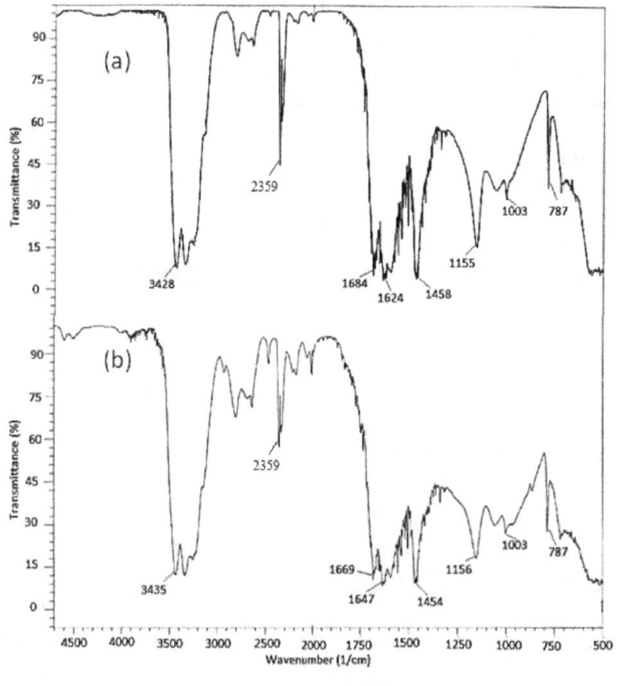

Figure 1: FT-IR spectra of urea. (a), control and (b), treated.

cm^{-1}; C-N-H bending at 1624 cm^{-1}; C-N stretching at 1458 cm^{-1} and 1003 cm^{-1}; NH_2 bending at 1155 cm^{-1} and out of plane NH_2 bending at 787 cm^{-1}. The spectrum was well supported by literature data [36]. The FT-IR spectrum of treated urea sample showed similar pattern of control IR absorption peaks except N-H stretching peak at 3435 cm^{-1}, C=O stretching peak at 1669 cm^{-1} and C-N-H bending peak at 1647 cm^{-1} (Table 1). The shifting of C=O stretching peak towards lower frequency and N-H stretching peak towards higher frequency were may be due to increase of conjugation effect in treated urea molecule. Due to conjugation, there may be increase in bond length of C=O and decrease in bond length of N-H bond of urea. It is already reported that the peak frequency (v) in IR spectra for any bond is directly proportional to its bond force constant (k) [37]. Also the bond force constant (k) is inversely related to average bond length (r) [38]. Hence, it is presumed that shifting of peak wavenumber corresponding to C=O and N-H bond could be due to change in corresponding bond length after biofield treatment. Data also exhibited that C-N-H bending frequency was shifted towards higher frequency as compared to control sample. It could be due to alteration in bond angle of C-N-H in urea after biofield treatment [39]. As there is occurrence of conjugation effect in treated sample due to biofield treatment, which may lead to increase stability in treated urea as compared to control. Apart from these peaks, small, sharp absorption bands in the region from 4000-3000 cm^{-1} and 1800-1600 cm^{-1} were appeared due to vapour phase water and the predominant CO_2 absorption band occurs as a doublet at 2359 cm^{-1} in both control and treated sample.

FTIR spectra of control and treated samples of thiourea are shown in Figure 2. The FT-IR spectrum of control thiourea sample showed NH_2 asymmetric and symmetric stretching peaks at 3363 cm^{-1} and 3169 cm^{-1} respectively. Other peaks were observed for C-N stretching at 1465 cm^{-1}; C=S asymmetric and symmetric stretching at 1412 cm^{-1} and 730 cm^{-1} respectively; C-N symmetric stretching peak at 1086 cm^{-1} and N-C-S bending at 621 cm^{-1}. The peaks in spectrum of control sample were well supported by literature data [40,41]. The FT-IR spectrum of treated thiourea sample showed similar peaks like in control sample except N-C-S bending peak at 660 cm^{-1} and NH_2 asymmetric stretching peak at 3387 cm^{-1} (Table 1). Hence, the shifting of NH_2 stretching peak towards higher frequency as compared to control (3363 cm^{-1}→3387 cm^{-1}) suggest that biofield treatment may reduce the bond length. As described earlier, it ultimately may cause some changes in bond force constant i.e. strengthening of bond which could provide more stability to the compound. Also the change in frequency of N-C-S bending peak suggests that there may be some alteration in bond angle of treated sample as compared to control after biofield treatment. The FT-IR spectra also showed small, sharp absorption bands in the region from 4000-3000 cm^{-1} and 1800-1600 cm^{-1} which were appeared due to vapour phase water and a doublet peak at 2359 cm^{-1} due to CO_2 absorption band in both control and treated sample.

The FT-IR spectra of control and treated samples of sodium carbonate are shown in Figure 3. The FT-IR spectrum of control sample was interpreted regarding the characteristic of IR absorption bands known for carbonate group [42]. IR spectrum of control sodium carbonate showed C-O in plane and out of plane bending peaks at 881 cm^{-1} and 701 cm^{-1} respectively and C-O stretching peak at 1445 cm^{-1}. The FT-IR spectrum of control sample also showed the peak at 2943 cm^{-1} i.e. O-H stretching frequency which could be due to water absorption by sample. The vapour phase water absorption is also evident due to appearance of small, sharp peaks in region of 4000-3000 cm^{-1}. The treated sample also showed O-H stretching peak at 3007 cm^{-1} other than small and sharp peaks in region of 4000-3000 cm^{-1} due to water absorption. The spectrum of treated sample showed similar peaks except C-O out of plane bending peak i.e. at 690 cm^{-1} [43]. The change in C-O bending peak as compared to control sample (Table 2) could be possible due to some change in bond angle of treated sample after biofield treatment.

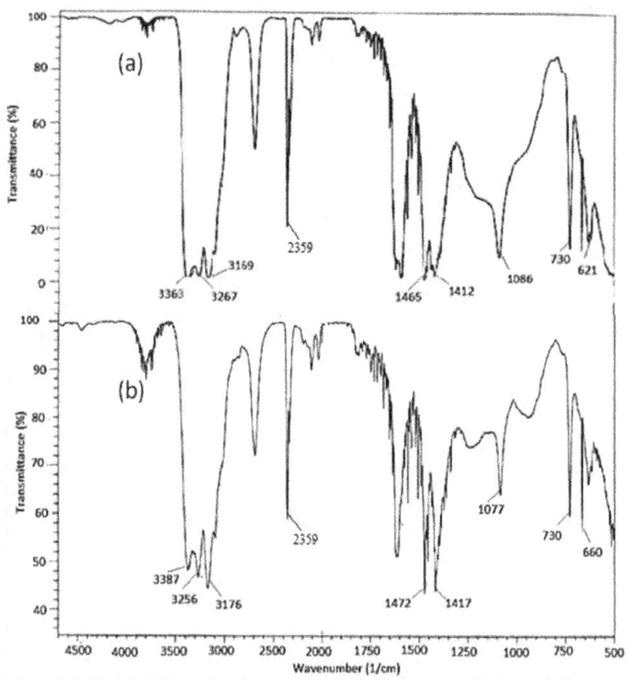

Figure 2: FT-IR spectra of thiourea. (a), control and (b), treated.

S. No.	Functional group	Wavenumber (cm^{-1})			
		Urea		Thiourea	
		Control	Treatment	Control	Treatment
1.	N-H stretching (in plane)	3428	3435	3169	3176
2.	N-H stretching (out of plane)	-	-	3363	3387
3.	C=O stretching	1684	1669	-	-
4.	C-N-H bending	1624	1647	-	-
5.	C-N stretching	1458, 1003	1454, 1003	1465, 1086	1472, 1077
6.	NH_2 bending (in plane)	1155	1156	-	-
7.	NH_2 bending (out of plane)	787	787	-	-
8.	C=S asymmetric stretching	-	-	1412	1417
9.	C=S symmetric stretching	-	-	730	730
10.	N-C-S bending	-	-	621	660

Table 1: Vibration modes observed in organic pharmaceutical compounds such as urea and thiourea.

FT- IR spectra of control and treated magnesium sulphate samples are shown in Figure 4. The spectrum of control sample showed O-H stretching peak at 3355 cm^{-1} and O-H bending peak at 1684 cm^{-1} which may be appeared due to absorption of water molecules by compound. IR spectra of treated sample also showed O-H stretching and bending peaks at 3279 cm^{-1} and 1660 cm^{-1} respectively. The absorption of vapour phase water is also evident by appearance of small peaks in region of 4000-3000 cm^{-1} and 1800-1600 cm^{-1} in both control and treated sample. Also a predominant CO_2 absorption band occurs as a doublet at 2359 cm^{-1} in both control and treated sample. Other peaks appeared in control sample were mainly due to presence of sulphate group. These were S-O asymmetric stretching peak at 1070 cm^{-1}; S-O symmetric stretching at 983 cm^{-1} and S-O bending at 621 cm^{-1} [44,45]. In case of treated sample spectra, similar peaks were observed. The only change observed was in S-O bending peak, which was shifted to higher frequency as compared to control sample (621→ 647 cm^{-1}) (Table 2). It may be due to alteration in bond angle S-O after biofield treatment.

UV-Vis spectroscopic analysis

The λ_{max} value corresponding to each control and treated samples are shown in Table 3. In UV spectra of control urea sample, the absorption peak was shown at 201 nm whereas in treated sample the

S. No.	Functional group	Wavenumber (cm^{-1})			
		Sodium carbonate		Magnesium sulphate	
		Control	Treatment	Control	Treatment
1.	O-H stretching	2943	3007	3355	3279
2.	C-O stretching	1445	1440	-	-
3.	C-O bending (in plane)	881	881	-	-
4.	C-O bending (out of plane)	701	690	-	-
5.	O-H bending	1772	1772	1684	1660
6.	S-O asymmetric stretching	-	-	1070	1077
7.	S-O symmetric stretching	-	-	983	985
8.	S-O bending	-	-	621	647

Table 2: Vibration modes observed in inorganic pharmaceutical compounds such as sodium carbonate and magnesium sulphate.

S. No.	Name of compound	Lambda max (nm)	
		Control	Treated
1.	Urea	201	220
2.	Thiourea	241, 202	241, 202
3.	Sodium carbonate	206	205.8
4.	Magnesium sulphate	358	360

Table 3: Maximum absorbance wavelength of control and treated sample of different pharmaceutical compounds.

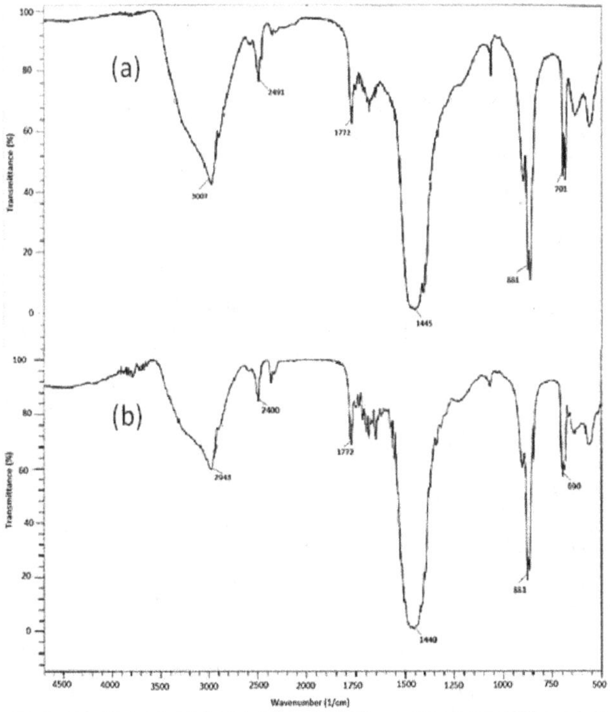

Figure 3: FT-IR spectra of sodium carbonate. (a), control and (b), treated.

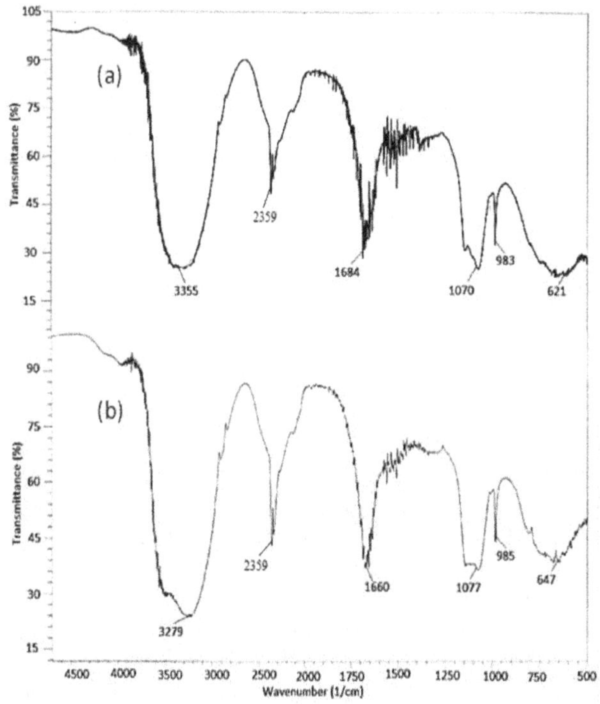

Figure 4: FT-IR spectra of magnesium sulphate. (a), control and (b), treated.

absorption peak was appeared at 220 nm. Shifting of λ_{max} towards higher wavelength in treated sample as compared to control may occur due to increase in conjugation effect in urea molecule. This result was also supported by FT-IR data. Further, it is assumed that conjugation in treated urea may lead to higher stability after biofield treatment [46]. In UV spectra of both control and treated samples of thiourea showed two absorption peaks (202 and 241 nm) and were well supported by literature data [38]. The UV spectra of control and treated samples of sodium carbonate also showed similar pattern of λ_{max} i.e. at 206 nm and which was well supported by literature data [47]. Similarly no change was found in λ_{max} in UV spectra of treated magnesium sulphate as compared to control. These observations suggest that biofield treatment might not make any alteration in chromophore groups present in thiourea, sodium carbonate and magnesium sulphate which are mainly responsible for absorption of light.

Our group previously reported the impact of biofield energy on physical, thermal and spectroscopic characteristics of various metals and powders [30-33]. The present study also showed the effect of biofield treatment on structural and bonding properties of different pharmaceutical compounds which are used in pharmaceutical industry because of their intrinsic pharmacological action or as intermediate agent. The change in IR frequencies and λ_{max} suggest that due to increase in conjugation effect or force constant between bonds (e.g., urea and thiourea), chemical stability of these compounds might increase after biofield treatment. The increase in stability can improve their shelf life and effectiveness [48] and make them more suitable to be used in pharmaceutical preparations.

Conclusion

Altogether, the results of present study showed that there has significant impact of biofield treatment on spectral properties of urea, thiourea, sodium carbonate, and magnesium sulphate. FT-IR data showed significant change in stretching frequencies in treated sample of urea which may be due to increased conjugation effect induced by biofield treatment and attribute to increased stability of treated urea sample. Similarly, a significant change was found in IR peak frequencies related to stretching and bending vibrations of treated samples of thiourea, sodium carbonate and magnesium sulphate which could be due to alteration in bond angle and bond strength after biofield treatment. UV spectroscopic result of urea was also supported by IR data, which suggest that biofield treatment may alter the conjugation effect within the molecule. Thus, it is postulated that biofield treatment can make some alteration at the atomic level, which could further affect the stability of the bonds and hence, the stability of compounds.

Acknowledgement

The authors would like to acknowledge the whole team of MGV Pharmacy College, Nashik for providing the instrumental facility.

The generous support of Trivedi Science, Trivedi Master Wellness and Trivedi Testimonials is gratefully acknowledged.

References

1. Stellman JM (1988) Encyclopedia of occupational health and safety. (4th edition) International Labor Organization, Geneva, Switzerland.
2. Sekhon BS, Gandhi L (2006) Medicinal uses of inorganic compounds-1. Resonance 11: 75-89.
3. Kurzer F, Sanderson PM (1956) Urea in the history of organic chemistry: Isolation from natural sources. J Chem Educ 9: 452-459.
4. Pan M, Heinecke G, Bernardo S, Tsui C, Levitt J (2013) Urea: a comprehensive review of the clinical literature. Dermatol Online J 19: 20392.
5. Schreiner PR1 (2003) Metal-free organocatalysis through explicit hydrogen bonding interactions. Chem Soc Rev 32: 289-296.
6. Angerer JK, Schaller KH (1988) Analyses of hazardous substances in biological materials. Fresenius' Journal of Analytical Chemistry 346: 828-829.
7. Pabst A (1930) On the hydrates of sodium carbonate. Am Mineral 15: 69-73.
8. http://medical-dictionary.thefreedictionary.com/sodium+carbonate
9. Buchel KH, Moretto HH, Woditsch P (2000) Industrial inorganic chemistry (2nd edition) Wiley-VCH, Germany.
10. Frakes MA, Richardson LE 2nd (1997) Magnesium sulfate therapy in certain emergency conditions. Am J Emerg Med 15: 182-187.
11. Panyachariwat N, Steckel H (2014) Stability of urea in solution and pharmaceutical preparations. J Cosmet Sci 65: 187-195.
12. Garland SN, Valentine D, Desai K, Li S, Langer C, et al. (2013) Complementary and alternative medicine use and benefit finding among cancer patients. J Altern Complement Med 19: 876-881.
13. Planck M (1956) Treatise on thermodynamics, (3rd edition) Longmans, Green, London (UK).
14. Einstein A (1905) Does the inertia of a body depend upon its energy-content. Ann Phys 18: 639-641.
15. Rivera-Ruiz M, Cajavilca C, Varon J (2008) Einthoven's string galvanometer: the first electrocardiograph. Tex Heart Inst J 35: 174-178.
16. Rubik B (2002) The biofield hypothesis: its biophysical basis and role in medicine. J Altern Complement Med 8: 703-717.
17. Trivedi MK, Tallapragada RM (2008) A transcendental to changing metal powder characteristics. Met Powder Rep 63: 22-28.
18. Trivedi MK, Tallapragada RM (2009) Effect of super consciousness external energy on atomic, crystalline and powder characteristics of carbon allotrope powders. Mater Res Innov 13: 473-480.
19. Dabhade VV, Tallapragada RMR, Trivedi MK (2009) Effect of external energy on atomic, crystalline and powder characteristics of antimony and bismuth powders. Bull Mater Sci 32: 471-479.
20. Trivedi MK, Patil S, Tallapragada RM (2013) Effect of biofield treatment on the physical and thermal characteristics of vanadium pentoxide powder. J Material Sci Eng S11: 001.
21. Trivedi MK, Patil S, Tallapragada RM (2014) Atomic, crystalline and powder characteristics of treated zirconia and silica powders. J Material Sci Eng 3: 144.
22. Patil S, Nayak GB, Barve SS, Tembe RP, Khan RR (2012) Impact of biofield treatment on growth and anatomical characteristics of Pogostemon cablin (Benth.). Biotechnology 11: 154-162.
23. Nayak G, Altekar N (2015) Effect of a biofield treatment on plant growth and adaptation. J Environ Health Sci 1: 1-9.
24. Trivedi MK, Patil S, Bhardwaj Y (2008) Impact of an external energy on *Staphylococcus epidermis* [ATCC-13518] in relation to antibiotic susceptibility and biochemical reactions-An experimental study. J Accord Integr Med 4: 230-235.
25. Trivedi MK, Patil S, Bhardwaj Y (2009) Impact of an external energy on *Enterococcus faecalis* [ATCC-51299] in relation to antibiotic susceptibility and biochemical reactions-An experimental study. J Accord Integr Med 5: 119-130.
26. Shinde V, Sances F, Patil S, Spence A (2012) Impact of biofield treatment on growth and yield of lettuce and tomato. Aust J Basic & Appl Sci 6: 100-105.
27. Lenssen AW (2013) Biofield and fungicide seed treatment influences on soybean productivity, seed quality and weed community. Agricultural Journal 8: 138-143.
28. Misra P, Dubinskii M (2002) Ultraviolet spectroscopy and UV lasers. Marcel Dekker. New York, USA.
29. Ovalles F, Gallignani M, Rondon R, Brunetto MR, Luna R (2009) Determination of sulfate for measuring magnesium sulfate in pharmaceuticals by flow analysis-Fourier transforms infrared spectroscopy. Lat Am J Pharm 28: 173-182.
30. Trivedi MK, Nayak G, Patil S, Tallapragada RM, Latiyal O (2015) Studies of the atomic and crystalline characteristics of ceramic oxide nano powders after bio field treatment. Ind Eng Manage 4: 161.
31. Trivedi MK, Patil S, Tallapragada RM (2013) Effect of biofield treatment on the physical and thermal characteristics of silicon, tin and lead powders. J Material Sci Eng 2: 125.

32. Trivedi MK, Patil S, Tallapragada RM (2012) Thought Intervention through bio field changing metal powder characteristics experiments on powder characterization at a PM plant. Proceeding of the 2nd International Conference on Future Control and Automation 2: 247-252.

33. Trivedi MK, Patil S, Tallapragada RM (2015) Effect of biofield treatment on the physical and thermal characteristics of aluminium powders. Ind Eng Manage 4: 151.

34. Sibilia JP (1996) A guide to materials characterization and chemical analysis. John Wiley and Sons Ltd, USA.

35. Barbara S (2004) Infrared Spectroscopy: Fundamentals and applications. Wiley-VCH, Germany.

36. Piasek Z, Urbanski T (1962) The infra-red absorption spectrum and structure of urea. B Pol Acad Sci-Tech X: 113-120.

37. Ghosh M, Dilawar N, Bandyopadhyay AK, Raychaudhuri AK (2009) Phonon dynamics of Zn (Mg,Cd)O alloy nanostructures and their phase segregation. J Appl Phys 106: 1-6.

38. EL-Mallawany RA (1989) Theoretical and experimental IR spectra of binary rare earth tellurite glasses-1. Infrared Phys 29: 781-785.

39. Pretsch E, Buhlmann P, Affolter C (2009) Structure Determination of Organic Compounds. (4th edition). Springer Verlag, Berlin, Heidelberg, Germany.

40. Ravi B, Jegatheesan A, Neelakandaprasad B, Sadeeshkumar C, Rajarajan G (2014) Optical and conductivity analysis of thiourea single crystals. Rasayan J Chem 7: 287-294.

41. Begum SA, Hossain M, Podder J (2009) An investigation on the growth and characterization of thiourea single crystal grown from aqueous solutions. J Bangladesh Acad Sci 33: 63-70.

42. Miller FA, Wilkins CH (1952) Infrared spectra and characteristic frequencies of inorganic ions. Anal Chem 24: 1253-1294.

43. Coates J (2000) Interpretation of infrared spectra, a practical approach. Encyclopedia of Analytical Chemistry. John Wiley and Sons Ltd, USA

44. Chaban GM, Huo WM, Lee TJ (2002) Theoretical study of infrared and raman spectra of hydrated magnesium sulfate salts. J Chem Phys 117: 2532-2537.

45. Sieranski T, Kruszynski R (2012) Magnesium sulfate complexes with hexamethylenetetramine and 1, 10-phenanthroline: Thermal, structural and spectroscopic properties. J Therm Anal Calorim 109: 141-152.

46. Filutowicz Z, Lukaszewski K, Pieszynski K (2004) Remarks on spectra-photometric monitoring of urea in dialysate. JMIT 8: 105-110.

47. Adams GE, Boag JW, Michael BD (1965) Reactions of the hydroxyl radical. Part 1.-Transient spectra of some inorganic radical-anions. Trans Faraday Soc 61: 1674-1680.

48. Blessy M, Patel RD, Prajapati PN, Agrawal YK (2014) Development of forced degradation and stability indicating studies of drugs-A review. J Pharm Anal 4: 159-165.

Relationship of Pb in House Dust and Ambient Air

Brian Gulson[1,2]* and Alan Taylor[3]

[1]*Graduate School of the Environment, Macquarie University, Sydney, Australia*
[2]*Commonwealth Scientific and Industrial Research Organisation (CSIRO), Earth Science and Resource Engineering, Sydney, Australia*
[3]*Department of Psychology, Macquarie University, Sydney, Australia*

Abstract

We evaluated the relationship of lead (Pb) using high precision Pb isotopes from ambient air particulates and dust fall accumulation in 59 residences in Sydney New South Wales Australia by the Petri Dish Dust method (PDD) to determine if the dust is a reliable indicator of exposure in cases where air Pb data may not be available. Over the period 1993-2002, Pb values in air samples were higher in winter whereas the Pb loadings for PDD values were slightly higher in spring and summer. These differences are probably the result of differences in sampling times of the air particulates (24-h) and PDD (~3 months).There was no seasonal or suburb effect for the isotopic ratios. Both air and PDD samples showed a strong increase in $^{206}Pb/^{204}Pb$ over time. PDD data were predicted by the air data ($p < 0.001$) and provide a useful adjunct in monitoring exposures.

Keywords: Dust particles; Sampling times; Exposures

Introduction

Exposure to house dust is the most important contributor to blood Pb (PbB) in young children [1-8]. The usual method of estimating exposure to dust is collection by vacuum cleaners [9,10] or surface wipes [9], summarized in US EPA [11]. These sampling methods have various limitations including lack of information about deposition rates, unless resampling over specific time periods is specifically undertaken. An alternative method makes use of collecting trays [12] or dishes [13-16] that monitor exposure for varying lengths of time. Material collected in these vessels is via airborne pathways and may derive from such sources as activities in the house (e.g., renovation, smoking), tracked in dust, and windblown through doors and windows. The exterior sources may be resuspended soil and dust which may contain a legacy of past leaded-gasoline use or leaded paint [17-19], a fact that is continuously misinterpreted by most people in the community (internationally) who think Pb is no longer an issue because Pb has been removed from gasoline and paint in many countries.

Ambient air has been monitored over decades mainly by environmental protection agencies and some researchers [11]. Although air Pb levels have decreased dramatically with cessation of the use of leaded gasoline in many countries (e.g., Thomas et al. [20]) the contribution of Pb in ambient air is still of importance in monitoring environmental exposures to the community and individuals in residences. For example, monitoring of Pb in Total Suspended Particulates (TSP) from High Volume (HV) air filters in Sydney by New South Wales Environment Protection Authority (EPA) showed the annual concentration decreasing from 0.75 µg/m^3 in 1991 to <0.1 µg/m^3 in 2000 [11], levels being so low for their laboratory methods that Pb measurements were discontinued from 2002. Nevertheless Pb has been monitored in Sydney $PM_{2.5}$ particulates over decades by Cohen and colleagues at ANSTO [21,22] using sensitive ion beam methods.

In spite of rapid reductions in air Pb associated with removal of Pb from gasoline and use of Pb in paint, air Pb may still be an important contributor to PbB especially in children. For example, Brink et al. [23] found that Pb measured in the US EPA's National Air Toxics Assessment was a significant predictor of PbB ≥10 µg/dl in children. Using data from the National Health and Nutrition Examination Surveys (NHANES) III and 9908 and air Pb data in TSP from the US EPA, Richmond-Bryant et al. [24] concluded that a larger relative public health benefit among children may be derived from decreases in air Pb at low air Pb exposures. In a US study of 3 urban neighborhoods which measured 23 trace elements from 24-hour $PM_{2.5}$ particulates in outdoor, indoor and personal samples, Adgate et al. [25] found that personal exposure is likely to be underestimated by outdoor central site monitors. Earlier, Tu and Knutson [26] concluded that a compliance with outdoor air quality standards did not ensure a satisfactory indoor situation.

Previous studies have usually evaluated associations between soil-exterior dust- house dust-paint and indoor air with personal monitors [27,28] but generally not exterior ambient air. Some of the studies evaluating the relationship of urban air to house dust include those of Angle and McIntire [29], Manton et al. [30], Laidlaw et al. [19], Rabinowitz et al. [31]. In this study we evaluated the association, using high precision Pb isotopes, between Pb in air collected from high volume air filters in Australia's largest city, Sydney, and house dust collected by the petri dish method to determine the usefulness of data from the petri dish method for other studies where air Pb data are unavailable. This is a follow-up to earlier papers which focused on Pb isotopic relationships on Sydney air [32,33] or elemental associations in $PM_{2.5}$ particulates [33,34].

Methods

Air filters

Particulates collected on filters ('air filters') were obtained from an ongoing air quality monitoring program carried out by the NSW EPA.

Filters were analyzed from samples collected in Sydney's Central Business District (CBD) and from a nearby suburb (Rozelle), situated about 5 km west of the CBD. Air filters from Rozelle have been analyzed for Pb isotopic ratios monthly from January 1991 till May 1996, and several were also analyzed from 1987 to 1989 (total n=138). Those from the CBD have been analyzed on a monthly basis only from April 1994 with four additional samples from 1992 and 1993 (n=36). To evaluate differences over a wider area, filters from a number of other locations in Sydney were occasionally analyzed.

***Corresponding author:** Brian L Gulson, Emeritus Professor, Graduate School of the Environment, Faculty of Science, Macquarie University, Sydney NSW 2109 Australia
E-mail: brian.gulson@mq.edu.au

A High Volume Air Sampler (HVAS) following Australian Standard AS 2724.3-1984 was used. Sampling was carried out continuously for 24 hours on a one-day-in-six cycle. Particles in the approximate size range 0.1 µm to 50 µm were collected. Rozelle samples were collected on a filter which was situated within the HVAS 1.13 m above ground level. The HVAS at Rozelle was located in a parkland setting several hundred meters from a major thoroughfare (>70,000 vehicles per day). The CBD HVAS was located 4m above ground on a street awning.

Preliminary measurements for 4 filters collected within 1 week showed small isotopic variations but which were not statistically significantly different (p 0.08- 0.7 in $^{207}Pb/^{206}Pb$ ratios for 4 different time intervals) so that ongoing sampling was restricted to filters that were collected in the first week of each month.

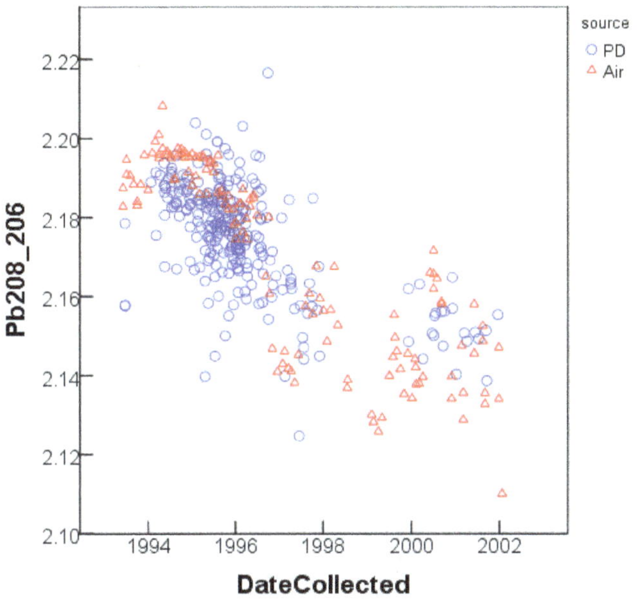

Figure 1: Measured $^{208}Pb/^{206}Pb$ ratios versus Date sampled for air filters and petri dish dust.

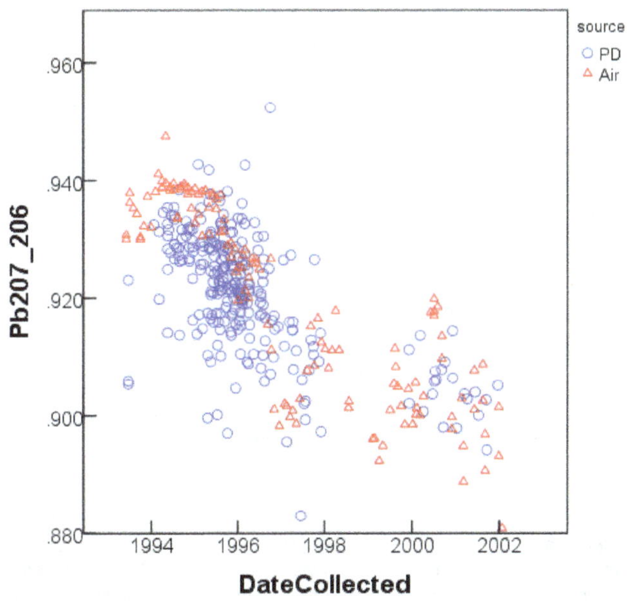

Figure 2: Measured $^{207}Pb/^{206}Pb$ ratios versus Date sampled for air filters and petri dish dust.

A smaller number of samples (n=13 each) were measured for $PM_{2.5}$ Teflon filters over the period 1998 to 2004 for 2 locations in Sydney: one from Mascot, a suburb close to the CBD and a high trafficked area, and the other, a relatively rural (background) setting at Richmond, 20km west of the CBD.

Personal air monitoring for a 24-hour period was undertaken within 12 houses in the inner Sydney suburbs. Although the air samples were collected by different methods and this would affect Pb values, it is not an issue for the Pb isotopic results. The methods for analysis of these filters are detailed in Chiaradia et al. [32], Cohen et al. [34] and Gulson et al. [14].

House dust

The use and advantages of petri dishes to provide ongoing dust fall accumulation was described in earlier publications [14,16]. The PDD collections were usually for a period of about 3 months. The dishes were placed in at least 3 locations: kitchen, main living area, child's bedroom. As early analyses showed only small differences between the different locations in a residence, thereafter the solution from each dish was combined for the appropriate time interval. Dust was collected from 56 houses with the number of collections ranging from 1 to 10 resulting in data for 261 dishes.

Statistical analyses

The analyses were based on 130 air measurements collected between the 5/6/1993 and 23/1/2002 and 261 PDD measurements collected between 23/6/1993 and 18/12/2001. Additional air measurements were available before and after the above periods but these were omitted from analyses so as to maximize the time overlap between the air and PDD samples.

Initial mixed model analyses with each of the Pb isotopic measures as dependent variables included the following independent variables: source (air or PDD), suburb (seven nested under source = PDD and five nested under source = air), season, and time at which the sample was collected, the last coded as fractional number of years since the date the first sample in the analysis subset was collected (5/6/1993). The initial analyses also included a quadratic term for time (time^2) and interactions between source and time and time^2 respectively. The model also included a random factor referred as location which allowed for the correlation between observations obtained at a common collection point (e.g., the living room of a house).

Results

Isotopic compositions

The data sets used in the statistical analyses are presented in Figures 1 to 3 and predicted isotopic results shown in Figures 4 to 6. Outputs of the results from the statistical analyses are given in the supplementary notes.

The dominant trend is a decrease in $^{208}Pb/^{206}Pb$ and $^{207}Pb/^{206}Pb$ ratios and a positive increase in $^{206}Pb/^{204}Pb$ ratios over time which is slightly sharper for air measurements than for PDD measurements. The steady increase in $^{206}Pb/^{204}Pb$ (or decrease in $^{207}Pb/^{206}Pb$ and $^{208}Pb/^{206}Pb$) over time in both types of samples probably reflects the gradual phasing out of Pb from gasoline and increasing contributions of Pb from natural materials such as soils and industrial materials.

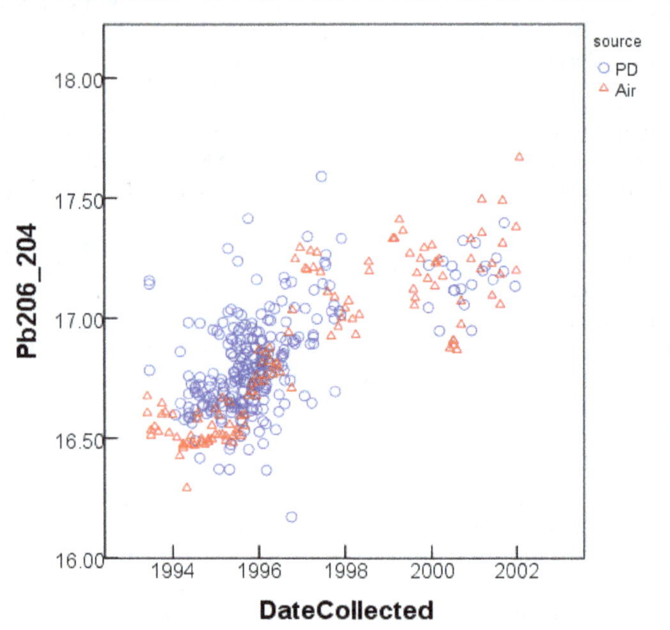

Figure 3: Measured $^{206}Pb/^{204}Pb$ ratios versus Date sampled for air filters and petri dish dust.

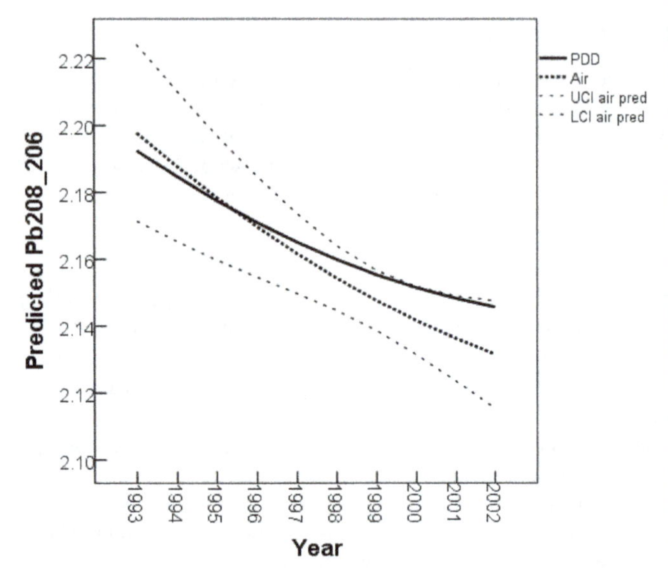

Figure 4: Predicted $^{208}Pb/^{206}Pb$ ratios versus Date sampled for air filters and PDD. The bold solid line is the curve fitted to the PDD measurements, the bold dotted line is the curve fitted to the air measurements and the light dashed lines are the upper and lower 95% confidence intervals for the prediction of the air measurements from the PDD measurements, derived as described above.

Initial analyses showed that while there was a curvilinear change in the measurements over time for each dependent variable, this did not vary with source, so only the main effect of time^2 was retained in the models the results of which are reported here. The interaction between linear time and source was significant, however, and was included in later models. Initial analyses also revealed three air observations the standardized residuals for which were near to 5. These data were excluded from subsequent analyses.

The general pattern of results was the same for each dependent variable (Supplementary material, which gives the ANOVA tables and the parameter estimates for the mixed model analyses). Suburb and season were non-significant (although slightly lower values of $^{207}Pb/^{206}Pb$ and $^{208}Pb/^{206}Pb$ were observed for autumn), while the main effect of time^2 and the interaction between source and time was significant. The proportion of variance of the dependent variables accounted for by all independent variables (calculated according to the method given by Snijders and Bosker, [35]) ranged from 0.53 to 0.59, while the unique variance accounted for by the source by time interaction ranged from 0.031 to 0.049.

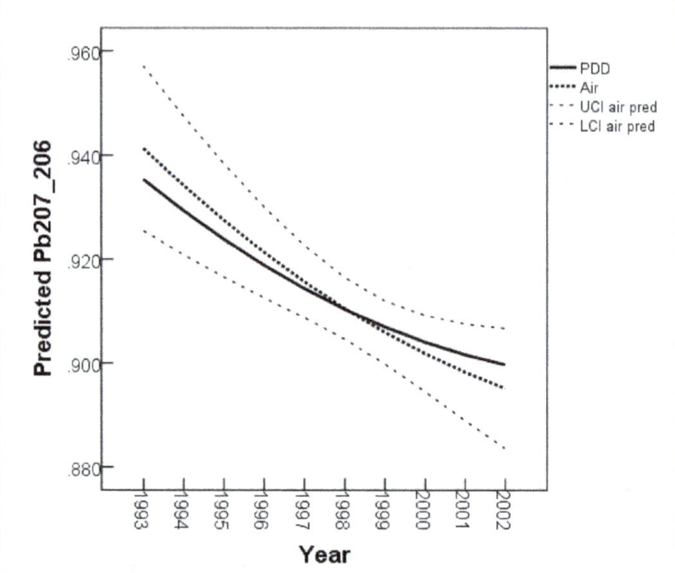

Figure 5: Predicted $^{207}Pb/^{206}Pb$ ratios versus Date sampled for air filters and PDD. The bold solid line is the curve fitted to the PDD measurements, the bold dotted line is the curve fitted to the air measurements and the light dashed lines are the upper and lower 95% confidence intervals for the prediction of the air measurements from the PDD measurements, derived as described above.

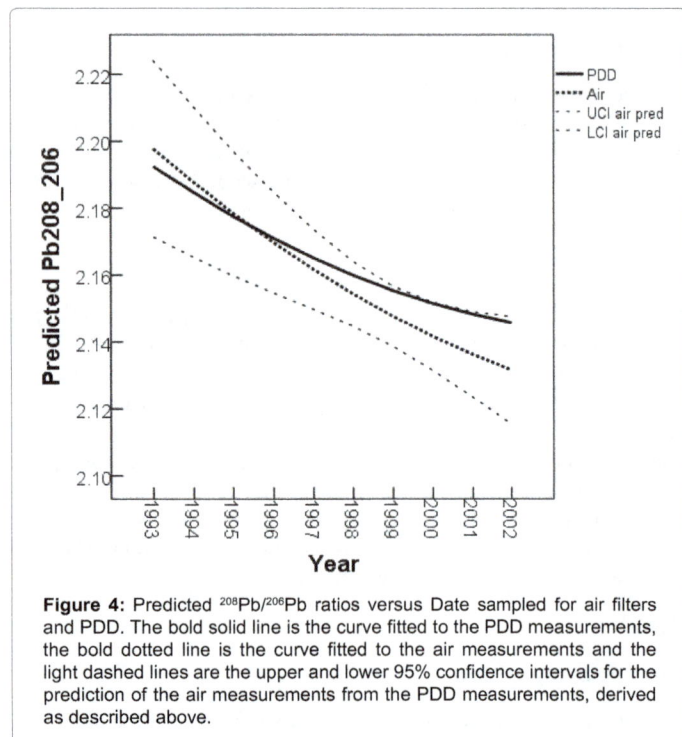

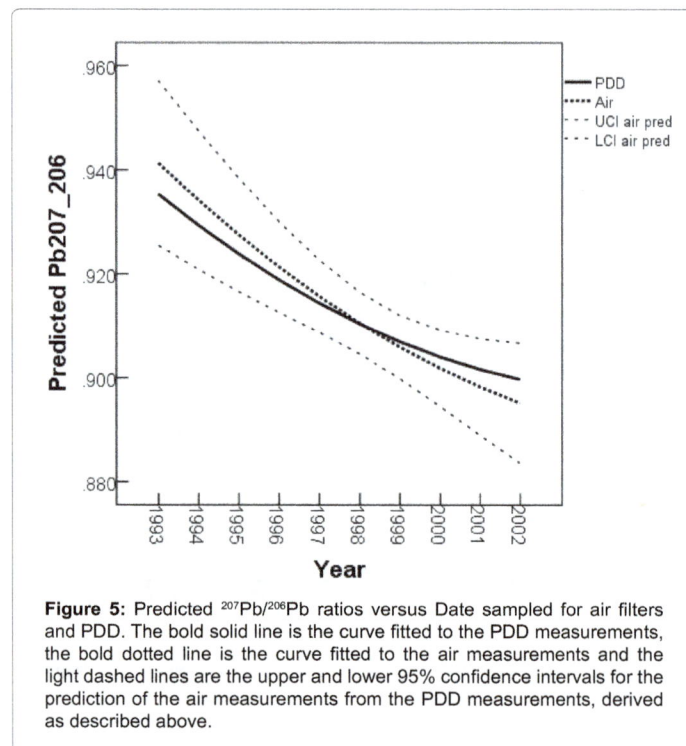

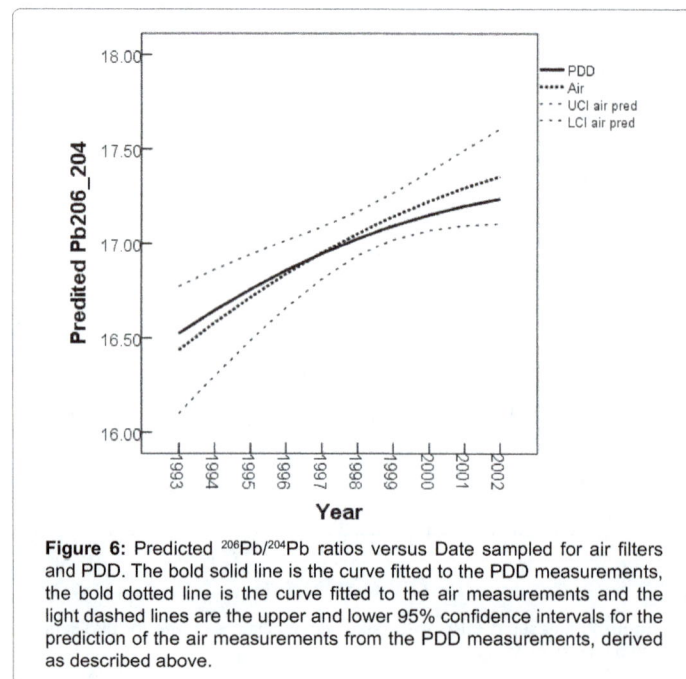

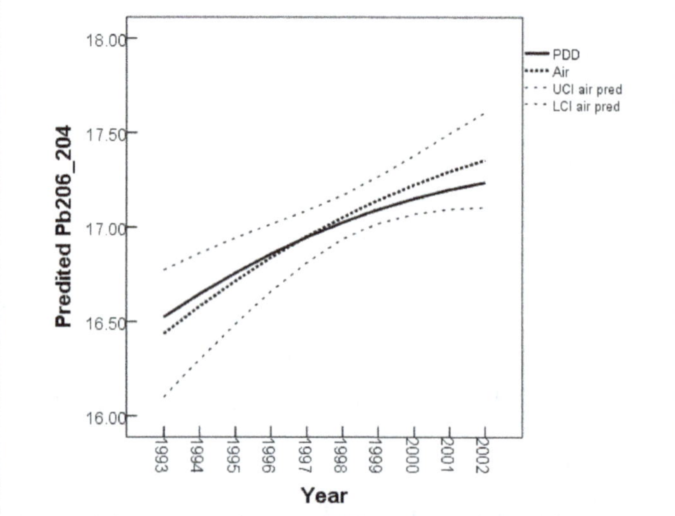

Figure 6: Predicted $^{206}Pb/^{204}Pb$ ratios versus Date sampled for air filters and PDD. The bold solid line is the curve fitted to the PDD measurements, the bold dotted line is the curve fitted to the air measurements and the light dashed lines are the upper and lower 95% confidence intervals for the prediction of the air measurements from the PDD measurements, derived as described above.

Dependent variable	B_{source}	$B_{(source \times time)}$
$^{208}Pb/^{206}Pb$	-0.005237	0.002161
$^{207}Pb/^{206}Pb$	-0.005998	0.001171
$^{206}Pb/^{204}Pb$	0.087706	-0.022941

Table 1: Regression coefficients used in predicting air measurements from PDD measurements.

The predicted values of the three different measurements over time for each source are shown in Figures 4 to 6. The bold solid line in each graph is the curve fitted to the PDD measurements, and the bold dotted line is the curve fitted to the air measurements.

The regression equation derived from the mixed model for each dependent variable provided a basis for making predictions from one source to the other, in particular from PDD measurements to air measurements. A prediction equation for each dependent variable was derived by equating the separate equations for the two sources with respect to time then rearranging and simplifying to give rise to an expression of the following form: for a given time, t,

$$air_t = PDD_t - B_{source} - B_{(source \times time)} \times t \quad (1)$$

Where B_{source} is the regression coefficient for the effect of source and $B_{(source \times time)}$ is the coefficient for the interaction between source and time. The coefficients for each dependent variable are shown in Table 1.

In order to estimate the variability of predictions derived from the present data, bootstrap sampling was used. Estimates of the two coefficients used in the above equation (the observed values of which are shown in Table 1) were obtained from analyses carried out on approximately 1000 bootstrap samples, using the mixed procedure and bootstrap procedures in Stata 13. Predictions of air measurements from the PDD measurements were obtained for each pair of coefficients, and the standard deviation of the predicted values was used as the standard error to construct 95% confidence intervals for the predictions.

The sampling with replacement was performed at the cluster level rather than at the level of individual measurements; that is, all the measurements for a given location, meaning a suburb or house, were either included in the bootstrap (possibly more than once), or not included at all. This method can be regarded as fairly stringent and gave rise to the conservative 95% confidence intervals shown in Figures 4 to 6.

Pb values

There were no significant differences in Pb loadings for PDD between suburbs but there were higher values in spring and summer compared with winter (Supplementary materials).

In contrast to the PDD, during the 4 years of detailed investigation of TSP in the Sydney suburb of Rozelle, Pb concentrations displayed a maximum in late autumn-start of winter (Figure 4) [32]. Similar trends have been observed in other Sydney suburbs (Figure 4) [34]. The higher Pb values in the autumn-winter months are probably related to thermal inversions typical of the cold season or to drier conditions during this period compared with summer. On the other hand the higher Pb contents in summer for the PDD may arise from the propensity of residents for open doors and windows in summer and the higher temperatures giving rise to dust which may then enter the residence. The decreasing Pb concentrations over time (p 0.008) reflect the decreasing use of Pb from gasoline, terminated in 2002.

Discussion

Comparison with other studies

Earlier studies of house dust did not include exterior air Pb, only reported Pb values (e.g., Succop et al. [28]) and rarely employed Pb isotopes. Furthermore, where Pb isotopes were measured, the dust sampling was by wiping material that had accumulated on surfaces and/or from door mats (e.g., Manton et al. [30]). In their investigation of aerosols, hand wipes, house dust, and a 24-hour duplicate diet Manton et al. [30] concluded that the Omaha aerosols represented a source distinct from the local household dust. In Shanghai, China, Liang et al. [36] found that children's blood lead levels were strongly correlated with air lead and mainly caused by coal consumption and fly ash after phasing out of leaded gasoline.

There are very few published isotopic data from Eastern Australia which would be relevant to compare with our analyses. Using a low volume sampler system powered by a diaphragm pump with a collection duration of from 1 to 3 months, Bollhöfer and Rosman [37,38] published data 2 samples in Sydney in March to May 1994 (2000) and 6 samples in 1994-95 and 6 in 1998 from Melbourne (2002). The sets of data from Bollhöfer and Rosman are encouragingly almost identical to our results for the same time periods even though the collection methods and laboratories were completely different [37]. The 1994 Sydney samples had $^{206}Pb/^{204}Pb$ ratios of 16.49 and 16.45 whereas those for Melbourne were slightly higher varying from 16.53 to 16.69 and in 1998 ranged from 16.8 to 17.3. Bollhöfer and Rosman [38] suggest that Pb isotopic ratios in Melbourne were slightly more radiogenic (e.g., higher $^{206}Pb/^{204}Pb$ ratio) during the summer compared with winter as we observed for summer-autumn but this is not clear from the data presented in their Table 1. Only one sample of gasoline from Western Australia was analyzed to evaluate the relationship between tetra-alkyl Pb and air in contrast to the numerous analyses for Sydney reported by Chiaradia et al. [32]. In the Sydney study, Chiaradia et al. [32] estimated that more than 90% of the Pb in HV air filters was derived from gasoline and gasoline Pb was also the primary source of surface soil and pavement dust.

Monthly sampling over a 15-month period (November 2010-January 2012) of dust deposition gauges placed in the rear yards of 5 houses in Sydney showed higher atmospheric Pb loadings in the summer/autumn with the lowest values in winter [15].

Dust from petri dishes in the houses described here over a 9-year time interval (1993-2002) showed higher Pb loadings in spring and summer compared with winter (Supplementary material). No difference between seasons for Pb loadings was observed by Laidlaw et al. [15], measured using the same petri dish method. In a longitudinal study in Sydney from 2001 to 2006 of 108 houses (1163 analyses) using PDD sampling over 6 monthly intervals, no seasonal effect for Pb loadings was observed [39]. The differences between our results and those of Laidlaw et al. [15] may arise from the different years of sampling and the limited number of 5 houses studied by Laidlaw et al. [15] compared with the 59 houses in our study.

Using vacuum cleaner dust from 82 houses in different Sydney suburbs, Chattopadhyay et al. [40] suggested that while air Pb levels decreased dramatically over decades, Pb concentrations in household dust remain unchanged due to accumulation of Pb from old paints and almost 80 years of leaded gasoline use [41,42].

Conclusions

The tracking of isotopic data for the air filters and house dust indicates that house dust provides an alternative indicator of exposure especially where air monitoring is not possible. The absence of seasonal or suburban effects on the isotopic data indicates applicability of house dust measurements over a wide area; the Sydney Basin area sampled in our study encompasses some 40 by 40 km. PDD is a simple, non-invasive sampling method compared with personal air monitoring and provides an integrated picture of air-dust relationships over say 3-monthly periods. PDD provide isotopic information that can be used in other environmental studies such as monitoring with mosses and lichens and lake sediments.

In trying to evaluate sources of metals in the environment now that Pb has been removed from gasoline in most countries and the important role of leaded paint has been more strongly publicized, it is worth reiterating a comment by Professor Roy Harrison "instead of having a few dominant sources we are now subject to a large number of relatively weak sources and therefore the source signatures become very obscure in the atmospheric signal" [33].

Acknowledgement

We wish to thank: the NSW EPA for allowing access to their air monitoring data and filters; CSIRO, US National Institute of Environmental Health Sciences (NO1-ES-05292), and Macquarie University for partial funding of the research; Karen Mizon, Michael Korsch, Massimo Chiaradia, Matthew James, and Nicole Pattison for analytical support; Bill Jameson and the late Katy Mahaffey for support and encouragement.

References

1. Buchet JP, Roels H, Lauwerys R, Bruaux P, Claeys-Thoreau F, et al. (1980) Repeated surveillance of exposure to cadmium, manganese, and arsenic in school-age children living in rural, urban, and nonferrous smelter areas in Belgium. Environ Res 22: 95-108.
2. Charney E, Sayre J, Coulter M (1980) Increased lead absorption in inner city children: where does the lead come from? Pediatrics 65: 226-231.
3. Duggan MJ, Inskip MJ (1985) Childhood exposure to lead in surface dust and soil: a community health problem. Public Health Rev 13: 1-54.
4. Harrison RM (1979) Toxic metals in street and household dusts. Sci Total Environ 11: 89-97.
5. Lanphear BP, Matte TD, Rogers J, Clickner RP, Dietz B, et al. (1998) The contribution of lead-contaminated house dust and residential soil to children's blood lead levels. A pooled analysis of 12 epidemiologic studies. Environ Res 79: 51-68.
6. Laxen DPH, Lindsay F, Raab GM, Hunter R, Fell GS, et al. (1987) The variability of lead in dusts within homes of young children. Lead in the Home Environment. Thornton, I, Culbard, E. (Editors). Science Reviews Ltd., Northwood, England, pp. 113-125.
7. Mielke HW, Laidlaw MA, Gonzales C (2010) Lead (Pb) legacy from vehicle traffic in eight California urbanized areas: continuing influence of lead dust on children's health. Sci Total Environ 408: 3965-3975.
8. Thornton I, Davies DJ, Watt JM, Quinn MJ (1990) Lead exposure in young children from dust and soil in the United Kingdom. Environ Health Perspect 89: 55-60.
9. Clark S, Bornschein RL, Pan W, Menrath W, Roda S (1995) An examination of the relationships between the U.S. Department of Housing and Urban Development floor lead loading clearance level for lead-based paint abatement, surface dust lead by a vacuum collection method and pediatric blood lead. Appl Occup Environ Hyg 10: 107-110.
10. Farfel MR, Lees PS, Rohde CA, Lim BS, Bannon D, et al. (1994) Comparison of a wipe and a vacuum collection method for the determination of lead in residential dusts. Environ Res 65: 291-301.
11. United States Environmental Protection Agency (2013) Office of Research and Development National Center for Environmental Assessment – RTP Division. Integrated Science Assessment for Lead. EPA/600/R-10/075F. Research Triangle Park, NC.
12. van Alphen M (1999) Atmospheric heavy metal deposition plumes adjacent to a primary lead-zinc smelter. Sci Total Environ 236: 119-134.
13. Bornschein RL, Succop P, Dietrich KN, Clark CS, Que Hee S, et al. (1985) The influence of social and environmental factors on dust lead, hand lead, and blood lead levels in young children. Environ Res 38: 108-118.
14. Gulson BL, Davis JJ, Mizon KJ, Korsch MJ, Bawden-Smith J (1995) Sources of soil and dust and the use of dust fallout as a sampling medium. Science Total Environ 166: 245-262.
15. Gulson B (2014) Comments on: Identification of lead sources in residential environments: Sydney Australia. By Laidlaw MAS, Zahran S, Pingitore N, Clague J, Devlin G, Taylor MP, 2013. Environmental Pollution 184, 238-246. Environ Pollut 185: 372-373.
16. Wlodarczyk J, Jardim K, Robertson R, Aldrich R, Toneguzzi R, et al. (1997) Measuring the amount of lead in indoor dust: long-term dust accumulation in petri dishes (a pilot study). NSW Public Health Bulletin 8: 92-93.
17. Laidlaw MAS, Filippelli GM (2008) Resuspension of urban soils as a persistent source of lead poisoning in children: a review and new directions. Appl Geochem 23: 2021-2039.
18. Laidlaw MAS, Zahran S, Mielke HW, Taylor MP, Filippelli GM (2012) Re-suspension of lead contaminated urban soil as a dominant source of atmospheric lead in Birmingham, Chicago, Detroit and Pittsburgh, USA. Atmos Environ 49: 302-310.
19. Zahran S, Laidlaw MA, McElmurry SP, Filippelli GM, Taylor M (2013) Linking source and effect: resuspended soil lead, air lead, and children's blood lead levels in Detroit, Michigan. Environ Sci Technol 47: 2839-2845.
20. Thomas VM, Socolow RH, Fanelli JJ, Spiro TG (1999) Effects of reducing lead in gasoline: An analysis of the international Experience. Environ Sci Technol 33: 3942-3948.
21. Cohen DD, Martin JW, Bailey GM, Crisp PT, Bryant E, et al. (1994) A twelve-month survey of lead in fine airborne particles in the major population areas of New South Wales. Clean Air 28: 79-88.
22. Australian Nuclear Science and Technology Organisation (ANSTO) (2013) Fine particle pollution: coastal NSW.
23. Brink LL, Talbott EO, Sharma RK, Marsh GM, Wu WC, et al. (2013) Do US ambient air lead levels have a significant impact on childhood blood lead levels: results of a national study. J Environ Public Health 2013: 278042.
24. Richmond-Bryant J, Meng Q, Davis A, Cohen J, Lu SE, et al. (2014) The influence of declining air lead levels on blood lead-air lead slope factors in children. Environ Health Perspect 122: 754-760.
25. Adgate JL, Mongin SJ, Pratt GC, Zhang J, Field MP, et al. (2007) Relationships between personal, indoor, and outdoor exposures to trace elements in PM(2.5). Sci Total Environ 386: 21-32.
26. Tu KW, Knutson EO (1988) Indoor outdoor aerosol measurement for two residential buildings in New Jersey. Aerosol Science and Technology 9: 71-82.
27. Davies DJ, Thornton I, Watt JM, Culbard EB, Harvey PG, et al. (1990) Lead intake and blood lead in two-year-old U.K. urban children. Sci Total Environ 90: 13-29.
28. Succop P, Bornschein R, Brown K, Tseng CY (1998) An empirical comparison of lead exposure pathway models. Environ Health Perspect 106 Suppl 6: 1577-1583.
29. Angle CR, McIntire MS (1979) Environmental lead and children: the Omaha study. J Toxicol Environ Health 5: 855-870.
30. Manton WI, Angle CR, Krogstrand KL (2005) Origin of lead in the United States diet. Environ Sci Technol 39: 8995-9000.
31. Rabinowitz M, Leviton A, Bellinger D (1985) Home refinishing, lead paint, and infant blood lead levels. Am J Public Health 75: 403-404.
32. Chiaradia M, Gulson BL, James M, Jameson CW, Johnson D (1997) Identification of secondary lead sources in the air of an urban environment. Atmos Environ 31: 3511-3521.
33. Gulson B, Korsch M, Dickson B, Cohen D, Mizon K, et al. (2007) Comparison of lead isotopes with source apportionment models, including SOM, for air particulates. Sci Total Environ 381: 169-179.
34. Cohen DD, Gulson BL, Davis JM, Stelcer E, Garton D, et al. (2005) Fine-particle Mn and other metals linked to the introduction of MMT into gasoline in Sydney, Australia: Results of a natural experiment. Atmos Environ 39: 6885-6896.
35. Snijders T, Bosker R (2012) Multilevel Analysis: An Introduction to Basic and Advanced Multilevel Modeling. London, Sage Publishers.
36. Liang F, Zhang G, Tan M, Yan C, Li X, et al. (2010) Lead in children's blood is mainly caused by coal-fired ash after phasing out of leaded gasoline in Shanghai. Environ Sci Technol 44: 4760-4765.
37. Bollhöfer A, Rosman KJR (2000) Isotopic source signatures for atmospheric lead: the Southern Hemisphere. Geochim Cosmochim Acta 64: 3251-3262.

38. Bollhöfer A, Rosman KJR (2002) The temporal stability of lead isotopic signatures at selected sites in the Southern and Northern Hemispheres. Geochim Cosmochim Acta 66: 1375-1386.

39. Gulson B, Mizon K, Taylor A, Korsch M, Davis JM, et al. (2014) Pathways of Pb and Mn observed in a 5-year longitudinal investigation in young children and environmental measures from an urban setting. Environ Poll 191: 38-49.

40. Chattopadhyay G, Lin KC, Feitz AJ (2003) Household dust metal levels in the Sydney metropolitan area. Environ Res 93: 301-307.

41. Clark CS, Bornschein R, Succop P, Roda S, Peace B (1991) Urban lead exposures of children in Cincinnati, Ohio. Chem. Speciation Bioavailability 3: 163-171.

42. Harney J, Trunov M, Grinshpun S, Willeke K, Choe K, et al. (2000) Release of lead-containing particles from a wall enclosure. AIHAJ 61: 743-752.

14

In vitro Metabolic Stability Study of New Cyclen Based Antimalarial Drug Leads Using RP-HPLC and LC-MS/MS

Apoorva V Rudraraju[1], Mohammad F Hossain[1], Anjuli Shrestha[1], Prince NA Amoyaw[1], Babu L Tekwani[2] and Faruk Khan MO[1*]

[1]*College of Pharmacy, Southwestern Oklahoma State University, 100 Campus Drive, Weatherford, Ok 73096, USA*
[2]*National Center for Natural Products Research, University of Mississippi, University, MS 38677, USA*

Abstract

Metabolic stability of the new antimalarial drug leads is determined using Human Liver Microsome (HLM) and specific cytochrome P450 enzyme (CYP2C8) taking the clinically used antimalarial drug chloroquine as a positive control. Experiment is done using standard methods. All the assays were conducted in 0.5 M phosphate buffer at pH 7.4. In general the metabolic reaction was initiated by adding 1 mM NADPH and 0.5 mg of enzyme. Incubations were done with time frequency of 0 hr, 1 hr, and 2 hrs at 37°C and the reactions were terminated by adding acetonitrile in the equal amounts of the assay mixture taken. The samples were centrifuged for 15 minutes at 10,000×g at 4°C and an aliquot of the supernatant was subjected to analysis using HPLC as well as LC-MS to confirm the masses of the drug and/or metabolite(s), if any. While chloroquine was found to be metabolized in a predictable manner by both HLM and CYP2C8, the drug leads were metabolically stable at similar experimental conditions. This study demonstrated that the new drug leads are worth conducting further preclinical evaluations.

Keywords: Metabolic stability; Preclinical study; Drug lead analysis; Drug discovery

Introduction

With the arrival of new combinational drug synthesis, the need for an expeditious evaluation of drug safety has become a paramount topic in drug discovery. The duration and impact of drug discovery revolving around drug absorption, distribution, metabolism and excretion will be evaluated according to their determined metabolic stability. The major organ for drug metabolism is liver that contains major drug metabolizing enzyme called Cytochrome P450 (CYP) system [1]. Drug metabolism refers to the susceptibility of compounds to biotransformation that depends on the presence of groups in the molecule those are open to enzyme catalyzed transformation [2]. Drug metabolism can be divided into two phases, Phase I and phase II. Phase I involves oxidation, reduction and hydrolysis reactions, which are catalyzed by the CYP and flavin containing monooxygenases (FMOs) whereas Phase II involves conjugation reactions catalyzed by metabolic enzymes like UDP-glucuronyltransferases (UGTs) and sulfotransferases [3,4].

Recognizing the metabolites of drugs is of paramount importance in drug discovery and development. The identification of drug metabolites in the early stages of the drug discovery is important in the development processes. The analytical tools like Liquid Chromatography-Mass Spectrometry (LC-MS) and HPLC play prominent role in these processes. Through this process of identification, the pharmacokinetic profiles can be assessed that are highly significant in detecting safety and efficacy of the drug leads before they are progressed to the clinical trials.

The investigation for metabolites take an advantage of the fact that majority of drug metabolites can be classified as predictable as they are formed from common accepted biotransformation reactions. However, there are many other illustrations of primary metabolites that are formed from uncommon reactions and are, therefore, not easily predictable. Molecular masses of predicted metabolites (m/z values) can be estimated based on mass shifts from the parent drug. For example, Chloroquine is the major antimalarial drug that used in the treatment and prophylaxis of malaria, the protonated molecular mass of the metabolite of chloroquine, desethylchloroquine (DCQ), is 292 m/z to that of the parent drug 320 m/z. Evaluation of some expected metabolites can be achieved by the acquisition of the complete MS spectrum using various MS instruments and also by Extracted Ion Chromatography (EIC) [4,5]. Using LC/MS, Clarke (2001) defined broadly the approaches for the identification of metabolites in biological matrices. After identifying the unknown metabolite ion peaks, further investigation to achieve detailed information on the pathways for the structure evaluation is done by multistage product ion scans [6,7].

The main purpose of the present study was to determine the metabolic stability of a series of newly discovered cyclen bisquinoline antimalarial drug leads [8] and related compounds (Figure 1) by RP-HPLC and LC-MS techniques. Compound B was shown to be highly effective antimalarial agent both *in vitro* and in vivo and was found to work by inhibiting β-hematin formation [8]. This lead structure fulfilled few important criteria for a new drug lead: 1) it is a 4-aminoquinoline derivative that are the most trusted class of antimalarials, 2) active against chloroquine-resistant as well as multidrug resistant isolates of *Plasmodium falciparum*, 3) simple pharmacophore structure that will afford low cost manufacturing, and 4) it also showed oral efficacy in mice model. All these features warranted further preclinical studies with these leads and related newly synthesized compounds as shown in Figure 1. Compounds A, C, D and E were also shown to have potent *in vitro* antimalarial activity and also serve as drug leads (manuscript is under preparation). The present study was thus aimed at determining the *in vitro* metabolic stability and identifying potential metabolites by HPLC and LC-MS techniques using HLM and CYP enzymes. The study was designed based on the known metabolic pathway of the related clinically used drug chloroquine (CQ) [9-12] and thus utilized both pooled human liver microsomes as well as specific isozyme CYP2C8 (Figure 1).

***Corresponding author:** Faruk Khan MO, College of Pharmacy, Southwestern Oklahoma State University, 100 Campus Drive, Weatherford, Ok 73096, USA E-mail: faruk.khan@swosu.edu

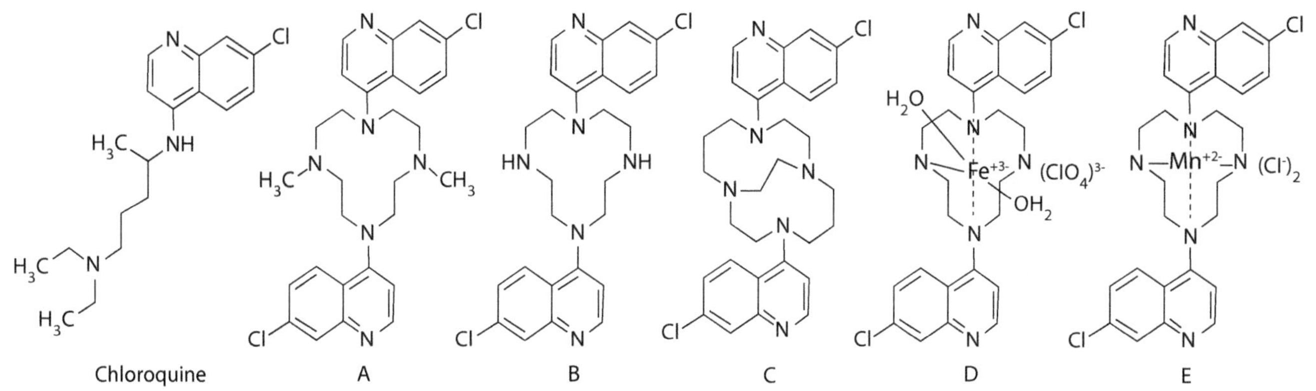

Figure 1: Chemical structures of chloroquine and the main drug leads A, B, C, D, and E.

Materials and Methods

Materials

Chloroquine diphosphate purchased from Pfaltz & Bauer. NADPH, anhydrous with assay 93-100% (HPLC), was purchased from Sigma Aldrich. 0.5 M Potassium phosphate buffer pH 7.4 was purchased from BD Gentest. Human Liver Microsome (HLM) was purchased from BD Bioscience and Corning. Human cDNA expressing CYP 450 and 2C8 were purchased from XenoTech and Corning. In the solvent system methanol Triethylamine, DMSO, and acetonitrile were purchased from Fisher Scientific. Dibasic anhydrous sodium phosphate and phosphoric acid were purchased from Fisher Scientific. Deionized water used was further purified by filtration and degassing. The drug leads mentioned in the Figure 1 were synthesized.

Chromatographic conditions

HPLC Method 1: The HPLC system used was Agilent 1100 series (Hewlett-Packard-Strasse 8, 76337 Waldbronn Germany), equipped with a pump, Diode array detector with UV lamp operated at variable wavelengths, autosampler, and thermostat. Data Acquisition was performed using Open Lab CDS Chem Station Edition software package with A.01.02 version implemented in the chromatographic system. Detection at 250 nm and 280 nm was standard. The stationary phase consisted of Waters X-Bridge C-18 column (4.6 mm×150 mm, 5.0 μm particle size, pore volume 0.76 cm^3/g) purchased from Waters Corporation (34 Maple Street, Milford, Massachusetts 01757-3696 U.S.A). It was operated at constant temperature of 25°C on both ends of the column.

The mobile phase consisted of a mixture of 0.04% formic acid in water (i), acetonitrile (ii) and methanol (iii) in a gradient elution mode as shown in Table 1 [9]. The flow rate was maintained at 1.0 ml/min.

LC-MS system used was Shimadzu prominence LC 20AT, equipped with Degasser (DGU- 20A5), Auto sampler (SIL-20AHT), Refractive index detector (RID-10A), Diode array detector (SPDM-20M20A), Fraction collector (FRC-10A) and a mass spectrometer (LC-MS 2020). Data Acquisition was performed by Lab Solutions Real Time Analysis Software, implemented in the chromatographic system.

HPLC Method 2: Chromatographic separation of CQ, drug lead B, and metabolites was successfully achieved on a Waters X-Bridge C-18 column (4.6 mm×250 mm, 5.0 μm particle size, part no. 186003117) purchased from Waters Corporation in an isocratic separation mode with mobile phase consisting of 0.1% of triethylamine in methanol and 0.02 M dibasic sodium phosphate (anhydrous) at pH 3.5 adjusted with Phosphoric Acid in the ratio of (60:40, v/v). The flow rate was maintained at 1.0 ml/min, the column oven temperature at 40°C and the effluent was monitored at 325 nm (based on the λ). The peak purity data were obtained using Photodiode Array (PDA) detector in the sample chromatograms. The method was found linear over the concentration range of 3.5–200.1 μg/ml (R^2=0.99) for drug lead B.

General procedure

The method used to determine the metabolism was a modified method obtained from BD Bioscience, which is summarized here. Metabolic stability study of a drug can be determined by treating it with the liver enzyme that correlates with the in vivo conditions. The assay mixture is prepared by the combination of substrate (0.01 mM to 0.5 mM), enzyme (0.5 mg), buffer (0.5 M), and 1 mM NADPH cofactor. The order of addition of the assay component also plays a major role in the stability study. Metabolism can be initiated by pre-warming or incubating the substrate, buffer, and the cofactor to 37°C and then adding liver microsomes or CYP2C8 to the mixture. It is customary that the assay mixture should be thoroughly mixed. After the preparation of the Total Assay Mixture (TAM), required volume of the sample is collected at different time intervals. The reaction is then terminated using acetonitrile in equal volumes as that of the sample solution collected. The samples are then transferred into the microcentrifuge tubes, vortexed for 2 minutes and then centrifuged at 10,000×g for 15 minutes. The purpose of centrifugation is to remove the protein. The sample solution is then separated from the protein pellet by taken the supernatants. These supernatant solutions are transferred into the HPLC auto sampler vials and are analyzed according to the analytical method.

Metabolic stability in HLM using HPLC Method 1

The HLM used was pooled and prepared from freshly frozen human tissues which were tested negative for pathogens using PCR. The HLM used in this study was comprised of 330 pmoles/mg of total P450 and 420 pmoles/mg cyt. b_5. It is a mixture of CYP1A2, 2A6, 2B6, 2C8, 2C9, 2C19, 2D6, 2E1, 3A4, 4A11, FMO, UGT1A1, UGT1A4, and UGT1A9 where the enzyme activity of CYP2C8, 3A4, and 2D6 was 82, 3200, and 110 pmoles/(mg×min) respectively. The volume of the enzyme per vial was 0.5 ml and the protein content was 20 mg/ml in 250 mM sucrose.

All the incubations were performed in duplicates. Therefore, all together there were 6 samples for each drug, one blank and 6 samples of the positive control. The total experiment with 5 drugs has about 65 samples. All the components used in the TAM are shown in Table

Time (Min)/solvent	0.04% formic acid in Water	Acetonitrile	Methanol
0	96	0	4
10	64	30	6
12	4	90	6

Table 1: Mobile phase for metabolic stability study: Gradient elution mode.

S. No.	Ingredients	Volume (in 1 mL of total assay mixture)	Final Concentration
1	Sample	10 µl	0.05 mM
2	Buffer (0.5 M)	200 µl	100 mM
3	DI Water	598 µl	-
4	NADPH	167 µl	1 mM
5	Enzyme	25 µl	0.5 mg

Table 2: Total assay mixture using HLM.

S. No.	Ingredients	Volume (TAM)	Final Concentration
1.	Sample	180 µl	0.5 mM
2.	Buffer	1275 µl	0.2 M
3.	NADPH	300 µl	1m M
4.	Enzyme	45 µl	25 pmoles/ml

Table 3: Total assay mixture using CYP 2C8.

S. No.	Ingredients	Sample		Standard	Blank
		Volume, 2 mL of TAM	Final Concentration	Volume, 2 mL of TAM	Volume, 2 mL of TAM
1	Substrate (1 mM)	20 µl	0.01 mM	20 µl	---
2	Buffer (0.5 M)	400 µl	100 mM	400 µl	400 µl
3	NADPH	200 µl	1 mM	---	200 µl
4	DI Water	1355 µl	---	1580 µl	1375 µl
5	Enzyme	25 µl	0.5 mg	---	25 µl

Table 4: Total assay mixture using HLM and CYP2C8.

2 with their volumes. All the ingredients in the TAM are taken in the order of their serial number (Table 2).

5 mM of the substrate stock solutions were prepared by dissolving in DMSO. The buffer used was 0.5 M potassium phosphate pH 7.4 (BD Biosciences Cat No. 451201). TAM was prepared by adding 10 µl of the sample, 200 µl of the buffer solution, 598 µl of water, and 167 µl of NADPH. It was then pre-incubated for 2 minutes in a shaking water bath and 25 µl of the enzyme was added to initiate the reaction. The capped tube was mixed thoroughly by inverting a couple of times. Immediately, at time t=0 hr, 300 µl of the TAM was taken into a centrifuge tube and the reaction was terminated by placing it on ice bath and adding 300 µl of acetonitrile. The remaining TAM was warmed in a shaking water bath at 37°C for 2 hrs and the same procedure is followed for the samples at time t=1 hr and t=2 hr. The incubation mixtures were then centrifuged for 15 minutes at 10,000×g at 4°C. About 500 µl aliquots of the supernatant were collected and were subjected to analysis using the HPLC conditions as shown in Table 1. After running HPLC the samples were frozen and then lyophilized to remove the solvent. The dried samples were then dissolved in methanol and water (50:50 v/v) for LC-MS analyses.

Metabolic stability in CYP2C8 using HPLC method 1

The CYP enzyme is comprised of a human CYP2C8 and human CYP-reductase co-expressed in *Escherichia coli.* The concentration of the CYP P450 is 1.0 nmol/ml and its protein concentration is 10.0 mg/ml. All the assays were conducted in 0.2 M phosphate buffer at pH 7.4. One blank for each sample A, B, C, D, and E was prepared by adding 180 µl of the sample in 1620 µl of the buffer. The total assay mixture contained 25 pmoles/ml of CYP2C8 and 0.5 mM of sample as shown in the table 3.

180 µl of Sample, 1275 µl buffer, and 300 µl of 1 mM NADPH was taken into a reaction flask and was pre-incubated for 2 minutes. The reaction was initiated by adding 45 µl enzyme. Incubations were done with increasing time (t=0 hr, 1 hr, 2 hrs) at 37°C. Immediately at time=0 hr, 300 µl of the assay mixture was pipetted into a centrifuge tube and 300 µl of acetonitrile was added to it to stop the reaction. All the samples were done in triplicates. Therefore, all together there were 9 samples for each drug, one blank and 9 samples of the positive control. The total experiment with 5 drugs had about 95 samples.

The remaining TAM was incubated for 1 hr and 2 hrs, same procedure of sampling was followed as mentioned in the HLM section. Then the samples were centrifuged for 15 minutes at 10,000×g at 4°C. An aliquot about 500 µl of the supernatant fraction were collected and subjected to analysis using the above HPLC conditions as shown in Table 1. The retention times were noted and the amounts of drugs metabolized were calculated according to their percentage peak area using the equation: % peak area of the metabolite/(% peak area of the drug+% peak area of the metabolite). After running HPLC the samples were frozen and then lyophilized to remove the solvent. The dried samples were then dissolved in methanol and water (50:50 v/v) and tested for their mass. The mass of the drug and the metabolite were determined by using LC-MS.

Metabolic stability in HLM and CYP2C8 using HPLC method 2

To verify the metabolic stability results using HPLC method 1, HPLC method 2 was carried out using drug lead B, and CQ as a positive control. For that we prepared sample, standard and blank for our drug lead B and CQ in duplicates as mentioned in the following Table 4. All the ingredients in the TAM were taken in the order of their serial number (Table 4).

Sample stock solutions of Chloroquine diphosphate and Cylen Bisquinoline Hydrochloride Salt were prepared by dissolving in water. 0.5 M potassium phosphate pH 7.4 (BD Biosciences Cat No. 451201) was used as a buffer. Sample solutions were prepared by adding 20 µl of substrate, 400 µl of the buffer solution, 200 µl of the NADPH solution and 1355 µl of water. Standard solutions were prepared by adding 20 µl of Substrate, 400 µl of the buffer solution and 1580 µl of water. Blank solutions were prepared by adding 400 µl of the buffer solution, 200 µl of the NADPH solution and 1375 µl of water.

All the above solutions were then pre-incubated for 5 minutes in a shaking water bath at 37°C. Reaction was initiated by adding 25 µl of the enzyme (from Corning HLM-452161 and CYP2C8-456252). These TAMs were mixed thoroughly by inverting a couple of times and returned to the water bath. Immediately, at time t=0 hr, 500 µl of the TAM was taken into a centrifuge tube and the reaction was terminated by placing it on ice bath and adding 500 µl of acetonitrile. The remaining TAMs were warmed in a shaking water bath at 37°C for 2 hrs and the same procedure was followed for the samples at time t=1 hr and t=2 hr. The incubation mixtures were then centrifuged for 15 minutes at 10,000×g at 4°C. About 500 µl aliquots of the supernatant were collected and subjected to analysis using the HPLC method 2. After running HPLC the samples were frozen and then lyophilized to remove the solvent. The dried samples were then dissolved in methanol and water (50:50 v/v) for LC-MS analyses.

Results and Discussion

HPLC Method 1

Results using HLM: Under the provided experimental conditions

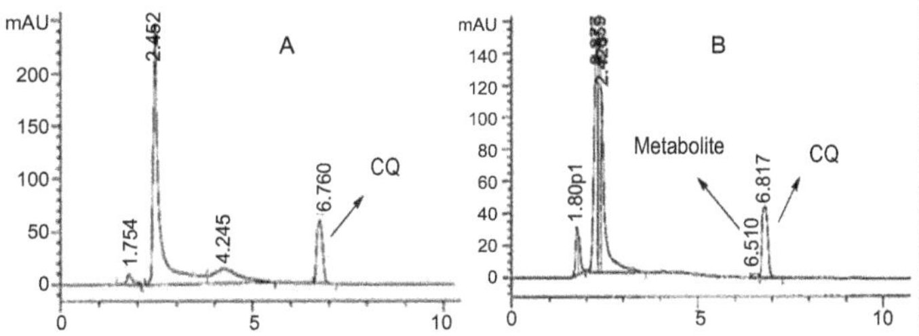

Figure 2: HPLC chromatogram of chloroquine at time A) 0 hour and B) 2 hour using HLM.

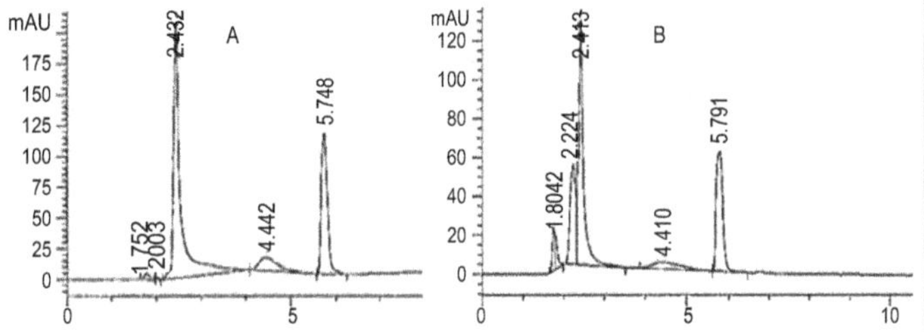

Figure 3: HPLC chromatogram of drug lead B at time A) 0 hour and B) 2 hour using HLM.

	Time t=1 hr			Time t=2 hr		
Drugs	**RT**	**Metabolism**	**MS- (m/z)**	**RT**	**Metabolism**	**MS- (m/z)**
A	8.136	NO	523	8.141	NO	523
B	5.737	NO	495	5.791	NO	495
C	4.869	NO	561	4.877	NO	561
D	5.818	NO	885	5.822	NO	885
E	5.822	NO	619	5.831	NO	619
CQ/DCQ	6.801/6.485	YES	320, 292*	6.817/6.510	YES	320, 292*

*Percentage of the metabolite formation was calculated according to the obtained percentage peak areas of the metabolite and the drug. The percent of CQ metabolized using HLM at time t=1 hr and 2 hrs was 5.12% and 6.146%, respectively.

Table 5: Data obtained from HPLC and MS for the drugs using HLM.

it was observed that CQ was metabolized in a predictable manner by HLM into the metabolite DCQ; contrarily all our experimental drug leads were metabolically stable. Not only was the metabolite for CQ identified in HPLC (Figure 2) but also in MS (Figure 7) reconfirming our results. Both drug (320 m/z) and metabolite masses (292 m/z) were observed in the LC-MS analysis of the samples. According to the area under curve ratios, CQ was converted to about 5.12% of its N-deethylated metabolite after 1 hour of incubation and 6.42% after 2 hours of incubation. On the other hand, the drug leads A (523m/z), B (495m/z), C (561m/z), D (885 m/z), and E (619 m/z) were found to be metabolically stable in both HPLC and LC-MS analyses. A sample chromatogram of CQ and compound B are shown in the Figures 2 and 3. The retention times and the peak areas of the drug leads A, B, C, D, E and CQ obtained at time t =1 hr and 2 hr are shown in the Table 5. At time t=0 min there was no metabolism with any of the drug samples. At time t=1 hr CQ showed some metabolism whereas the drug leads were stable. There was no evidence of any metabolite mass of any of the drug leads when they were analyzed by LC-MS. The molecular ion peaks (m/z) were similar to those of their molecular weights (Figures 2 and 3) (Table 5).

Results using CYP2C8: As shown in Figure 4, about 10.33% of CQ was transformed into DCQ after 1 hour of incubation and 12.26% after 2 hours of incubation, whereas the drug lead B was stable as shown in Figure 5. The drug leads A (523 m/z), B (495 m/z), C (561 m/z), D (885 m/z), and E (619 m/z) were metabolically stable at similar experimental conditions. The retention times of the drugs A, B, C, D, E and CQ obtained at time t=0 were 8.136, 5.971, 4.869, 5822, 5.817 and 6.998, respectively, and no metabolite peak was evident. The retention times of the drug leads and the CQ at time t=1 hr and t=2 hr is shown in the Table 6. All the samples were estimated in triplicates and the results shown are the average of the three trials (Figures 4 and 5) (Table 6).

With these results, it was confirmed that CYP2C8 has an important role in the metabolism of chloroquine and the drug leads were metabolically stable. The experiment was extended up to 24 hours for the drug leads and consequently no metabolism was observed. The drug masses were confirmed by MS (Figures 6-9).

HPLC Method 2

Results using HLM: Under provided experimental conditions using HPLC method 2, similar results were obtained, CQ was metabolized by HLM into the metabolite DCQ and contrarily the drug lead B was metabolically stable. The metabolite was identified by HPLC and also by LC-MS reconfirming the results shown in method 1. Both drug (320 m/z) and metabolite (292 m/z) masses were observed in the LC-MS analysis of the samples. According to the area under curve ratios, CQ was converted to about 4.29% of its N-deethylated metabolite after 1

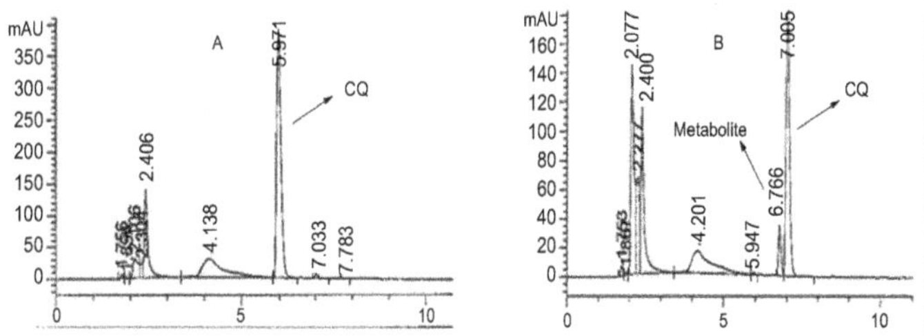

Figure 4: HPLC chromatogram of chloroquine at time A) 0 hour and B) 2 hour using CYP2C8.

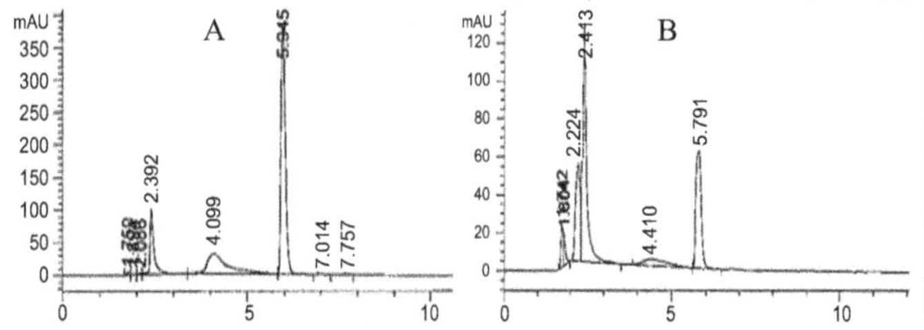

Figure 5: HPLC chromatogram of drug B at time A) 0 min and B) 2 hour using CYP2C8.

	Time t= 1 hr			Time t= 2 hr		
Drugs	**RT**	**Metabolism**	**MS- (m/z)**	**RT**	**Metabolism**	**MS-(m/z)**
A	8.126	NO	523	8.136	NO	523
B	5.973	NO	495	5.944	NO	495
C	4.875	NO	561	4.869	NO	561
D	5.819	NO	885	5.822	NO	885
E	5.821	NO	619	5.817	NO	619
CQ/DCQ	7.017/6.784	YES	320, 292*	7.011/6.772	YES	320, 292*

*The percent of CQ metabolized using CYP2C8 at time t=1hr and 2hrs was 12.26% and 10.33%, respectively.

Table 6: Data obtained from HPLC and MS for the drugs using CYP2C8.

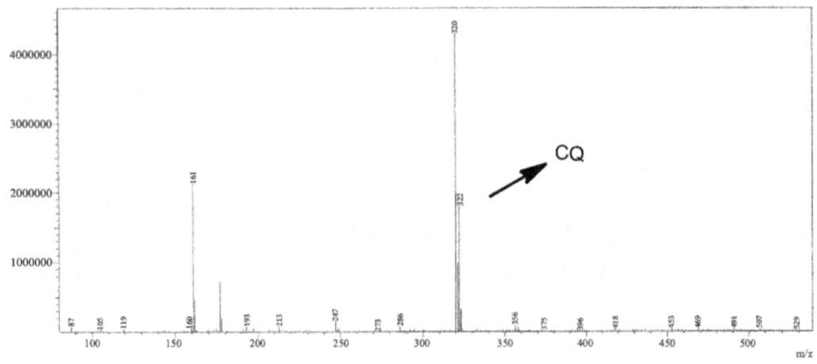

Figure 6: MS spectrum showing chloroquine standard.

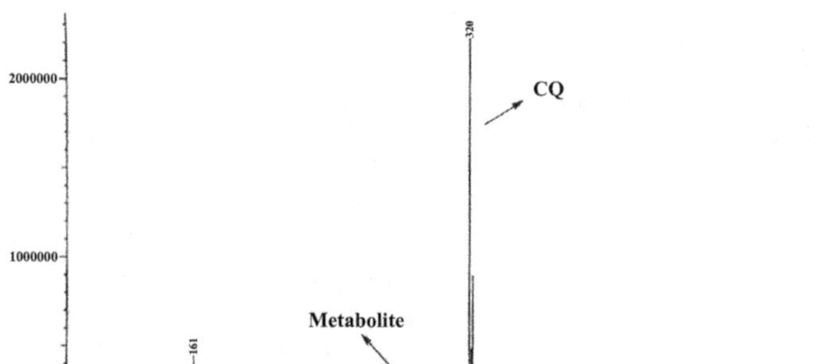

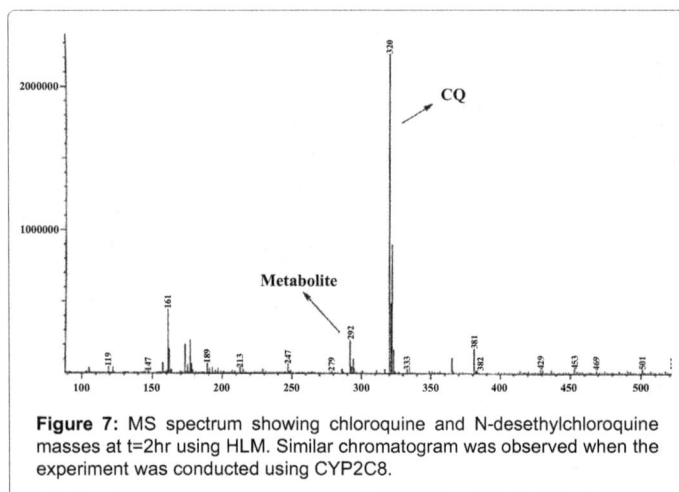

Figure 7: MS spectrum showing chloroquine and N-desethylchloroquine masses at t=2hr using HLM. Similar chromatogram was observed when the experiment was conducted using CYP2C8.

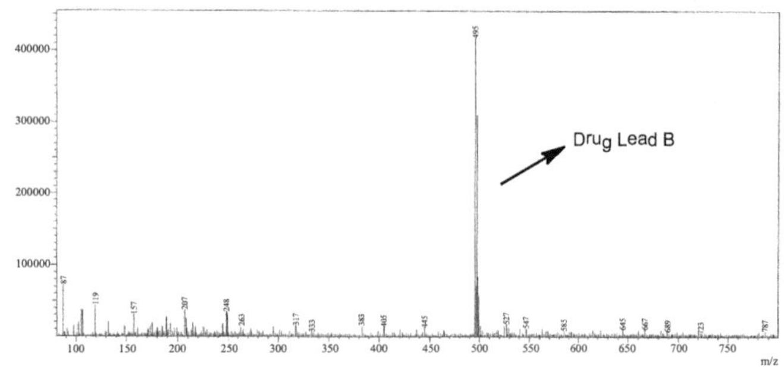

Figure 8: MS spectrum showing drug lead B standard.

hour of incubation and 6.61% after 2 hours of incubation. On the other hand, the drug leads B (495 m/z) was found to be metabolically stable in both HPLC and LC-MS analyses. A sample chromatogram of CQ and compound B are shown in Figure 10C and Figure 11C, respectively. The retention times and the peak areas of the drug lead B and CQ obtained at time t=1hr and 2hr are shown in the Table 7.

Results using CYP2C8: Under provided experimental conditions using HPLC method 2, similar results were obtained, CQ was metabolized by CYP2C8 into the metabolite DCQ and contrarily the drug lead B was metabolically stable. Using CYP2C8 about 2.23% of CQ was transformed into DCQ after 1 hour of incubation and 2.89% after 2 hours of incubation. The drug lead B (495 m/z) was metabolically stable at similar experimental conditions (Figure 11D). The retention times of the drug lead B and the CQ at time t=1 hr and t=2 hr are shown in the Table 8. All the samples were estimated in duplicates and the results shown are the average of the two trials (Figures 10 and 11) (Table 8).

According to the results obtained from HPLC method I and II, it was confirmed that both HLM and CYP2C8 has an important role in the metabolism of chloroquine whereas, all drug leads were metabolically stable to these enzyme.

Discussion

The *in vitro* metabolism of the new drug leads compared to the positive control chloroquine in the presence of HLM and cDNA expressing CYP2C8 have been demonstrated in this study. In the pharmacokinetic studies of CQ, Frisk-Holmberg [6,7] reported the single dose kinetics of CQ and its major metabolite DCQ in healthy subjects and identified that CQ is dealkylated into two main metabolites, N-desethylchloroquine (DCQ) and N-bis-desethylchloroquine (BDCQ). Identification of metabolites is done by LC-MS. Mass spectrometers help in separating the compounds from each other using the difference in mass-to-charge ratio (m/z) of ionized compounds.

In blood and plasma, the concentration of DCQ was detected up to 20 to 50% of those of the parent compound CQ, whereas the conversion to BDCQ was <10% [6,7]. It is known that CYPs are major drug metabolizing enzyme system. Ofori- Adjei and Ericsson [13] estimated that 30 to 50% of an administered dose of chloroquine is metabolized by the liver via cytochrome P450 enzymes. Identification of the major enzymes, associated with drug's metabolism therefore started by evaluating and identifying the role of CYP isoforms. Xue-Qing Li et al. [14] identified human CYPs that metabolize anti-parasitic drugs using three approaches, (a) Relative Activity Factor (RAF), (b) correlation analysis of their activity, and (c) inhibitors diagnostics. They reported that the major CYPs involved in the metabolism are CYP isoforms 2C8, 3A4, and 2D6. The RAF approach is used significantly to bridge the gap between liver microsomes and recombinant systems. It also estimates the individual P450 contributions to drug metabolism [15,16]. Using the same approach, Projean et al. [9] also reported that CYPs 2C8, 3A4, and 2D6 are the main isoforms catalyzing DCQ formation. This was the first investigation of CQ metabolism in HLM and they identified that CYP2C8 and CYP3A4 contributes collectively

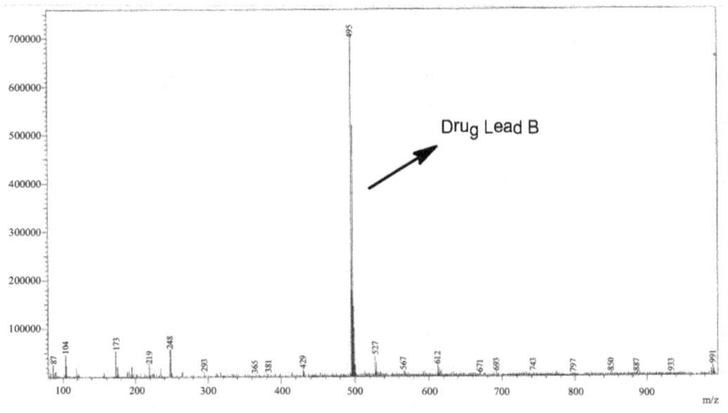

Figure 9: MS spectrum showing drug lead B mass at t=2 hr using HLM. Similar chromatogram was observed when the experiment was conducted using CYP2C8.

	Time t=1 hr			Time t=2 hr		
Drugs	**RT**	**Metabolism**	**MS- (m/z)**	**RT**	**Metabolism**	**MS- (m/z)**
B	13.135	NO	495	13.10	NO	495
CQ/DCQ	5.460/4.480	YES	320, 292*	5.470/4.508	YES	320, 292*

*The percent of CQ metabolized using HLM at time t=1 hr and 2 hrs was 4.29% and 6.61%, respectively

Table 7: Data obtained from HPLC and MS for the drug using HLM.

	Time t=1 hr			Time t=2 hr		
Drugs	RT	Metabolism	MS- (m/z)	RT	Metabolism	MS- (m/z)
B	13.145	NO	495	13.129	NO	495
CQ/DCQ	5.486/4.533	YES	320,292*	5.432/4.517	YES	320, 292*

*The percent of CQ metabolized using CYP2C8 at time t=1 hr and 2 hrs was 12.26% and 10.33%, respectively

Table 8: Data obtained from HPLC and MS for the drugs using CYP2C8.

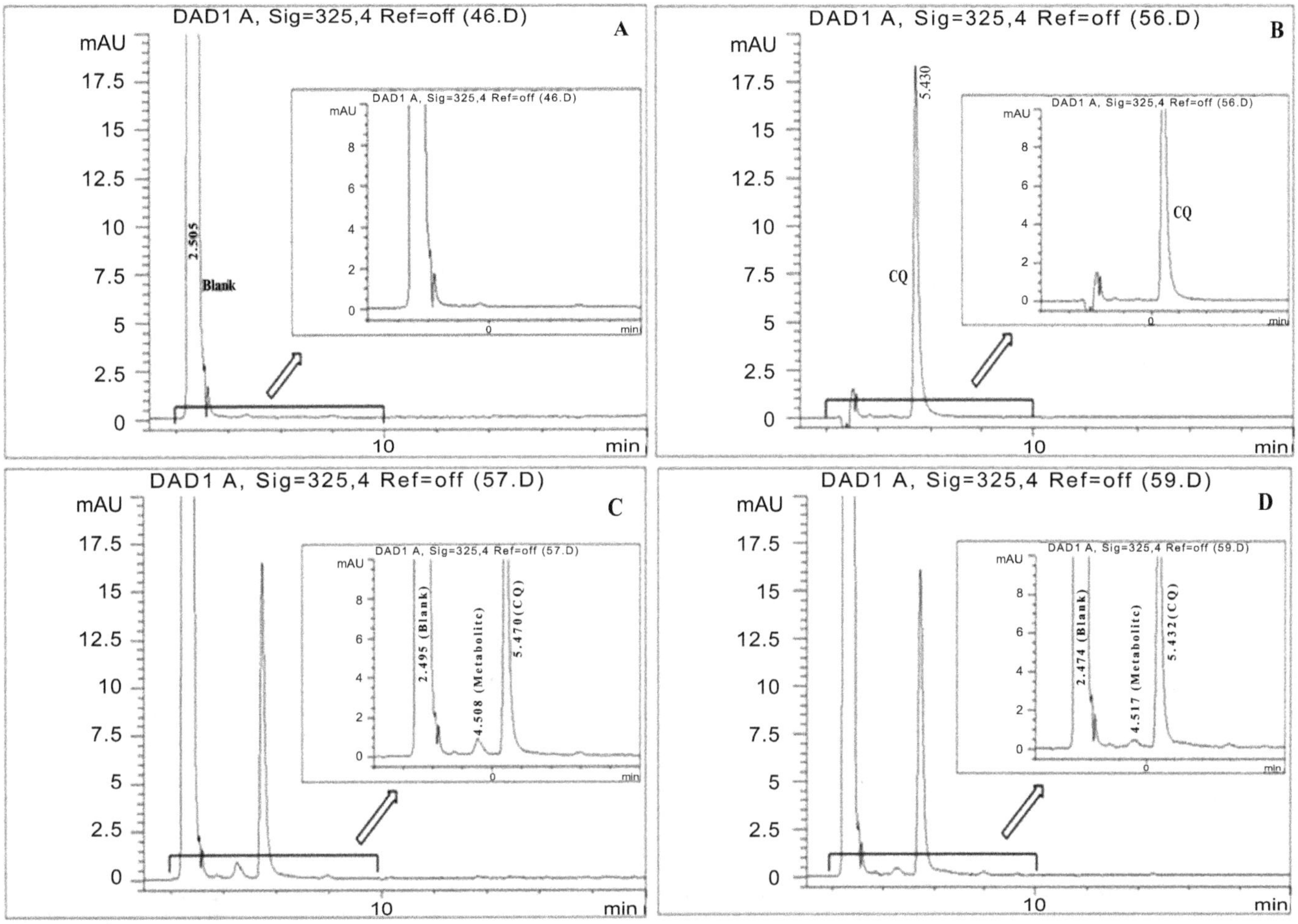

Figure 10: HPLC chromatograms of chloroquine. A) Blank (NADPH, Enzyme & Buffer), B) CQ (Control without Enzyme), C) CQ in HLM (t=2 hr), and D) CQ in CYP2C8 (t=2 hr).

more than 80% of the total CQ N-desethylation over a wide range of concentrations. Their study suggested that, in humans, the therapeutic concentration of CQ that is metabolized into DCQ is primarily via CYP2C8 and CYP3A4, whereas 2D6 plays a significant role when CQ concentrations are low. This study also shows that CYPs, 2C8, and 3A4 constitute low-affinity and high-capacity systems, whereas CYP2D6 has higher affinity but a significantly lower capacity. Among these CYP2C8 represents approximately 5% of the total hepatic CYPs and metabolizes drugs that are amides or weak bases with two hydrogen bond acceptors [11,17,18]. Kim et al. [10] also identified that CYP2C8 and 3A4/5 are involved in metabolism of CQ into desethylchloroquine in human liver microsomes. This was also consistent with the previous data obtained from Ducharme and Farinotti [19]. Previous studies show that the relative content of CYP 3A4, 2C8 and 2D6 in 60 human samples are 29%, 18%, and 2%, respectively [20]. Based on these literature evidence, HLM and CYP2C8 were obvious selection to study the *in vitro* metabolism stability of the new drug leads due to the similarity of the structures and activity of the lead compounds with chloroquine. While chloroquine was metabolized by both HLM and CYP2C8 considerably, all these new drug leads were metabolically stable.

In the Figure 10, the HPLC chromatograms clearly showed that CQ was metabolized and there was an additional peak before the parent peak which was absent in blank and control sample, partially confirms the metabolism of chloroquine. Moreover, when the samples were tested with LC-MS (Figure 7), positive control chloroquine sample at time t=2 hrs gave two ion peaks, one at 320 m/z and another at 292 m/z, which confirms the metabolism of the drug.

In the Figure 11, the HPLC chromatograms clearly showed that drug lead B was not metabolized but the appearance of an additional peak before the parent peak in blank and control sample confirms the presence of impurity in drug lead B. The spectrum (Figure 9) obtained

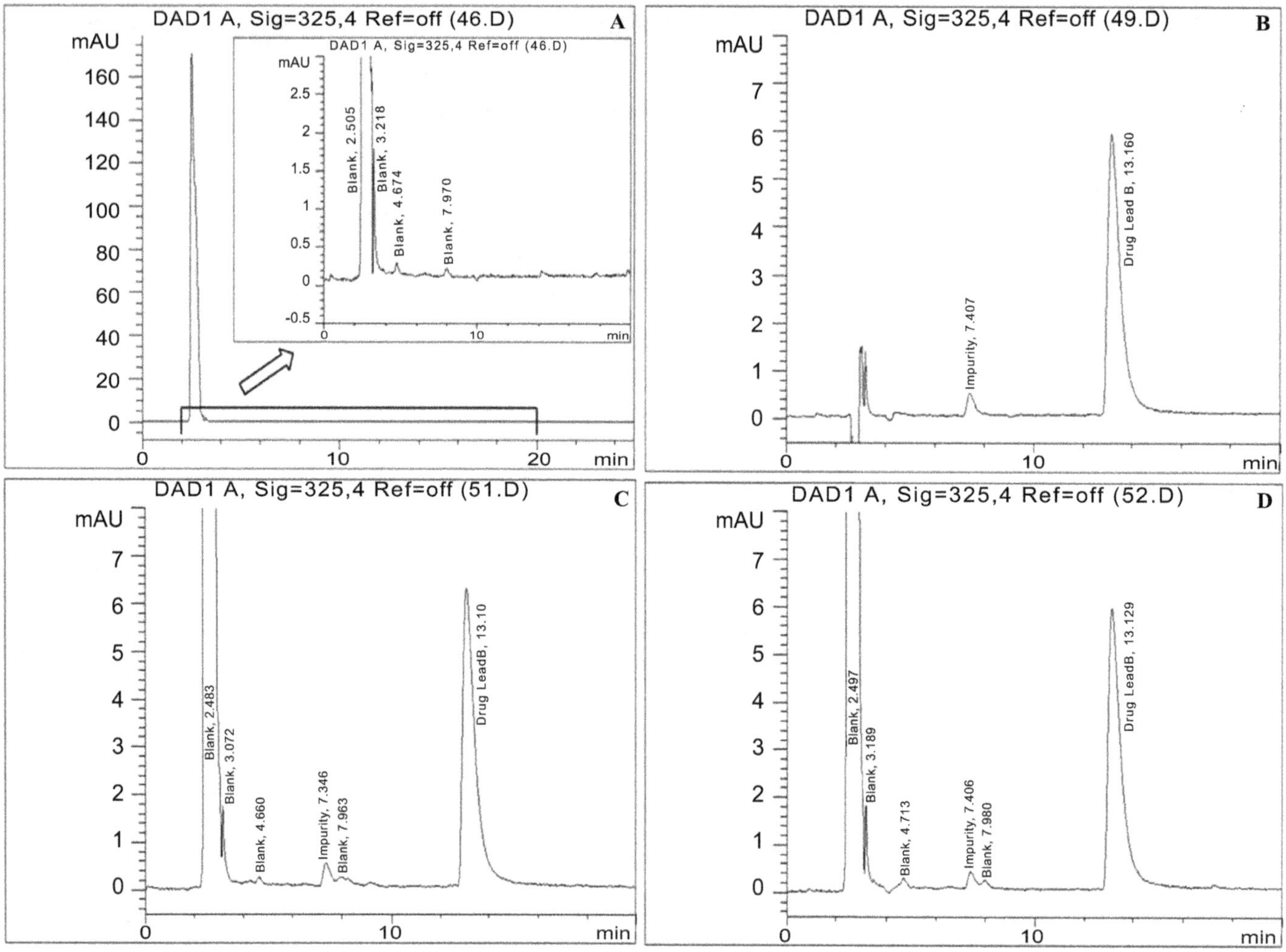

Figure 11: HPLC chromatograms of drug lead B. A) Blank (NADPH, Enzyme & Buffer), B) Drug lead B (Control without Enzyme), C) Drug lead B in HLM (t=2 hr), and D) Drug lead B in CYP2C8 (t=2 hr).

from LC-MS showed the ion peak of the drug lead B at time t=2 hrs as 495 m/z, which confirms its molecular weight. Henceforth there is no metabolism of the drug lead B.

As discussed earlier, CYP enzymes provide information regarding the metabolites. Nebert and Russell [21] mentioned that there are 270 different CYP gene families, and 18 of those gene families were recorded in mammals that encrypt 57 CYP genes in humans. Out of all CYP enzymes, CYP2C8, 3A4, and 2D6 are involved in Chloroquine metabolism. This was proven in our experiment where CYP2C8 contributed to more than 10% of the total CQ *N*-desethylation over a 100 μM concentration [22]. It has low affinity and high capacity in metabolizing the chloroquine. Further investigations would be conducted utilizing more CYP isozymes at different concentration ranges and time intervals to identify any possible metabolic pathways for these drug leads in near future as a part of our continuous preclinical analysis. This study demonstrated a convenient method development and also a metabolic stability of the drug leads against most important CYP isozyme that is involved in the metabolism of standard drug chloroquine.

Acknowledgment

This project was supported by the National Institute of General Medical Sciences of the National Institutes of Health through Grant Number 8P20GM103447.

References

1. Casarett LJ, Doull J, Klaassen CD (2001) Parkinson A: Biotransformation of Xenobiotics. Casarett and Doull's Toxicology: The Basic Science of Poisons, McGraw-Hill Medical Publishing Division, New York, USA.
2. Ariens EJ, Simonis AM (1982) Optimization of pharmacokinetics-an essential aspect of drug development by metabolic stabilization. Strategy in drug research. Keverling Buisman JA, editor. Elsevier Scientific Publishing Company, Amsterdam.
3. Smith DA, Jones BC, Walker DK (1996) Design of drugs involving the concepts and theories of drug metabolism and pharmacokinetics. Med Res Rev 16: 243-266.
4. Baranczewski P, Stańczak A, Sundberg K, Svensson R, Wallin A, et al. (2006) Introduction to in vitro estimation of metabolic stability and drug interactions of new chemical entities in drug discovery and development. Pharmacol Rep 58: 453-472.
5. Tiller PR, Romanyshyn LA (2002) Liquid chromatography/tandem mass spectrometric quantification with metabolite screening as a strategy to enhance the early drug discovery process. Rapid Commun Mass Spectrom 16: 1225-1231.
6. Frisk-Holmberg M, Bergqvist Y, Domeij-Nyberg B (1983) Steady state disposition of chloroquine in patients with rheumatoid disease. Eur J Clin Pharmacol 24: 837-839.
7. Gustafsson LL, Walker O, Alván G, Beermann B, Estevez F, et al. (1983) Disposition of chloroquine in man after single intravenous and oral doses. Br J Clin Pharmacol 15: 471-479.

8. Khan MO, Levi MS, Tekwani BL, Khan SI, Kimura E, et al. (2009) Synthesis and antimalarial activities of cyclen 4-aminoquinoline analogs. Antimicrob Agents Chemother 53: 1320-1324.

9. Projean D, Baune B, Farinotti R, Flinois JP, Beaune P, et al. (2003) In vitro metabolism of chloroquine: identification of CYP2C8, CYP3A4, and CYP2D6 as the main isoforms catalyzing N-desethylchloroquine formation. Drug Metab Dispos 31: 748-754.

10. Kim KA, Park JY, Lee JS, Lim S (2003) Cytochrome P450 2C8 and CYP3A4/5 are involved in chloroquine metabolism in human liver microsomes. Arch Pharm Res 26: 631-637.

11. Lewis DF (2004) 57 varieties: the human cytochromes P450. Pharmacogenomics 5: 305-318.

12. Ducharme J, Farinotti R (1996) Clinical pharmacokinetics and metabolism of chloroquine. Focus on recent advancements. Clin Pharmacokinet 31: 257-274.

13. McChesney EW, Banks WF Jr, Sullivan DJ (1965) Metabolism of chloroquine and hydroxychloroquine in albino and pigmented rats. Toxicol Appl Pharmacol 7: 627-636.

14. Ofori-Adjei D, Ericsson O (1985) Chloroquine in nail clippings. Lancet 2: 331.

15. Li XQ, Björkman A, Andersson TB, Gustafsson LL, Masimirembwa CM (2003) Identification of human cytochrome P(450)s that metabolise anti-parasitic drugs and predictions of in vivo drug hepatic clearance from in vitro data. Eur J Clin Pharmacol 59: 429-442.

16. Crespi CL (1995) Xenobiotic-metabolizing human cells as tools for pharmacological and toxicological research. Advances in Drug Research 26: 179-235.

17. Störmer E, von Moltke LL, Shader RI, Greenblatt DJ (2000) Metabolism of the antidepressant mirtazapine in vitro: contribution of cytochromes P-450 1A2, 2D6, and 3A4. Drug Metab Dispos 28: 1168-1175.

18. Pelkonen O, Turpeinen M, Hakkola J, Honkakoski P, Hukkanen J, et al. (2008) Inhibition and induction of human cytochrome P450 enzymes: current status. Arch Toxicol 82: 667-715.

19. Musana AK, Wilke RA (2005) Gene-based drug prescribing: clinical implications of the cytochrome P450 genes. WMJ 104: 61-66.

20. Ducharme J, Farinotti R (1997) Rapid and simple method to determine chloroquine and its desethylated metabolites in human microsomes by high-performance liquid chromatography with fluorescence detection J Chromatogr B Biomed Sci Appl 698: 243-250.

21. Shimada T, Yamazaki H, Mimura M, Inui Y, Guengerich FP (1994) Interindividual variations in human liver cytochrome P-450 enzymes involved in the oxidation of drugs, carcinogens and toxic chemicals: studies with liver microsomes of 30 Japanese and 30 Caucasians. J Pharmacol Exp Ther 270: 414-423.

22. Nebert DW, Russell DW (2002) Clinical importance of the cytochromes P450. Lancet 360: 1155-1162.

Gas Chromatography-Mass Spectrometric Method for Simultaneous Separation and Determination of Several Pops with Health Hazards Effects

Nagwa ABO EL-Maali* and Asmaa Yehia Wahman

Department of Chemistry, Faculty of Science, Assiut University, Assiut, Egypt

Abstract

The hazards effect of persistent organic pollutants, POPs, on the human health has lead us to modify the ASTM method D-5175 to enhance both sensitivity and selectivity of their simultaneous separation. As their separation is difficult- due to the similarity in their chemical and physical properties- that lead to co-elution in extraction, we proposed a validated method for their simultaneous determination using liquid/liquid microextraction followed by GC/MS in the SIM mode. The method is advantageous since the time needed for the chromatographic analysis of all analytes is less than 17 min. Method Detection Limits (MDLs) and Limit of Detection (LODs) reached sub- ppb levels and in many cases are lower than those achieved in the standard test method ASTM D-5175 for many analytes. Besides, three pesticides namely: Hexachlorocyclopentadiene, p,p'-DDE and trifluralin have been added to the method with good accuracy and precision. Application to several environmental samples has been successfully assessed and supported by proficiency testing samples provided from Absolute Standards®, Inc.

Keywords: GC-MS; POPs; Validation; Application to environmental matrices

Introduction

The importance of the Persistent Organic Pollutants (POPs) with their health effects has lead to looking for accurate and reliable methods for their determinations [1-6] using chromatographic techniques in many matrices viz. ground water [2], human serum [3,4] water and drinking water [5,6], fruits and vegetables [1,7] and tap water [8].

Organochlorines (OCs) are a lipophilic class of chemicals that include OC pesticides and other persistent organic pollutants, such as Polychlorinated Biphenyls (PCBs). It is well known that environmental and/ or dietary exposure to OCs results in the bioaccumulation of these chemicals in the human body especially, in adipose tissue, serum and breast milk [9,10].

Despite of the long-term adverse effects on humans, animals and environment [11], recent studies in East-Asian countries have reported elevated concentration of OCPs in various environmental media suggesting that same OCPs are still being used [12]. OC-exposure has been linked with a number of children diseases such as asthma, abnormalities of the productive tract, diabetes, and growth and neurobehavioral disorders [13,14]. In Spain [15,16], the level of chemical contamination by OCs of the population of the canary Island has been evaluated although they're banned in Spain in the late 1970s. In US, PCBs exposures are encountered by the general public by eating contaminated food or living near a previously operating PCB factory hazardous waste site [17], although they are banned in the United States in 1977. PCBs have been classified as probable human carcinogenic and are listed in the top 10% of EPA's most toxic chemical [18]. Of the 209 PCBs congeners four non-ortho and eight mono-ortho congeners are currently recognized by the World Health Organization (WHO) as "dioxin like" in their toxic effects [19]. Routine analysis of OCs in environmental in different matrices has been achieved through GC/MS and different extraction techniques [20-23].

Therefore, the aim of the present work is to validate and enhance the sensitivity of the ASTM method D5175 for the determination of OCPs viz. Alachlor, Aldrin, Dieldrin, Endrin, Heptachlor, Heptachlorepoxide, Hexachlorobenzene, lindane, Methoxychlor , PCBs congeners namely: PCB 28, 52,118,138, and 180 cited in Table 1 in the presence of the new analytes viz. Hexachlorocyclopentadiene, p,p'-DDE and the organofluorine pesticides trifluralin in other matrices viz. waste water and transformer oils with new levels lower than those cited in the literature.

Experimental

Chemicals and reagents

Organochlorine pesticides: Alachlor, aldrin, dieldrin, endrin, heptachlor, heptachlorepoxide, hexachlorobenzene, hexachlorocyclopentadiene, lindane, methoxychlor, p,p'-DDE and the organofluorine one- trifluralin -with purity higher than 96.0% and PCBs 28, 52, 118, 138 and 180 with purity higher the 96.0% were acquired from Sigma, reference standards are acquired from AccuuStandards®, Inc Lot # 209111013 and AbsoluteStandards®, Inc Lot # 032409. Proficiency testing samples are from AbsoluteStandards®, Inc Lot # 091608. Sodium chloride, Sigma, sodium thiosulfate, Merck. Methanol, Hexane and acetone (HPLC grade) were from Sigma. Ultrapure water used was from Milli-Q system model: Milli-Q Gradient A10, Elix 3UV and Tank 60L, Serial NO: F7AN24007K F7BN90274I, USA.

Preparation of standards

Standard solution, Stock: These solutions prepared from pure standard materials of each PCBs and Pesticides (1000 μg/ml).

- By accurately weighting about 1.0 mg of pure material. Dissolve the material in 1 ml of methanol absolute in 1.5 ml vials; the weight is used without correction to calculate the concentration of the stock.

***Corresponding author:** Nagwa ABO EL-Maali, Department of Chemistry, Faculty of Science, Assiut University, 71516-Assiut, Egypt, E-mail: nelmaali@live.com

Analyte Name	Chemical structure	MCL, µgL^{-1}, EPA	Health effect
Alachlor	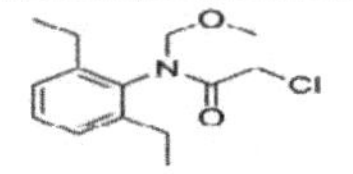	2.000	Eye, liver, kidney or spleen problems; anemia; increased risk of cancer
Aldrin	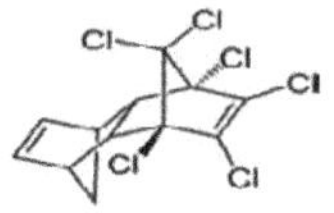	0.030*	Both aldrin and dieldrin are highly toxic to humans, the target organs being the central nervous system and the liver. Severe cases of both accidental and occupational poisoning and a number of fatalities have been reported
Dieldrin	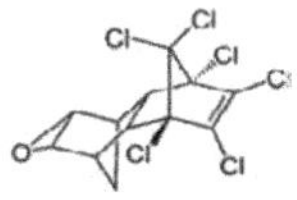	0.030*	
Endrin	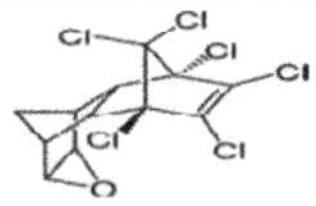	2.000	Liver problems
Heptachlor	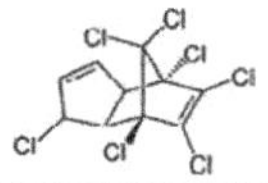	0.400	Liver damage; increased risk of cancer
Heptachlor epoxide	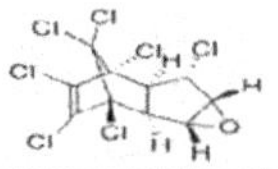	0.200	Liver damage; increased risk of cancer
Hexachlorobenzene	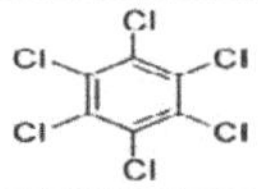	1.000	Liver or kidney problems; reproductive difficulties; increased risk of cancer
Hexachlorocyclopentadiene	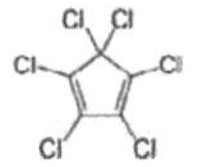	50.000	Kidney or stomach problems
Lindane	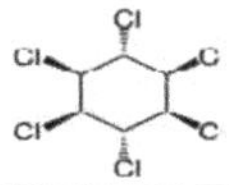	0.200	Liver or kidney problems
Methoxychlor	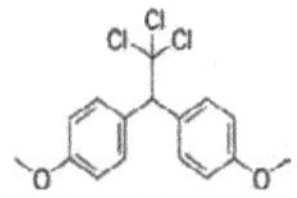	40.000	Reproductive difficulties
p,p'-DDE	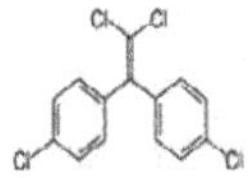	1.000*	Exposure to technical-grade DDT, an increased risk for pancreatic cancer could not be excluded. Pesticide applicators are exposed primarily to p,p'-DDT, whereas it is the p,p'-DDE metabolite to which the general population is exposed in the diet or drinking-water.
PCB 28	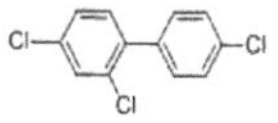	0.500	Skin changes; thymus gland problems; immune deficiencies; reproductive or nervous system difficulties; increased risk of cancer
PCB 52	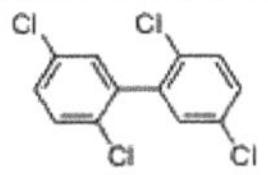		
PCB 118	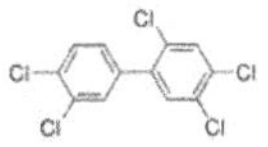		
PCB 138	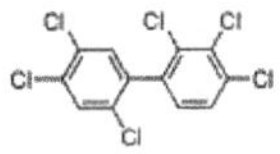		
PCB 180	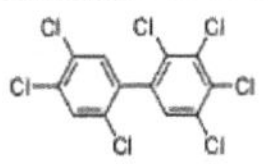		
Trifluralin	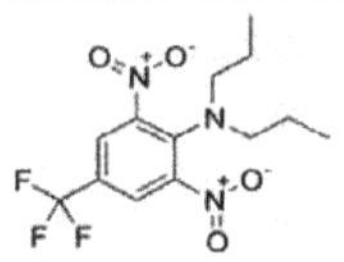	20.000*	In a study in the USA, the use of trifluralin was associated with an increased risk for non-Hodgkin lymphoma. In contrast, a study of ovarian cancer in Italy did not suggest an association with trifluralin exposure. In both studies, the numbers of exposed subjects were small. A larger study in the USA showed no association with leukaemia (20). IARC concluded that there is inadequate evidence in humans for the carcinogenicity of trifluralin (20).

Table 1: Structure, Maximum Contaminant Level (MCL) and health effects for Pesticides (OCP, OFP) and polychlorinated biphenyls (PCBs) under investigation. *MCL from WHO.

- Store standard solution in freezer and protect from light. Stock standard solution should be checked frequently for assign degradation or evaporation, especially prior to preparation calibration standard from them.

- Store standard solution must be replaced if comparison with checked standard indicates a problem.

Standard solution, secondary dilution: Use the stock standard solution to prepare secondary dilution standard solution in methanol and check frequently for singe of degradation evaporation especially just before preparing calibration standard.

Sample preparation and collection

- When sampling from a water tap, open the tap and allow the system to flush until the water temperature has stabilized (usually about 10 min). Adjust the flow to about 500 ml/min and collect samples from the flowing stream.

- When sampling drinking, surface, well , and waste water a sampling water apparatus model Easy-Load® Masterflex® USE 15, 24 TUBING Model:7518-12 part No: 4,813,855 Assembled in USA is used.

Sample preservation and storage

In 1 L empty bottle, add 8 mL of 1 M sodium thiosulfate just prior to sample collection. The samples must be chilled to 4°C at the time of collection and maintained at that temperature until the analyst is prepared for the extraction process. Store samples and extracts at 4°C until analysis has been completed. Extract all samples as soon as possible after collection. Results of holding time studies suggest that all analytes were stable for 14 days when stored under these conditions.

Instrumentation

GC separation was performed using Gas Chromatograph from Agilent Technologies Model 7890A equipped with temperature programming capability, splitless injector, capillary column, and Mass Quadrupole Spectrometry detector Model 5975B. A computer data system is MSD Chem Station E.0201.1177 used for measuring peak areas and heights.

Gas chromatograph parameters

The analytical columns used were DB-1701P (30 m × 0.25 mm × 0.25 μm), Agilent Part No.122-7732 as a primary column and DB-5ms (30 m × 0.25 mm × 0.25 μm), Agilent Part No.122-5532 as a secondary one, the oven temperature was set at 60°C for 0.50 min, increased to 140°C at 120°C/min, 260°C at 11°C/min then to 260°C for 5.5 min. The volume of the injected sample was 1 μL in split less mode. The injector temperature was set at 250°C. Helium (99.999%, purity) was used as carrier constant flow, 1 $mLmin^{-1}$.

Mass spectrometer parameter

The mass spectrometer was operated in electron impact (70 eV of ion energy), with 4.0 min solvent delay, SIM acquisition mode, mass quadruple and mass source kept at 150°C and 230°C.

Data analysis

Analysis of data is done using Microsoft® Office Excel 2003 (11.5612.5606) Part of Microsoft Office Professional Edition 2003, Product ID: 73931-640-0000106-75603.

Extraction procedure

Stored samples are removed from the fridge and allow to equilibrate to room temperature. To 35 ml of each sample, add 6 g NaCl in the separating funnel. Recap and dissolve the NaCl by inverting and shaking several times (approximately 30 sec). Remove the cap, add 2 ml of n-hexane recap and shake vigorously by hand for 2 min, inverting the separating funnel while shaking. Stand the separating funnels upright and allows the water and hexane phases to separate. Transfer 0.5 ml of hexane layer into an auto sampler vial; inject 1 μl portions into the gas chromatograph for analysis.

For transformer oil samples, solid phase extraction procedure using AccuBOND II FLORISIL Cartridges provided from Agilent Part No. 188-2460 was applied using the following method: weigh 0.2 gm of the transformer oil sample, pass through the FLORISIL Cartridges followed by flush five times with 2 mL aliquots of n-Hexane, the eluent was collected in a 10 mL volumetric flask then completed to the mark and mix thoroughly prior to the GC/MS analysis.

Results and Discussion

GC-MS separation

To confirm the retention times of the POPs – under investigation- OCPs, OFP and PCBs, a mass range of 50-500 m/z was scanned. Thereafter the SIM mode was applied to monitor the mixture. The selected ions (m/z) used for confirmation and quantification are cited in Table 2. POPs- under investigation- are eluted from the column in the following order: Hexachlorocyclopentadiene, Hexachlorobenzene, Trifluralin, lindane, PCB 28, Heptachlor, PCB 52, Aldrin, Alachlor, Heptachlorepoxide, p,p'-DDE, Dieldrin, PCB 118, PCB 138, Endrin, PCB180 and Methoxychlor (Figure 1). It is worthy to mention that the time needed for the chromatographic analysis is less than 17 min.

Optimization of the extraction procedure

In order to achieve the highest recoveries for the compounds under investigation, different organic solvents have been tried for this purpose, among them hexane and dichloromethane. With hexane only 2 ml for 2 min gave rise to best recovery all the analytes under investigation while using dichloromethane it needs 6.3 ml for 6 min is required.

Analyte	Primary Ion	Secondary Ion(s)
Alachlor	45.0	146.1, 160.1, 188.1, 224.1
Aldrin	66.0	44.0, 79.0, 91.0, 262.0
Dieldrin	79.0	44.0, 108.1, 262.9, 276.9
Endrin	81.0	67.0, 263.0, 79.0, 53.0, 261.0, 265.0
Heptachlor	100.0	272.0, 65.0, 237.0, 102.0
Heptachlorepoxide	81.0	44.0, 236.8, 262.9, 352.9
Hexachlorobenzene	284.0	286.0, 282.0, 249.0, 288.0
Hexachlorocyclopentadiene	237.0	239.0, 235.0, 272.0, 95.0
Lindane	181.0	183.0, 111.0, 219.0, 109.0
Methoxychlor	227.1	44.0, 113.7, 152.1, 212.1
p,p'-DDE	246.0	105.0, 176.0, 210.0, 318.0
PCB 28	256.0	257.0, 258.0, 259.0, 260.0
PCB 52	292.0	255.0, 257.0, 290.0, 294.0
PCB 118	325.9	254.0, 256.0, 323.9, 327.9
PCB 138	359.9	287.9, 289.9, 357.9, 361.9
PCB 180	393.9	323.9, 395.9, 397.9
Trifluralin	306.0	43.0, 264.0, 248.1, 290.0

Table 2: Characteristic ions for the investigated pesticides and PCBs.

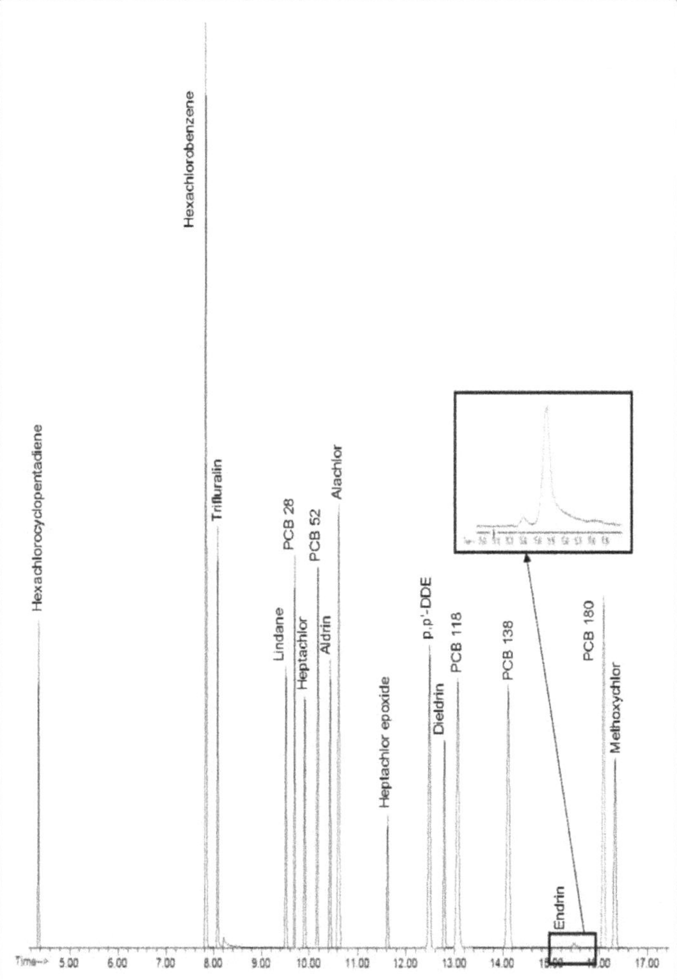

Figure 1: Hexane spiked with all analytes under investigation.

Method validation

Method validation is performed to provide evidence that the method is fit for the purpose for which it is used. Since the key challenges in the validated methods is that only well-characterized reference materials with well documented purities should be used during method validation activities, all steps are validated using reference materials, this includes specificity, accuracy, linearity, precision, range, detection limit, quantitation limit and robustness.

In order to assess these parameters, the method was therefore tested for Linearity, range Table 3. The analytical method demonstrated initial and extended validation as being capable of providing mean recovery values at each spiking level within the range 70-120%, spiked recovery experiments are performed (Figure 2), In order to check the precision of the proposed method, a minimum of 5 replicates is performed Table 4 summarizes these data.

Method Detection Limit (MDL) and Limit of Detection (LOD) - for the analytes under investigation - are cited in Table 5. Under normal conditions, reproducibility of data is tested in order to be sure that the method is robust. By changing the pH of the extract and the oven temperature, laboratory reproducibility as RSD% was found to be ≤ 20%, for all compounds indicating that the method is robust.

Quality control

Validation are supported and extended by method performance verification during analysis through analytical quality control AQC. AQC data are used to validate the extension of the method to new analytes viz Hexachlorocyclopentadiene, p,p'-DDE and trifluralin, new matrices viz. waste water and transformer oils and also to new concentration levels. Minimum quality control requirements are checked for POPs determination these include.

Analysis of laboratory reagent blanks (LRB)

All glassware and reagent interferences are taken under control by checking the extract and the reagent for any source of contamination within the retention time window of all analytes under investigation.

Initial demonstration of capability

This has been checked at different spiking levels which have been selected at a concentration level about ten times the estimate detection limit or at the maximum contaminant level for each analyte. For all aliquots analyzed, the recovery value for each analyte falls in the range 70-120%.

Analysis of laboratory fortified blanks (LFB)

Table 6 illustrates the spiking concentration of each analyte in the LFB sample with the calculated accuracy as percent recovery (%R). The recovery of all analytes under investigation fall inside the control limits (X ± 3S); where X is the mean percent recovery and S is the standard deviation of the percent recovery.

Analysis of laboratory fortified sample matrix (LFM)

To assess analytes recovery, a known spike of Aldrin, Alachlor , Heptachlorepoxide, Dieldrin, Methoxychlor, p,p' - DDE, PCB 28, PCB 52, PCB 118, PCB 138 and PCB 180 is added to waste water matrix as shown in Figure 3.

Analysis of reference materials (QCS) and proficiency testing

In order to assure the correct execution of the whole procedure for each individual sample and the correct injection of each final sample extract in the GC system, the use of one or more quality control (QC-) standards is utilized. These compounds are added at different steps of the procedure e.g., to the samples prior to extraction as surrogate standard or to the final sample extract just before injection as instrument internal standards. Analysis of QC samples provided from

Analyte	Calibration Equation	R^2
Alachlor	y=945.77x	0.999
Aldrin	y=1281.3x-1279.1	0.993
Dieldrin	y=841.31x-840.35	0.993
Endrin	y=15.60x-21.671	0.986
Heptachlor	y=901.89x-3608.8	0.989
Heptachlor epoxide	y=876.49x-4385.6	0.994
Hexachlorobenzene	y=6065.70x-23753	0.995
Hexachlorocyclopentadiene	y=2341.40x-64838	0.963
Lindane	y=1086.50x-607.36	0.993
Methoxychlor	y=1392.8x-27039	0.995
p,p'-DDE	y=2547.7x	0.998
PCB 28	y=3235.3x	0.997
PCB 52	y=8424.3x	0.995
PCB 118	y=2480.7x	0.997
PCB 138	y=2034.4x	0.996
PCB 180	y=939.48x	0.997
Trifluralin	y=550.82x-415.15	0.991

Table 3: Calibration equation and R^2 for pesticides and polychlorinated biphenyls.

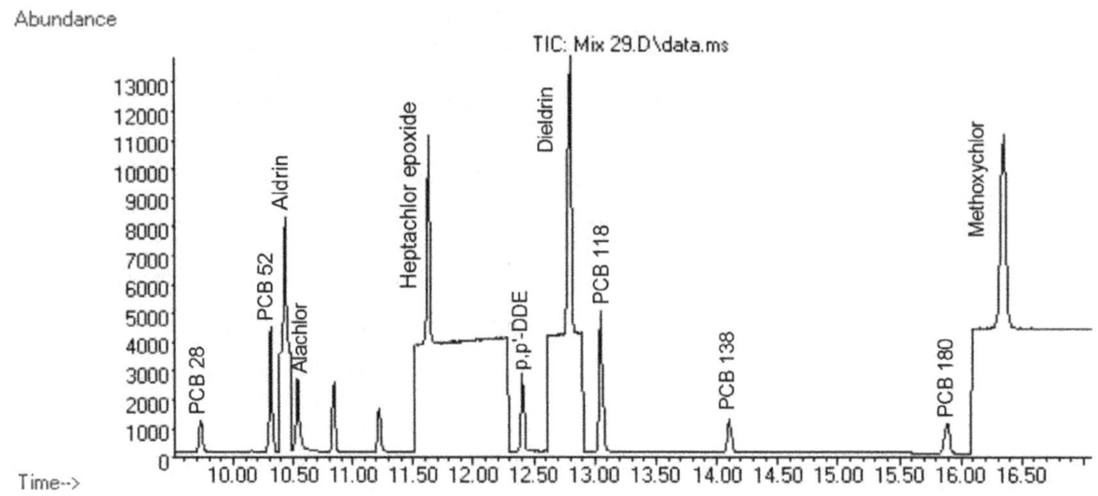

Figure 2: Extract of Reagent Water Spiked with, 3.966 µg/L Aldrin, 2.368 µg/L of Alachlor, 10.416 µg/L of Heptachlorepoxide, 12.800 µg/L of Dieldrin, 20.956 µg/L Methoxychlor, 1.413 µg/L p,p' - DDE, 0.498 µg/L PCB 28, 0.436 µg/L PCB 52, 1.830 µg/L PCB 118, 0.619 µg/L PCB 138 and 0.115 µg/L PCB 180.

Analyte	MDL, µg/L			PQL, µg/L		
	Reagent Water	Tap Water	Waste Water	Reagent Water	Tap Water	Waste Water
Alachlor	0.017	No [A]	No [A]	0.053	No [A]	No [A]
Aldrin	0.079	0.029	0.132	0.169	0.093	0.421
Dieldrin	0.109	0.030	0.139	0.346	0.096	0.443
Endrin	1.003	No [A]	No [A]	2.006	No [A]	No [A]
Heptachlor	0.759	No [A]	No [A]	2.416	No [A]	No [A]
Heptachlor epoxide	0.107	0.099	0.163	0.342	0.316	0.521
Hexachlorobenzene	0.010	No [A]	No [A]	0.0201	No [A]	No [A]
Hexachlorocyclopentadiene	3.762	No [A]	No [A]	7.523	No [A]	No [A]
Lindane	0.081	No [A]	No [A]	0.161	No [A]	No [A]
Methoxychlor	5.000	No [A]	No [A]	10.000	No [A]	No [A]
p,p'-DDE	0.006	0.005	0.012	0.019	0.017	0.037
PCB 28	0.020	0.007	0.061	0.056	0.021	0.194
PCB 52	0.013	0.006	0.029	0.042	0.020	0.092
PCB 118	0.014	0.008	0.031	0.045	0.025	0.099
PCB 138	0.012	0.008	0.077	0.038	0.027	0.245
PCB 180	0.010	0.006	0.025	0.031	0.018	0.081
Trifluralin	1.007	No [A]	1.078	2.014	No [A]	No [A]
[A]No analysis conducted.						
*Analytes Provided from Sigma-Aldrich.						
**Analytes Provided from Absolute Standard Lot # 032409						

Table 4: Accuracy and Precision for Pesticides and PCBs in reagent, tap and waste water.

Analyte	MDL, µg/L			PQL, µg/L		
	Reagent Water	Tap Water	Waste Water	Reagent Water	Tap Water	Waste Water
Alachlor	0.017	No [A]	No [A]	0.053	No [A]	No [A]
Aldrin	0.079	0.029	0.132	0.169	0.093	0.421
Dieldrin	0.109	0.030	0.139	0.346	0.096	0.443
Endrin	1.003	No [A]	No [A]	2.006	No [A]	No [A]
Heptachlor	0.759	No [A]	No [A]	2.416	No [A]	No [A]
Heptachlor epoxide	0.107	0.099	0.163	0.342	0.316	0.521
Hexachlorobenzene	0.010	No [A]	No [A]	0.0201	No [A]	No [A]
Hexachlorocyclopentadiene	3.762	No [A]	No [A]	7.523	No [A]	No [A]
Lindane	0.081	No [A]	No [A]	0.161	No [A]	No [A]
Methoxychlor	5.000	No [A]	No [A]	10.000	No [A]	No [A]
p,p'-DDE	0.006	0.005	0.012	0.019	0.017	0.037
PCB 28	0.020	0.007	0.061	0.056	0.021	0.194
PCB 52	0.013	0.006	0.029	0.042	0.020	0.092
PCB 118	0.014	0.008	0.031	0.045	0.025	0.099
PCB 138	0.012	0.008	0.077	0.038	0.027	0.245
PCB 180	0.010	0.006	0.025	0.031	0.018	0.081
Trifluralin	1.007	No [A]	1.078	2.014	No [A]	No [A]
[A]No analysis conducted						

Table 5: Method detection limit (MDL) and practical quantitation limit (PQL).

Accu Standard*, and Absolute Standards* Inc., are shown in Figure 4 and Table 7 respectively. In addition sharing in a proficiency testing program provided from Absolute Standards*, Inc., was successfully achieved as shown in Figure 5 and Table 8 indicating that our results meet the performance criteria for the provided QC sample datasheets.

Qualifying results with uncertainty data

Measurement uncertainty is a quantitative indicator of the confidence in the analytical data and describes the range around a reported or experimental result within which the true value can be expected to lie within a defined probability (confidence level). Uncertainty ranges must take into consideration all sources of error. To determine the uncertainty associated with analytical results, the available sufficient data derived from method validation /verification, inter-laboratory studies (e.g., proficiency tests provided from Absolute Standards*, Inc proficiency testing provider) and in-house quality control tests provided from Accuu Standard* are applied to estimate the uncertainties. Uncertainty associated with repeatability of measurements for these true samples in the main elements of the uncertainty budget. The expanded uncertainty is calculated and cited in Table 9 as follows:

$$u = K\frac{SD}{\sqrt{n}}$$

Where K: is the coverage factor (it has a value of 2 at 95% confidence level); SD: is the standard deviations; n: is the number of measurements.

Since uncertainty tends to be greater at lower levels, especially as the LOQ is approached. It was therefore necessary to generate

Analyte	Spiking Level, ppb	Recovery (%R) Relative Standard Deviation	
		R%	RSD%
Alachlor	0.800	77.318	55.902
Aldrin	7.500	74.915	0.982
Dieldrin	5.000	78.802	0.533
Endrin	7.500	92.003	2.627
Heptachlor	10.000	98.948	0.719
Heptachlorepoxide	15.000	75.746	2.428
Hexachlorobenzene	15.000	92.125	0.570
Hexachlorocyclopentadiene	20.00	115.899	0.550
Lindane	10.000	91.890	3.293
Methoxychlor	20.000	104.355	0.477
p,p'-DDE	0.020	70.775	13.402
PCB 28	1.500	69.283	22.619
PCB 52	2.048	99.953	25.132
PCB 118	0.027	75.892	21.876
PCB 138	0.014	118.874	20.267
PCB 180	0.020	109.254	14.171
Trifluralin	5.000	97.896	0.210

Table 6: Laboratory fortified blank for pesticides and polychlorinated biphenyls.

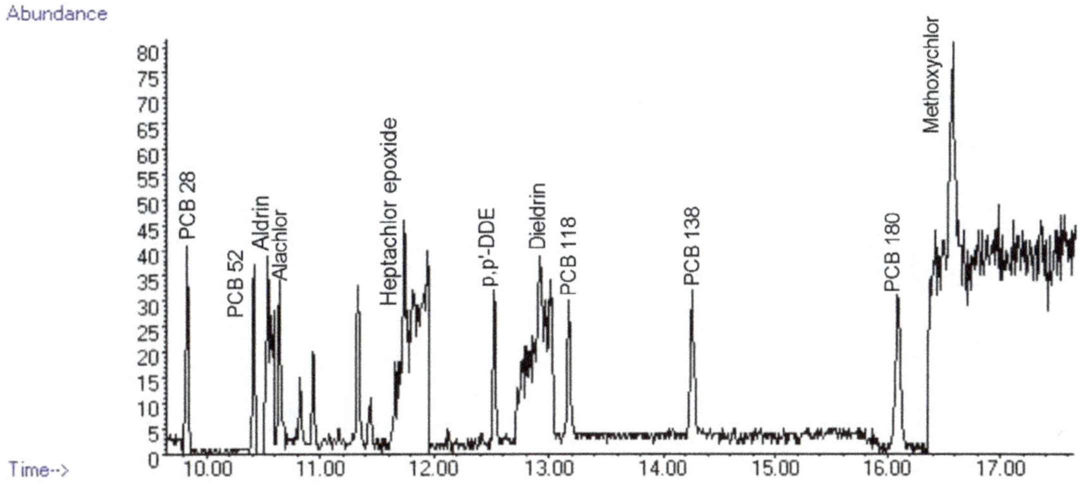

Figure 3: Extract of waste water spiked with 1.060 µg/L Aldrin, 0.044 µg/L of Alachlor, 4.784 µg/L of Heptachlorepoxide, 0.898 µg/L of Dieldrin, 15.556 µg/L Methoxychlor, 0.010 µg/L p,p' - DDE, 0.015 µg/L PCB 28, 0.003 µg/L PCB 52, 0.009 µg/L PCB 118, 0.012 µg/L PCB 138 and 0.038 µg/L PCB 180.

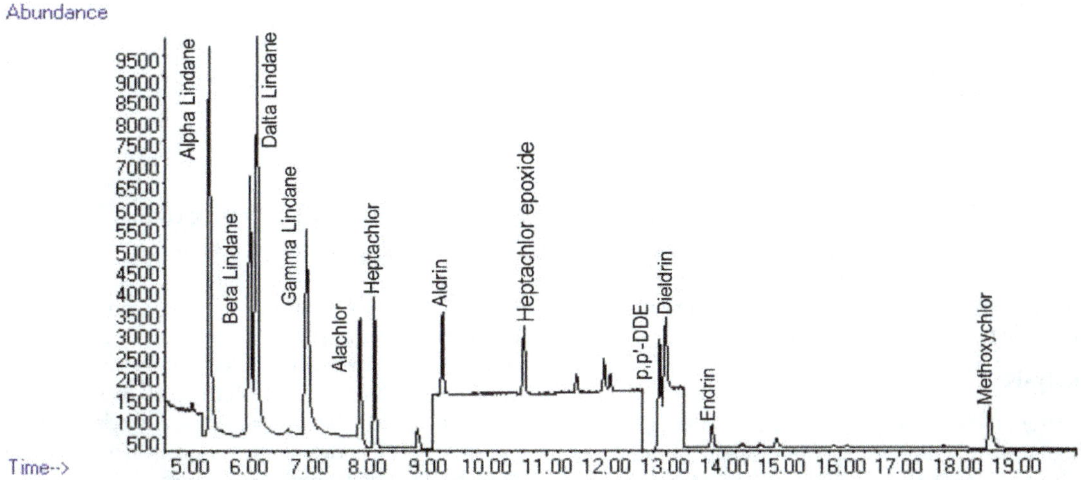

Figure 4: Extraction of Quality Control Sample provided from AccuStandard®, Inc Lot # 209111013- with 1.800 µg/L Aldrin, 3.550 µg/L of Alachlor, 2.175 µg/L of Dieldrin, 47.265 µg/L of Endrin, 10.000 µg/L of Heptachlor and Heptachlorepoxide, 7.518 µg/L Lindane, 23.647 µg/L Methoxychlor and 0.721 µg/L p,p'-DDE.

Analyte	True Concentration, ppb	Observed Concentration, ppb	Accuracy	
			%R	%Bias
Aldrin	2.000	2.218	110.887	10.887
Dieldrin	5.000	3.940	78.802	-21.198
Endrin	7.500	6.900	92.003	-7.997
Heptachlor	7.500	7.222	96.287	-3.713
Heptachlor epoxide	15.000	11.362	75.746	-24.254
Hexachlorobenzene	15.000	13.819	92.125	-7.875
Hexachlorocyclopentadiene	40.000	35.119	87.798	-12.202
Lindane	5.000	4.416	88.327	-11.673
Methoxychlor	20.000	20.871	104.355	4.355
Trifluralin	10.000	9.628	96.277	-3.723

Table 7: Quality control sample for pesticides provided from AbsoluteStandards®, Inc., Lot # 032409.

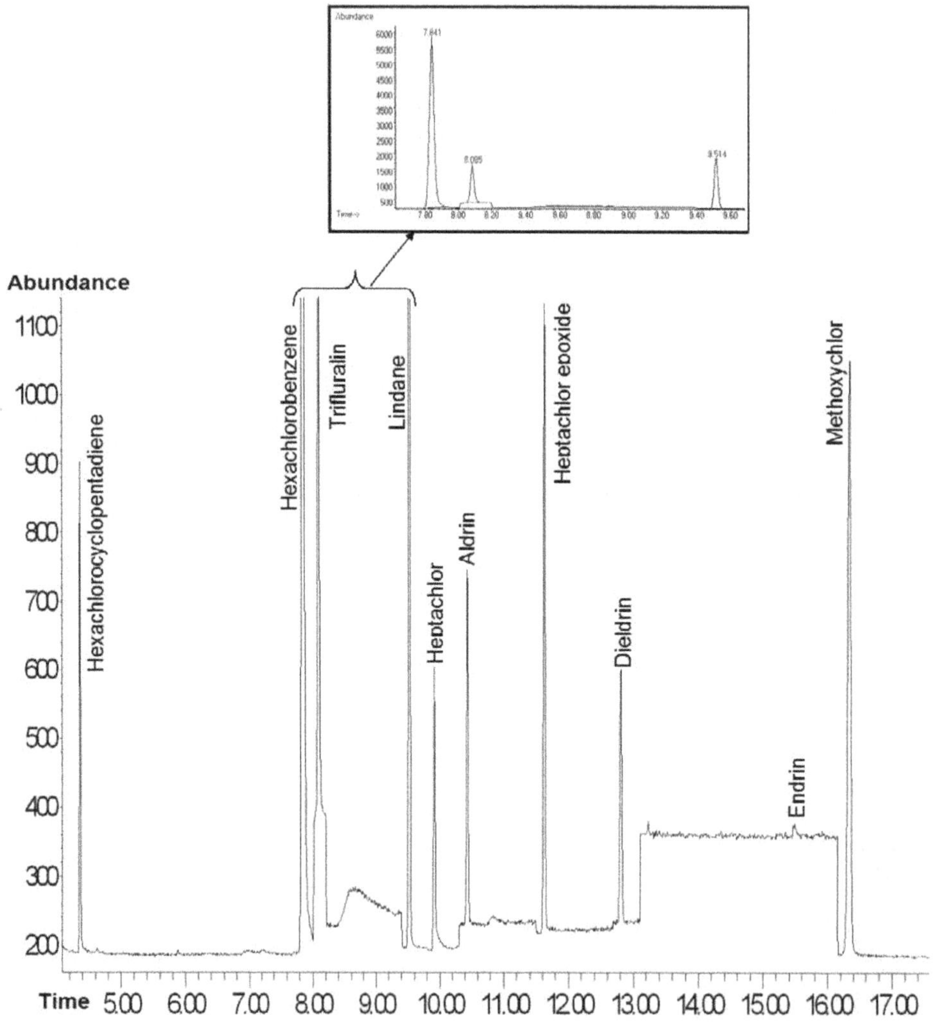

Figure 5: Extract of Practice PT Sample provided from AbsoluteStandard®, Inc, Lot # 091608.

uncertainty data for a range of concentrations if typical uncertainty is to be provided for a wide range of analytes data.

Real samples analysis

Nile River water (Assiut, Egypt), ground water (Assiut), waste water (Zenar, Assiut), tape water from our laboratory and Transformer oils (Cemex, Assiut) were analyzed using the proposed method. Chromatograms are shown in Figure 6 and data are cited in Table 10.

According to the MCL for water provided in Table 1 and the International permissible concentration of PCB's in Transformer oil as cited in the Environmental Protection Agency, EPA, USA regulations that is: >50 ppm= Non-PCB transformer, 50-500 ppm = PCB-contaminated transformer and ≥ 500 ppm= Repeat the reclassification process until the transformer can be classified as to non-PCB or a PCB-contaminated status; or remove the transformer from service, it is clear that for water samples a contamination with Aldrin and Dieldrin is noticeable while some of the transformer oils are to be considered as PCB-contaminated transformers.

Component	Method	Reported value	Assign Value	Accepted limits	
				Low	High
Aldrin	Modified ASTM D 5175-03	1.241	1.75	0.743	2.20
Dieldrin	Modified ASTM D 5175-03	1.330	1.40	0.874	1.85
Endrin	Modified ASTM D 5175-03	2.254	2.30	1.610	2.99
Heptachlor	Modified ASTM D 5175-03	3.644	3.36	1.850	4.87
Heptachlor epoxide	Modified ASTM D 5175-03	5.915	4.75	2.610	6.89
Hexachlorobenzene	Modified ASTM D 5175-03	4.697	3.81	1.750	4.81
Hexachlorocyclopentadiene	Modified ASTM D 5175-03	27.726	22.6	4.280	31.80
Lindane	Modified ASTM D 5175-03	2.070	2.40	1.320	3.48
Methoxychlor	Modified ASTM D 5175-03	19.654	25.0	13.80	36.30
Propachlor	Modified ASTM D 5175-03	---------	2.60	1.510	3.58
Trifluralin	Modified ASTM D 5175-03	2.280	4.00	2.120	5.02

Table 8: Proficiency Testing Evaluation Report for Organochlorine Pesticides mixture in water supply provided from Absolute Standard®, Inc., USA.

Analyte	Conc., ppb	Reagent Water		Tap Water		Waste Water	
		n[A]	u[B]	n[A]	u[B]	n[A]	u[B]
Alachlor	0.089	6	± 0.004	7	± 0.002	7	± 0.007
	6.375	4	± 2.361	4	± 0.213	7	± 0.311
	12.800	7	± 3.311	4	± 0.213	7	± 1.296
Aldrin	0.160	6	± 0.014	5	± 0.008	7	± 0.032
	10.200	4	± 0.588	4	± 1.120	7	± 0.460
	20.480	3	± 0.803	7	± 1.947	7	± 0.865
Dieldrin	0.293	7	± 0.026	7	± 0.007	7	± 0.033
	12.750	4	± 1.190	4	± 4.809	5	± 1.737
	25.600	5	± 2.967	No[C]	No[C]	5	± 3.584
Endrin	2.000	5	± 0.065	No[C]	No[C]	No[C]	No[C]
Heptachlor	5.020	6	± 0.369	No[C]	No[C]	No[C]	No[C]
Heptachlor epoxide	0.280	7	± 0.026	7	± 0.024	7	± 0.039
	12.750	4	± 1.511	7	± 0.553	7	± 1.208
	25.600	5	± 5.871	5	± 1.671	7	± 4.738
Hexachlorobenzene	5.00	5	± 0.041	No[C]	No[C]	No[C]	No[C]
Hexachlorocyclopentadiene	20.060	7	± 0.228	No[C]	No[C]	No[C]	No[C]
Lindane	2.010	7	± 0.051	No[C]	No[C]	No[C]	No[C]
Methoxychlor	15.008	7	± 2.518	7	± 1.518	5	± 1.156
p,p'-DDE	0.020	7	± 0.001	4	± 0.002	7	± 0.003
	14.329	7	± 1.047	7	± 0.703	5	± 0.353
	28.672	5	± 1.397	7	± 5.992	7	± 1.194
PCB 28	0.009	7	± 0.021	7	± 0.001	7	± 0.015
	6.141	7	± 1.015	5	± 0.553	5	± 0.162
	12.288	7	± 3.162	7	± 2.894	5	± 0.541
PCB 52	0.008	6	± 0.003	7	± 0.001	7	± 0.007
	8.188	7	± 0.653	4	± 0.798	7	± 0.239
	16.384	7	± 0.972	5	± 1.144	4	± 0.799
PCB 118	0.027	7	± 0.003	5	± 0.002	7	± 0.008
	6.899	4	± 0.914	7	± 0.415	7	± 0.505
	13.824	5	± 2.015	7	± 2.222	7	± 0.878
PCB 138	0.014	5	± 0.003	5	± 0.003	7	± 0.019
	10.235	5	± 0.114	7	± 0.646	4	± 0.495
	20.480	7	± 0.895	7	± 3.514	7	± 0.564
PCB 180	0.020	6	± 0.003	7	± 0.002	7	± 0.006
	20.470	7	± 2.247	7	± 0.769	5	± 0.466
	40.960	5	± 2.491	7	± 13.262	7	± 1.256
Trifluralin	2.014	7	± 0.061	No[C]	No[C]	No[C]	No[C]

[A] Number of measurements

Table 9: Expanded uncertainty for pesticides and polychlorinated biphenyls in reagent, tap and waste water.

Conclusion

The validation and application of GC MS method in the Selected Ion Monitoring (SIM) mode for the simultaneous determination of the pesticides and PCBs has been evaluated in this study. The optimal conditions of extraction techniques have been obtained. The established method can be applied to determine the concentration of the pesticides in real water samples and transformer oils. The recoveries in water are

Matrices/Analytes	Concentration, µg/L								
	Alachlor	Aldrin	Dieldrin	p,p'-DDE	PCB 28	PCB 52	PCB 118	PCB 138	PCB 180
Waste Water1	---	---	---	---	0.006	0.004	0.002	0.007	---
Waste Water2	---	---	---	0.003	---	0.001	---	---	---
Waste Water3 (Zennar)	---	0.071	---	---	0.038	---	---	---	---
Water Irrigation (Grape Residue)	---	0.453	---	---	---	---	---	---	---
Tap w ater	0.067	---	---	---	---	---	---	---	---
Ground water	0.052	---	---	---	---	---	---	---	---
El ibrahiemia w ater	0.067	---	0.073	---	---	---	---	---	---
Naga Hammady water	0.046	---	---	---	---	---	---	---	---
Nile w ater1	0.055	---	0.294	---	---	---	---	---	---
Nile w ater2	0.035	---		---	---	---	---	---	---
Nile w ater3	0.026	---	---	---	---	---	---	---	---
Concentration, mg/L									
Transformer Oil-1	---	---	---	---	---	53.830	---	---	3.730
Transformer Oil-2	---	---	---	---	---	---	---	---	8.300
Transformer Oil-3	---	---	---	---	---	55.230	---	---	5.100
Transformer Oil-4	---	---	---	---	---	26.900	---	---	---
Transformer Oil-5	---	---	---	---	---	100.900	---	---	---
Transformer Oil-6	---	---	---	---	---	33.130	---	---	14.560
Transformer Oil-7	---	---	---	---	---	---	---	---	17.460
Transformer Oil-8	---	---	---	---	---	54.460	---	---	15.700
Transformer Oil-9	---	---	---	---	---	---	---	---	2.530
Transformer Oil-10	---	---	---	---	---	---	---	---	0.960
Transformer Oil-11	---	---	---	---	---	---	---	---	0.960
Transformer Oil-12	---	---	---	---	---	---	---	---	5.600
Transformer Oil-13	---	---	---	---	---	---	---	---	10.900
Transformer Oil-14	---	---	---	---	---	---	---	---	11.530
Transformer Oil-15	---	---	---	---	---	15.230	---	---	16.600
Transformer Oil-16	---	---	---	---	---	---	---	---	6.030

[1,2,3] are different samples from different areas

Table 10: Application in different Matrices.

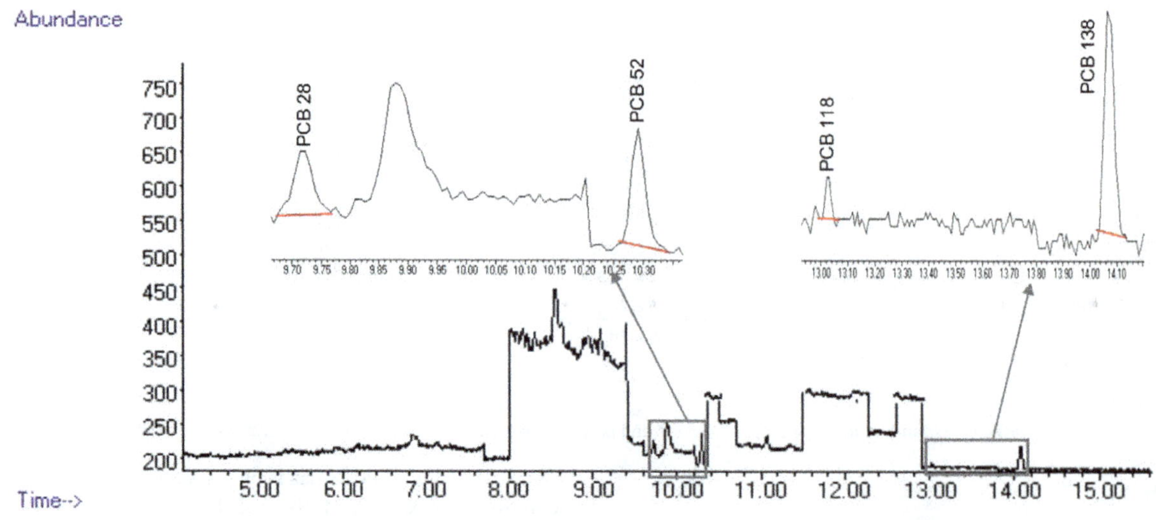

Figure 6: Analysis of real waste water sample.

Analyte	Range, µg/L (proposed method)	Range, µg/L (ASTM method)	MCL, µg/L (EPA/WHO*)
Alachlor	0.050-12.800	0.500-37.500	2.000
Aldrin	0.080-40.000	0.040-1.420	0.03*
Dieldrin	0.040-40.000	0.100-7.500	0.03*
Endrin	2.006-15.045	0.100-7.500	2.00
Heptachlor	0.480-80.000	0.040-1.410	0.40
Heptachlorepoxide	0.040-320.000	0.040-1.420	0.20
Hexachlorobenzene	0.020-321.600	0.010-0.370	1.00
Hexachlorocyclopentadiene	7.523-320.960	----	50.00
Lindane	0.161-20.100	0.040-1.390	0.20
Methoxychlor	10.000-320.000	0.200-15.000	40.00
p,p'-DDE	0.014-28.672	----	1.00*
PCB 28	0.006-12.288	0.500-50.000	0.5
PCB 52	0.008-16.384		
PCB 118	0.027-13.824		
PCB 138	0.010-20.480		
PCB 180	0.020-40.960		
Trifluralin	2.014-20.140	----	20.00*

Table 11: Comparison betw een concentration range in the proposed method and ASTM D5175.

from 70% to 120%. Adequate repeatability, good linearity and the low detection limits prove the capability and credibility for the validation of method by analyzing proficiency testing samples provided from AbsolueStandards®, Inc.

Comparing the data produced from our proposed method , using the universal detector MS in the SIM mode, with those from the ASTM D 5175 method obtained with electron capture detector, Table 11 gave evidence that our proposed method may solve many environmental pollution problems for ultra- trace pollutants since it can reach sub-ppb and ppt levels.

Acknowledgements

The authors would like to thank the Management of Supporting Excellence Unit, Egyptian Ministry of Higher Education for supporting this work through the grant provided (Project CEP1-043-ASSU, 2014).

References

1. Štěpán R, Hajšlová J, Kocourek V, Tichá J (2004) Uncertainties of gas chromatographic measurement of troublesome pesticide residues in apples employing conventional and mass spectrometric detectors. Anal Chim Acta 520: 245- 255.
2. Linsinger TP, Führer M, Kandler W, Schuhmacher R (2001) Determination of measurement uncertainty for the determination of triazines in groundwater from validation data. Analyst 126: 211-216.
3. Moreno Frías M, Garrido Frenich A, Martínez Vidal JL, Mateu Sánchez M, Olea F, et al. (2001) Analyses of lindane, vinclozolin, aldrin, p,p'-DDE, o,p'-DDT and p,p'-DDT in human serum using gas chromatography with electron capture detection and tandem mass spectrometry. J Chromatogr B Biomed Sci Appl 760: 1-15.
4. Martínez Vidal JL, Moreno Frías M, Garrido Frenich A, Olea-Serrano F, Olea N (2002) Determination of endocrine-disrupting pesticides and polychlorinated biphenyls in human serum by GC-ECD and GC-MS-MS and evaluation of contributions to the uncertainty of the results. Anal Bioanal Chem 372: 766-775.
5. Ratola N, Santos L, Herbert P, Alves A (2006) Uncertainty associated to the analysis of organochlorine pesticides in water by solid-phase microextraction/ gas chromatography-electron capture detection--evaluation using two different approaches. Anal Chim Acta 573-574: 202-208.
6. Quintana J, Martí I, Ventura F (2001) Monitoring of pesticides in drinking and related waters in NE Spain with a multiresidue SPE-GC-MS method including an estimation of the uncertainty of the analytical results. J Chromatogr A 938: 3-13.
7. Yenisoy-Karakaş S (2006) Validation and uncertainty assessment of rapid extraction and clean-up methods for the determination of 16 organochlorine pesticide residues in vegetables. Anal Chim Acta 571: 298-307.
8. Picó Y, Rodriguez R, Mañes J (2003) Capillary electrophoresis for the determination of pesticide residues. Trends Anal Chem 22: 133- 151.
9. Luzardo OP, Mahtani V, Troyano JM, Alvarez de la Rosa M, Padilla-Pérez AI, et al. (2009) Determinants of organochlorine levels detectable in the amniotic fluid of women from Tenerife Island (Canary Islands, Spain). Environ Res 109: 607-613.
10. Luzardo OP, Goethals M, Zumbado M, Alvarez-León EE, Cabrera F, et al. (2006) Increasing serum levels of non-DDT-derivative organochlorine pesticides in the younger population of the Canary Islands (Spain). Sci Total Environ 367: 129-138.
11. Haraguchi K, Koizumi A, Inoue K, Harada KH, Hitomi T, et al. (2009) Levels and regional trends of persistent organochlorines and polybrominated diphenyl ethers in Asian breast milk demonstrate POPs signatures unique to individual countries. Environ Int 35: 1072-1079.
12. Wong MH, Leung AO, Chan JK, Choi MP (2005) A review on the usage of POP pesticides in China, with emphasis on DDT loadings in human milk. Chemosphere 60: 740-752.
13. Feeley M, Brouwer A (2000) Health risks to infants from exposure to PCBs, PCDDs and PCDFs. Food Addit Contam 17: 325-333.
14. Ribas-Fito N, Gladen BC, Brock JW, Klebanoff MA, longnecker MP (2006) Prenatal exposure to 1,1-dichloro-2,2-bis (p-chlorophenyl)ethylene (p,p'-DDE) in relation to child growth. Int J Epidemiol 35: 853-856.
15. Boada LD, Lara PC, Alvarez-León EE, Losada A, Zumbado ML, et al. (2007) Serum levels of insulin-like growth factor-I in relation to organochlorine pesticides exposure. Growth Horm IGF Res 17: 506-511.
16. Zumbado M, Goethals M, Alvarez-León EE, Luzardo OP, Cabrera F, et al. (2005) Inadvertent exposure to organochlorine pesticides DDT and derivatives in people from the Canary Islands (Spain). Sci Total Environ 339: 49-62.
17. Hopf NB, Ruder AM, Succop P (2009) Background levels of polychlorinated biphenyls in the U.S. population. Sci Total Environ 407: 6109-6119.
18. http://www.epa.gov/epawaste/hazard/tsd/pcbs/index.htm
19. World Health Organization (2003) Polychlorinated Biphenyls: Human Health Aspects, Concise International Chemical Assessment Document 55.
20. Serôdio P, Nogueira JMF (2004) Multi-residue screening of endocrine disrupters chemicals in water samples by stir bar sorptive extraction-liquid desorption-capillary gas chromatography–mass spectrometry detection. Anal Chim Acta 517: 21-32.
21. Muir D, Sverko E (2006) Analytical methods for PCBs and organochlorine pesticides in environmental monitoring and surveillance: a critical appraisal. Anal Bioanal Chem 386: 769-789.
22. (1991) U. S. Environmental of Organic Compounds in Drinking Water, EPA 600/4-88/039, U. S. Government Printing Office, Washington, DC.
23. (2011) Annual Book of ASTM Standards.

Evaluation of Carbonylated Proteins in Hepatitis C Virus Patients

Mahmoud Mohamed Alou-El-Makarem[1]*, Moussa Madany Moustafa[1], Mohamed Abdel-Aziz Fahmy[1], Aamer Mohamed Abdel-Hamed[1], Khaled Nagy El-fayomy[3] and Medhat Mohamed Abdel-Salam Darwish[2]

[1]Department of Medical Biochemistry, Al Azhar University, Cairo, Egypt
[2]Faculty of Medicine, Al Azhar University, Damietta, Cairo, Egypt
[3]Department of Internal Medicine, Al Azhar University, Damietta, Cairo, Egypt

Abstract

Carbonylated proteins are irreversible posttranslational oxidative modifications, which may interfere with the normal homeostasis of cell growth inducing liver cirrhosis and risk of malignancy.

Objective: To determine plasma levels of carbonylated proteins and total antioxidant capacity and evaluate their role in HCV hepatitis patients and HCV induced liver cirrhosis before and after antiviral therapy; interferon and ribavirin.

Methods: This study included twenty chronic hepatitis C patients with cirrhotic changes, twenty chronic hepatitis C patients without cirrhotic changes and before taking interferon therapy and Fifteen chronic hepatitis C patients without cirrhotic changes and after taking antiviral therapy; interferon (PEG-IFN α2a 180-μg/week and Ribavirin 800 mg capsule one time daily for 24 weeks). Twenty male healthy individuals were included as the control group (age, and body mass index matched). All patients were taking liver support supplements containing vitamins; C, E, folic acid and carotenoids.

Results: There was a highly significant increase (p value 0.00001) in plasma carbonylated protein level in cirrhotic patients (44.9 ± 5.63 nM/dL) as compared to the control group (22.3 ± 3.35 mM/L). TAC in serum in cirrhotic patients was significantly decreased to 0.765 ± 0.249 mM/L as compared to all other groups. These patients were taking antioxidants vitamins (vitamin C, carotenoids and vitamin E), and other supplements known to have antioxidant effects (silimaryn, trace elements), which did not increase their TAC.

Conclusions: Carbonylated proteins may play a role in HCV induced liver cirrhosis. The currently used antioxidants did not increase the antioxidant capacity of plasma. New antioxidants as well as inducers of antioxidant enzymes may be helpful in increasing TAC and prevention of formation of carbonylated proteins and liver cirrhosis.

Keywords: Carbonylated proteins; Total antioxidant capacity; HCV hepatitis; Liver cirrhosis; Interferon; Ribavirin

Introduction

Egypt has a high prevalence of Hepatitis C Virus (HCV), and its complications as liver cirrhosis and hepatocellular carcinoma. Hepatitis C virus infection represents a major health issue worldwide due to its burden of chronic liver disease and extra hepatic manifestations. HCV hepatitis is treated with antiviral therapy, which includes interferon and ribavirin. The therapy usually extends for 24 weeks or more. Yet, in Egypt, liver cirrhosis is still a complication of HCV hepatitis, even in patients who received the antiviral therapy, which may suggest that additional factors contribute for the cirrhosis. HCV variants with reduced susceptibility to interferon can occur naturally, even before treatment begins [1].

HCV causes oxidative stress by a variety of processes, such as activation of prooxidant enzymes, weakening of antioxidant defenses, organelle damage, and metals unbalance. A focal point, in HCV-related oxidative stress onset, is the mitochondrial failure. Mitochondria have a central role in energy production, metabolism, and metals homeostasis, mainly copper and iron. Furthermore, mitochondria are direct viral targets, because many HCV proteins associate with them. They are the main intracellular free radicals producers and targets [2].

Oxidative stress is a condition of oxidant /antioxidant disequilibrium where there is overproduction of Reactive Oxygen Species (ROS) on one side and a deficiency of enzymatic and non-enzymatic antioxidants on the other side. Examples of non-enzymatic antioxidants are; bilirubin, vitamin C, vitamin E, β- carotene, and flavonoids. The resulting oxidative stress damages various cellular component including lipids, proteins and nucleic acids inhibiting their normal function. ROS damage to cellular component plays an important role in numerous diseases including cancer, metabolic syndrome, atherosclerosis, cardiovascular diseases, neurological disorders, Alzheimer disease, diabetes mellitus, aging, autoimmune diseases, and chronic inflammatory diseases [3].

Oxidative stress induces reactions in cellular and blood lipids and proteins. ROS can mediate intra- and intermolecular cross-linking of peptides and proteins and fragmentation of polypeptide chains. In addition, protein carbonyls are formed because of the oxidative modifications of proteins. Carbonyl groups are introduced into proteins by two distinct mechanisms: oxidative (direct) and non-oxidative (indirect). Oxidative mechanisms, which are metal catalyzed, involve the direct reaction of certain reactive oxygen species (e.g., hydrogen peroxide and lipid hydro peroxides) with protein side chains. Non-oxidative carbonylation of proteins involves the reaction of the nucleophilic centers in cysteine, histidine or lysine residues with reactive carbonyls (RCOs). RCOs are carbonyl-containing malondialdehyde, acrolein and carbohydrates (e.g., glyoxal, methylglyoxal) [4].

Damage of proteins by oxidative stress could be involved in inflammation-related carcinogenesis. Detection of elevated levels of protein carbonyls in blood or tissues indicates generally a disease associated dysfunction and defective immunological responses and macrophages functions [5].

***Corresponding author:** Mahmoud Mohamed Alou-El-Makarem, Department of Medical Biochemistry, Al Azhar University, Cairo, Egypt
E-mail: aref48@mail.com

The aim of this work was to determine the plasma level of carbonylated proteins and total antioxidant capacity and evaluate their role in HCV hepatitis patients and HCV induced liver cirrhosis. In addition, the effect of antiviral therapy; interferon and ribavirin on their levels in plasma was evaluated.

Subjects and Methods

Patients and controls

This study included 55 male patients with chronic hepatitis C from the internal medicine department, Damietta, Egypt. The patients were already diagnosed by viral markers.

Twenty male healthy individuals were included as the control group (age, and body mass index matched; Table 1).

Patients were divided into 3 groups:

Group 1: Twenty chronic hepatitis C patients with cirrhotic changes diagnosed by Sonography.

Group 2: Twenty chronic hepatitis C patients without cirrhotic changes and before taking interferon therapy.

Group 3: Fifteen chronic hepatitis C patients without cirrhotic changes and after taking antiviral therapy; interferon (PEG-IFN α2a 180-µg/ week and Ribavirin 800 mg capsule one time daily for 24 weeks).

All patients were taking liver support supplements containing vitamins; C, E, folic acid and carotenoids.

Exclusion criteria

Subjects suffering from any systemic disease or autoimmune disease were excluded from the study. In addition, subjects with positive HBV or have taken HBV vaccine, smokers and obese subjects with BMI >30 kg/M^2 were excluded from the study.

Only patients with HCV hepatitis were included in this study.

Written medical consent was signed by all participants, and the protocol of the work was approved by the Ethical Committee of Al-Azhar University (Table 1).

All the patients were taking orally liver supporting drugs: Silymarin; 400 mg, vitamin C 250 mg, vitamin E; 20 mg, selenium36.6µg, zinc7.3 mg and Glutathione 250 mg, daily.

		Weight (kg)	Height (cm)	BMI (kg/M^2)
Control	Mean ± SD	88.7 ± 7.5	177.85 ± 6.7	28.02 ± 1.8
Group 1	Mean ± SD	84.2 ± 10.7	176.85 ± 7.9	26.81 ± 2.1
Group 2	Mean ± SD	83.8 ± 9.7	178.5 ± 9.8	26.29 ± 2.5
Group 3	Mean ± SD	85.4 ±7.3	182 ± 9.7	25.8 ± 2.4

Table 1: Weight, height and BMI of patients and control group.

		Control	Group 1	Group 2	Group 3
Mean ± SD		3.9 ± 0.38	2.04 ± 0.567	4.04 ± 0.377	3.8 ± 0.376
Control	t-test P-value		12.417 0.0003*	-0.835 0.409	1.133 0.265
HCV before interferon	t-test P-value		-13.1 0.0002*	----------	----------
HCV after interferon	t-test P-value		---------	1.914 0.065	----------

*Highly significant

Table 2: Serum albumin in different groups (g/dL).

Blood samples

Blood samples were obtained from patients and healthy controls after 12 hours of fasting.

Two mL in polypropylene tubes, left to clot for 20 minutes at 37°C, centrifuged at 3000 g for 10 minutes and serum was separated and used for determination of ALT, AST, alkaline phosphatase, and albumin.

Four mL in polypropylene tubes containing sodium citrate as an anticoagulant for determination of carbonylated protein level and TAC in plasma. Blood was centrifuged for 10 minutes at 3000 g immediately after collection and plasma was removed and stored at <-80°C till assay.

Methods

Serum albumin, Alanine Transaminase (ALT), Aspartate Transaminase (AST) and Alkaline Phosphatase (ALP) were assayed on Roche/Hitachi 902 chemistry auto analyzer, by using kits supplied by Roche Diagnostic, Germany [6].

Total plasma antioxidative capacity was determined by the reaction of antioxidants in the sample with a defined amount of exogenously provided hydrogen peroxide (H_2O_2). The antioxidants in the sample eliminate a certain amount of H_2O_2, The residual H_2O_2 was determined calorimetrically by an enzymatic reaction which involves the conversion of 3, 5, dichloro-2-hydroxy benzene sulphonate to a colored product which was measured at 505 nm [7].

Carbonylated proteins in plasma were measured using Cayman's Protein Carbonyl Colorimetric Assay Kit. The kit utilized the DNPH reaction as described by Levine et al. [8]. After incubation of DNPH with samples, trichloroacetic acid 10% (w/v) was added and samples were centrifuged at 11,000 g in a cooling centrifuge. The precipitates were washed with a mixture of ethanol/ethyl acetate, 1:1 (v/v) three times and then dissolved in guanidine hydrochloride 6M (pH 2.3). The absorbance was then measured at 360 nm and the concentration of CO groups was calculated using the molar extinction coefficient (e) of 22,000 cm^{-1} M^{-1}.

Protein concentration of the samples was determined by the Bradford's assay [9] using a microplate reader (Bio-Rad 3550; Bio-Rad, Hercules, CA) at 595 nm and BSA as standard.

Statistical analysis

Results were expressed as Mean ± Standard Deviation (SD). Comparison between groups was done using Student's t test with significance defined as $p \leq 0.05$.

Results

Serum albumin

Serum albumin level was 3.9 ± 0.38 g/dl in the control group. Serum albumin levels were 2.04 ± 0.567, 4.04 ± 0.377 and 3.8 ± 0.376 g/dl in-group 1, 2 and 3 respectively (Table 2). There was a highly significant decrease in albumin level in cirrhotic patients (group 1) compared to all other groups.

Plasma total antioxidant capacity

Total plasma antioxidant capacity levels were 1.55 ± 0.272, 2.04 ± 0.290 and 2.63 ± 0.193 mmol/liter in the control group, group 2 and group 3 respectively. Plasma TAC in chronic HCV hepatitis patients before and after antiviral therapies were therefore, significantly increased. Plasma TAC in group 1 (cirrhotic patients) was significantly decreased to 0.765 ± 0.249 mmol/liter as compared to all other groups (Table 3).

	Control	Group 1	Group 2	Group 3
Mean ± SD	1.55 ± 0.272	0.765 ± 0.249	2.04 ± 0.290	2.63 ± 0.193
Control	t-test P-value	9.545 0.0004*	-5.453 0.002*	-13.067 0.0003*
HCV before interferon	t-test P-value	-14.88 0.0002*	-----------	-----------
HCV after interferon	t-test P-value	----------	-6.861 0.002*	-----------

*Highly significant

Table 3: Total anti-oxidant capacity in plasma (mmol/litre).

	Control	Group 1	Group 2	Group 3
Mean ± SD	22.3 ± 3.35	44.9 ± 5.63	27.22 ± 2.92	24.9 ± 2.83
Control	t-test P-value	-15.414 0.00001*	-4.916 0.003*	-2.43 0.21
HCV before interferon	t-test P-value	12.475 0.0003*	-----------	----------
HCV after interferon	t-test P-value	-----------	2.312 0.027	-------------

*Highly significant

Table 4: Plasma carbonylated proteins in different groups (nM/dL).

		AST U/dl	ALT U/dl	AST/ALT (AAR)	ALP U/L
Control	Mean ± SD	25.8 ± 4.3	25.1 ± 3.7	1.03 ± 0.1	79.7 ± 22.3
Group 1	Mean ± SD	24.8 ± 3.7	25.05 ± 3.1	0.99 ± 0.06	89.45 ± 27.8
Group 2	Mean ± SD	124.5 ± 27.4*	131.35 ± 20.8*	0.97 ± 0.04	218.45 ± 50.3*
Group 3	Mean ± SD	25.8 ± 3.7	27.87 ± 3.8	0.93 ± 0.08	88 ± 22.2

*Statistical significant when compared to the control group (p-value ≤ 0.01)

Table 5: Serum aminotransferases, alkaline phosphatase (ALP) and AST/ALT ratio (AAR) in control group and patients.

Plasma carbonylated proteins

Carbonylated proteins level in plasma of HCV patients with cirrhosis was 44.9 ± 5.63 nM/dL. Carbonylated proteins levels in control group, and group 3 were 22.3 ± 3.35, and 24.9 ± 2.83 mM/L respectively. There was a highly significant increase in carbonylated protein level in cirrhotic patients compared to all other groups. Carbonylated proteins levels in-group 2 was 27.22 ± 2.92 nM/dL, showing a significant increase as compared to control group and group 3 (Table 4).

Serum aminotransferases, alkaline phosphatase and AST/ALT ratio (AAR) in control group and patients

Serum levels of AST, ALT and ALP in HCV hepatitis patients were significantly high (AST; 124.5 IU/dl, ALT; 131.35 IU/dl, ALP; 218.45 ± 50.3 U/L) when compared to the control group (AST; 25.8 ± 4.3 IU/dl, ALT; 25.1 ± 3.7 IU/dl, 79.7 ± 22.3 U/L). The levels of serum aminotransferases and ALP in patients taking interferon and ribavirin therapy, decreased to near control values. Interferon therapy in these patients decreased the inflammatory process in hepatocytes. AAR ratio in cirrhotic patients in this study was 0.99 (24.8/25.05). The ratio was 1.03 ± 0.1, 0.97 ± 0.04 and 0.93 ± 0.08 in control group, group 2 and group 3 respectively. There was no statistical significance between the values of all groups (Table 5).

Discussion

Serum albumin

HCV hepatitis is an acute to chronic inflammation of hepatocytes, which may develop into liver cirrhosis. The levels of serum albumin were significantly decreased in patients with HCV hepatic cirrhosis (2.04 g/dl) compared to the control group (3.94 g/dl). The changes in the level of albumin were not significant in HCV hepatitis patients without cirrhosis. The low level of serum albumin and cirrhotic changes in liver parenchyma may lead to ascites and edema in lower limbs. In humans, albumin synthesis takes place only in the liver. Albumin is not stored by the liver but is secreted into the portal circulation as soon as it is manufactured. Under physiological conditions, albumin may have significant antioxidant potential. This may be related to the abundance of sulfhydryl (-SH) groups on the albumin molecule [10].

Albumin therapy is recommended in refractory ascites not responsive to diuretics. Indications for albumin therapy are linked to the antioxidant activity of albumin. Albumin exchange has emerged as promising liver support therapies for liver failure and other toxic syndromes. They are designed to remove a broad range of blood-borne toxins and to restore normal functions of the circulating albumin by replacing defective forms of albumin and albumin molecules saturated with toxins with normal albumin [11].

Serum aminotransferases and alkaline phosphatase

The most commonly used markers of hepatocyte injury are AST (cytosolic and mitochondrial forms), and ALT (cytosolic). As markers of hepatocellular injury, AST and ALT also lack some specificity because they are found in other tissue. HCV is one of the risk factors for hepatocellular carcinoma. AST/ALT ratio (AAR) may be used as a possible surrogate marker for identifying patients at high risk for developing hepatocellular carcinoma. AAR > 1.4 might be a useful tool to identify candidates at high risk for HCC. This ratio in cirrhotic patients in this study was 0.99 (24.8/25.05). The ratio was 1.03 ± 0.1, 0.99 ± 0.06, 0.97 ± 0.04 and 0.93 ± 0.08 in control group, group 1, group 2 and group 3 respectively. There was no statistical significance between the values of all groups. AAR in all patients was lower than the cut value of suspecting malignancy [12].

Alkaline phosphatase in serum of HCV hepatitis patients without antiviral therapy, and using only hepatic supportive drugs was significantly increased. inflammation of the bile ducts occurs frequently in chronic hepatitis C, and decreases in response to IFN therapy [13].

Carbonylated proteins

Protein carbonyl group can be generated directly by amino acid oxidation or indirectly by forming adduct with lipid peroxidation product. Carbonylation is an irreversible post-translation modification that often leads to loss of protein function. Carbonylated proteins are a stable marker of severe oxidative stress because damage to the protein structure is irreversible and may cause an inhibition of their enzymatic activity or an increased susceptibility to proteolysis [14].

Detection of elevated levels of protein carbonyls in blood or tissues, indicates generally a disease associated dysfunction and defective immunological responses and macrophages functions [15].

The carbonylated modification of protein may lead to proteins that are recognized as non-self by the immune system. The resultant antibody will cross react with normal tissue proteins so initiating autoimmune diseases [16]. Chemicals, which increase ROS production and oxidative stress damage to protein, may lead to dys-morphogenesis and teratogenesis in fetus [17].

ROS interfere with the expression of number of genes, signal transduction pathway, and are thus instrumental in the process of carcinogenesis. The abnormal behaviors of neoplastic cells can be traced to an alteration in cell signaling mechanisms such as cytoplasmic receptor for tyrosine kinas, altered level of growth factor, proteins of transcription apparatus, protein involved in cell cycle. All these protein can be oxidized and carbonylated by ROS. Permanent modification of genetic materials and protein responsible for DNA replication and repair and control of cell cycle by oxidative damage represents the first step in carcinogenesis. The significantly high levels of carbonylated proteins in HCV cirrhotic patients may suggest that they are at risk of hepatocellular carcinoma. These protein oxidations should be treated before they are initiated at the stage of HCV hepatitis [18].

Total anti-oxidant capacity

Assay of Total anti-oxidant capacity measures a complex of non-enzymatic antioxidants present in blood, which include exogenous antioxidants such as ascorbic acid, α tocopherol, β carotene and polyphenols. It also measures the endogenous antioxidants such as reduced glutathione, uric acid, and bilirubin. All patients were taking antioxidants including Silymarin and ascorbic acid, as routine since they were diagnosed as hepatitis patients even before the cause of hepatitis was determined for about three months before enrolled in this study. Silymarin flavonolignans used by patients in this study as polyphenol antioxidant are rapidly absorbed [19]. Silymarin displays anti-inflammatory effects on T lymphocytes in vitro and modest nonspecific immunomodulatory effects in vivo [20]. Ascorbic acid inhibits intracellular ROS generation and reduces the ethanol-induced inflammation in hepatocytes [21].

Taking antioxidants may explain the high level of blood total antioxidants in HCV hepatitis patients as compared to the control group. Supplementation of exogenous antioxidants did not increase antioxidant status in cirrhotic patients. The low levels of TAC in patients with liver cirrhosis may be either due to decrease in the level of plasma albumin or in the synthesis of glutathione and other endogenous antioxidants. In addition, it may be due to increased ROS production in the body.

New antioxidants as well as inducers of antioxidant enzymes; Mn/superoxide dismutase and catalase, may be helpful in increasing TAC and prevention of formation of carbonylated proteins. Drugs, which can increase specifically proteolysis of carbonylated proteins, may be tried in liver cirrhosis patients [22].

Conclusion

Carbonylated proteins are irreversible posttranslational modification, which may interfere with the normal homeostasis of cell growth inducing liver cirrhosis and risk of malignancy [23-25]. There was a highly significant increase (p value 0.00001) in carbonylated protein level in cirrhotic patients (44.9 ± 5.63 n M/dL) as compared to the control group (22.3 ± 3.35 mM/L) [26-28].

TAC in plasma in cirrhotic patients was significantly decreased to 0.765±0.249 mM/L as compared to all other groups [29]. These patients were taking antioxidants vitamins (vitamin C, carotenoids and vitamin E), and other supplements known to have antioxidant effects (silimaryn, trace elements), which did not increase their TAC [30,31].

References

1. Pawlotsky JM (2011) Treatment failure and resistance with direct-acting antiviral drugs against hepatitis C virus. Hepatology 53: 1742-1751.
2. Arciello M, Gori M, Balsano C (2013) Mitochondrial dysfunctions and altered metals homeostasis: new weapons to counteract HCV-related oxidative stress. Oxid Med Cell Longev 2013: 971024.
3. Liochev SI (2013) Reactive oxygen species and the free radical theory of aging. Free Radic Biol Med 60: 1-4.
4. Dkhar P, Sharma R (2011) Amelioration of age-dependent increase in protein carbonyls of cerebral hemispheres of mice by melatonin and ascorbic acid. Neurochem Int 59: 996-1002.
5. Thanan R, Oikawa S, Yongvanit P, Hiraku Y, Ma N, et al. (2012) Inflammation-induced protein carbonylation contributes to poor prognosis for cholangiocarcinoma. Free Radic Biol Med 52: 1465-1472.
6. Withold W, Rick W (1994) Evaluation of an immunoradiometric assay for bone alkaline phosphatase mass concentration in human sera. Eur J Clin Chem Clin Biochem 32: 91-95.
7. Koracevic D, Koracevic G, Djordjevic V, Andrejevic S, Cosic V (2001) Method for the measurement of antioxidant activity in human fluids. J Clin Pathol 54: 356-361.
8. Levine RL, Garland D, Oliver CN, Amici A, Climent I, et al. (1990) Determination of carbonyl content in oxidatively modified proteins. Methods Enzymol 186: 464-478.
9. Bradford MM (1976) A rapid and sensitive method for the quantitation of microgram quantities of protein utilizing the principle of protein-dye binding. Anal Biochem 72: 248-254.
10. Nicholson JP, Wolmarans MR, Park GR (2000) The role of albumin in critical illness. Br J Anaesth 85: 599-610.
11. Rozga J, Piątek T, Małkowski P (2013) Human albumin: old, new, and emerging applications. Ann Transplant 18: 205-217.
12. Tseng PL, Wang JH, Hung CH, Tung HD, Chen TM, et al. (2013) Comparisons of noninvasive indices based on daily practice parameters for predicting liver cirrhosis in chronic hepatitis B and hepatitis C patients in hospital and community populations. Kaohsiung J Med Sci 29: 385-395.
13. Banner BF, Allan C, Smith L, Savas L, Bonkovsky HL (1996) Effect of interferon therapy on bile duct inflammation in hepatitis C. Virchows Arch 428: 253-259.
14. Hopps E, Caimi G (2013) Protein oxidation in metabolic syndrome. Clin Invest Med 36: E1-8.
15. Bozinovski S, Vlahos R, Zhang Y, Lah LC, Seow HJ, et al. (2011) Carbonylation caused by cigarette smoke extract is associated with defective macrophage immunity. Am J Respir Cell Mol Biol 45: 229-236.
16. Madian AG, Regnier FE (2010) Proteomic identification of carbonylated proteins and their oxidation sites. J Proteome Res 9: 3766-3780.
17. Madian AG, Myracle AD, Diaz-Maldonado N, Rochelle NS, Janle EM, et al. (2011) Determining the effects of antioxidants on oxidative stress induced carbonylation of proteins. Anal Chem 83: 9328-9336.
18. Han Y, Chen JZ (2013) Oxidative stress induces mitochondrial DNA damage and cytotoxicity through independent mechanisms in human cancer cells. Biomed Res Int 2013: 825065.
19. Zhu HJ, Brinda BJ, Chavin KD, Bernstein HJ, Patrick KS, et al. (2013) An assessment of pharmacokinetics and antioxidant activity of free silymarin flavonolignans in healthy volunteers: a dose escalation study. Drug Metab Dispos 41: 1679-1685.
20. Adeyemo O, Doi H, Rajender Reddy K, Kaplan DE (2013) Impact of oral silymarin on virus- and non-virus-specific T-cell responses in chronic hepatitis C infection. J Viral Hepat 20: 453-462.
21. Abhilash PA, Harikrishnan R, Indira M (2013) Ascorbic acid is superior to silymarin in the recovery of ethanol-induced inflammatory reactions in hepatocytes of guinea pigs. J Physiol Biochem 69: 785-798.
22. Robbins D, Zhao Y (2011) The role of manganese superoxide dismutase in skin cancer. Enzyme Res 2011: 409295.
23. Bergmeyer HU, Horder M, Rej R (1986) International Federation of Clinical Chemistry (IFCC) Scientific Committee, Analytical Section: approved recommendation (1985) on IFCC methods for the measurement of catalytic concentration of enzymes. Part 3. IFCC method for alanine aminotransferase (L-alanine: 2-oxoglutarate aminotransferase, EC 2.6.1.2). J Clin Chem Clin Biochem 24: 481-495.
24. Choi J (2012) Oxidative stress, endogenous antioxidants, alcohol, and hepatitis C: pathogenic interactions and therapeutic considerations. Free Radic Biol Med 52: 1135-1150.
25. Johnston DE (1999) Special considerations in interpreting liver function tests. Am Fam Physician 59: 2223-2230.
26. Doumas BT, Watson WA, Biggs HG (1971) Albumin standards and the measurement of serum albumin with bromcresol green. Clin Chim Acta 31: 87-96.

27. Hull MC, Morris CG, Pepine CJ, Mendenhall NP (2003) Valvular dysfunction and carotid, subclavian, and coronary artery disease in survivors of hodgkin lymphoma treated with radiation therapy. JAMA 290: 2831-2837.

28. Mirici-Cappa F, Caraceni P, Domenicali M, Gelonesi E, Benazzi B, et al. (2011) How albumin administration for cirrhosis impacts on hospital albumin consumption and expenditure. World J Gastroenterol 17: 3479-3486.

29. Seronello S, Montanez J, Presleigh K, Barlow M, Park SB, et al. (2011) Ethanol and reactive species increase basal sequence heterogeneity of hepatitis C virus and produce variants with reduced susceptibility to antivirals. PLoS One 6: e27436.

30. Suski JM, Lebiedzinska M, Bonora M, Pinton P, Duszynski J, et al. (2012) Relation between mitochondrial membrane potential and ROS formation. Methods Mol Biol 810: 183-205.

31. Wahlefeld AW, Herz G, and Bernt E (1972) Modification of Malloy-Evelyn method for a simple, reliable determination of total bilirubin in serum. Scand J Clin Lab Invest 29:11-12.

Migraine Headaches: Feverfew or Chamomile Leaves?

Snezana Agatonovic-Kustrin[1]*, David Babazadeh Ortakand[2] and David W Morton[2]

[1]Faculty of Pharmacy, Universiti Teknologi MARA (UiTM), Bandar Puncak Alam, Selangor 42300, Malaysia
[2]School of Pharmacy and Applied Science, La Trobe Institute of Molecular Sciences, La Trobe University, Bendigo 3550, Australia

Abstract

The purpose of this study was to compare and analyse active components in feverfew and chamomile using High Performance Thin Layer Chromatography as the analytical method. Both plants belong to the same Asteraceae family and feverfew is sometimes mistaken for German chamomile due to similar flowers. Feverfew leaves have been traditionally used in the treatment of migraine, with Parthenolide regarded as the primary active ingredient. On the other hand, bisabolol and chamazulene have anti-inflammatory properties, and are the main active components in German chamomile essential oil which is obtained by steam distillation of flower heads.

Bisabolol and chamazulene were present in higher concentrations in flowers and in leaves from flowering German chamomile. Parthenolide was present in higher concentration in leaves. Parthenolide and chamazulene are both terpenoids, derived from the same sesquiterpene precursor, farnesyl diphosphate, via two different biosynthetic pathways. Our study of Feverfew and German chamomile suggests that the Parthenolide pathway is favoured in leaves, while formation of matricin and bisabolol is favoured in flowers.

Anti-inflammatory activity of chamazulene and the presence of Parthenolide could explain and justify the use of chamomile in the treatment and prevention of migraine.

Keywords: Feverfew; German chamomile; HPTLC fingerprinting; Parthenolide; Bisabolol; Chamazulene

Introduction

Feverfew, also known as wild chamomile, has been traditionally used in the treatment of headache and migraine. Since clinical trials have confirmed its effectiveness against headaches and migraine [1]. Feverfew is recommended as a migraine prophylactic [2]. The most well-studied and abundant group of active compounds in feverfew are the sesquiterpene lactones produced by superficial leaf glands, with Parthenolide considered to be the main active ingredient [3,4]. Parthenolide can also be found in other Asteraceae species, such as German Chamomile [5].

Due to similar flowers, Feverfew (*Chrysanthemum parthenium/ Tanacetum parthenium L.*) is sometimes mistaken for German Chamomile (*Matricaria Recutita*) [6]. Although German chamomile belongs to the same Asteraceae family as Feverfew, its use in migraine prophylaxis is not evidenced. Pharmacological effect of German Chamomile is mainly associated with its essential oils extracted from the flowers. The main active constituents in essential oil are (-)-α-bisabolol and its oxides, chamazulene [7], hydrophilic polyphenols, flavonoids (e.g., apigenin) [8] and their glucosides [9]. The essential oil of both German and Roman chamomile is a light blue in colour due to the presence of terpenoid chamazulene (5%). Bisabolol comprises 50% of German chamomile's essential oil. Chamazulene and bisabolol are present in higher concentrations in German than in Roman chamomile. Since Parthenolide, alpha-bisabolol, matricin and chamazulene, all belong to sesquterpenes, a class of terpenes that consist of three isoprene units, the aim of this study was to compare High Performance Thin Layer Chromatographic (HPTLC) fingerprints of Feverfew and German chamomile and to quantify chamazulene, bisabolol and Parthenolide in fresh leaf and flower heads. Parthenolide content of dried leaf deteriorates on storage, and many commercial feverfew preparations have been shown to contain Parthenolide in concentrations well below the stated content [10] or even absent in some formulations [11].

Research comparing of Feverfew and German chamomile regarding its biologically active compound is lacking; furthermore evidence of Parthenolide content in German chamomile leaf extracts is limited. The current study aims to compare the levels of bisabolol, chamazulene and Parthenolide in Feverfew and German Camomile extracts, both from flower heads and leaves, and rationalize the use of German chamomile leaves in anti-migraine therapy.

Materials and Methods

Chemicals

German Chamomile, Feverfew and Calendula plants were purchased from the local nursery (Bendigo Wholefoods and Nursery, Bendigo, Australia) and grown in growth chambers for 5 months at a constant temperature of 21°C using a 12 hour day/ night cycle. Standardised supercritical carbon dioxide (CO_2) extract of Feverfew and German Chamomile were donated by FLAVEX Naturextrakte GmbH (Rehlingen, Germany), (97%) Aldrich (Milawauki, USA), Parthenolide were purchased from from Sigma-Aldrich (Munich, Germany), chamazulene and bisabolol from Phytolab (Vestenbergsgreuth, Germany). Glacial acetic acid and concentrated sulphuric acid were obtained from (Merck, Darmstadt, Germany). Anisaldehyde was purchased from ACROS organics (New Jersey, USA). Separations were performed on 20 × 10 cm normal phase Silica gel 60 F254 HPTLC plates. All solvents and chemicals used were of analytical grade. Acetone and acetic acid were purchased from Merck (Kilsyth, Victoria, Australia), hexane from BDH (Poole, England) ethyl acetate from Sigma-Aldrich (Munich, Germany), and Milli-Q (Millipore) water was used.

Standard solutions

0.1 mg/mL standard solutions of bisabolol chamazulene and

***Corresponding author:** Snezana Agatonovic-Kustrin, Faculty of Pharmacy, Universiti Teknologi MARA (UiTM), Bandar Puncak Alam, Selangor 42300, Malaysia, E-mail: snezana@puncakalam.uitm.edu.my

Parthenolide were prepared using absolute ethanol (Thermo Fisher Sci., Scoresby, Australia) and stored at 4°C to prevent degradation.

Extraction

10 g of fresh leaves were picked and frozen. Frozen leaves were macerated into a fine paste in a mortar and pestle, placed into 22 × 80 mm cellulose extraction thimble (Whatman, Little Chalfont, UK), and refluxed in a Soxhlet apparatus for 12 hours using 100 mL of absolute ethanol. The resulting extracted solution was concentrated to approximately 10 mL using a rotary evaporator (Buchi Rotavapor Model R-200), transferred into a 25 mL volumetric flask and adjusted to volume using absolute ethanol. All solutions were stored at 4°C to minimise degradation.

High performance thin layer chromatography

HPTLC plates were pre-washed before use with a blank run of ethanol, dried and activated, by heating in an oven at 110°C for 15 minutes. Samples were sprayed as 8 mm wide bands using a 100 µL HPTLC syringe (Hamilton, Bonaduz, GR, Switzerland) with a semi-automatic sample applicator (Linomat 5, Camag, Muttenz, Switzerland), 8 mm from the lower edge, with 10 mm distance from each side, and a distance of 4 mm between each tracks (15 tracks in total).

Anisaldehyde reagent for post-chromatographic derivatization: The anisaldehyde spray reagent solution was freshly prepared before use by combining anisaldehyde with a refrigerated solution of glacial acetic acid /concentrated sulphuric acid (Merck, Darmstadt, Germany) in the ratio of 0.5:50:1.

HPTLC plate development and visualisation: Chromatographic plates were developed in an Automated Multiple Development Chamber (AMD 2, Camag, Muttenz, Switzerland) using hexane: ethyl acetate: acetic (20:10:1) acid as the mobile phase. Images of the developed plates for both standards and analysed extracts, were recorded using a TLC-visualiser (Camag, Muttenz, Switzerland) equipped with a 12-bit Charged Couple Device (CCD) digital camera and winCATS software (Camag, Muttenz, Switzerland) under UV light at 366 nm and 256 nm, and white light above and below the plate. Developed plates were photographed before and after derivatization by spraying with anisaldehyde solution.

WinCATS image capturing parameters were fixed to ensure high quality images and reproducibility between plates. Quantitative HPTLC analysis was performed using VideoScan Digital Image Evaluation software (Camag, 2003) and set to recognise fluorescent bands.

Method validation

Linearity of the method for quantification of bisabolol, chamazulene and Parthenolide was assessed by plotting chromatographic peak area versus applied amounts of standards in µg over a range of: 0.2-6.0 µg for bisabolol; 0.1-6.0 µg for chamazulene; and 0.1-6.0 µg for Parthenolide.

Regression analysis was performed using the least-squared method. Specificity was assessed by the capacity of the optimised mobile phase to separate applied standards. Repeatability of the method (intra assay precision) was assessed by applying replicates at two different concentrations within the linearity range. Variance between replicates was expressed as a Relative Standard Deviation (%RSD). The detection limit (LOD) was calculated by multiplying the standard deviation of multiple measurements (n=5) by 3 and then dividing by the slope of the calibration curve [12]. The Limit of Quantification (LOQ) [12] was determined from multiple measurements (n=3) of the chromatographic response of a single sample mixture using Equation 1.

$$LOQ = \frac{10 \times Sd}{Slope} \quad (1)$$

Results and Discussion

A simple HPTLC method was developed for simultaneous determination of bisabolol, chamazulene and Parthenolide using a mixture of hexane: ethyl acetate: acetic (20:10:1) acid as the mobile phase. Bands were characterised by their R_f values and colour after derivatization by spraying with anisaldehyde solution. Bisabolol, chamazulene and Parthenolide were observed after derivatization as purple, brown and navy blue coloured bands under visible light (Figure 1) and R_f values of 0.75 ± 0.009, 0.86 ± 0.005 and 0.67 ± 0.005 respectively.

Parthenolide is present in higher concentrations in leaves than in flower heads, while chamazulene and bisabolol are found in larger quantities in flowers. Furthermore, when compared Chamomile extracts, chamazule and bisabolol are found in leaves extracted from flowering German Chamomile plants while their content was bellow limit of detection in leaves obtained from non-flowering German and Roman chamomile plants (Figure 2). Parthenolide is found in leaves only and could not be detected in flower head extracts

Digitized images of plant extracts and standards from the plates were converted to chromatograms before analysis (Figure 3).

Linearity of the method was achieved in the concentration range of 0.5-4.0 µg for bisabolol; 0.6-5.0 µg for chamazulene, and 0.3-2.0 µg for Parthenolide (Figure 4). The correlation coefficients for each analytes were greater than 0.98 (Table 1). Since the slopes were not significantly different from unity, the method did not show proportional error. Small y-intercept values suggested unbiased method.

The sensitivity of the method was determined by calculating LOD and LOQ. The LOD values were 0.30, 0.22 and 0.10 µg, while the LOQ values were found to be 1.00, 0.78 and 0.36 µg for bisabolol, chamazulene and Parthenolide respectively.

Good repeatability of the method was confirmed from the coefficient of variation of five replicates at two different concentrations of standards within their calibration ranges (Table 2) with averaged coefficients of variation of 4.1%, 3.0% and 3.2% for bisabolol, chamazulene and Parthenolide respectively. The relatively low values of %RSD show that the method is precise and can be used to quantify bisabolol, chamazulene and Parthenolide in plant extracts.

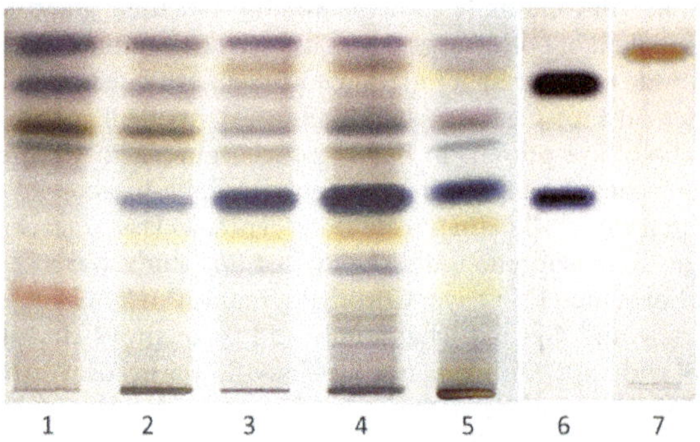

Figure 1: HPTLC fingerprints of chamomile and feverfew extracts and marker compounds. Mobile phase, hexane: ethyl acetate: acetic acid (20:10:1); derivatization with anisaldehyde reagent, white light under (tracks 1-6) and under 254 nm (track 7); from left to right: track 1, Chamomile supercritical flower head extract; track 2, Chamomile soxhlet leaf extract; track 3, Feverfew supercritical plant extract; track 4, Feverfew soxhlet leaf extract; track 5, Feverfew flower extract; track 6, parthenolide and bisablol; track 7, chamazulene;

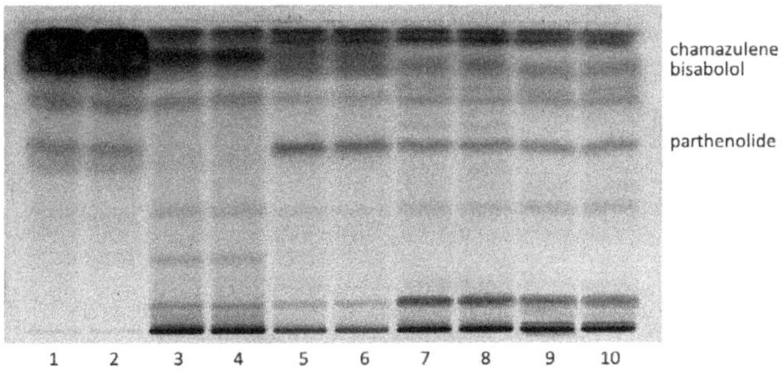

Figure 2: HPTLC fingerprints of chamomile extracts. Mobile phase, hexane: ethyl acetate: acetic acid (20:10:1); derivatization with anisaldehyde reagent, white light under; from left to right: tracks 1 and 2, Chamomile flower heads; tracks 3 and 4 Chamomile supercritical flower head extract; track 5 and 6, German Chamomile soxhlet leaf extract from flowering plant; track 7 and 8, Chamomile Roman Chamomile soxhlet leaf extract; tracks 9 and 10, german Chamomile soxhlet leaf extract.

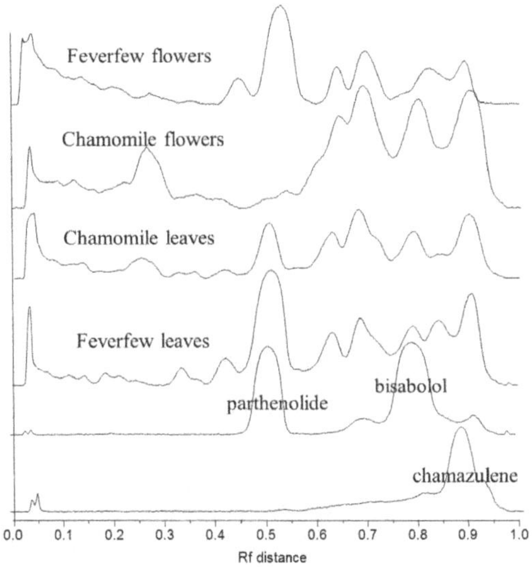

Figure 3: Superimposed chromatograms of feverfew and chamomile extracts with parthenolie, bisbolol and chamazulene chromatograms.

The highest concentration of Parthenolide was found in Feverfew leaves>Feverfew flowers>Chamomile leaves>Chamomile flowers. The highest amount of bisabolol was found in Chamomile flowers>Chamomile leaves>Feverfew flowers≅Feverfew leaves. The highest concentration of chamazulene was found in Chamomile flowers>Chamomile leaves>Feverfew flowers>Feverfew flowers. Furthermore, Parthenolide cannot be quantified in Chamomile flowers.

The average bisabolol, chamazulene and Parthenolide concentrations in chamomile and feverfew extracts are given in Table 2.

Although the flowers are of great botanical importance, they are only a minor source of natural products that are used. Chamomile is among most important examples of plants whose flowers are used. The rational for use of Chamomile flowers in the treatment of headache and migraine can be explained with the antioxidant activity of bisabolol. It has been suggested that the formation of free radicals in the brain and the resultant oxidative damage may be involved in the pathogenesis of migraine [13] so it may be that the beneficial effects of Chamomile is due to the presence of free radical scavenging antioxidant compounds present in flowers.

Parthenolide and chamazulene are sesquiterpenes derived from the same sesquiterpene precursor, farnesyl diphosphate (FPP). The initial folding and cyclisation of FPP to sesquiterpenes is possible via six alternative reaction pathways [14], depending on the enzyme present in the tissue. Such tissue-specific synthesis of terpenoids is a widely reported phenomenon in a range of plant genera [15]. Different reactions lead to the formation of Parthenolide and matricin [16]. Our study suggests that the Parthenolide pathway is favoured in leaves, while formation of matricin (i.e., chamazulene) is favoured in flowers since higher concentrations of Parthenolide were found in leaves and higher concentrations of chamazulene were found in flowers.

Chamazulene is a thermal decomposition product from matricin, responsible for the dark blue coloration of the oil. Matricin becomes inflammatory through conversion to chamazulene [17]. Chamazulene is a natural profen, e.g., Nonsteroidal anti-inflammatory drug, an inhibitor of cyclooxygenase-2(COX-2), but not of cyclooxygenase-

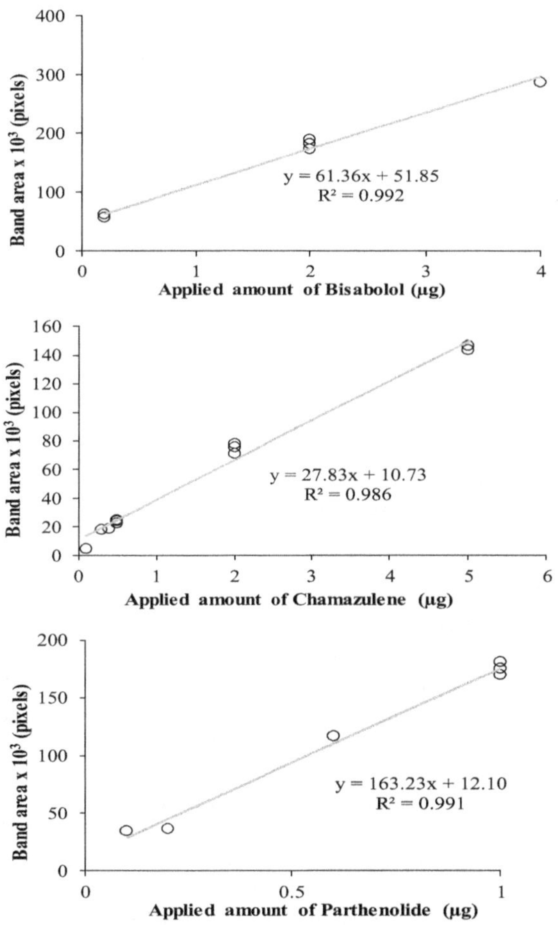

Figure 4: Calibration curves relating standard concentration to peak areas.

Standard	Applied (μg)	Mean theoretical value (μg)	Mean recovery (%)	RSD (%)
Bisabolol	2.00	2.09	104.39	5.30
	5.00	4.19	83.88	3.01
Chamazulene	2.00	2.23	111.43	4.49
	5.00	4.68	93.65	1.51
Parthenolide	1.11	1.10	109.67	5.35
	1.48	1.49	74.70	1.08

Table 1: Accuracy and precision of the method (n=5).

1(COX-1) [17]. Therefore matricin becomes inflammatory through conversion to chamazulene. Selective cyclooxygenase-2 inhibitors have been studied in acute migraine treatment based on the theory of upregulation of COX-2 in the inflammation involved in migraine pathophysiology [18].

The neurotransmitter serotonin (5-hydroxytryptamine [5-HT]) plays a role in the development of migraine attacks [19]. This conclusion is supported by evidence indicating that migraine patients tend to have low levels of serotonin in their brains [20]. Although the exact mechanism that links abnormal 5-HT neurotransmission to the manifestation of head is not fully understood, a deficit on 5-HT descending pain inhibitory system is probably the most implicated in migraine pathophysiology. It has been suggested that Parthenolide may be a low-affinity antagonist at 5HT2A receptors [21]. Drugs that are serotonin receptor antagonists are used in migraine prevention (e.g., methysergide) [22]. *In vitro* studies have shown that Parthenolide, as well as other sesquiterpene lactones, inhibits serotonin release [10,23].

Assuming that molecules with similar geometry will have similar biological activities, we wanted to compare Parthenolide and chamazulene molecules with serotonin molecule. Molecular similarity is a pivotal concept in drug discovery and the cornerstone of Structure-Activity Relationship (SAR) and structural clustering analysis. The rationale behind this approach is that structurally similar molecules are likely to have similar physicochemical properties [24], similar interactions with target receptors or enzymes (Active Analog Principle) [25] and should exhibit the same (or similar) biological activities.

The structure similarity can be calculated from the 2D structure fingerprint and expressed as Tanimoto coefficient [26] or from the 3D shape/pharmacophore superposition. The threshold of the shape similarity is 80%, and of the feature (pharmacophore) similarity is 50%. The most common 2D similarity calculation algorithms utilize molecular fingerprints as the similarity measure and capture the molecular connectivity/substructure/chemical feature information and analytical metrics are used to compare the implicit relationship between the two compounds, with the Tanimoto coefficient [26]. The 3D superposition algorithms are designed to find spatial similarity between molecules. 3D similarity measurements usually involve geometrical information of predefined objectives from the 3D molecular conformations, which include pharmacophores, molecular shapes, and molecular fields. During the 3D similarity calculation, 3D superposition is optimized by molecular shape/feature overlap and represented by the shape/feature similarity score. Superposition of the molecules was done using on line PubChem 3D Viewer (https://pubchem.ncbi.nlm.nih.gov/vw3d/vw3d.cgi) and similarity score was calculated using. The online PubChem services provide fast molecular similarity measure by 3D shape overlap and are expressed as similarity shape score. The target-independent structure of the PubChem database, without any previous information on the physical or biological properties of the compounds, can be used to compare molecules [27]. Highly similar molecules may or may not have highly similar chemical or biological properties. Two molecules that are similar to a third molecule may be similar to each other or may not be similar.

Atoms of the one molecule are matched or assigned to the atoms of the other molecule and structures are superimposed. 3D alignment method requires consideration of the degrees of freedom that are related to the conformational flexibility of the molecules and is aimed to determine the alignment where similarity measure is at a maximum. The 3D superposition algorithms are designed to find spatial similarity between molecules following the paradigm that a necessary condition

	Supercritical (CO_2)		Soxhlet (Ethanol)		
	Chamomile flowers	**Feverfew plant**	**Chamomile leaves**	**Feverfew flowers**	**Feverfew leaves**
Bisabolol	2.15	1.30	0.53	0.47	0.42
Chamazulene	5.35	3.35	1.71	0.66	1.58
Parthenolide	0.49	0.59	0.38	0.33	0.75

Table 2: Average concentrations (n=3) of bisabolol, chamazulene and Parthenolide (µg/µL) found in Chamomile and Feverfew extracts.

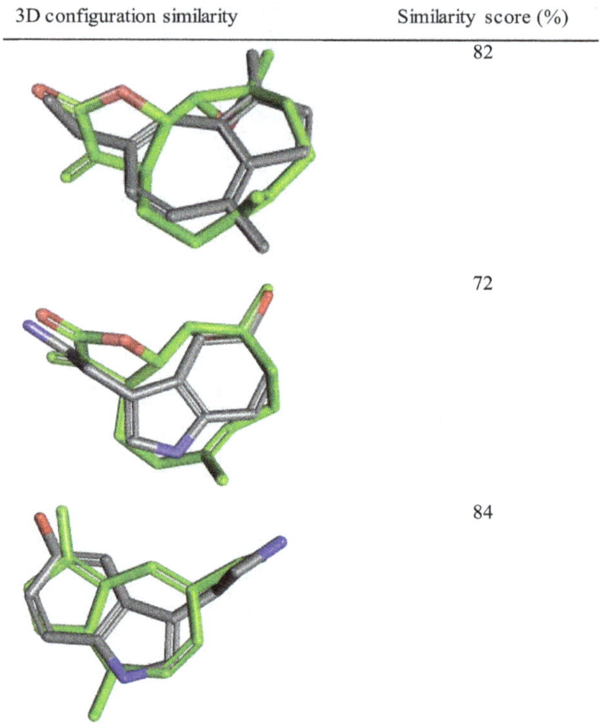

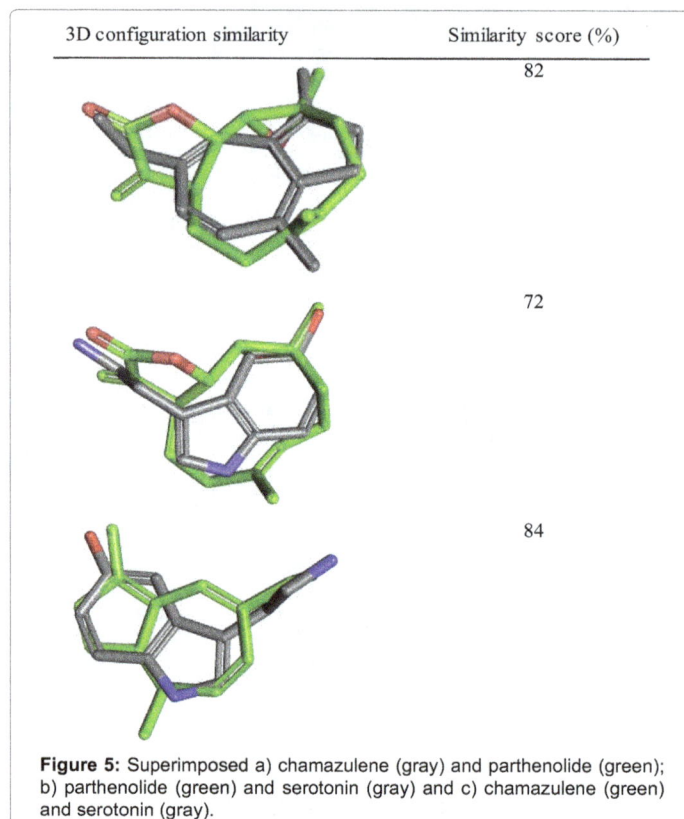

Figure 5: Superimposed a) chamazulene (gray) and parthenolide (green); b) parthenolide (green) and serotonin (gray) and c) chamazulene (green) and serotonin (gray).

for functional similarity is similar molecular geometry. To our surprise chamazulene has shown higher similarity (84%) to serotonin than Parthenolide (72%), while chamazulene and Parthenolide have 82% similarity score to each other (Figure 5). As a rule of the thumb, similarity greater than 80% provides 50% chance that the molecules will have similar activities [28].

Conclusion

Parthenolide is present in higher concentrations in leaves than in flower heads. Chamazulene and bisabolol are mainly found in flowers, with more observed in Chamomile flowers than in Feverfew flowers. Parthenolide and chamazulene are both terpenoids, both derived from the same sesquiterpene precursor farnesyl diphosphate via two possible biosynthetic pathways. Our study suggests that the Parthenolide pathway is favoured in leaves, while formation of matricin and bisabolol is favoured in flowers. Chamazulene is a natural profen (a non-steroidal anti-inflammatory drug). Selective cyclooxygenase-2 inhibitors have been studied in acute migraine treatment based on the theory of upregulation of COX-2 in the inflammation involved in migraine pathophysiology. The anti-inflammatory activity of chamazulene, and the presence of Parthenolide in both chamomile and feverfew, could explain and justify their use in the treatment and prevention of migraine. However, further research into the use of fresh German chamomile leaves in anti-migraine therapy is needed.

References

1. Vogler BK, Pittler MH, Ernst E (1998) Feverfew as a preventive treatment for migraine: a systematic review. Cephalalgia 18: 704-708.
2. Khan SI, Abourashed EA, Khan IA, Walker LA (2003) Transport of parthenolide across human intestinal cells (Caco-2). Planta Med 69: 1009-1012.
3. Piela-Smith TH, Liu X (2001) Feverfew extracts and the sesquiterpene lactone parthenolide inhibit intercellular adhesion molecule-1 expression in human synovial fibroblasts. Cell Immunol 209: 89-96.
4. Rios J, Passe MM (2004) Evidenced-based use of botanicals, minerals, and vitamins in the prophylactic treatment of migraines. J Am Acad Nurse Pract 16: 251-256.
5. Chaves JS, Da Costa FB (2008) A proposal for the quality control of *Tanacetum parthenium* (feverfew) and its hydroalcoholic extract. Revista Brasileira de Farmacognosia 18: 360-366.
6. Smith RM, Burford MD (1993) Comparison of flavanoids in feverfew varieties and related species by principal components analysis. Chemometr Intell Lab Syst 18: 285-291.
7. Pino JA, Bayat F, Marbot R, Aguero J (2002) Essential oil of chamomile *Chamomilla recutita* (L.) Rausch. from Iran. J Essent Oil Res 14: 407-408.
8. Máday E, Szöke E, Muskáth Z, Lemberkovics E (1999) A study of the production of essential oils in chamomile hairy root cultures. Eur J Drug Metab Pharmacokinet 24: 303-308.
9. Mulinacci N, Romani A, Pinelli P, Vincieri FF, Prucher D (2000) Characterization of *Matricaria recutita* L. Flower extracts by HPLC-MS and HPLC-DAD analysis. Chromatographia 5: 301-307.
10. Marles RJ, Kaminski J, Arnason JT, Pazos-Sanou L, Heptinstall S, et al. (1992) A bioassay for inhibition of serotonin release from bovine platelets. J Nat Prod 55: 1044-1056.
11. Groenewegen WA, Heptinstall S (1986) Amounts of feverfew in commercial preparations of the herb. Lancet 1: 44-45.
12. Robinson KA (1986) Chemical Analysis. Little, Brown and Company, Boston.
13. Cordero MD, Cano-García FJ, Alcocer-Gómez E, De Miguel M, Sánchez-Alcázar JA (2012) Oxidative stress correlates with headache symptoms in fibromyalgia: coenzyme Q_{10} effect on clinical improvement. PLoS One 7: e35677.
14. Dickschat JS (2011) Isoprenoids in three-dimensional space: the stereochemistry of terpene biosynthesis. Nat Prod Rep 28: 1917-1936.
15. Croteau R, Johnson MA (1984) Biosynthesis of terpenoids in glandular trichomes. In: Rodriguez E, Healey PL, Mehta L (Eds.), Biology and Chemistry of Plant Trichomes. Plenum Press, New York. pp: 133-185.
16. Gutta P, Tantillo DJ (2006) Theoretical studies on farnesyl cation cyclization: pathways to pentalenene. J Am Chem Soc 128: 6172-6179.
17. Ramadan M, Goeters S, Watzer B, Krause E, Lohmann K, et al. (2006) Chamazulene carboxylic acid and matricin: a natural profen and its natural prodrug, identified through similarity to synthetic drug substances. J Nat Prod 69: 1041-1045.
18. Loo CY, Tan HJ, Teh HS, Raymond AA (2007) Randomised, open label, controlled trial of celecoxib in the treatment of acute migraine. Singapore Med J 48: 834-839.
19. Hamel E (2007) Serotonin and migraine: biology and clinical implications. Cephalalgia 27: 1293-1300.
20. Panconesi A (2008) Serotonin and migraine: a reconsideration of the central theory. J Headache Pain 9: 267-276.
21. Weber JT, O'Connor MF, Hayataka K, Colson N, Medora R, et al. (1997) Activity of Parthenolide at 5HT2A receptors. J Nat Prod 60: 651-653.
22. Murphy JJ, Heptinstall S, Mitchell JR (1988) Randomised double-blind placebo-controlled trial of feverfew in migraine prevention. Lancet 2: 189-192.
23. De Weerdt CJ, Bootsma HP, Hendriks H (1996) Herbal medicines in migraine prevention Randomized double-blind placebo-controlled crossover trial of a feverfew preparation. Phytomedicine 3: 225-230.
24. Raevsky OA, Trepalin SV, Trepalina HP, Gerasimenko VA, Raevskaja OE (2002) SLIPPER-2001 -- software for predicting molecular properties on the basis of physicochemical descriptors and structural similarity. J Chem Inf Comput Sci 42: 540-549.
25. Maggiora GA, Johnson MA (1990) Concepts and applications of molecular similarity. American Chemical Society. Meeting. John Wiley, New York.
26. Baldi P, Nasr R (2010) When is chemical similarity significant? The statistical distribution of chemical similarity scores and its extreme values. J Chem Inf Model 50: 1205-1222.
27. Cincilla G, Thormann M, Pons M (2010) Structuring Chemical Space: Similarity-Based Characterization of the PubChem Database. Molecular Informatics 29: 37-49.
28. Taketo O, Akira K (2011) Metabolomics. Handbook of Molecular and Cellular Methods in Biology and Medicine, Third Edition. CRC Press. pp: 471-484.

18

Verification of Phase Diagrams by Three-Dimension Computer Models

Lutsyk VI[1,2]* and Vorob'eva[1]

[1]*Institute of Physical Materials Science, Siberian Branch of Russian Academy of Sciences, Russia*
[2]*Buryat State University Ulan-Ude, Russia*

Abstract

Present paper is the survey of the works, dedicated to the elimination of contradictions in the publications, which describe the calculated and/or experimental results of investigations of the three-component systems phase diagrams. Special approach to the construction of phase diagrams in the form of their assembling from the surfaces and the phase regions into the three-dimensional (3D) computer model as the effective tool of the detection of the incorrect interpretation of the obtained experiment or of errors in the thermodynamic calculations of the phase diagrams fragments, caused by a deficiency in the initial information, is proposed. 3D computer models of Au-Ge-Sn, Au-Ge-Sb, Ag-Au-Bi, Ag-Sb-Sn, Au-Bi-Sb T-x-y diagram are considered.

Keywords: Phasediagrams; Computer simulation; Three-dimensional Visualization

Introduction

Such known programs as Lukas Program, ThermoCalc, ChemSage, FACTSage, MTDATA, PanEngine, PANDAT are created for the calculations of phase equilibria. Thanks to them it became possible the use of more realistic models of the thermodynamic properties of phases, the calculation of phase diagrams in the complex two-component systems and the systems with the large number of components. The CALPHAD-method, which makes possible to generalize and to refine within the framework information about the phase equilibria and the thermodynamics of phases for one model, is most claimed today. It is effective means for decreasing the volumes of the experiments, necessary for understanding of phase transformations in the alloys and the ceramics. The CALPHAD ideology became the powerful means of theoretical studies and obtaining of adequate information about the phase equilibria. Thermodynamic properties and phase diagrams for the technologically important multi-component materials can be predicted with its use. The reliable thermodynamic descriptions of two-component systems are the basis of the data bases with such characteristics. However, the CALPHAD application is limited by a deficiency in reliable thermodynamic data and the weak possibilities of the visualization of three-dimensional objects. In addition to this, using of thermodynamic methods of the states diagram calculating is hindered by the need of evaluating the thermodynamic properties of phases (in the absence of experimental data) and the agreement of experimental data of phase equilibria with the thermodynamic models. The innovation technology of assembling the space models of multidimensional phase diagrams from the entire totality of the geometric images corresponding to them is proposed: "To decode the diagrams topology the schemes of uni- and invariant states had been elaborated. This sort of schemes with phase's routes designations makes possible to calculate the number of phase regions, surfaces and to know a type of every surface (plane, ruled or unruled surface). Detailed analysis of T-x-y diagrams geometrical constructions had been carried out with their aid, and their computer models had been designed" [1].

3D models of T-x-y diagrams: Approaches, the principles of the construction

Basic principle of the design of the three-dimensional (3D) computer model of the ternary system T-x-y diagram is the assembling of three-dimensional objects of its surfaces and phase regions [2]. The 3D model constructing is fulfilled in several stages: 1) the two-dimensional (2D) table and then the 3D scheme of uni- and invariant states (planes of invariant reactions and ruled surfaces), 2) the prototype (unruled surfaces), 3) transformation of the prototype into the real system model. It should be noted that, as many geometrically simple diagrams are already described in the known monographs (for instance, [3,4], therefore the Reference book of 3D and 4D models of virtual T-x-y and T-x-y-z diagrams of basic topological types is created (look about it, for instance, in [5]. And to construct the real system T-x-y diagram of a simple topology it suffices to take from the Reference book the finished model (or the combination of two-three simple models). The 3D model is obtained after the input of the concrete coordinates of base points (corresponding to invariant transformations in the binary and ternary system) and correction of the curvature of T-x-y diagram lines and surfaces. For instance, the liquidus of the Au-Ge-Sn=A-B-C phase diagram with six binary compounds (incongruently melting R1, R2 of the variable and R4=$AuSn_2$, R5=$AuSn_4$ of the constant concentration (R5 is decomposed below 49.8°C), R3=AuSn is the stoicheometry congruently melting compound, R6=Au_5Sn exists below 179.3°C) is the result of triangulation by the quasi-binary Ge-AuSn section, and each subsystem (Au-Ge-AuSn and AuSn-Ge-Sn) is the double combination of the classical topological type of the T-x-y diagram with a incongruently melting binary compound. But both subsystems are formed by two incongruently compounds (Figure 1). And the sub-solidus is complicated by the allotropy of tin (transformation of 2 modifications at low, ~13°C, temperatures). First step of the 3D model design is the analysis of the T-x-y diagram geometric structure using the uni- and invariant states scheme. This is the usual Sheil' phase reactions scheme [6], the trajectories of a change in the concentrations of the interacting phases in which are written below each three-phase reaction [2] (Tables 1 and 2). The first (with highest temperature) and second (with lowest one) points are accepted as the base points for each trajectory. They obtain designations (for instance, Table 1 and Figure 1): binary eutectic "e" and peritectic "p" with the subscripts e_{AB}, e_{BC}, p_{AR1}, etc; concentrations of appropriate solid phases "A_B" and "B_A" as participants of, for instance, the eutectic reaction L→A+B; liquid concentrations in ternary eutectic "E", quasi-

***Corresponding author:** Lutsyk VI, Institute of Physical Materials Science, Siberian Branch of Russian Academy of Sciences, Russia
E-mail: vluts@ipms.bscnet.ru

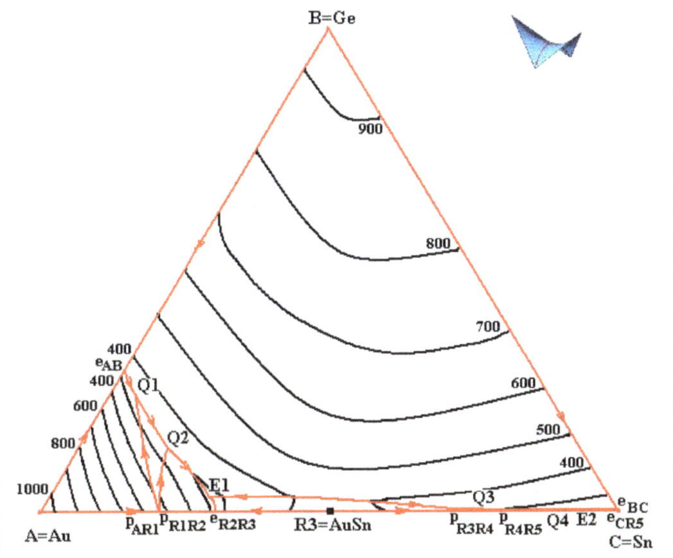

Figure 1: X-Y projection of the Au-Ge-Sn=A-B-C T-x-y diagram 3D computer model.

peritectic "Q", peritectic "P", etc invariant reactions; concentrations of appropriate solid phases "A_{Q1}", "B_{Q1}", "$R1_{Q1}$" (corresponding, for instance, to the reaction Q1: L+A→B+R1). Superscript "0" is assigned to based points of the temperature scale (on the lower face of the trigonal prism of the T-x-y diagram). These trajectories together with the appropriate lines of binary systems form the contours of the diagram surfaces [1]: curves $e_{AB}Q_1$, $p_{AR1}Q_1$, corresponding to the liquid L, belong to inner contours of the liquidus surfaces (liquidus – q_A, q_B, q_{R1}, etc), and appropriate them curves A_BA_{Q1} and $A_{R1}A_{Q1}$ belong to the solidus surfaces (*s*olidus - s_A, s_B, s_{R1}, etc). The binary combinations of such lines are the directing curves for three ruled surfaces (ruled - superscript "r") on the boundaries of the three-phase region. Since these curves belong to the contours of liquidus (q) or solidus (s), than ruled surfaces on the boundaries of the region L+A+B are designated as q^r_{AB}, q^r_{BA}, s^r_{AB}. If both the directing curves of a ruled surface are located on the contour of solidus surface, such ruled surface of "solidus type" is designated by the letter "s". For instance, the solidus type ruled surface s^r_{AB} has directing curves A_BA_{Q1} and $A_{R1}A_{Q1}$, which belong to contours of solidus surfaces s_A and s_B. But if one of the directing curve (e_{AB}Q1) belongs to the liquidus surface (q_A), then the ruled surface is named as "the liquidus type" (q^r_{AB}). As a result, according to the uni- and invariant states schema of the Au-Ge-Sn system (Table 1), one should expect besides six invariant four-phase transformations (two – eutectic E1, E2 and four quasi-peritectic Q_1, Q_2, Q_3, Q_4), indicated in the quasi-peritectictoid Q_5: R_2+R_3→B+R_6 and the eutectoid E_3: R_5→B+C+R_4 interactions, and also the polymorphous transformation E_4: C→C_1+B+R_4 in the sub-solidus. These "derived" from the uni- and invariant states scheme solid-phase invariant transformations Q_5, E_3, E_4 are confirmed by the isopleths [7]. Thus, according to the uni- and invariant states scheme, the Au-Ge-Sn=A-B-C T-x-y diagram consists of 173 surfaces (8 - liquidus and 8 - solidus, 38 - solvus, 2 - transus, 81 ruled surfaces and 9 horizontal (isothermal) planes-complexes, corresponding to invariant reactions in the ternary system, and each of them is divided into 4 simplexes. Surfaces are borders of 65 phase regions (8 two-phase L+I with liquid (I=A, B, C, R1-R5), 10 one-phase (I=A, B, C, C1, R1-R6), 20 two-phase I+J without liquid, 13 three-phase L+I+J with liquid, 14 three-phase I+J+K without liquid). Further the tabular (2D) scheme is transferred into the graphic 3D form. Since the horizontal (isothermal) plane or the complex, which consists of four simplexes, corresponds to each four-phase transformation, then, it is possible to construct, for instance, for the Au-Ge-Sn system the quadrangle $A_{Q1}B_{Q1}Q_1R_1Q_1$ and other 8 complexes of different types for reactions Q_2-Q_5, E_1-E_4 (Table 1). Then it makes possible to draw three lines $e_{AB}Q_1$, A_BA_{Q1}, B_AB_{Q1} as the directing ones and to obtain the prototype of the three-phase region L+A+B. The preliminary contours of other three-phase regions are depicted analogously. The 3D uni- and invariant states scheme is so constructed. Next step is to obtain the prototype of the T-x-y diagram by designing of the liquidus, solidus, solvus, transus surfaces. And last step includes the refinement of the curvature of the directing curves of ruled surfaces, closing the contours of unruled surfaces by curves of binary systems and correcting their isothermal lines. In this stage the 3D model of the real system T-x-y diagram is formed. Finished T-x-y diagram 3D model allows to construct any arbitrarily assigned sections and to calculate mass balances of the coexisting phases in all stages of the crystallization for any arbitrarily assigned concentration [8]. Furthermore, the option of the determination of conditions for changing the type of three-phase reaction in any three-phase region is provided in 3D models [9]. Moreover 3D computer models of phase diagrams are an effective tool for the verification of those experimentally constructed isothermal sections and isopleths for checking the correctness of the interpretation of data, obtained from the experiment and the thermodynamic calculation [10,11]. The quality of each model depends on completeness and authenticity of initial data. But, independent of the initial information, even primitive initial model is capable to carry out such very important function as searching of contradictions in the description of phase diagrams geometric structures or data, obtained from the different publications. Such possibilities of 3D models can be seen on the examples of the using of the metal systems T-x-y diagrams - the bases of the creation of the materials, promising as the lead-free solders. Initial data for these 3D models are taken from the special atlases of phase diagrams for Pb-free solders [7,12]. Their authors carried out for each system thorough selection and agreement of experimental data, which was being accompanied by the necessary thermodynamic calculations in the CALPHAD-technology. The additional experiment was fulfilled in the absence or the doubtfulness of data. Final result for each phase diagram of ternary system was represented in the form of the table of invariant reactions, x-y projection of liquidus and two-three isothermal sections and isopleths. Nevertheless, in spite of so scrupulous selection of represented data, they are not always deprived of contradictions. It should be noted that the 3D computer models construction is accompanied by the re-designation of components and compounds. For instance, Au, Ge, Sn, low-temperature modification of tin, six binary compounds are denoted in the Au-Ge-Sn T-x-y diagram 3D model, correspondingly, as A, B, C, C_1, R_1, R_2, etc. This makes it possible to avoid confusion in the designations, accepted in the CALPHAD, where it is taken into consideration for the designation of phases their crystal structure. For instance, compounds $AuSn_4$ and AuSn are denoted in atlases of phase diagrams for lead-free solders in the volume, devoted to binary systems as AUSN4 и AU1SN, correspondingly, but in other volume of the same series as PTSN4_TYPE and NIAS_TYPE. Both germanium and low-temperature modification of tin are designated by the same symbol DIAMOND_A4 (of their structure prototype).

Au-Ge-Sb: "Template" for the Au-Ge-Sb T-x-y diagram can be taken from the Reference book and, after input the coordinates (concentration-temperature) of base points, after the correction of the curvature of lines and surfaces, to obtain the 3D model of the real Au-Ge-Sb T-x-y diagram with the compound $AuSb_2$ (Figure 2). In this case the comparison of the 3D model isopleths 17 at % Ge (Figure 3b) with that published (Figure 3a) makes it possible to reveal immediately the

Table 1: Uni- and invariant states scheme of the Au-Ge-Sn=A-B-C system.

$A>B>p_{AR1}>p_{R1R2}>e_{BR3}>R3>e_{AB}>Q1>Q2>p_{R3R4}>Q3>e_{R2R3}>E1>p_{R4R5}>>Q4>C>e_{BC}>e_{CR5}>>E2>p^{R6}_{R2R3}>Q5>e^{R5}_{CR4}>E3>C1>e^{c}_{C1R4}>e^{c}_{BC1}>E4$

A-C	A-B-C	B-C	A-B
$L+A \rightarrow R1$			
$p_{AR1}Q1,\ A_{R1}A_{Q1},\ R1_{A}R1_{Q1}$			
$L+R1 \rightarrow R2$			$L \rightarrow A+B$
$p_{R1R2}Q2,\ R1_{R2}R1_{Q2},\ R2_{R1}R2_{Q2}$	$Q1: L+A \rightarrow B+R1$	$L \rightarrow B+R3$	$e_{AB}Q1,$
$A+B+R1$		$e_{BR3}Q3(E1),$	$A_{B}A_{Q1},$
$A_{Q1}A^{0}_{Q1},\ B_{Q1}B^{0}_{Q1},\ R1_{Q1}R1^{0}_{Q1}$		$B_{R3}B_{Q3(E1)},$	$B_{A}B_{Q1}$
$L \rightarrow B+R1$	$Q2: L+R1 \rightarrow B+R2$	$R3_{B}R3_{Q3(E1)}$	
$Q1Q2,\ B_{Q1}B_{Q2},\ R1_{Q1}R1_{Q2}$			
$B+R1+R2$	$L \rightarrow B+R2$		
$B_{Q2}B^{0}_{Q2},\ R1_{Q2}R1^{0}_{Q2},\ R2_{Q2}R2^{0}_{Q2}$	$Q2E1,\ B_{Q2}B_{Q1},\ R2_{Q2}R2_{E2}$		
$L+R3 \rightarrow R4$			
$p_{R3R4}Q3,\ R3_{R4}R3_{Q3},\ R4_{R3}R4_{Q3}$	$Q3: L+R3 \rightarrow B+R4$		
$B+R3+R4$			
$B_{Q3}B^{0}_{Q3},\ R3_{Q3}R3^{0}_{Q3},\ R4_{Q3}R4^{0}_{Q3}$			
$L \rightarrow R2+R3$	$E1: L \rightarrow B+R2+R3$		
$e_{R2R3}E1,\ R2_{R3}R2_{E1},\ R3_{R2}R3_{E1}$	$L \rightarrow B+R4$	$L \rightarrow B+C$	
	$Q3Q4,\ B_{Q3}B_{Q4},\ R4_{Q3}R4_{Q4}$	$e_{BC}E2,$	
$L+R4 \rightarrow R5$		$B_{C}B_{E2},$	
$p_{R4R5}Q4,\ R4_{R5}R4_{Q4},\ R5_{R4}R5_{Q4}$	$B+R2+R3$	$C_{B}C_{E2}$	
	$B_{E1}B_{Q5},\ R2_{E1}R2_{Q5},\ R3_{E1}R3_{Q5}$		
$L \rightarrow B+R5$	$Q4: L+R4 \rightarrow B+R5$		
$Q4E2,\ B_{Q4}B_{E2},\ R5_{Q4}R5_{E2}$			
$L \rightarrow C+R5$			
$e_{CR5}E2,\ C_{R5}C_{E2},\ R5_{C}R5_{E2}$	$E2: L \rightarrow B+C+R5$		
$R2+R3 \rightarrow R6$			
$R2_{R6}R2_{Q5},\ R3_{R6}R3_{Q5},\ p^{R6}_{R2R3}Q5$	$B+C+R5$	$B+R4+R5$	
	$B_{E2}B_{E3},\ C_{E2}C_{E3},\ R5_{E2}R5_{E3}$	$B_{Q4}B_{E3},$	
$Q5: R2+R3 \rightarrow B+R6$		$R4_{Q4}R4_{E3},$	
$R5 \rightarrow C+R4$	$B+R3+R6$	$R5_{Q4}R5_{E3}$	
$e^{R5}_{CR4}E3,\ C_{R5}C_{E3},\ R4_{R5}R4_{E3}$	$B_{Q5}B^{0}_{Q5},\ R3_{Q5}R3^{0}_{Q5},\ R6_{Q5}R6^{0}_{Q5}$		
$B+R2+R6$	$E3: R5 \rightarrow B+C+R4$		
$B_{Q5}B^{0}_{Q5},\ R2_{Q5}R2^{0}_{Q5},\ R6_{Q5}R6^{0}_{Q5}$			
$C \rightarrow C1+R4$	$B+C+R4$	$C \rightarrow B+C1$	
$e^{c}_{C1R4}E4,\ C1_{R4}C1_{E4},\ R4_{C1}R4_{E4}$	$B_{E3}B_{E4},\ C_{E3}C_{E4},\ R4_{E3}R4_{E4}$	$e^{c}_{BC1}E4,$	
$B+C1+R4$	$E4: C \rightarrow B+C1+R4$	$B_{C1}B_{E4},$	
$B_{E4}B^{0}_{E4},\ C1_{E4}C1^{0}_{E4},\ R4_{E4}R4^{0}_{E4}$		$C1_{B}C1_{E4}$	

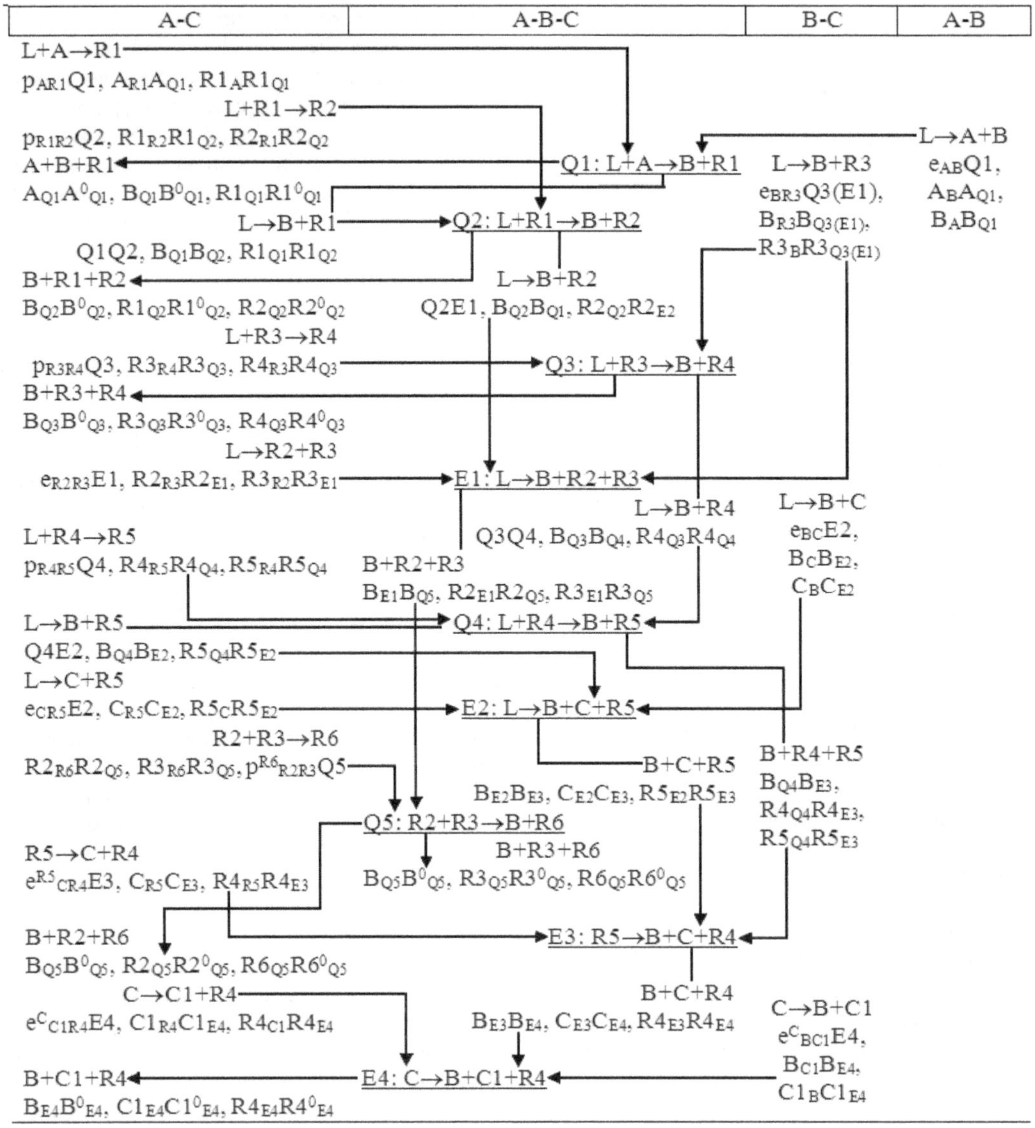

error of the latter: the intersection of the liquidus curve and the ruled surface section on the border of the regions L+Ge and L+Ge+Sb. It is worthwhile to note that the analogous section, parallel to side Au-Sb, but at other distance from Ge (not 17 at %, but 15 at %), in other paper, devoted to this system [13], is also constructed. The sections of liquidus and ruled surface are passed closely, but they do not intersect each other in this section.

Ag-Au-Bi: Despite the fact that the information, placed into the atlas, was thoroughly checked (according to its authors), the description of the Ag-Au-Bi system is contradictory: in the binary system Au-Bi, cited according to the data [14-19], the compound Au₂Bi decomposes at the temperature 110˚C, whereas the solid solution on its basis in the ternary system isopleths exists also at 0˚C (Figures 4c and 4d). To understand these contradictions, it is necessary to discuss the papers [14-18], the information from which in the form of an isothermal section and four isopleths were included into the atlas. The Ag-Au-Bi=A-B-C T-x-y diagram has a simple geometric construction. It is formed by binary systems: with continuous series of solid solutions (Ag-Au=A-B), an eutectic (Ag-Bi=A-C), and an eutectic-peritectic (Au-

Bi=B-C) with the binary incongruently melting compound Au₂Bi=R. Liquidus consists of three fields of primary crystallization of: bismuth, solid solution based on the binary compound Au₂Bi=R and the solid solution Ag(Au)=A(B). Liquidus surfaces intersect in the curves, connecting the binary eutectics e_{AC} and e_{CR} and the peritectic p_{BR} (Figure 4) with the point Q, corresponding to the invariant quasi-peritectic reaction L+Ag(Au)→Bi+Au₂Bi or L+A(B)→C+R. The compound Au₂Bi exists up to 50˚C according to the 2005 year paper [14] (Figure 5a). This compound decomposes at 110˚C (Figure 5b) according to the 2006 year work [16], and these data, as the most reliable, were chosen for the atlas [12]. However, the solid solution based on Au₂Bi occurs in isopleths of below by temperatures from 450K [17] and to 0˚C [12], in the phase regions with the lowest temperatures. But if this compound decomposes in the binary system Au-Bi at 110˚C, then the solid solution on its basis must indeed decompose, also, in the Ag-Au-Bi ternary system. This problem may be resolved by means of the 3D model too. In order to demonstrate this contradiction graphically, two versions of the 3D computer model of the T-x-y diagram were constructed: in one version the compound Au₂Bi decomposes in the

Table 2: Uni- and invariant states scheme of the Au-Bi-Sb=A-B-C system. $A>C>p_{CR2}>p_{AR1}>e_{AR2}>Q>B>e_{BR1}>E>Y$

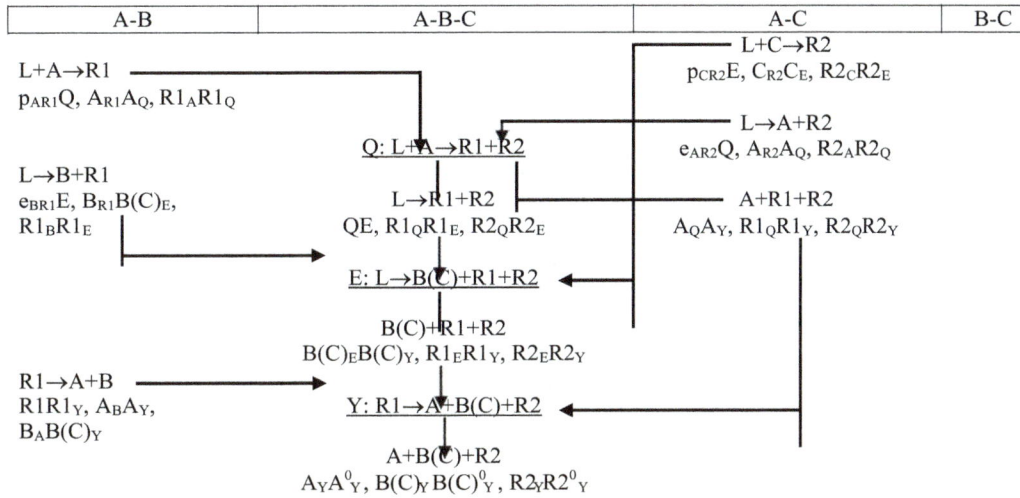

A-B	A-B-C	A-C	B-C
		L+C→R2 $p_{CR2}E$, $C_{R2}C_E$, $R2_CR2_E$	
L+A→R1 $p_{AR1}Q$, $A_{R1}A_Q$, $R1_AR1_Q$			
	Q: L+A→R1+R2	L→A+R2 $e_{AR2}Q$, $A_{R2}A_Q$, $R2_AR2_Q$	
L→B+R1 $e_{BR1}E$, $B_{R1}B(C)_E$, $R1_BR1_E$	L→R1+R2 QE, $R1_QR1_E$, $R2_QR2_E$	A+R1+R2 A_QA_Y, $R1_QR1_Y$, $R2_QR2_Y$	
	E: L→B(C)+R1+R2		
	B(C)+R1+R2 $B(C)_EB(C)_Y$, $R1_ER1_Y$, $R2_ER2_Y$		
R1→A+B $R1R1_Y$, A_BA_Y, $B_AB(C)_Y$	Y: R1→A+B(C)+R2		
	A+B(C)+R2 $A_YA^0_Y$, $B(C)_YB(C)^0_Y$, $R2_YR2^0_Y$		

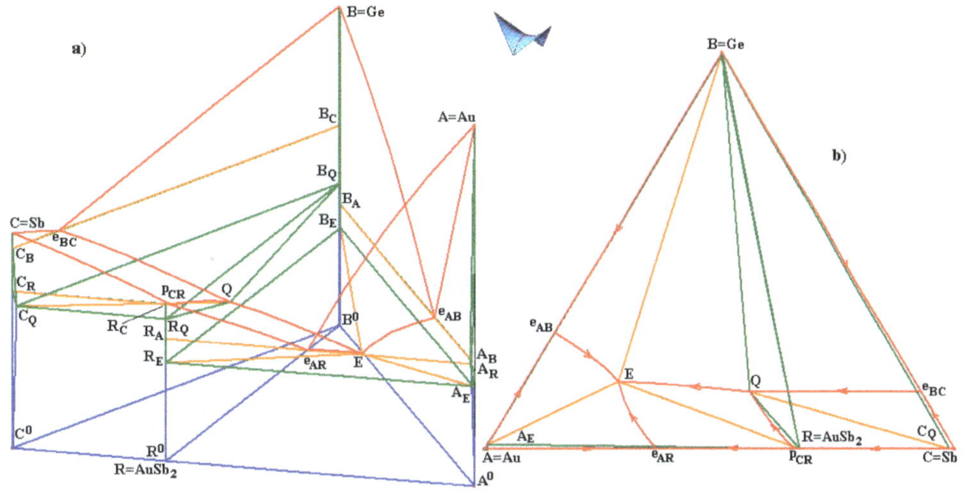

Figure 2: The Au-Ge-Sb=A-B-C T-x-y diagram 3D computer model (a), its x-y projection (b).

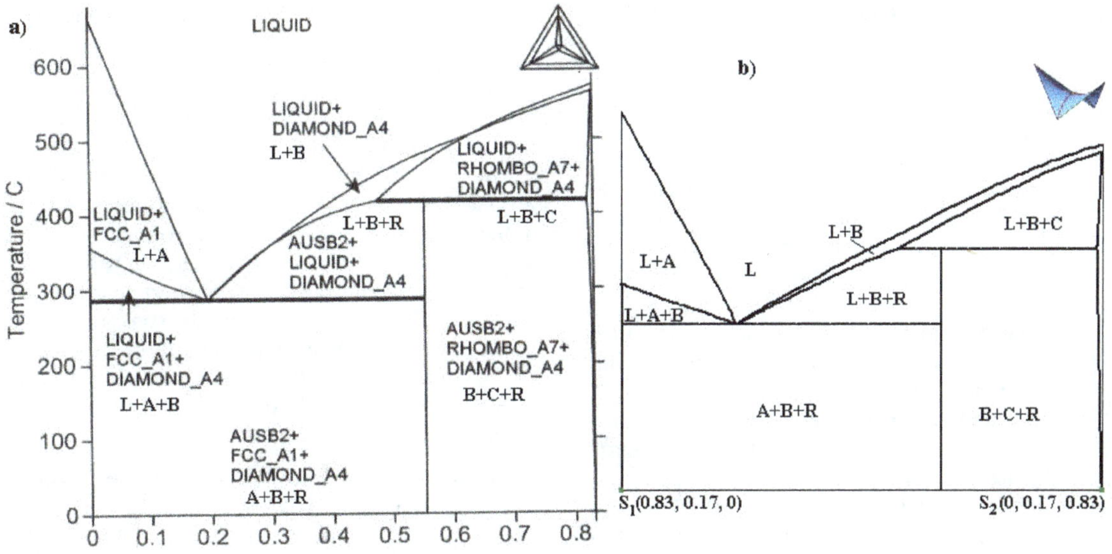

Figure 3: Isopleth 17 at% Ge of the Au-Ge-Sb system

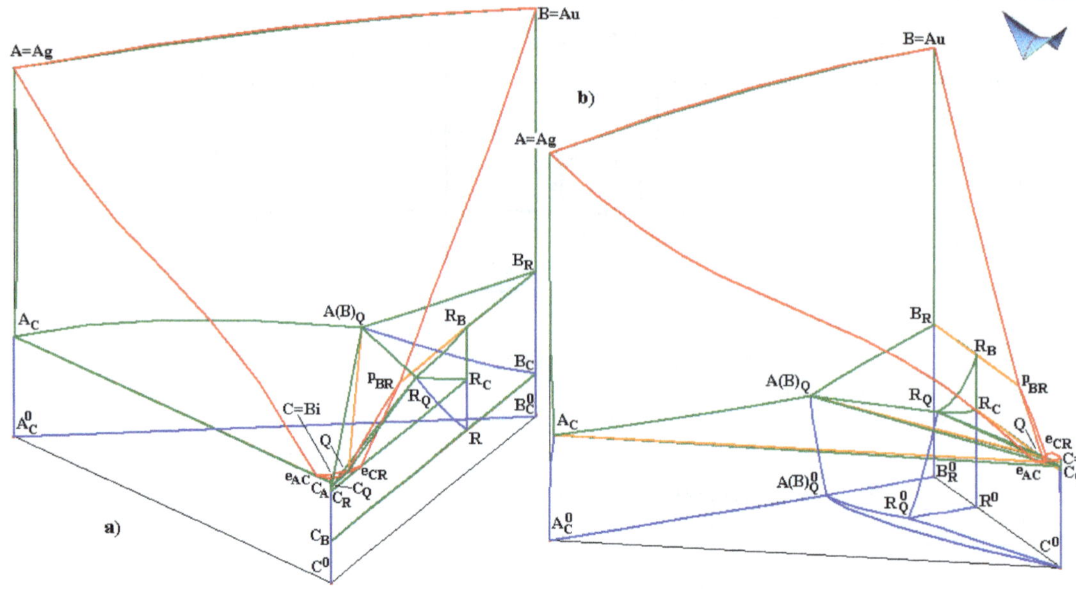

Figure 4: Designed by the data of [12] 3D computer model of the Ag-Au-Bi=A-B-C T-x-y diagram: with (a) and without (b) decomposition of the solid solution based on the compound R=Au_2Bi.

binary system at 110°C and, correspondingly, the solid solution on its basis in the ternary system does not exist below by this temperature too (Figure 4a), in other version a decomposition is absent (Figure 4b). Both versions of the T-x-y diagram differ in the solvus surfaces and the boundaries of the three-phase region A(B)+C+R. If the solid solution on the basis of the compound R decomposes, then the process R→A(B)+C is closed within the three-phase region with boundaries given by the directing curves R_QR, $A(B)_QB_R$, C_QC_R (Figure 4a). If the solid solution based on the compound R in the ternary system does not decompose, then the region A(B)+C+R below the horizontal plane Q is bounded by three ruled surfaces with the directing curves $R_QR^0_Q$, $A(B)_QA(B)^0_Q$, C_QC^0 (Figure 4b). Both versions of the T-x-y diagram consists of 28 surfaces (3 liquidus and 3 solidus, 6 solvus, 12 ruled surfaces and 4 horizontal planes, corresponding to simplexes of the complex of the invariant transformation Q: L+A(B)→C+R) and 14 phase regions. Border solution on the compound R base has the linear region of the homogeneity. Bordering it surfaces of two solvus $v_{R\_A(B)}$ (R_BRR_Q or $R_BR_QR^0_QR^0_B$), v_{RC} (R_CRR_Q or $R_CR_QR^0_QR^0_C$) and solidus s_R ($R_BR_QR_C$) have a triangular shape (Figure 4b). A conclusion about the temperature boundaries of the compound Au_2Bi existence cannot be made from the phase reactions scheme of [14]. It corresponds to the uni- and invariant states scheme, which does not consider its decomposition. Note also that phase reactions are written in the scheme of [14] incorrectly: the peritectic reaction L+Au→Au_2Bi and the eutectic one L→Au_2Bi+Bi are changed by the places. Moreover the eutectic reaction is written as L→(Ag)+(Bi). Isopleths 20 at% Ag (Figure 5a-c), 50 at% Bi, 85 at% Bi, Ag:Au=1:4 and the isothermal section T=230°C are shown in papers [12,17]. There is no contradictions of the 3D model and the published sections, including the isothermal section at $T<T_Q$ in both versions and the high-temperature part of the T-x-y diagram (higher than reaction Q at T_Q=251.9°C). If the compound Au_2Bi decomposes, the versions of the 20 at % Ag isopleth, published in [12] (Figure 4c) and [17,18] (Figure 5d-f), are reproduced by the 3D model, where the section S_1S_2 of the three-phase region, designated in [12] as FCC1_A1+RHOMBO_A7+AU2BI_C15, is bounded by curves 6-8 and 7-9 (Figure 5e). However, in the case of the based on R solid solution decomposition the vertical plane S_1S_2 intersects the ruled surfaces q^{rR}_{BC} and s^{rR}_{BC} (to notate the boundaries of the three-phase region with a phase reaction R→A(B)+C without liquid, the superscript "R" is added) through curves 7-8 and 6-8. The last are the boundaries of the three-phase region Ag(Au)+Bi+Au_2Bi=A(B)+C+R (Figure 5f). Since the three-phase region Ag(Au)+Bi+Au_2Bi in the published in [11,17] isopleths exists up to 0°C, then it is assumed that the temperature of the Au_2Bi compound decomposition after addition of the third component (Ag) reduces due to the formation of the solid solution Ag(Au). However, the participation of the compound in the invariant reaction L+Ag(Au)→Bi+Au_2Bi at 251.9°C and without the decomposition of the compound Au_2Bi→Ag(Au)+Bi at 110°C suggests that the three-phase region Ag(Au)+Bi+Au_2Bi exists only within the temperature interval 251.9-110°C. The two-phase region Ag(Au)+Bi is lower than this region under the ruled surface $A(B)_QC_QC_BB_C$ in all concentration diapason. Because at ultralow temperatures, according to the third law of thermodynamics, the continuous solid solution Ag(Au) should undergo decomposition into components [20,21] this two-phase region should be replaced by the three-phase region Ag+Au+Bi.

Ag-Sb-Sn: Analogous contradictions are in the description of the Ag-Sb-Sn T-x-y diagram. From one side, the binary compound Sb_2Sn_3 in the Ag-Sb-Sn T-x-y diagram, represented in [22-24], exists at temperatures up to 0°C. With another side, this compound decomposes at 242.4°C in [12]. Its existence is limited by the same temperature in the analogous system Ni-Sb-Sn too. Consequently, the conditions for the compound Sb_2Sn_3 existence require additional experimental study. But now the 3D model is designed in two versions. The first one corresponds to [22-24] with the polymorphous transformations in the eutectoid reaction, and the 3D model consists of 99 surfaces and 62 phase regions. In the different version, constructed according to the data of [12], one additional eutectoid reaction precedes the polymorphous transformation, and the 3D model consists of 109 surfaces and 66 phase regions.

Au-Bi-Sb: The results of experimental study and thermodynamic correlation using the CALPHAD-technology of the Au-Bi-Sb=A-B-C system with the compounds Au_2Bi=R_1 and $AuSb_2$=R_2, are presented in [25,26]. Earlier six isopleths had been shown in [27], which then were also constructed in [26]. The authors of atlas [12] after the analysis of

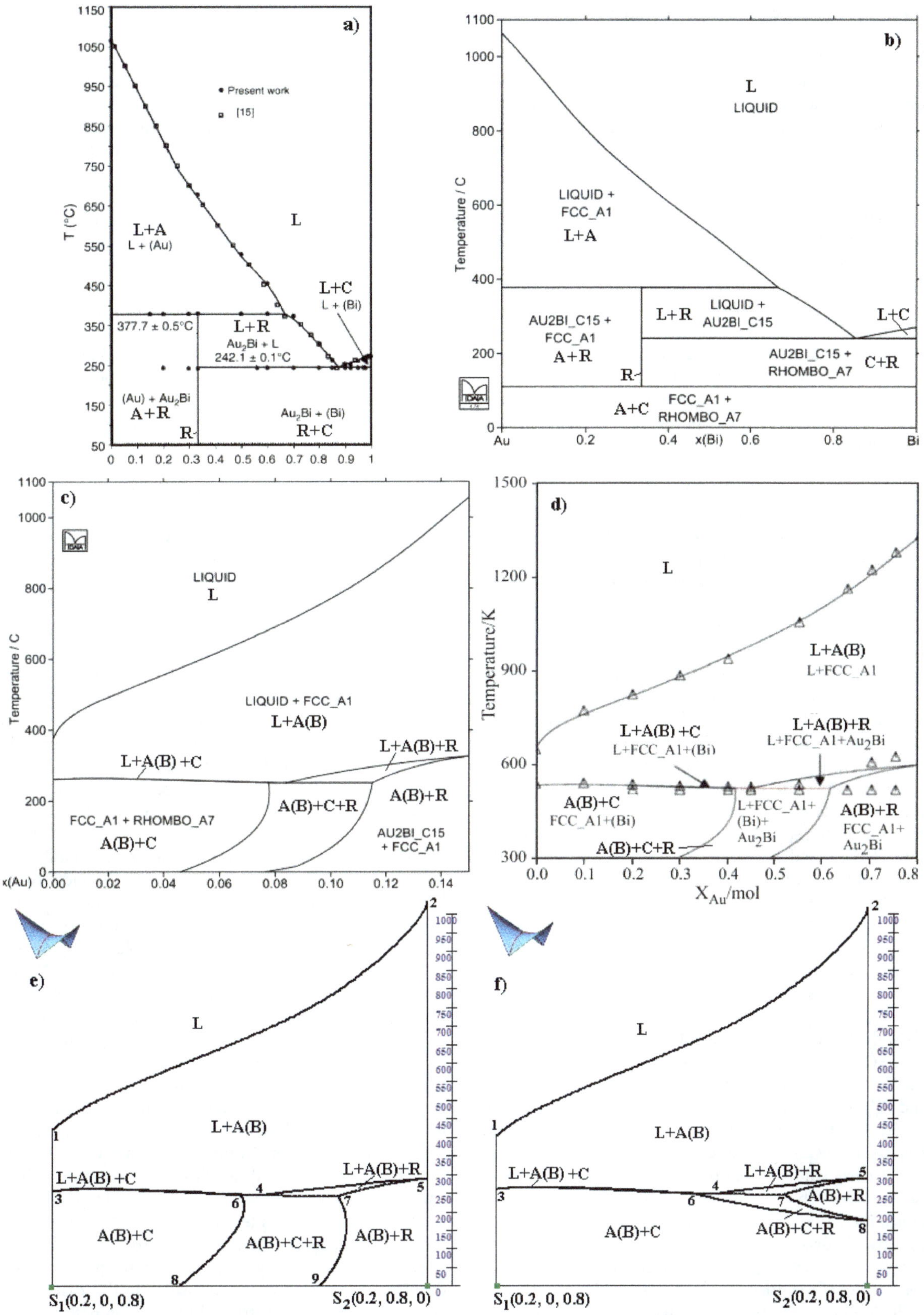

Figure 5: Binary system Au-Bi according to the data of [12] (a) and [17] (b): isopleth $S_1(0.2, 0.8, 0)$-$S_2(0.2, 0, 0.8)$ of the Ag-Au-Bi; [12] (c): [17] (d): 3D model, designed according to the data of [12] with (e) and without (f) decomposition of the solid solution based on the compound R=Au_2Bi.

these publications preferred the results of the data of [26], but correcting and obtaining "the better agreement in comparison with the experimental results". The region of the homogeneity of the solid solution Bi(Sb)=B(C) is shown in [26] in the isothermal section 300°C. Analogous section is used for the illustration of the phase diagrams calculation results by the method of the convex hulls [28]. However, the binary system Bi-Sb adjoins not with the solid solution B(C) region, but with the two-phase region Bi(Sb)+$AuSb_2$=B(C)+R_2, containing the compound with the third component (gold), in the same section, published in the atlas and in [25]. So, versions of the Au-Bi-Sb T-x-y diagram 3D computer model were also used for explaining this contradiction in iso- and the polythermal sections, constructed by the different authors. Each version is designed according to the data of concrete publication and reproduces all given there sections and projections. This makes it possible to compare the sections of diagrams, obtained from different works. According to the atlas data, two invariant reactions – quasi-peritectic Q: L+A→R_1+R_2 at 296°C and eutectic E: L→B(C)+R_1+R_2 at 239.2°C with participation of the solid solution of bismuth with antimony B(C) takes a place in the Au-Bi-Sb system. However, the Au_2Bi=R_1 compound decomposes in the binary system at 110°C and it is absent in the atlas sections at temperatures lower than 110°C. Consequently, it is necessary to add into the phase reactions scheme the invariant decomposition Y: R_1→A+B(C)+R_2 (Table 2). Formal enumeration of surfaces and phase regions gives according to the uni- and invariant states scheme (Table 2) 4 liquidus and 4 solidus surfaces, 12 surfaces of solvus, the cupola of the disintegration of the solid solution B(C), 3 complexes of the invariant reactions (Q at 296°C, E at 239.2°C, Y at ~110°C), divided each into 4 horizontal triangular simplexes, 15 ruled surfaces or borders of 5 three-phase regions with liquid and 12 ruled surfaces as borders of 4 three-phase regions without liquid. Concentration coordinates (0, 0.948, 0.052) of the vertex $B(C)_E$ of the complex, which corresponds to the eutectic reaction in the ternary system, are indicated in the atlas [12], as if this point belongs to the Bi-Sb=B-C binary system. Since this is impossible, the point $B(C)_E$ was displaced in the computer model inside the prism to position (0.010, 0.948, 0.042). Because of this the solidus $s_{B(C)}$ surface ceased to belong to edge B-C. Surfaces of solidus (s_{R1}) and solvus (v_{R1A}, $v_{R1\_B(C)}$, v_{R1R2}), corresponding to the stoichiometric compound R1, which does not exist below 110°C, practically coincide with the vertical line R1 within the temperature interval 377.5-110°C, where 377.5°C is the temperature of the peritectic p_{AR1}. Border solution based on the compound R_2 has the linear homogeneity region. The part of the solidus s_{R2} ($R_{2A}R_{2Q}R_{2E}R_{2C}$) corresponding to it takes the triangular shape $R_{2C}R_{2A}R_{2Q}$, and the line $R_{2Q}R_{2E}$ coincides with the line $R_{2C}R_{2E}$ (Figure 6a). Analogously the surfaces of solvus v_{R2A} ($R_{2A}R_{2Q}R_{2Y}R_2^0$), $v_{R2\_B(C)}$ ($R_{2C}R_{2E}R_{2Y}R_2^0$), v_{R2R1} ($R_{2Q}R_{2E}R_{2Y}$) have in the same plane the configurations with the triangular fragments v_{R2A} ($R_{2A}R_{2Q}R_{2Y}$), $v_{R2\_B(C)}$ ($R_{2C}R_{2E}R_{2Y}$), v_{R2R1} ($R_{2Q}R_{2E}R_{2Y}$), corresponding to the solidus s_{R2}. But since in the Au-Sb=A-C system the compound R_2 is stoichiometric, and points $R_2{}^{\circ}{}_A$ and $R_2{}^{\circ}{}_C$, practically coincide with the vertical line in the point R_2, they are supplemented in the ternary system by the point, denoted in the uni- and invariant states scheme (Table 2) as $R_2{}^0{}_Y$, they all obtains the same designation R_2°, corresponding to the prism base (Figure 6a). So, the Au-Bi-Sb T-x-y diagram consists of 60 surfaces and 24 phase regions, including six surfaces (s_{R1}, $v_{AB(C)}$, v_{AR1}, v_{R1A}, $v_{R1\_B(C)}$, v_{R1R2}), which practically degenerated in the vertical lines A and R_1 and four surfaces ($s^r{}_{AR1}$, $q^{rR1}{}_{A\_B(C)}$, $v^r{}_{AR1\_(Q)}$, $v^r{}_{AR2\_(Y)}$) practically coincided with the A-B and A-C edges of the prism. Since this 3D model corresponds to the basic principles of geometric thermodynamics (the phase rule and the law of adjoining phase regions), then it makes it possible to explain the reasons for contradictions in the publications of different authors. The curve of the solid solution Bi(Sb) disintegration on Bi and Sb is depicted at temperatures lower than 200°C in the published in [12] the Bi-Sb binary system phase diagram (Figure 7). This curve obviously assigns the solid solution disintegration cupola in the Au-Bi-Sb ternary system. Track of a section can be seen in the atlas [12] in the 20 at % Bi isopleth (Figure 8a). However the two-phase region Bi+Sb is denoted in this section as the three-phase : Bi+Sb+$AuSb_2$=B+C+R2 (or RHOMBO_A7+RHOMBO_A7+AUSB2 in the designations of [12]). The two-phase region Bi(Sb)+$AuSb_2$=B(C)+R2 (RHOMBO_A7+AUSB2) is denoted above it. Thus, the phase regions, which contain the compound with gold, adjoin with the binary system, formed by bismuth and antimony. Perhaps, the authors [12] were mistaken in the designations of these phase regions, and the compound $AuSb_2$=R_2 should be removed, after renaming the two-phase region to the single-phase (the solid solution of bismuth with antimony), and the three-phase to the two-phase (disintegration of this solid solution). But the single-phase region RHOMBO_A7 would then be adjacent to three-phase regions Bi(Sb)+Au_2Bi+$AuSb_2$=B(C)+R1+R2(RHOMBO_A7+AU2BI_C15+AUSB2) and L+Bi(Sb)+$AuSb_2$=L+B(C)+R2 (L+RHOMBO_A7+AUSB2is not designed in [12], violating the law of adjoining phase regions. As was mentioned above, authors of [12] preferred the results of the work [26]. It is necessary to note that the binary system Bi-Sb is examined in [26] at temperatures above 500 K – above the binodal curve and the temperature 400 K is the minimum in all isopleths. The disintegration cupola of the solid solution Bi(Sb) is absent in the sections, published in [26], and it is possible only to assume that it would appear at temperatures below 400 K. However, there is no contradiction in the designations of phase regions, because the single-phase region of the solid solution Bi(Sb) adjoins directly to the Bi-Sb system as, for instance, in the 20 at % Bi isopleth (Figure 8c). On the other hand, the binodal curve assigns the region of the disintegration of the solid solution in the Bi-Sb binary system, and it can be seen the track of its section not only in isopleths in the atlas [12] (Figure 8a), but also in the paper [25], moreover with the same contradictory designations of phase regions near the Bi-Sb system. To explain the reasons of the contradiction in determining phase regions and to understand, whether the authors considered the decomposition of the solid solution Bi(Sb)=B(C) in the binary system Bi-Sb, on the one hand in the publications [12,25], and in [26,27] on the other hand, three versions of the Au-Bi-Sb T-x-y diagram 3D model were designed according to the data of [12] (Figure 6), [25] and [26]. In contrast to [12], the compounds Au_2Bi=R1 and $AuSb_2$=R_2 are treated as stoichiometric in [25], and there is no disintegration of the solid solution Bi(Sb) in [26]. If "no", then isopleths in [26] (Figure 8c) and [27] are accurate. And they, like the corresponding 3D model isopleth (Figure 8d), depicts the boundaries of the solid solution Bi(Sb)=B(C) region as traces of sections of the surfaces of solidus and solvus (curves 5-6 and 6-18). They are very close to the temperature axis and are hard to see, but they are present in the sections. If "yes", and the solid solution decomposes in the system Bi-Sb, then a curve corresponds to it in the T-x diagram (Figure 7). It generates the disintegration cupola in the ternary system, which is present in the isopleths of [12] (Figure 8a) and [25]. In that case errors in the designations of phase regions in these papers are connected with the fact that the sections of the solidus $s_{B(C)}$ and solvus $v_{B(C)\_}R_2$ surfaces - curves 5-6, 6-20 (Figure 8b) are lost in the appropriate sections of the 3D models. At the same time, it should be noted that the curve of solvus (the curve 6-18 in Figure 8d) of the model section cannot be very close to the temperature axis, when the disintegration cupola is present, which in this case must be assumed, also, in the section, constructed in [26] Figure 8c. They should move away from the axis in order "to give a place" for the trace of the section of the disintegration cupola, as in Figure 8b. Thus, the final version of the T-x-y diagram 3D computer model consists of 60 surfaces and 24

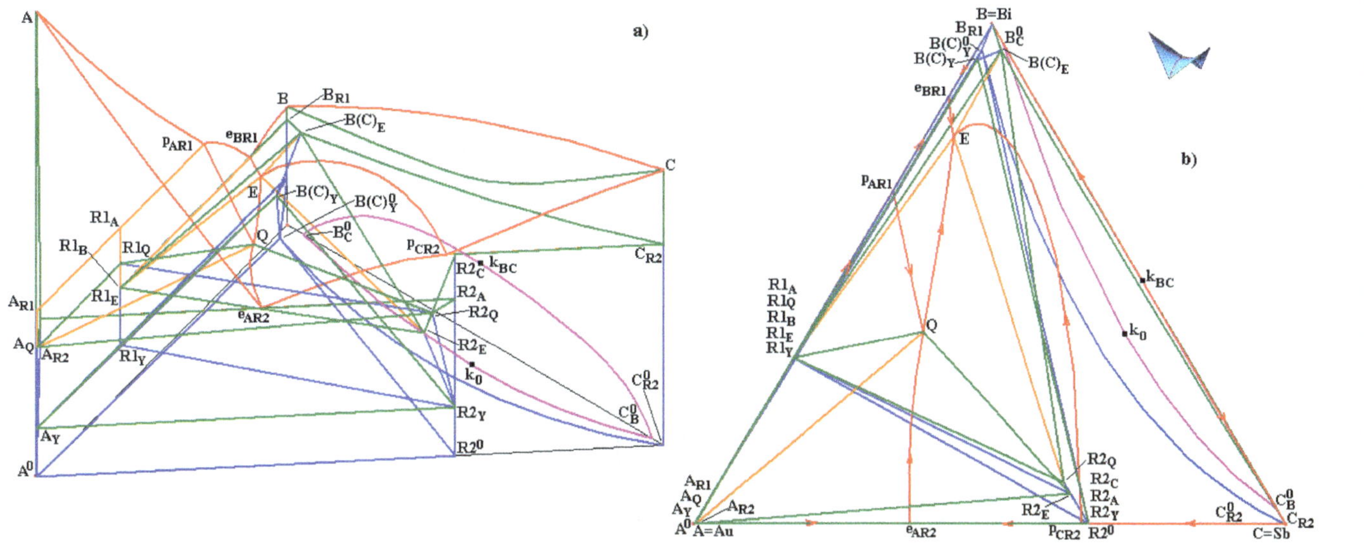

Figure 6: 3D computer model of the Au-Bi-Sb=A-B-C T-x-y diagram, designed according to data [12], but with the added surfaces of solidus $s_{B(C)}$ and solvus $v_{B(C)\_R2}$ (a); x-y projection (b).

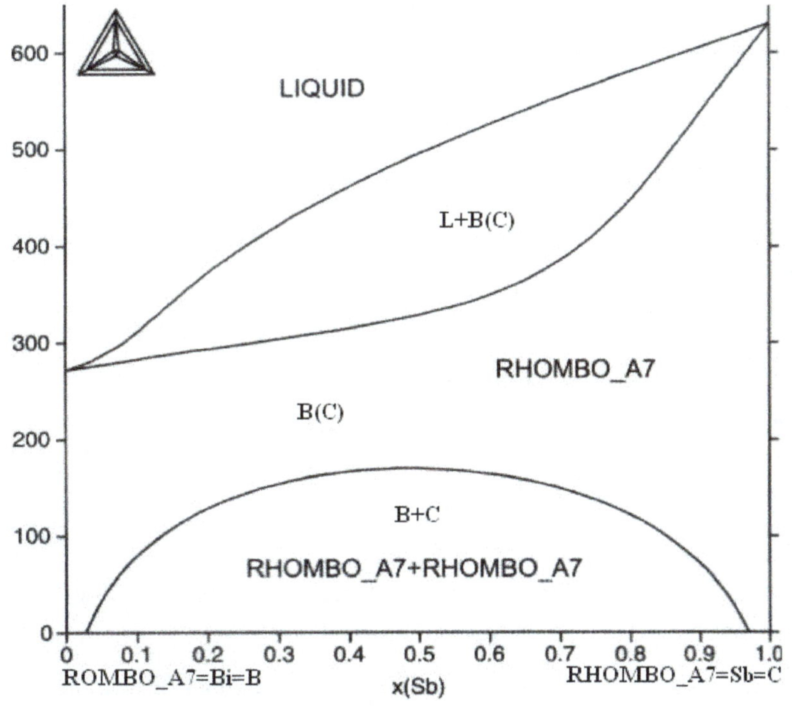

Figure 7: Binary system Bi-Sb [12] (the axis of temperature is marked in °C).

phase regions and reproduced the data of [12], but with the added surfaces of solidus $s_{B(C)}$ and solvus $v_{B(C)\_}R_2$. Traces of these surfaces are shown by dashed lines on the 20 at% Bi isopleth (Figure 8b).

Conclusion

It is convenient to use the 3D computer models of T-x-y diagrams, designed according to the data of the different authors, for the agreement of the sections and for searching of contradictions in calculations or incorrect interpretation of experiment. Despite the fact that the binary system Au-Bi is well studied [14,16,19] and it shown in the publications [16,19] that the solid solution on basis of the compound Au_2Bi does not exist in the ternary system at temperatures below 110°C, the analysis of isopleths, carried out with the aid of two versions of the Au-Ag-Bi T-x-y diagram 3D computer model, constructed according to the data [12,17], showed that the decomposition this solid solution was not considered in [12] and [17]. Therefore the low-temperature part of these sections must be corrected. The correct form of section is shown in the Figure 5f. In order to be confident in the correctness of the Au-Sb-Sn system description in [22-24], it is necessary to explain, actually the compound Sb_2Sn_3 is decomposed at 242.2°C as in [12], or the

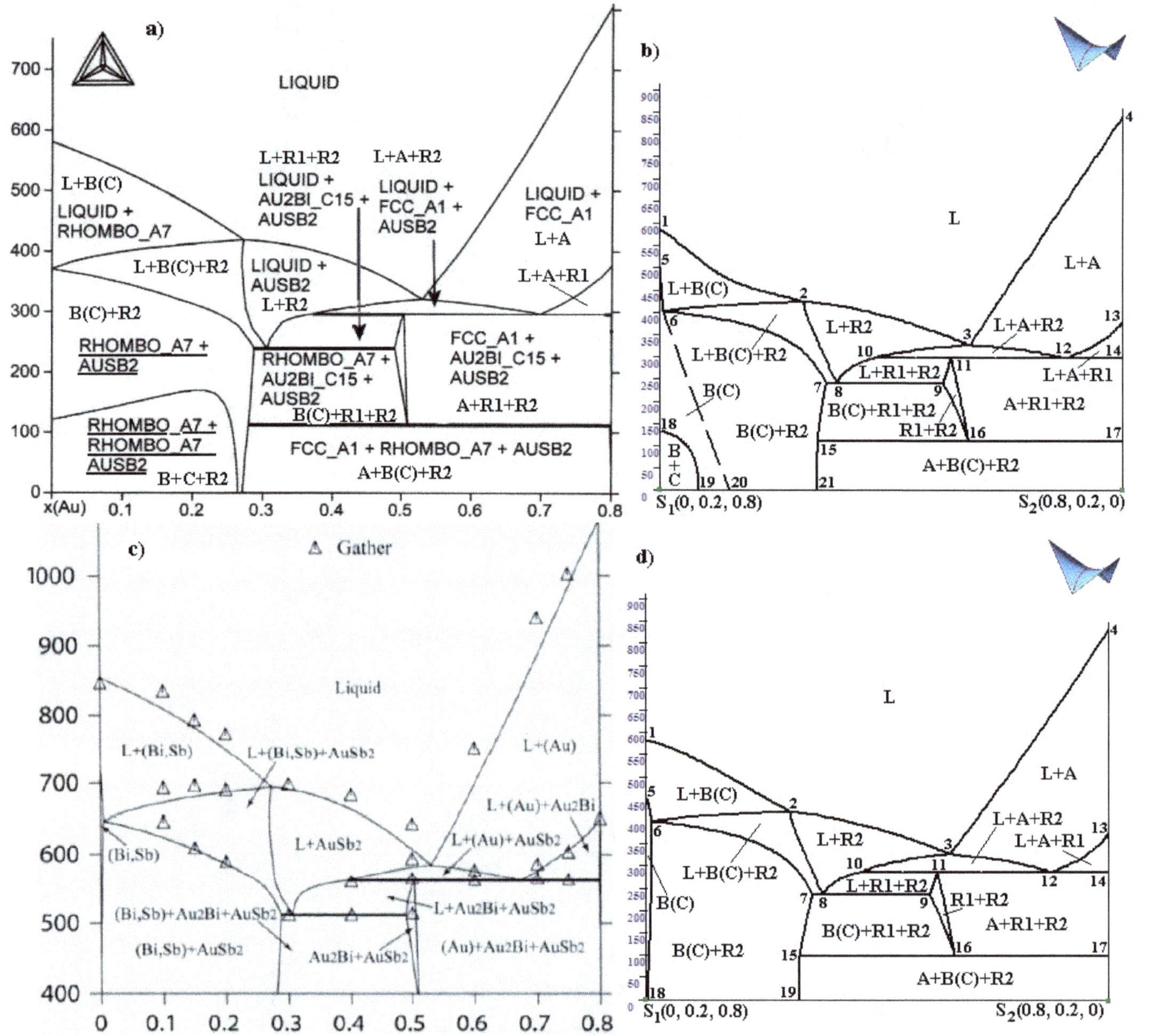

Figure 8: Isopleth $S_1(0, 0.2, 0.8)$-$S_2(0.8, 0.2, 0)$ of the Au-Bi-Sb: [12] (a): [26] (c): and 3D models according to data [12] (b): (the axis of temperature is marked in ˚C) and [26] (d): (sections of the added surfaces of solidus $s_{B(C)}$ and solvus $v_{B(C)\_R2}$. are drawn with dashed lines).

temperature boundaries of its existence are stretched to 0˚C as it shown in [22] Fragment of the 17 at % Ge isopleth is not correctly in Figure 8 The analysis, carried out with the aid of three versions of the Au-Bi-Sb T-x-y diagram 3D computer model for publications [12,25,26], showed that: a) sections of the solidus $s_{B(C)}$ and solvus $v_{B(C)\_}R_2$ surfaces in isopleths of the 3D computer models by curves 5-6 and 6-20 (Figure 8b), 5-6 and 6-18 (Figure 5d), are missed in sections of the diagram in [12] (Figure 8a) and in [25] b) phase regions Bi(Sb) and Bi+Sb are missed in the same sections in [12] (Figure 8a) and in [25];c) the cupola of the solid solution Bi(Sb) disintegration to bismuth and antimony (Figure 8c), which is assigned by the binodal of the Bi-Sb binary system [12], is lost in [26] (Figure 6). Most preferable is the version of the T-x-y diagram 3D model, which is reconstructed according to the data of [12], but with the added surfaces of solidus $s_{B(C)}$ and solvus $v_{B(C)\_}R_2$ (Figure 6). At the same time, it is possible only to speak, that the structure of the ternary system phase regions near the binary system Bi-Sb at temperatures below 150˚C is contradictory and requires an additional experimental study.

Acknowledgements

This work was been performed under the program of fundamental research SB RAS (Project 0336-2016-0006), it was partially supported by the RFBR (Projects 15-43-04304, 17-08-00875) and the RSF (Project 17-19-01171).

References

1. Lutsyk V, Vorob'eva (2009) From topology to computer model: ternary systems with polymorphism. Abstracts of the international conference on phase diagram calculations and computational thermochemistry (CALPHAD XXXVIII), Prague, Czech Republic, p: 66.

2. Lutsyk VI, Vorob'eva (2010) Computer models of eutectic-type T-x-y diagrams with allotropy. Two inner liquidus fields of two low-temperature modifications of the same component. J Therm Anal Calorim 101: 25-31.
3. Rhines FN (1956) Phase Diagrams in Metallurgy: Their development and application. McGraw-Hill Book Company, MC. New York-Toronto-London.
4. Prince A (1966) Alloy Phase Equilibria. Elsevier Publ Comp Amsterdam-London, New York.
5. Lutsyk VI, Vorob'eva, Zelenaya AE (2015) 3D reference book on the oxide systems space diagrams as a tool for data mining. Solid State Phenomena 230: 51-54.
6. Lukas HL, Henig ET, Petzow G (1986) 50 Years Reaction Scheme after Erich Scheil. Z Metallkd 76: 360-367.
7. Dinsdale A, Kroupa A, Watson A (2012) COST Action MP0602 - Handbook of high-temperature lead-free solders - Atlas of phase diagrams 1: 218.
8. Lutsyk VI, Vorob'eva (2016) 3D model of the T-x-y diagram of the Bi-In-Sn system for designing microstructure of alloys. Rus J Inorg Chem 61: 188-207.
9. Lutsyk VI, Vorob'eva, Shodorova SY (2015) Determining the conditions for changes of the three-phase rreaction type in a V-Zr-Cr systems. Rus J Phys Chem A 89: 2331-2338.
10. Lutsyk VI, Vorob'eva, Shodorova SY (2016) Verification of the T-x-y diagram of the Ag-Au-Bi system using a 3D computer model. Rus J Inorg Chem 61: 858-866.
11. Lutsyk VI, Vorob'eva (2015) Three-dimensional model of phase diagram of Au-Bi-Sb system for clarification of thermodynamic calculations. Rus J Phys Chem A 89: 1715-1722.
12. Dinsdale A, Watson A, Kroupa A (2008) Atlas of phase diagrams for lead-free soldering compiled by COST 531 1: 277.
13. Wang J, Leinenbach C, Roth M (2009) Thermodynamic description of the Au-Ge-Sb ternary system. J Alloys Compd 485: 577-582.
14. Zoro E, Dichi E, Servant C, Legendre B (2005) Phase equilibria in the Ag-Au-Bi ternary system. J Alloys Compd 400: 209-215.
15. Zoro E, Boa D, Servant C, Legendre B (2005) Enthalpies of mixing of the liquid phase in the ternary system Ag-Au-Bi. J Alloys Compd 398: 106-112.
16. Servant C, Zoro E, Legendre B (2006) Thermodynamic reassessment of the Au-Bi system. CALPHAD 30: 443-448.
17. Zoro E, Servant C, Legendre B (2007) Thermodynamic assessment of the Ag-Au-Bi system. CALPHAD 31: 89-94.
18. Zoro E, Servant C, Legendre B (2007) Thermodynamic assessment of the Ag-Au-Bi and Ag-Au-Sb systems. J Therm Anal Calorim 90: 347-353.
19. Chevalier PY (1988) Thermodynamic evaluation of the Au-Bi system. Thermochim Acta 130: 15-24.
20. Zakharov AM (1990) State diagrams of binary and ternary systems.
21. Fedorov PP (2010) Third law of thermodynamics as applied to phase diagrams. Rus J Inorgan Chem 55: 1722-1739.
22. Chen SW, Chen PY, Wu HJ, Chiu CN, Huang YC, et al. (2008) Phase equilibria of Sn-Sb-Ag ternary system (I): experimental. Metall Mater Trans A 39: 3191-3198.
23. Gierlotka W, Huang YC, Chen SW (2008) Phase equilibria of Sn-Sb-Ag ternary system (II): calculation. Metall Mater Trans 39: 3199-3209.
24. Chen SW, Chen CC, Gierlotka W, Zi AR, Chen PY, et al. (2008) Phase equilibria in the Sn-Sb binary system. J Electron Mater 37: 992-1002.
25. Manasijevic D, Minic D, Zivkovic D, Zivkovic Z (2008) Experimental study and thermodynamic calculation of Au-Bi-Sb system phase equilibria. J Phys Chem Solids 69: 847-851.
26. Wang J, Meng FG, Liu HS, Liu LB, Jin ZP (2007) Thermodynamic modeling of the Au-Bi-Sb ternary system. J Electron Mater 36: 568-577.
27. Prince A, Raynor GV, Evans DS (1990) Phase diagrams of ternary gold alloys, London: the Institute of Metals.
28. Voskov AL, Dzuban AV, Maksimov AI (2015) TernAPI program for the calculation of ternary phase diagrams with isolated miscibility gaps by the convex hull method. Fluid Phase Equilibria 388: 50-58.

Short and Robust HPLC-UV Method to Determine Serum Ribavirin Concentration without Evaporation Step

Abdul Rafiq Khan[1]*, Ali Al-Othaim[1], Khalid Muhammed Khan[2], Shazia Mrtaza[1], Sara Altraif[1], Waleed Tamimi[1], Waqas Jamil[3] and Ibrahim Altraif[1]

[1]*National Guard Health Affairs, Department of Pathology and Laboratory Medicine, Biochemical Metabolic Laboratory, Riyadh, Kingdom of Saudi Arabia*
[2]*Haji Ibrahim Jamal Research Institute of Chemistry, University of Karachi, Karachi, Pakistan*
[3]*University of Jamshoro, Jamshoro, Sindh, Pakistan*

Abstract

Objective: Measurement of ribavirin (RBV) is important for therapeutic drug monitoring in hepatitis C patient. A simple and fast high performance liquid chromatography (HPLC) method developed and validated to measure ribavirin concentration in serum samples without an evaporation step.

Design and method: About 500 µl serum sample, 50 µl internal standard and 20 mM ammonium acetate buffer (pH=8.5) were mixed for 30 seconds and centrifuged. The supernatant was transferred into preconditioned phenyl boronic acid cartridges for solid phase extraction. All cartridges were washed two times with 1 mL of 20 mM ammonium acetate buffer under vacuum not exceeding 10 psi. Ribavirin and internal standard were eluted with 300 µl of 3% formic acid. An aliquot of 100 µl was injected into HPLC system.

Results: The method was linear in the range of 0.1-8.0 mg/l with 0.05 mg/l as limit of detection. The correlation coefficient of method comparison was 0.975 with p value of 0.116 showing a good reproducibility of results. The mean accuracy was checked at three different concentrations and found to be between 107-110% for all three levels. The extraction efficiency was 65.5% for ribavirin in the range of 0.1-8.0 mg/l and 71.2% for internal standard at 50 mg/l. The intra assay precisions were determined at 0.5, 2.5, and 5.0 mg/l and % CV were found to be 2.2%, 5.0%, 4.5% respectively. The injection reproducibility at three levels was 5.5%, 6.1% and 3.3%. Removal of gravity flow and evaporation step made this method faster and easy for routine analysis of ribavirin samples.

Conclusion: The newly developed HPLC method was faster, accurate and sensitive. It applied for determining serum ribavirin level in hepatitis C patient in our hospital.

Keywords: Ribavirin; HPLC; Analysis

Introduction

Ribavirin (RBV) is a guanosine ribonucleoside analog which has spectrum of antiviral activity against DNA and RNA viruses [1,2]. Chronic hepatitis C virus (HCV) is treated with α-2-interferon along with ribavirin analogs. Combination therapy has greatly improved the rates of biochemical and virological response compared to the patients treated with interferon alone [3-6]. Ribavirin absorbs quickly and maximum plasma concentration achieve within 1.5 hour after oral administration. It slowly releases from kidney and gains steady state concentration after 4 weeks. Ribavirin produces significant side effects like hemolytic anemia and varying biological responses. Therefore, monitoring of ribavirin concentration is important to reduce adverse effects, to drive dose modification and to optimize management of HCV-infected patients receiving combination treatment [7-9].

Analytical methods for the determinations of ribavirin reviewed in 2007 [10]. Ribavirin was analyzed by tandem mass spectrometry in monkey and rat brain [11,12]. Capillary electrophoresis and high performance liquid chromatographic methods with UV detection were used for the determination of ribavirin in human plasma [13,14]. A high performance liquid chromatographic method with solid phase extraction was also reported using phenylboronic acid cartridges for sample clean up followed by evaporation [15]. Extraction of RBV from plasma was performed using a novel method based on ultrafiltration in one step that allows direct injection into the high-performance liquid chromatography without any prior steps of dryness or reconstitution [16]. The ultrafiltration technique is not commonly available in most of the lab and it limits the number of injection to 200-300 per column. The HPLC column gives high pressure after 200 samples with reduced resolution and bad peak shape. After reviewing most of the published methods for the determination of ribavirin in serum sample, we observed that all of these methods either involved the evaporation of purified serum sample after liquid - liquid extraction or after solid phase extraction which consumed time and labor especially when ribavirin extracted with 3% formic acid. After loading serum sample on solid phase cartridges, the washing and elution process involved the gravity flow of solvents which did not work for several samples. The third problem was direct injection of elution solvent into the HPLC column which gave a high solvent front merged with ribavirin peak.

In this study we have developed a faster cleaning method of serum sample by using phenylboronic acid cartridges with vacuum elution of washing and extraction solvent. The final elution solvent (3% formic acid) was directly injected into HPLC system without any evaporation step. The vacuum manifold was used under low vacuum to make extraction process faster without any blockage of cartridges which was observed many times when solvent passed under gravity force from PBA cartridges. A precise and low flow rate was used on 3 µ, 3 mm × 150 mm C_{18} column which completely resolved ribavarin peak with solvent front which was not possible with many published HPLC columns, mobile

***Corresponding author:** Abdul Rafiq Khan, National Guard Health Affairs, Department of Pathology and Laboratory Medicine, Biochemical Metabolic Laboratory, Riyadh, Kingdom of Saudi Arabia
E-mail: khanab@ngha.med.sa (or) abdulrafiq_khan@yahoo.com

phases and flow rates. There was no single HPLC method available so far which used vacuum manifold for sample cleaning and extraction as well as direction injection of elution solvent into HPLC column. A precise low flow rate was used to completely resolved ribavirin peak with solvent front. We have also evaluated the effect of interferences due lipemic, icteric and hemolysis on assay results of which was not reported in any of published chromatographic methods before. The method is fully validated according to FDA method validation guideline and successfully applied for the pharmacokinetic study and determination of Ribavirin trough level of hepatitis C patient treated with ribavirin along with interferon in National Guard Health Affairs, department of pathology and laboratory medicine, Riyadh, Kingdom of Saudi Arabia. The developed method fully validated by using standard method validation guideline as described below.

Materials and Methods

Reagent and chemicals

Ribavirin (Catalogue # R9644), internal standard (N^6-Methyladenosine 5-monophosphate sodium, Catalogue # M2780), formic acid and sodium dihydrogen phosphate were purchased from Sigma-Aldrich of highest purity. Bond elute phenylboronic acid (PBA, 100 mg, 1 ml) cartridges were obtained from Agilent, HPLC grade methanol and acetonitrile were purchased from Fisher Scientific. HPLC column 150 mm × 3 mm with 5 µm particle size was obtained from chromsystem Germany (Part 38130). Purified water was obtained from Millipore water purification system. Drug free blood samples were provided from our blood bank unit.

Instrumentation

High Performance Liquid Chromatographic system was Waters 2790 separation module which composed of binary pump and an auto-sampler. A tunable dual wavelength 2487 UV detector was used for monitoring both analytes and data was processed by using Waters empower 2 software. Vacuum manifold was obtained from supelcosil for sample preparation. An injection volume of 100 µl was loaded into the HPLC system.

Calibration and quality control preparation

For stock standard solution, accurately weighted 100 mg of ribavarine standard was transferred into 100 ml volumetric flask. The material was dissolved in HPLC grade water by sonication for 5 minutes and filled up to the mark. The stock standard solution (1000 mg/l) was further diluted to prepare working standard solution of 5, 25, 50, 100, 200 and 400 mg/l in water. Six Calibration standards of 0.1, 0.5, 1.0, 2.0, 4.0 and 8.0 mg/l were prepared by spiking 200 µl of each of these working standard solutions in 10 ml of ribavirin drug free serum in six different 15 ml glass tubes. The six levels of spiked serum calibration standards were mixed for 30 minutes, aliquoted into 500 µl and stored at -20°C until used. Stock standard solution (1000 mg/l) for quality control preparation was prepared by separately weighing about 100 mg of ribavirin and dissolved into 100 mL of HPLC grade water by 5 minutes sonication. Working quality control solutions of 25, 125, 250 mg/l were prepared in water by diluting stock solution. Three quality controls in serum samples (0.5, 2.5, 5.0 mg/l) were made by spiking 200 µl of each working quality controls solutions into 10 mL of ribavirin free serum sample separately. The three level of spiked serum quality controls were mixed, aliquoted into 500 µl and stored at -20°C until used.

Internal standard and reagent preparation

For stock internal standard solution, 50 mg of internal standard were accurately weighted and transferred into 100 ml volumetric flask. The content of flask was dissolved in HPLC grade water and made up to the mark. The working internal standard solution was prepared by transferring 10 mL of stock internal standard solution into 100 ml flask and volume was made up to the mark with HPLC grade water to get the final concentration of 50 mg/l. The working standard solution was aliquoted into 5 ml and stored at -40°C until used. Mobile phase A was made with 20mM buffer of KH_2PO_4 adjusted to the pH 3.0 with orthophosphoric acid and filtered with 0.45 µm. Mobile phase B was made by mixing 90% of HPLC grade acetonitril with 10% of HPLC grade water. Precipitation and extraction reagents were 20 mM ammonium acetate (pH 8.5 with ammonia) and 3% formic acid in HPLC grade water respectively.

Extraction procedure

The extraction of sample was performed by transferring 500 µl of ribavirin free serum, six level calibration standards, three levels quality controls and patient samples into separate 1.5 ml eppendorf centrifuged tubes. 50 µl of working internal standard and 500 mL of ammonium acetated were added in each tube, vortexed for 5 min and centrifuged at 10000 RPM for 5 min. The supernatants were loaded into phenylboronic acid cartridges (PBA) which were previously washed with 1 ml methanol followed by 1 ml of 20mM ammonium acetate (pH 8.5) by using vacuum manifold at the vacuum not exceeding 10 mm Hg. The SPS cartridges were drained by vacuum followed by washing with 1 ml of 20 mM ammonium acetate (pH 8.5) three times. After third washing, all the cartridges were dried under vacuum for 5 minutes. The ribavarin and internal standard were eluted from the cartridges with 300 µl of 3% formic acid at a vacuum not exceeding 10 mm Hg. The eluents were transferred into HPLC vials and 100 µl injected into HPLC systems.

HPLC analysis

A 100 µl of blank, calibration, quality controls and patient samples were injected into C_{18} HPLC column (150 mm × 3.0 mm, 3 µm, Chromsystem Germany) where ribavirin and internal standard were separated using mobile phase A (20mM buffer of KH_2PO_4 adjusted to the pH 3.0 with orthophosphoric acid) and mobile phase B (90% of HPLC grade acetonitril with 10% of HPLC grade water). The gradient elution was started with 0.5 ml/min of 97% mobile phase A and 3% of mobile phase B for first 10 minutes and then changed into 50% mobile phase A and B at for next 5 minute. The initial conditions were turned on for 5 min before the next sample injection. The tunable UV detector was operated at 230 nm to monitor ribavirin and changed to 262 nm after 15 min to monitor internal standard. The linear equations for the relationship between the peak areas of ribavirin and their concentration were determined by least-squares method using internal standard quantification mode of calculation. Quality controls and patient samples were quantified by six point calibration standard curve based on peak area toward concentration in serum sample. Chromatogram of blank without internal standard addition was free from all interferences at the typical retention time of ribavirin and internal standard.

Results and Discussion

Method development

Bond elut phenylboronic acid (PBA) cartridges were used because its boronat group had strong affinity to cis-diol containing compounds like ribavarin. The internal standard N^6-Methyadenosine 5-monophosphate was selected to have similar cis-diol group which provided better recovery. The UV detection was performed at 230 nm and 262 nm which were the wavelength of maximum absorbance of ribavarin and internal standard respectively. The saturation and washing of PBA cartridges were carried out as described in the general protocol of extraction method from Agilent. Both ribavarin and internal standard were eluted with 300 µl of 3% formic acid without any organic solvent which saved cost. The final elution solvent (3% formic acid) was directly injected on five different HPLC columns (supelcosil C_{18} 150 mm × 2.1 m, 3 µm, Hypersil BDS C_{18} 150 mm × 2.1 m × 3 µm, Atlantic T3 150 mm × 4.6 m × 3 µm, Zorbax C_{18} 250 mm × 4.6 m × 5 µm and Chromsystem C_{18} HPLC column 150 mm × 3 mm × 3 µm) with different mobile phase of different composition and flow rate. Ribavarin and internal standard were well resolved from solvent front only on Chromsystem C_{18} HPLC column (150 mm × 3 mm, 3 µm). The low flow rate was selected in order to have well separated peak of ribavarin from solvent front and to overcome the back pressure problem after 500-600 injections. An aliquot of 100 µl was loaded on HPLC column in order to have better signal to noise ratio at low concentration of ribavarin which was very useful if low sample volume (250 µl) was available for some patients. Waters tunable absorbance detector provided a facility to detect both molecules on two different wavelength and different retention time in a same injection.

Specificity and selectivity

Six different pool of serum samples were collected from different sources and analyzed by the assay procedure in duplicate with and without adding internal standard. The chromatograms of these blank samples showed no interfering peak from other component of blood at the retention time of ribavirin and internal standard. These blank samples were analyzed to demonstrate that the developed method was selective and specific for the analysis of ribavirin in serum samples. Chromatograms of a blank sample, serum calibration standard and a patient sample are shown in Figures 1-3 respectively.

Linearity and range

Six different amount of ribavirin was spiked in drug free serum ranging from 0.1-8.0 mg/l. After thorough mixing, these calibration

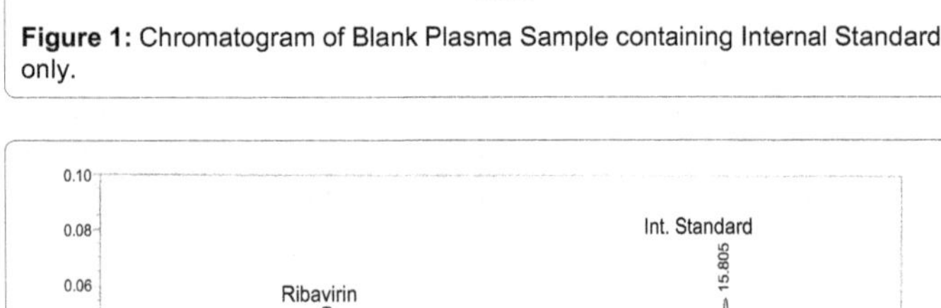

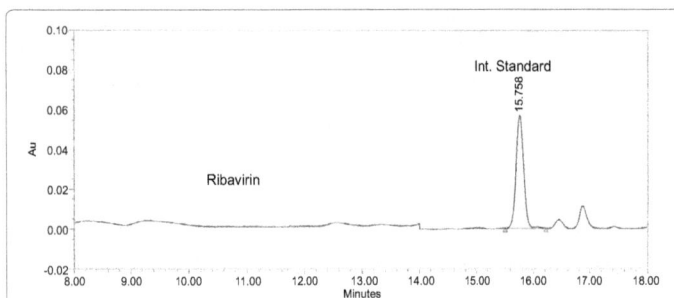

Figure 1: Chromatogram of Blank Plasma Sample containing Internal Standard only.

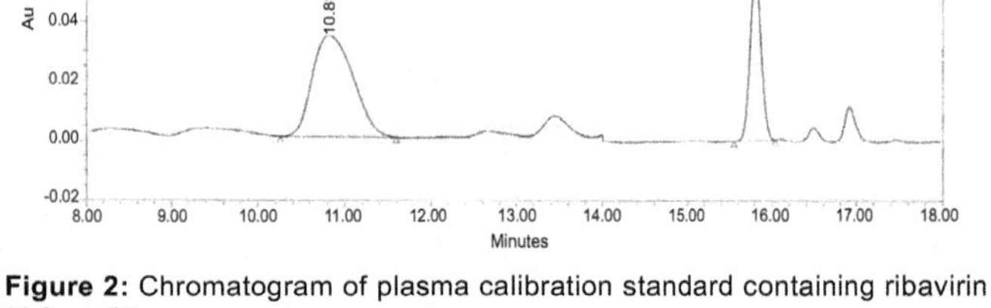

Figure 2: Chromatogram of plasma calibration standard containing ribavirin (4.0 mg/L).

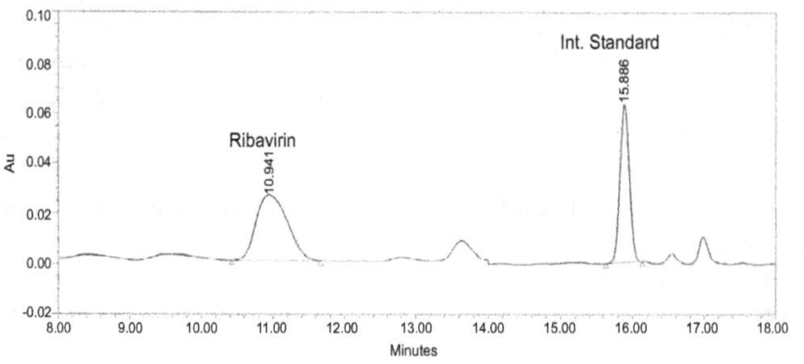

Figure 3: Chromatogram of patient sample receiving ribavirin (2.5 mg/L).

serums samples were extracted and analyzed in duplicate by the assay procedure. A linear calibration curve was obtained by plotting the peak area ratio of ribavirin to the peak area of internal standard against the corresponding six concentrations. Linear regression analysis was performed and results were calculated (Table 1).

Limit of detection and lower limit of quantification (LOD and LLOQ)

A serum sample of 0.05 mg/l and 0.1 mg/l were prepared by spiking ribavirin standard solution into drug free serum samples and analyzed 10 times by the assay procedure. The recovery and precision of the results were calculated and it was found that limit of detection was 0.05 mg/l whereas lower limit of quantification was 0.1 mg/l Table 1 according to FDA method validation guideline.

Accuracy and precision

Known amount of ribavirin at three different concentrations 0.5 mg/l, 2.5 mg/l and 5.0 mg/l levels were spiked in drug free serum samples and analyzed 5 times on two different days by newly developed HPLC method. Mean accuracy of 110.1%, 108.3% and 107.7% were obtained on 0.5 mg/l, 2.5 mg/l and 5.0 mg/l respectively. The intra assay precision determined at 0.5, 2.5, and 5.0 mg/l were found to be 2.2%, 5.0% and 4.5% respectively. The injection reproducibility were found to be 5.5 %, 6.1% and 3.3% at three levels calculated from the results of 5 replicate injections of extracted quality control samples on same day (Table 1).

Comparison and extraction efficiency

In an experiment, about 40 patient sample of varying ribavirin concentration were analyzed on different days by developed HPLC-UV method and results were compared to tandem mass spectrometry method. Statistical analysis was performed by comparing the mean values of both methods and bias. No significant bias and differences in patient results was found as indicated from the statistical analysis of two mean values (Table 1).

Extraction efficiency of the method was checked by preparing six calibration standards in serum sample followed by extraction procedure. Six calibration standards were also prepared in water and analyzed by the assay procedure without extraction. The experiment was performed twice on two different days. The mean recovery of ribavirin and internal standard, calculated from the speak area of ribavirin and internal standard in both serum (with extraction) and water sample (without extraction), were 65.5 and 71.2% respectively (Table 1).

Ruggedness and interference study

The developed method was used for the analysis of ribavirin patient samples in routine laboratory. Therefore the ruggedness was also checked in order to see the effect on retention time from day to day

Linearity	Range (mg/L)		Slope		Intercept		R²	
	0.1 - 8.0		0.310		-.006		0.999	
Sensitivity	LOD (mg/L)	% Recovery	% CV		LLOQ (mg/L)	% Recovery	% CV	
	0.05	122	19.2		0.1	110	5.5	
Accuracy (N=10)	Level 1 (0.5 mg/L)		Level 2 (2.5 mg/L)		Level 3 (5.0 mg/L)			
	110.1%		108.3%		107.3%			
Precision (N=10) % CV	Level 1 (0.5 mg/L)		Level 2 (2.5 mg/L)		Level 3 (5.0 mg/L)			
	2.2		5.0		4.5			
Reproducibility (N=10) CV	Level 1 (0.5 mg/L)		Level 2 (2.5 mg/L)		Level 3 (5.0 mg/L)			
	5.5		6.1		3.3			
Comparison	Slope	Intercept	Bias		p Value		R²	
	0.953	0.056	-0.03		0.116		0.975	
Extraction Efficiency	% Recovery of Ribavirin				% Recovery of Internal Standard			
	65.5				71.2			
System Suitability Test Evaluation								
Components	Retention Time (Min)		Theoretical Plates		Tailing Factors		Resolution	
Ribavirin	19.614		29160		1.2		33	
Internal Standard	32.708		17294		1.1			

Table 1: Validation data for Linearity, Accuracy, Precision, Comparison, and Extraction, Efficiency and System Suitability.

use. To determine the ruggedness of the chromatographic methodology developed, experimental conditions were purposely altered and chromatographic characteristics were evaluated. The pH of the mobile phase was adjusted to 2.5 and 3.5. The normal pH for the method was 3.0. Similarly the concentration of elution reagent (3% formic acid) and mobile phase composition were slightly changed and effects on retention time and results were monitored. In all these changes the percent variation in retention time was less than 5.0% for both ribavarin and internal standard with no significant affect in the results of quality control and patient samples.

For checking the effect of Billirubin, Hemolysis and Lipemia, a pool of patient sample was spiked with increasing amount of billirubine, heymolysat and lipid 0 µl, 10 µl, 20 µl, 30 µl and 40 µl and analyzed in duplicate. There is no significant difference in results was found as shown in Figure 4.

System suitability

The peaks of ribavarin and internal standards were well separated from each other and from other adjacent peaks whereas theoretical plates and tailing factors for both analytes were calculate by Empower software and summarized in Table 1.

Stability studies

Three patient samples were initial analyzed and then stored at 4°C and -20°C. After 5 day and 15 day, these samples were again analyzed by the assay procedure with freshly prepared calibration standard in order to assess the stability. The calculated recovery showed that the patient samples were stable at both temperature tested up to 15 days (Table 2).

Application

The developed HPLC method was successfully applied for serum ribavirin baseline pharmacokinetic profile and monitoring of trough ribavirin concentration from HCV patient treated with combination therapy after 4, 8, 12 and 24 weeks.

Conclusion

In conclusion, the present analytical method development provides a simple, faster and robust analysis of ribavirin in human serum sample which could be used for the monitoring of ribavirin concentration in patient with HCV infection in order to enhance the effectiveness of interferon therapy with ribavirin, optimized therapy and reduce adverse events particularly ribavirin related anemia. The present method offers advantages of speediness, simplicity, sensitivity and accuracy, and is applicable to the selective determination of ribavirin with shorter and faster extraction using vacuum and without evaporation step. This method has also evaluated for the most common interferences due to lipemic, icteric and hemolysis of serum samples and found to be free from the effects.

Days	At 4°C			At -20°C		
	Patient 1 (mg/L)	Patient 2 (mg/L)	Patient 3 (mg/L)	Patient 1 (mg/L)	Patient 2 (mg/L)	Patient 3 (mg/L)
Initial	0.469	1.258	2.622	0.469	1.258	2.622
5 Days	0.479	1.195	2.479	0.497	1.127	2.393
15 Days	0.485	1.200	2.275	0.426	1.273	2.532

Table 2: Stability Studies.

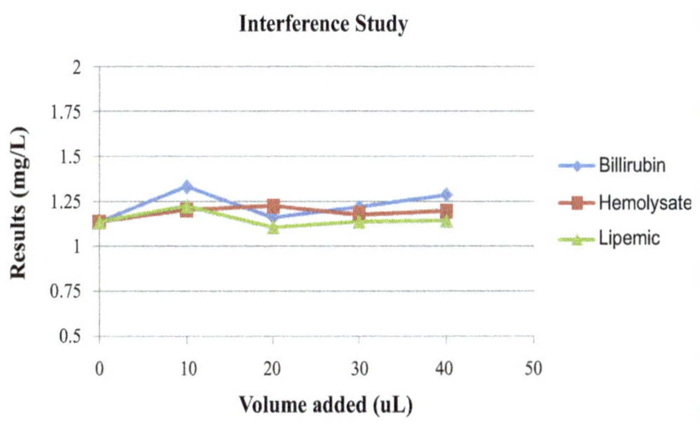

Figure 4: Inference Study.

References

1. Liu LX, Wang Y (2008) Advances of rabivirin clinical study. Mod J Integr Tradit Chin West Med 17: 5542-5543.

2. Reddy K, Nelson DR, Stefan Z (2009) Ribavirin: Current role in the optimal clinical management of chronic hepatitis C. J Hepatology 50: 402-411.

3. Davis GL (1999) Combination therapy with interferon Alfa and ribavirin as retreatment of interferon relapse in chronic hepatitis. C Semin Liver Dis 19: 49-55.

4. Heathcote J (2000) Antiviral therapy for patients with chronic hepatitis. C Semin Liver Dis 20: 185-199.

5. Hutchison JG, Poynard T (1999) Combination therapy with interferon plus ribavirin for the initial treatment of chronic hepatitis C. Semin Liver Dis 19: 57-65.

6. Poynard T, Marcellin P, Lee SS, Niederau C, Minuk GS, et al. (1998) Randomised trial of interferon alpha2b plus ribavirin for 48 weeks or for 24 weeks versus interferon alpha2b plus placebo for 48 weeks for treatment of chronic infection with hepatitis C virus. Lancet 352: 1426-1432.

7. Marina N, Nuria C, Belén R, Miguel AB, Pablo B, et al. (2005) Impact of ribavirin exposure on early virological response to hepatitis C therapy in HIV-infected patients with chronich hepatitis C. Antivir Ther 10: 657-662.

8. Rendón AL, Núñez M, Romero M, Barreiro P, Martín-Carbonero L, et al. (2005) Early monitoring of ribavirin plasma concentrations may predict anemia and early virologic response in HIV/hepatitis C virus-coinfected patients. J Acquir Immune Defic Syndr Early monitoring of ribavirin plasma. 39: 401-405.

9. Bosch M, Sanchez AJ, Rojas F, Ojeda C (2007) Ribavirin: Analytical determinations since the origin until today. J Pharm Biomed Anal 45: 185-193.

10. Li WK, Luo SY, Li SY (2007) Simultaneous determination of ribavirin and ribavirin base in monkey plasma by high performance liquid chromatography with tandem mass spectrometry. J Chromatogr B 846: 57-68.

11. Zironi E, Gazzotti T, Lugoboni B (2011) Development of a rapid LC–MS/MS method for ribavirin determination in rat brain. J Pharm Biomed Anal 54: 889-892.

12. Breadmore MC, Theurillat R, Thormann W (2004) Determination of ribavirin in human serum and plasma by capillary electrophoresis. Electrophoresis 25: 1615-1622.

13. Loregian A, Scarpa MC, Pagni S (2007) Measurement of ribavirin and evaluation of its stability in human plasma by high-performance liquid chromatography with UV detection. J Chromatogr B 856: 358-364.

14. Larrat S, Stanke F, Plages A, Zarski JP, Bessard G, et al. (2003) Ribavirin quantification in combination treatment of chronic hepatitis C. Antimicrob Agents Chemother 47: 124-129.

15. Svensson JO, Bruchfeld A, Schvarcz R, Ståhle L (2000) Determination of ribavirin in serum using highly selective solid-phase extraction and high-performance liquid chromatography. Ther Drug Monit 22: 215-218.

16. Morello J, Rodríguez-Nóvoa S, Cantillano ALR, González-Pardo G, Jiménez I, et al. (2007) Measurement of ribavirin plasma concentrations by high-performance liquid chromatography using a novel solid-phase extraction method in patients treated for chronic hepatitis C. Ther Drug Monit 29: 802-806.

20

Effect of Template on the Structure of Carbon Nanotubes Grown by Catalytic Chemical Vapor Deposition Method

Hekmat F[1], Sohrabi B[2]* and Rahmanifar MS[1]

[1]*Faculty of Basic Science, Shahed University, Tehran, Iran*
[2]*Department of Chemistry, Surface Chemistry Research Laboratory, Iran University of Science and Technology, Iran*

Abstract

Carbon nanotubes (CNTs) have been synthesized on an anodized aluminum oxide (AAO) template with acetylene by using catalytic chemical vapor deposition (CVD). It was found that the structure of CNTs can be mostly depends on the quality of the catalyst deposition in the pores of AAO template. Straight CNTs were observed when Ni catalyst deposited only in the bottom of holes as nanoparticles, but when Ni catalyst fulfilled the holes of AAO template as nanowires (NWs), coiled CNTs observed. The characterization of as-prepared materials was examined by SEM and Raman spectroscopy. In addition, it is realized that the nanostructure of the AAO template strongly affected the properties of as-grown CNTs. In the other word, based on the obtained results the diameter and pitch of coiled CNTs tightly depend on the size of metallic catalyst and in the results it depend on the pore diameter of AAO template.

Keywords: Nanostructures; Chemical vapor deposition; Electron microscopy; Raman spectroscopy

Introduction

Due to their unique mechanical, electrical, and thermal properties, carbon nanotubes (CNTs) have attracted particular attention among the known allotropes of carbon [1-4]. Although straight CNTs, including single walled CNTs (SWCNTs) and multi walled CNTs (MWCNTs) were commonly observed, different shapes of CNTs such as coiled, branched, and toroidal were reported [5] shortly after their discovering by Iijima [6]. It has been demonstrated that the nanostructure of CNTs can strongly affected the characteristic properties of different shapes of CNTs, so it can limit their applications. For example, unlike straight CNTs, coiled ones can show semi metallic properties beside the typical electrical characteristics of straight CNTs, such as exhibition metallic and semi conductive properties [7,8]. Almost simultaneously with claim that there is possibility about the existence of coiled CNTs with minimized energy confirmed by theoretical calculations [9-11] coiled carbon fibers were experimentally synthesized, principally from decomposition of hydrocarbons in the presence of metallic catalyst [12-18]. It is known that coiled CNTs are MWCNTs with incomplete crystalline structures created when paired pentagon-heptagon atomic rings arrange themselves periodically within the hexagonal carbon network. We have previously reported production of a novel porous coiled CNT/Ni-NW nanostructure by ambient thermal-CVD [19] successfully. Our investigations show that this material being a sufficient structure for improvement of electrode materials for advanced energy storage systems. On the other hand, there is a great interest for growing well aligned CNTs and specially vertically aligned CNTs (VA-CNTs) for developing field emission devices and a large variety of microelectronic devices, including nanotube sensors, optoelectronic systems, batteries, and supercapacitors [20-23]. One of the common methods for growing VA-CNTs is based on using a porous filter such as AAO template. By using this kind of nano template for growing CNTs, control the shape of CNTs being easy. So, the characteristic properties of CNTs especially their aspect ratios are tightly control with the nanostructure of template, including diameter and length of nano channels. Ever since the discovery of carbon nanotubes, a variety of methods have been reported for producing different shapes of CNTs, among which may be mentioned the following main methods: arc discharge, laser ablation, and CVD [21]. In the last few years, CVD method attracted much attention as the most utilized method, because it is a cost effective method for large-scale production of CNTs on a wide range of substrates. Recently, CVD method is given special emphasis as the most used method for synthesis different shapes of CNTs, specially coiled and aligned straight ones [22-24]. In this study, acetylene decomposition was performed over Ni catalysts supported on AAO template to investigate the role of catalyst on the structure of as-grown CNTs. To evaluate the effect of nanostructure of AAO template on the appearance of as grown CNTs, different condition of anodizing template, were investigated.

Experimental

Chemicals

High-purity aluminum foil (99.9995%, Merck, Germany), Perchloric acid (60%, Merck, Germany), ethanol (96%, Jonoob, Iran), oxalic acid 2-hydrate (99%, Panreac Quimica SA, E.U.), Choromic anhydride (KANTO Chemical Co. INC, Japan), phosphoric acid (85%, Merck, Germany), copper sulfate (Merck, Germany), chloridric acid (36%, Merck, Germany), hexahydrate nickel sulfate (Merck, Germany), heptahydrate nickel chloride (Merck, Germany), boric acid (Merck, Germany) were used as received. Finally, double ionized water was obtained from an OES water purification system (Oklahoma, USA).

Equipments

DC power supplier (MP6003, Megatek, Germany) was used in order to apply electric field in preparation of AAO template. Electro analyzer system SAMA 500 (Iran) was used for electrochemical deposition of Ni catalytic particles in the pores of AAO template. The samples investigation were done by using scanning electron microscopy (TESCAN, VEGA, Czech Republic), field-emission scanning electron microscope (TESCAN, Mira II and 3 LMU, Czech Republic) and Raman spectroscopy (BRUKER, SENTERRA, Germany).

***Corresponding author:** Sohrabi B, Department of Chemistry, Surface Chemistry Research Laboratory, University of Science and Technology, Iran
E-mail: sohrabi_b@yahoo.com

Preparation of template and growing CNTs

As illustrated in Figure 1, a porous AAO template was obtained by two-step anodization of a 3 × 3 × 0.3 mm³ high purity (99.9995%, Merck, Germany) aluminum foil which degreased with acetone by using ultrasonication and then annealed at 500°C for 5 hr. Before proceeding Aluminum by anodizing, the sample must pretreated using electropolishing for smoothing and brightening the surface of metal. In first step of anodization, the sample was anodized in an aqueous solution of oxalic acid (0.3 M) under a DC constant voltage (45 V) at low temperature (below 5°C) for 20, 22, and 25 hr. After the first step of anodization, the anodized surface layer was removed by electrochemical etching step which occurred in a mixture of chromic acid (1.8 wt %) and phosphoric acid (6 wt %) at 75°C for 3 hr. In the second step, anodization was done for 4 hr under the same conditions as mentioned in the first step. In order to facilitate the uniform electrodeposition of Ni which acts as catalyst in growing CNTs, immediately at the end of the second anodization, the voltage was dropped from 45 to 14 V at the rate of 0.5 V min^{-1}. The remaining Al substrate was mostly removed in a saturated copper sulphate aqueous solution in chloridric acid at room temperature [24]. Finally, the AAO template was etched in a 1M aqueous solution of phosphoric acid at a room temperature for 40 min for the further thin the barrier layer and widen the pore size. Catalytic Ni particles deposited in the nanochannels of as prepared AAO template from Watt bath which contained from hexahydrate nickel sulphate (330 g l^{-1}), heptahydrate nickel chloride (45 g l^{-1}) and boric acid (35 g l^{-1}) at 23°C and pH=2. The electrochemical deposition of Ni-nanoparticles was performed by using AAO template which sputtered with a layer of Au, as working electrode, platinum as counter electrode, and Ag/AgCl as reference electrode, and all of the potentials refer to the reference electrode. Ni nanoparticles were electrochemically deposited in nanochannels of AAO template applying at scan rate of 50 mV s^{-1} in the watt bat contained $NiSO_4.6H_2O$, H_3BO_3, and $NiCl_2.7H_2O$. The deposition time was selected by considering the deposition rate of Ni nanoparticles in the holes of AAO template which is about 8.4 nm. s^{-1} [25]. So, we can easily control the mass loading of Ni particles which deposited in the pores of AAO template, by adjusting the deposition time. The deposition of Ni nanoparticles were carried out in the potential range between -1.1 to 0 V applying 5 cycles for deposition Ni nanoparticles and 60 cycles for growing Ni nanowires (Ni-NWs) which fulfilled the holes of AAO template. Along with fulfilling pores with Ni-NWs, the color of surface changes to silver, so the Ni deposition must be stopped after the first change of color. Figure 2 shows the cyclic voltammograms recorded for deposition Ni nano catalysts in the pores of AAO template at 23°C and pH=2. CNTs were grown by using a catalytic pyrolysis of acetylene as hydrocarbon source and a mixture of hydrogen and argon as carrier gas. In the first step, catalytic Ni particles were reduced in a carrier gas (contains 10% H_2 and 90% Ar) at 500°C for 1 hr. In the second step, which started immediately after heating up to 650°C, CNTs were grown by catalytic decomposition of 10% C_2H_2 and 20% H_2 in an Ar carrier gas at a total flow rate of 150 sccm during 40 min. After growing CNTs, the furnace stay at 650°C for 10 min without any hydrocarbon source, then the furnace was cooled down slowly in presence of Ar. To have free standing VA-CNTs, surrounding AAO template was removed by using KOH solution in room temperature Figures 3a and 3b illustrate the preparation steps of growing VA-CNTs and coiled CNTs by using the CVD method, respectively. Our previous studies show that coiled carbon nanotubes grow directly on an anodized aluminum oxide (AAO) template which one side of it is covered with conducting Au layer and the other side is filled by using catalytic Ni nanowires.

Results and Discussion

Influence of the nanostructure of catalytic Ni particles on the structure of CNTs

The Figure 4 show the SEM images of two AAO based templates after deposition of Ni nano catalysts in their pores as nanoparticles and NWs, respectively. As it can be seen in Figure 4a, the catalyst particles deposited uniformly at the bottom of pores and in Figure 4b, the catalytic Ni-NWs were deposited in the pores of AAO template and filled the holes. Figures 4c and 4d show SEM images of as-grown straight VA-CNTs on the supported Ni nanoparticles by acetylene decomposition at 650°C before and after removing AAO nano template by using KOH aqueous solution in room temperature, respectively. Cross sectional and plane views of AAO/Ni-NW based template after growing coiled CNTs on the Ni-NWs were illustrated in Figures 4e and 4f, respectively. Raman spectroscopy is a method used to obtain suitable information about the structure of as-prepared two kinds of carbon nanotubes. Figure 5 represents the Raman spectra of grown CNTs on the AAO template using Ni nanoparticles and Ni-NWs as catalysts. The multi-walled structure of grown CNTs is identified by the sharp G bands which appeared at 1500-1600 cm^{-1} region of the wave number and attributed to the graphite E^{2}g optical mode and also by the one observed at the wave number of 1200-1300 cm^{-1} is known as D bands which corresponds to the disordered sp^2 carbon of graphite [26]. The ID/IG ratio of the as-grown coiled and straight CNTs is ~1.45 and ~0.865, respectively. It reveals that the crystallinity of the CNTs grown using Ni nanoparticles as catalyst is higher than that of coiled CNTs grown using Ni-NWs, and it should be noted that increasing the deposition time of Ni particles to fulfill the pores of AAO template can increase the possibility of creating pentagon-heptagon atomic rings within the hexagonal carbon network, which are known as defects [27,28]. Thermogravimetric analysis (TGA) was performed to precisely determine the content of coiled CNTs in the coiled CNT/Ni-NW electrode. The results of this analysis has previously been presented by using our group.

Influence of the pore diameter of AAO template on the nano-structure of coil CNTs

In the CVD process, the size of nano pores which were fulfilled by Ni-NWs play an important role for growing coiled CNTs. SEM observations indicated that the diameter and pitch of as-grown coiled CNTs (Figure 6) were strongly affected by the size of metallic nano catalysts, which was controlled significantly by the pore size

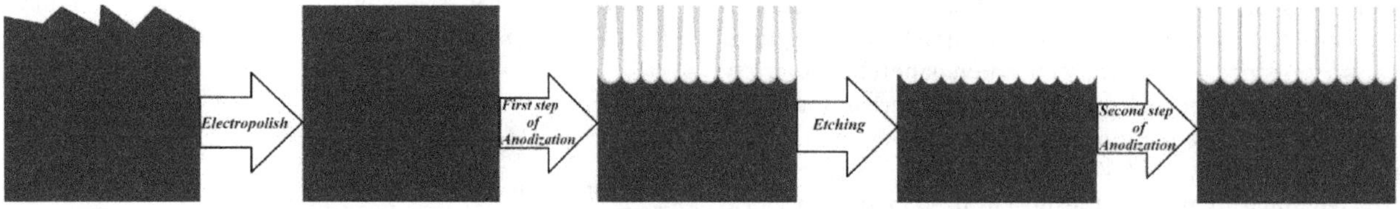

Figure 1: A schematic diagram of experimental setups for the anodization process.

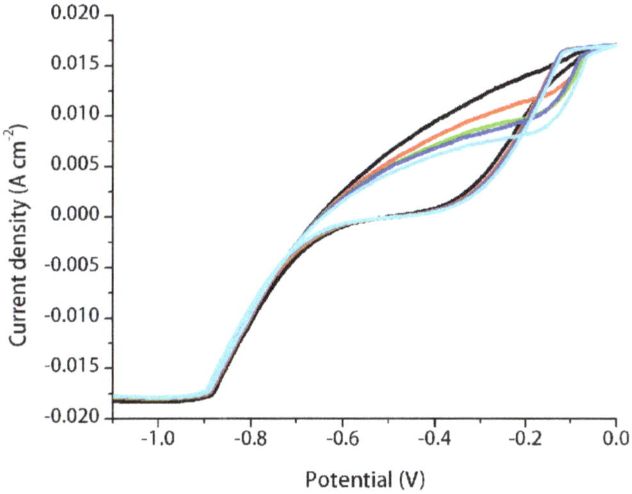

Figure 2: The Ni nano catalysts deposited in the pores of AAO template in Watt bath in scan rate of 50 mVs^{-1}.

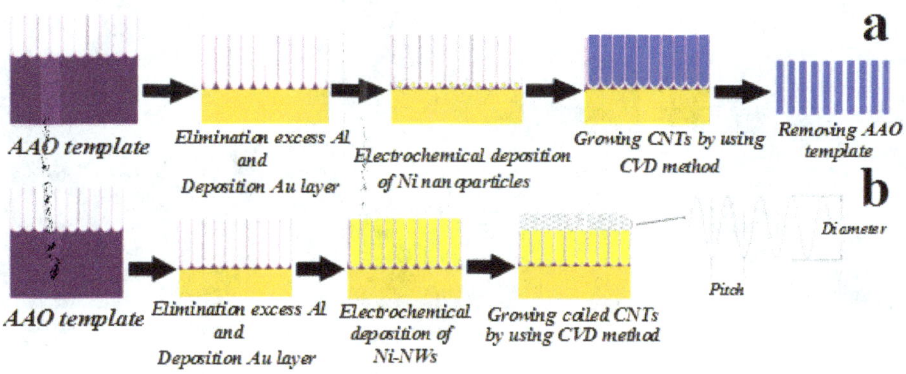

Figure 3: A schematic illustration of growing (a) VA-CNTs and (b) coiled CNTs, by using a catalytic CVD method.

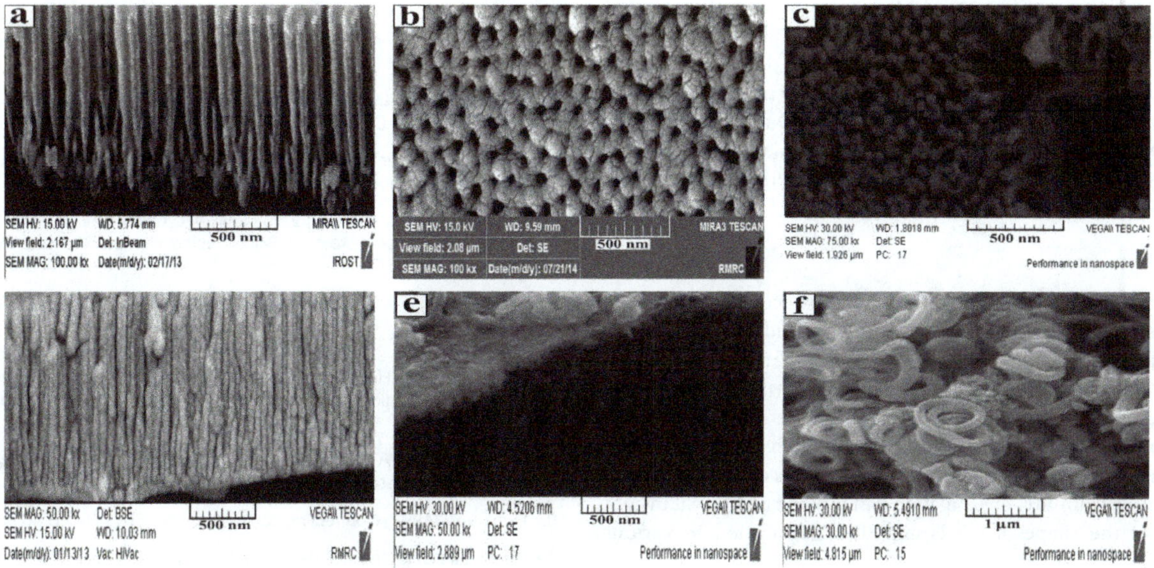

Figure 4: SEM images showing (a) AAO template after deposition Ni nanoparticles in the bottom of pores, (b) and (c) CNTs grown at 650°C for 40 min in nanochannels of AAO template before and after removing AAO template, respectively, (d) Ni-NWs deposited in the nanoholes of AAO template, (e) side view and surface morphology of AAO/Ni nanowire based template after growing coiled CNTs.

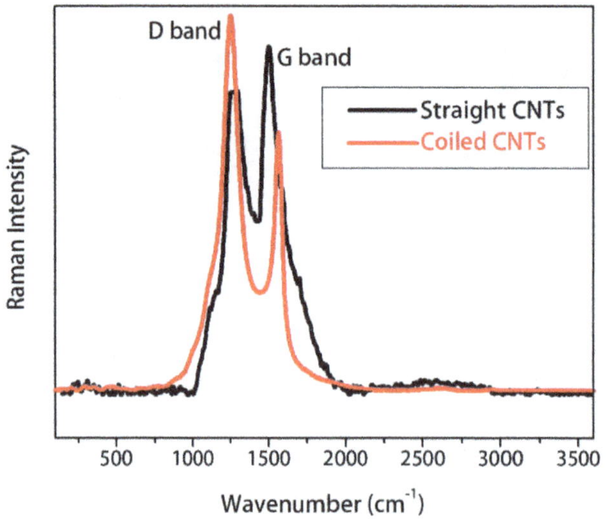

Figure 5: Raman spectrum of the coiled and straight CNTs.

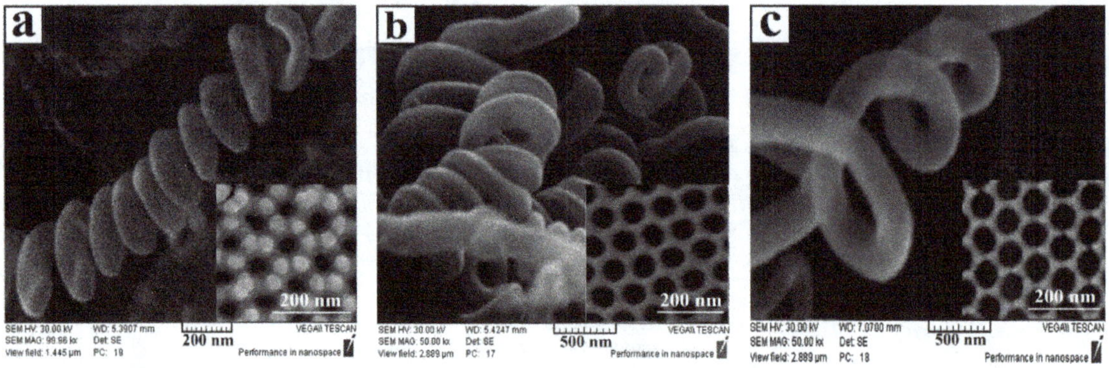

Figure 6: SEM images of coiled CNTs which were grown on the AAO template with pore size of (a) ~50 nm, (b) ~65 nm, (c) ~80 nm.

of AAO template [29]. As it can be seen in Figure 6 diameter, and pitch of coil CNTs were increased by increasing the pore size of AAO template. The insets of Figures 6a and 6c show the surface morphology of AAO templates. By increasing the diameter of pores and also the diameter of Ni-NWs from ~50 to ~80 nm, the diameter of coiled CNTs were increased from ~100 to ~400 nm, respectively. Based on the above results we can claim that the diameter and pitch of coiled CNTs are tightly depend on the pore diameter of AAO template [30].

Conclusion

In summary, two different types of CNTs were successfully fabricated on an AAO template by using CVD method. This study reveals that the structure of metallic catalyst and the size of nanoparticles of metallic catalyst are important parameters in growing CNTs which can strongly affect the microstructure of MWCNTs and also their alignment on the substrate. On the other hand, the diameter and pitch of coil CNTs can significantly controlled by the diameter of nano pores of AAO template. Therefore, results from this investigation can provide additional insight into control of the shape of CNTs and their application in various industrials.

Acknowledgements

We are grateful for financial support from Iran National Science Foundation (INSF), Iran (Grant No. 93/36575).

References

1. Patole SP, Kim HI, Jung JH, Patole AS (2011) The synthesis of vertically-aligned carbon nanotubeson an aluminum foil laminated on stainless steel. Carbon 49: 3522-3528.
2. Bai JB (2003) Growth of nanotube/nano fibre coils by CVD on an alumina substrate. Mater Lett 57: 2629-2633.
3. Beguin F, Presser V, Balducci A, Frackowiak E (2014) Carbons and Electrolytes for Advanced Supercapacitors. Adv Mater 26: 2219-2251.
4. Ismagilov RR, Shvets PV, Zolotukhin AA, Obraztsov AN (2013) Growth of a Carbon Nanotube Forest on Silicon using Remote Plasma CVD. Chem Vap Depo 19: 332-337.
5. Zhang M, Li J (2009) Carbon nanotube in different shapes. Mater Today 12: 12-18.
6. Iijima S (1991) Helical microtubules of graphitic carbon. Nature 354: 56-58.
7. Lau KT, Lu M, Hui D (2006) Coiled carbon nanotubes Synthesis and their potential applications in advanced composite structures. Compos. Part B Eng 37: 437-448.
8. Csató A, Szabó A, Fonseca A (2012) Synthesis and characterization of coiled carbon nanotubes. Catal Today 181: 33-39.
9. Dunlap BI (1992) Connecting carbon tubules. Phys Rev B 46: 1933-1936.
10. Amelinckx S, Zhang XB, Bernaerts D, Zhang XF (1994) Formation mechanism for catalytically grown helix-shaped graphite nanotubes. Science 265: 635-639.
11. Zhong OY, Su ZB, Wang CL (1997) Coil formation in multishell carbon nanotubes: competition between curvature elasticity and interlayer adhesion. Phys Rev Lett 78: 4055-4058.

12. Motojima S, Hasegawa I, Kagiya S, Momiyama M (1993) Preparation of coiled carbon fibers by pyrolysis of acetylene using a Ni catalyst and sulfur or phosphorus compound impurity. Appl Phys Lett 62: 2322-2323.

13. Zhang XB, Zhang XF, Bernaerts D, Van G (1994) The Texture of Catalytically Grown Coil-Shaped Carbon Nanotubules. Europhys Lett 27: 141-146.

14. Iwanaga H, Kawaguchi M, Motojima S (1993) A Growth Mechanism of Coiled Whiskers of Silicon Nitride and Carbon. Jpn J Appl Phy 32: 105-115.

15. Itoh S, Ihara S, Kitakami J (1993) Toroidal from of carbon C60. Phys Rev B 47: 1703-1704.

16. Itoh S, Ihara S, Kitakami J (1993) Helically coiled cage forms of graphitic carbon. Phys Rev B 48: 5643-5647.

17. Itoh S, Ihara S (1993) Toroidal forms of graphitic carbon II Elongated tori. Phys Rev B 48: 8323-8327.

18. Liu L, Zhou K, He P, Chen T (2013) Synthesis and microwave absorption properties of carbon coil–carbon fiber hybrid materials. Mater Lett 110: 76-79.

19. Hekmat F, Sohrabi B, Rahmanifar MS, Vaezi MR (2014) Super capacitive properties of coiled carbon nanotubes directly grown on nickel nanowires. J Mater Che A 2: 17446-17453.

20. Chen H, Roy A, Baek JB, Zhu L, Qu J (2010) Controlled growth and modification of vertically-aligned carbon nanotubes for multifunctional applications. Mater Sci Eng R 70: 63-91.

21. Prasek J, Drbohlavova J, Chomoucka J, Hubalek J (2011) Methods for carbon nanotubes synthesis. J Mater Chem. 21: 15872-15884.

22. Quinton BT, Barnes PN, Varanasi CV (2013) A Comparative Study of Three Different Chemical Vapor Deposition Techniques of Carbon Nanotube Growth on Diamond Films. J Nano mater 356259: 1-9.

23. Khavrus VO, Weiser M, Fritsch M, Leonhardt A (2012) Application of Carbon Nanotubes Directly Grown on Aluminum Foils as Electric Double Layer Capacitor Electrodes. Chem Vapor Depos 18: 53-60.

24. Chemical Vapour Deposition Precursors Processes and Applications (2009) RSC publishing.

25. Yen JH, Leu IC, Wu MT, Lin CC (2004) Density Control for Carbon Nanotube Arrays Synthesized by ICP-CVD Using AAO/Si as a Nano template. Electro chem Solid St 7: 29-31.

26. Chhowalla M, Teo KBK, Ducati C (2001) Growth process conditions of vertically aligned carbon nanotubes using plasma enhanced chemical vapor deposition. J Appl Phys 90: 5308-5317.

27. Shin YS, Hong JY, Ryu DH (2007) The Role of H2 in the Growth of Carbon Nanotubes on an AAO Template. J Korean Phys Soc. 50: 1068-1072.

28. Lu W, Qu L, Henry K (2009) High performance electrochemical capacitors from aligned carbon nanotube electrodes and ionic liquid electrolytes. J Power Sources 189: 1270-1277.

29. Wen S, Mho S, Yeo IH (2006) Improved electrochemical capacitive characteristics of the carbon nanotubes grown on the alumina templates with high pore density. J Power Sources 163: 304-308.

30. Hu W, Yuan L, Chen Z (2002) Fabrication and Characterization of Vertically Aligned Carbon Nanotubes on Silicon Substrates Using Porous Alumina Nano templates. J Nanosci Nanotech 2: 203-207.

21

Crystal Structure, Spectral Characterization and Biologically Studies of Mononuclear Transition Metal Complexes Derived from New N_2O_2 Type Ligand

Samina K Tadavi[1], Jamatsing D Rajput[1], Suresh D Bagul[1], Jaiprakash N Sangshetti[3], Amar A Hosamani[2] and Ratnamala S Bendre[1]*

[1]School of Chemical Sciences, North Maharashtra Jalgaon, Maharashtra, India
[2]Solid State and Structural Chemistry Unit, Indian Institute of Science, Bangalore, Karnataka, India
[3]YB Chavan College of Pharmacy, Dr. Rafiq Zakaria Campus, Aurangabad, Maharashtra, India

Abstract

The new mononuclear metal complexes viz Mn(III), Co(II), Ni(II) and Cu(II) have been synthesized by using tetradentate N2O2 donor symmetric schiff base 6,6'-((1E,1'E)-(1,2-phenylenebis(azanylylidene))bis(methanylylidene)) bis(5-isopropyl-2-methylphenol) HL and employing with corresponding metal chloride or acetate salts. After successful synthesis of compounds were thoroughly characterized by Elemental analysis, FT-IR, Uv-visible, NMR spectroscopy, LC-MS spectrometry, SEM analysis, Magnetic susceptibility measurement, Molar conductance, ESR spectroscopy. X-ray single crystal structure of HL schiff base has been determined. The synthesized compounds have been screened for their antimicrobial and antioxidant activities, which show significant results.

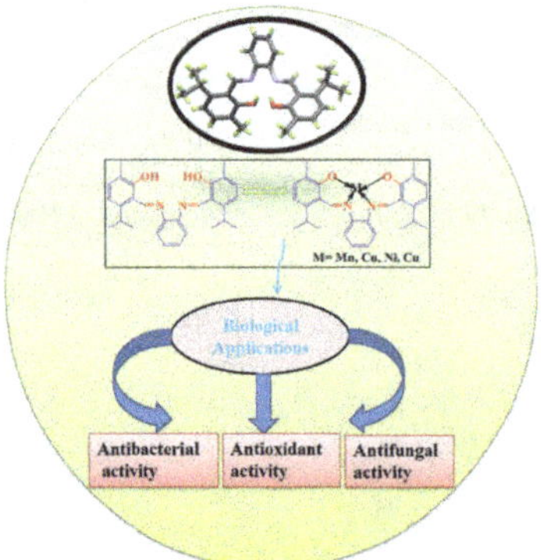

Keywords: Schiff base; Mononuclear transition metal complexes; ESR; Antimicrobial activity; Antioxidant activity

Introduction

Schiff bases (RHC=NR) are a class of organic compounds typically formed by condensation of a primary amine and an aldehyde and they are considered as privilege ligand [1,2]. Schiff base ligands with N_2O_2 donor atoms are well known to coordinate with a variety of metal ions. They have attracted much interest in recent years, due to ease of synthesis, their stability under a variety of oxidative and reductive conditions and their structural versatility associated with various applications. These compounds have various applications in industries as dyes, drug synthesis [3] bioinorganic chemistry [4] electrochemistry [5] dioxygen uptake and catalysis [6-10]. Tetra-dentate N_2O_2 donors schiff bases derived from o-phenylenediamine and salicyaldehyde have been widely studied in solid state. They have been studied for a variety of applications including biological, clinical and analytical. The previous work has shown that some drugs showed increased activity when administered as metal chelates rather than as organic compounds [11-13]. A search through literature reveals that no work has been done on the transition metal complexes of the symmetrical schiff base derived from o-phenylenediamine and 2-hydroxy-6-isopropyl-3-methyl benzaldehyde. In this paper, we report synthesis of a new type of tetradentate Schiff base ligand formed by the simple condensation of o-phenylenediamine with 2-hydroxy-6-isopropyl-3-methyl benzaldehyde and its Mn(III), Co(II), Ni(II) and Cu(II) mononuclear complexes. The prepared ligand and complexes were characterized using Elemental analysis, FT-IR, UV-visible, NMR, LC-MS, ESR and crystal structure and further screened for antimicrobial and antioxidant activity.

Experimental

All the solvents and chemicals were of commercial reagent grade and used as received without further purification. FT-IR spectra were recorded as KBr pellets on a SHIMADZU FT-IR- 8400 spectrometer from range 4000 to 400 cm^{-1}. The electronic spectra were recorded in DMF solutions on the UV 2400 series spectrophotometer. ^{1}H and ^{13}C-NMR spectra were precise with a BRUKER AVANCE III (400 MHz) spectrometer and proton chemical shifts are recorded in ppm relative to tetramethylsilane as an internal standard using $CDCl_3$ as solvent and the LC-MS spectra have been carried out with Waters Micromass Q-Tof Micro instrument. The elemental analyses were carried out with a Thermo Finnigan elemental analyzer. The X-band ESR spectra of

***Corresponding author:** Bendre RS, School of Chemical Sciences, North Maharashtra University, Maharashtra, India, E-mail: bendrers@gmail.com copper complex are recorded on a JES-FA200 EPR spectrometer with DPPH as standard. Magnetic susceptibilities are measured at room temperature on a guoy balance using $Hg[Co(NCS)_4]$ as reference. SEM-EDS analyses were performed on SEM JEOL JSM 6360 and JEOL JSM 5400, Japan. Molar conductance of metal complexes was calculated in DMF at room temperature on Systronic Conductivity Bridge.

Synthesis of Schiff base (L)

The ligand was prepared by dropwise addition of a warm ethanolic solution of o-phenylenediamine (1 mmol) to constantly stirring warm solution of 2-hydroxy-6-isopropyl-3-methyl benzaldehyde (2 mmol) in ethanol. The resulting solution was refluxed for 3 h, formation of orange precipitate. The obtained product was filtered, washed with cold ethanol and recrystallized from ethanol and dried at room temperature. Yield: 77 %Colour: Orange. Anal. calcd for $C_{28}H_{32}N_2O_2$ (%): C 78.47, H 7.53, N 6.54; Found: C 78.31, H 8.16 N 6.61. FT-IR (KBr pellet, cm^{-1}) ν_{max}: 3412 (OH), 1601 (C=N), 1460 (C=C), 1253 (C-O). UV-vis (DMF) λ_{max} (nm): 268, 341. ^{1}H- NMR ($CDCl_3$, 400 MHz) (δ, ppm): 14.16(s, 2H, -OH), 7.39-7.37 (m, 2H, Ar), 7.32-7.30 (m, 2H, Ar), 7.20 (d, 2H, J=8Hz), 6.74 (d, 2H, Ar), 3.48-3.45 (m, 2H, $-2(CH_3)$), 2.17 (s, 6H, $2(CH_3)$), 1.19-1.24 (m, 12H, $4(CH_3)$). ^{13}C- NMR ($CDCl_3$, 400 MHz) (δ, ppm): 162.31, 160.88, 147.99, 142.74, 134.72, 127.34, 124.05, 121.16, 114.95, 28.09, 24.07, 15.60. MS (m/z): calcd 428.57; obsv 429.4.

General procedure for the synthesis of mononuclear metal complexes

The complexes were prepared by dropwise addition of (1 mmol) warm ethanolic solution of ligand to the corresponding suitable ethanolic metal acetate or chloride salts (2 mmol) in basic condition under inert atmosphere nitrogen gas. The reaction mixture was refluxed till completion of reaction, the precipitate was obtained and collected product was washed with cold ethanol and then diethyl ether.

Mononuclear Mn(III) complex [1]: Yield: 71%. Colour: reddish brown. Anal. calcd for $C_{30}H_{33}MnN_2O_4$ (%): C 66.66, H 6.15, N, 5.18; Found: C 64.93, H 5.91, N 5.97. FT-IR (KBr, pellet cm^{-1}) ν_{max}: 1591 (C=N), 1276 (C-O), 1382 (C=C), 530 (M-O), 480 (M-N). UV-Vis (DMF) λ_{max} (nm): 268, 342, 471. LC-MS (m/z): calcd 540.53; obsv 481.3. μ_{eff}: 4.82 B.M. Conductance ($\Lambda_{M,}$ Ω^{-1} cm^2 mol^{-1}) in DMF: 9.60.

Mononuclear Co(II) complex [2]: Yield: 65%. Colour: brown. Anal. calcd for $C_{28}H_{30}CoN_2O_2$ (%): C 69.27 H 6.23, N 5.77; Found: C 68.29, H 5.14, N 5.58. FT-IR (KBr pellet, cm^{-1}) ν_{max}: 1541 (C=N), 1205 (C-O), 1382 (C=C), 530 (M-O), 480 (M-N). UV-Vis (DMF) λ_{max} (nm): 264, 506, 620. LC-MS (m/z): calcd 485.48; obsv 485.3 μ_{eff}: 3.80 B.M. Conductance (Λ_M, Ω^{-1} cm^2 mol^{-1}) in DMF: 14.35.

Mononuclear Ni(II) complex [3]: Yield: 80%. Colour: reddish orange. Anal. calcd for $C_{28}H_{30}N_2NiO_2$ (%): C 69.31, H 6.23, N 5.77; Found: C 70.34, H 6.03, N 6.35. FT-IR (KBr pellet, cm^{-1}) ν_{max}: 1560 (C=N), 1234 (C-O), 1452 (C=C), 474 (M-O), 432 (M-N). UV-Vis (DMF) λ_{max} (nm): 267, 307, 386, 490, 547. LC-MS (m/z): calcd 485.23; obsv 485.41. Conductance (Λ_M, Ω^{-1} cm^2 mol^{-1}) in DMF: 15.64.

Mononuclear Cu(II) complex [4]: Yield: 78%. Colour: Shining dark green. Anal. calcd for $C_{28}H_{30}CuN_2O_2$ (%): C 68.62, H 6.17, N 5.72, Found: C 66.79, H 5.18, N 5.67. FT-IR (KBr, cm-1) ν_{max}: 1600 (C=N), 1240 (C-O), 1404 (C=C), 550 (M-O), 455 (M-N). UV-Vis: (DMF) λ_{max} (nm): 264, 444, 632. LC-MS (m/z): calcd 490.10; obsv 490.40. μ_{eff}: 1.75 B.M. Conductance (Λ_M Ω^{-1} cm^2 mol^{-1}): 12.67. $g_{||}$=2.2317, g=2.0472.

Bioassay

Protocol for antibacterial activity: The antibacterial activity of the compounds was performed by enumerating the viable number of cells upon in the nutrient broth containing various concentrations of compounds. The viable number is represented by colony count method. The test organisms used on which the antibacterial activity was performed were *Escherichia Coli, Bacillus subtilis* (NCIM-2063), *Pseudomonas aeruginosa* (NCIM-2036) and *Staphylococcus aureus* (NCIM-2901). The minimum inhibitory concentration (MIC μg/mL) values were determined. The lowest MIC values show the potent activity. Ciprofloxacin and Ampicillin were used as standard drugs. All the experiment were carried out in triplicate and the averaged results are considered [14].

Protocol for antifungal activity: The antifungal activity was evaluated against different fungal strains, such as *Candida albicans* (NCIM 3471), *Fusarium oxysporum* (NCIM1332), *Aspergillus flavus* (NCIM 539), *Aspergillus niger* (NCIM 1196), and *Cryptococcus neoformans* (NCIM 576). Minimum inhibitory concentration (MIC, μg/mL) values of all the compounds were determined using the standard agar dilution method as per CLSI guidelines. By using miconazole and fluconazole as standard antifungal drugs. All the experiments are performed in triplicate and average results are considered [15].

Protocol for antioxidant activity: DPPH (2,2-diphenyl-1-picrylhydrazyl) radical scavenging activity was evaluated according to the reported method [16,17]. Ascorbic acid was used as standard. Each sample and positive control (ascorbic acid) were performed in triplicate. Scavenging activity versus concentration in ppm were plotted to estimate a 50% reduction of its initial value (EC_{50}). Lower absorbance of the reaction mixture indicates higher free radical scavenging activity. The capability to scavenge the DPPH radical was calculated using the following equation:

DPPH radical scavenging activity (%)=$\text{Absorbance}_{(control)} - \text{Absorbance}_{(standard)} / \text{Absorbance}_{(control)} \times 100$

Where, $\text{Absorbance}_{(control)}$:Absorbance of DPPH radical+methanol $\text{Absorbance}_{(standard)}$:Absorbance of DPPH radical+extract/standard.

Results and Discussion

The symmetrical Schiff base was prepared by simply condensation of 2-hydroxy-3-methyl-benzaldehyde [18] with o-phenylenediamine in 2:1 molar ratio in ethanolic solution (Figure 1). The Mn(III),

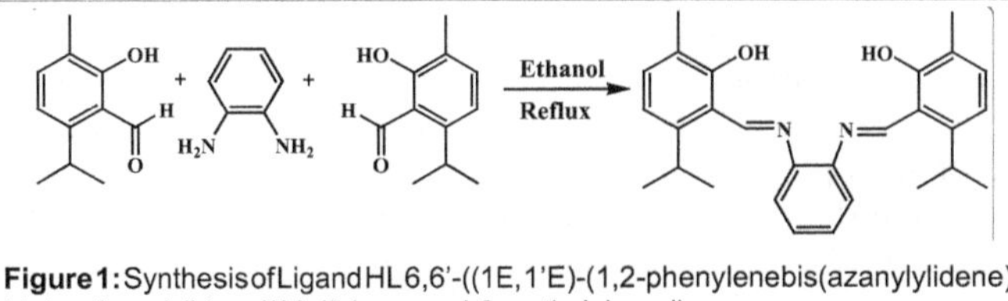

Figure 1: Synthesis of Ligand HL 6,6'-((1E,1'E)-(1,2-phenylenebis(azanylylidene))bis(methanylylidene))bis(5-isopropyl-2-methylphenol).

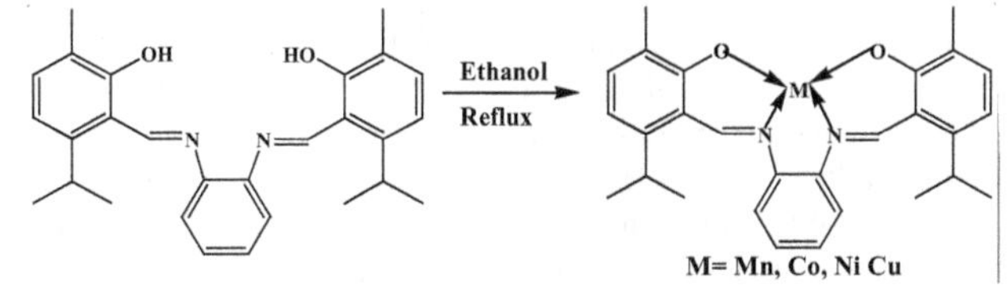

Figure 2: Synthesis of mononuclear metal (II) complexes with 6,6'-((1E,1'E)-(1,2-phenylenebis(azanylylidene))bis(methanylylidene))bis(5-isopropyl-2-methylphenol).

Compounds Name	*E. coli* (MIC μg/mL)	*P. aeruginosa* (MIC μg/mL)	*B. subtilis* (MIC μg/mL)	*S. aureus* (MIC g/mL)
Carvacrol	72	97	56	129
Carvcrol aldehyde	95	120	75	193
Ligand	102	49	63	77
Mn(II)	111	101	81	115
Co(II)	168	170	230	99
Ni(II)	63	93	74	56
Cu(II)	74	85	102	74
Ciprofloxacin	25	25	50	50
Ampicillin	100	100	250	250

Table 1: The antibacterial activity representation.

Compounds Name	*C. albicans* (MIC μg/ mL)	*A. flavus* (MIC μg/ mL)	*A. niger* (MIC μg/ mL)	*C. neoformans* (MIC μg/ mL)
Carvacrol	125	125	125	100
Carvacrol aldehyde	*	125	*	225
Ligand	100	150	150	*
Mn(II)	125	225	125	*
Co(II)	100	100	*	150
Ni(II)	150	*	100	125
Cu(II)	150	*	*	*
Miconazole	25	12.5	25	25
Fluconazole	12.5	6.25	12.5	6.25

Table 2: The antifungal activity representation.

Co(II), Ni(II) and Cu(II) mononuclear complexes were synthesized by the direct reaction of ligand with an equimolar amount of metal(II) acetate/ chloride salts in ethanol in 1:1 molar ratio (Figure 2). All the compounds are soluble in polar solvents. The spectral analyses agree well with the proposed structure of the complexes. All the compounds were subjected to *in vitro* antibacterial, antifungal and antioxidant activities.

Biological activity

As per the literature survey carvacrol possess good antibacterial, antifungal and antioxidant activities [19]. Our antibacterial data illustrated in Table 1, that the antibacterial activity decreases upon formylation of carvacrol to carvacrol aldehyde as compared to the carvacrol. While schiff base ligand possesses the superior activity as compared to the carvacrol against *Pseudomonas aeruginosa, Bacillius subtilis and Staphylococcus aureus.* All the metal complexes presented better antibacterial activity against the *Staphylococcus aureus* as compared to carvacrol (Table 2). The Ni(II) and Co(II) complexes exhibited better antibacterial activity than the ligand against all the microbes and standard ampicillin. The antifungal activity results indicated that carvacrol aldehyde possesses good antifungal activity against the *Aspergillus flavus* and *Cryptococcus neoformans.* The schiff base ligand shows the better activity in contrast to carvacrol for the fungal strain *Candida albicans* and moderate to *Aspergillus flavus* and *Aspergillus niger.* All the complexes show good activity against the *Candida albicans..* The Co(II) and Ni(II) complexes showed better activity against *Candida albicans, Aspergillus flavus and Aspergillus niger, Cryptococcus neoformans* respectively. The better antibacterial and antifungal activities of these complexes as compared to the ligand may be explained on the basis of chelation theory [20]. The process of scavenging DPPH-free radicals has been used to assess the antioxidant activity of specific compounds [21]. DPPH is a stable free radical that can accept an electron or hydrogen radical and get converted to a stable, diamagnetic molecule. DPPH has an odd electron and has strong absorption band at 517 nm. When this electron becomes paired off, the absorption decreases stiochometrically with respect to the number of electrons or hydrogen atoms taken up. Such a change in the absorbance by this reaction has been extensively adopted to test the capacity of several molecules to act as free radical scavengers (Table 3). Hence, more rapidly the absorbance decreases, the more potent is the antioxidant activity of the compound [22]. Our result indicate that carvacrol aldehyde possesses the strong activity as compared to the carvacrol while the Schiff base ligand possess lesser activity as compared to both. However upon complexation with metal ions *viz* Mn(III) and Cu(II) the activity was improved significantly. All the compounds showed comparable or better activity to that of standard ascorbic acid. The Mn(III) and Cu(II) complex showed significantly higher antioxidant activity followed by Ni(II) and Co(II) complexes at different concentrations.

X-ray crystallographic analysis

Yellow colored crystal of suitable size 0.23 × 0.22 × 0.21 mm^3 and mounted on 'Bruker APEX-II CCD' diffractometer equipped with graphite monochromated Mo Kα radiation in the wavelength of 0.71073 Å at room temperature. The summary of crystallographic parameters, data collection and refinement is given in Table 4 and the additional details concerning the data collections, structure solution and refinement are included in the supporting data. The CCDC No. for Ligand is 1443334. The single crystal of the Schiff base demonstrated the monoclinic system having $P2_1$ space group with the two molecules in the unit cell. The ORTEP diagram with numbering and packing diagram are shown in the Figure 3. The selected bond lengths and bond angles are depicted in Table 6 While, the hydrogen bonding parameters are given in Table 5. The crystal structure of the symmetric Schiff base ligand represented, the N1-C15 and C_8-N_2 distances as 1.286(8) and 1.277(8) A° respectively for C=N double bonding. The two C_4-O_1 and C_{21}-O_2 having bond distances 1.339(8) and 1.349(8)A° respectively illustrated the C-O phenolic single bond. The intramolecular hydrogen bonding formed by the H_1-O_1-N_2 and H_2-O_2-N_1, with distances of 2.617 and 2.579 A° has resulted in the formation of a five membered chelating ring. There is no solvent molecule appear in the structure. From this crystal structure study, it is confirmed that the symmetric Schiff base having two imine nitrogen atoms and two phenolic oxygen atoms (N_1, N_2, O_1, O_2) has been successfully synthesized and suggests that this ligand can easily act as tetradentate donor [23,24].

Spectral characterization

FT-IR: The IR spectroscopy is used to determine the characteristic peaks (OH, C=N, C-O) in ligand before and after complexation. The IR spectra of the ligand and complexes are compared, the band at 1601 cm^{-1} is characteristic of the azomethine nitrogen atom present in the schiff base ligand is found to shift to lower frequency region 1546-1556 cm^{-1} in all the complexes and illustrated the involvement of the azomethine nitrogen atom in coordination. The peak at 3412 cm^{-1} appear for the phenolic -OH group. The absence of the broad band at 3500-3100 cm^{-1}. indicated the absence of coordinated water molecule to complexes. The C-O peak appearing in the ligand at 1298 cm^{-1} is shifted to 1340-1388 cm^{-1} in all complexes. The involvement of oxygen and nitrogen atoms in coordination with metal is further definited by the appearance of weak and low frequency bands in the range 400-600 cm^{-1} corresponding to the M-O and M-N respectively [25].

Electronic spectra: The absorption value of high energy 265-280 nm and 340- 490 nm are assigned to the intraligand $\pi \rightarrow \pi^*$ and $n \rightarrow \pi^*$ transitions respectively. The electronic spectra of copper consist of weak d-d transition in the visible region as a broad band at 632 nm attributed to square planar complexes of similar tetradentate schiff base ligands with o-phenylenedimaine, The nickel complex shows the absorption at 547 nm indicating that the complex is diamagnetic nature and possesses square planar geometry. The absorption at 506 nm and 620 nm shows cobalt complex possess the square planar geometry [26].

Mass: In the mass spectra of the ligands and complexes the molecular ion peak is significantly more abundant than other fragment ions. The proposed fragmentations are equivalent with the empirical formula of proposed structure of the ligands and complexes.

NMR: In the ^1H NMR spectrum of the ligand in DMSO shows the following signals: 4.16 δ singlet are assigned to the phenolic -OH group and the values 7.39-7.37, 7.32-7.30, 7.20, 6.74 δ are assigned to the aromatic region, the peak at multiplet 3.48-3.45 δ and 1.19-1.24 δ are attributed to the 2-CH and of four methyl group of isopropyl group respectively and singlet peak at 2.17 δ assigned for methyl group.

SEM: Scanning electron micrograph (SEM) have been currently used to determine the morphology and the grain size of the metal complexes. The SEM photograph of the complexes are illustrated in the Figure 4 for all the metal complexes the SEM photograph were taken in the different scale range from 1 μm to 50 μm. From the SEM photograph it was noted there is a uniform matrix in all the metal complexes. The photograph of Mn(II) complex shows the flakes like morphology while Co(II) complex exhibit the rod like morphology. The Ni(II) and Cu(II) displays the fibre like morphology.

Conductivity measurement: The molar conductivity values of the complexes were in the range 15-25 Ω^{-1} cm^{-1} mol^{-1}. This value indicated that the complexes having non-electrolytic in nature [27].

Magnetic susceptibility: The Mn(III) complex shows the distorted square planer geometry having magnetic moment 4.85 B.M. The magnetic moment of the Co(II) complexes exhibit at 2.81 BM [28] consistent with Square planar geometry Ni(II) complexes are square planar and diamagnetic in nature [29] while the magnetic moment of Cu(II) complex exhibited in the range 1.75 representing that copper has square planar geometry showing one unpaired electron [30,31].

ESR spectra: The ESR spectra of the CuL complex in DMF solution at 77 K at liquid nitrogen temperature exhibits in the perpendicular region, three of four hyperfine features are well resolved while the fourth one is overlapped by *g* features [32]. The values of ESR parameters G_{avg}, G, α^2, *f* for Cu(II) complex calculated are 2.22, 2.040, 2.1, 5.75, 186 × 10^{-4}, 0.79, 118, respectively. From the calculated values it illustrated that

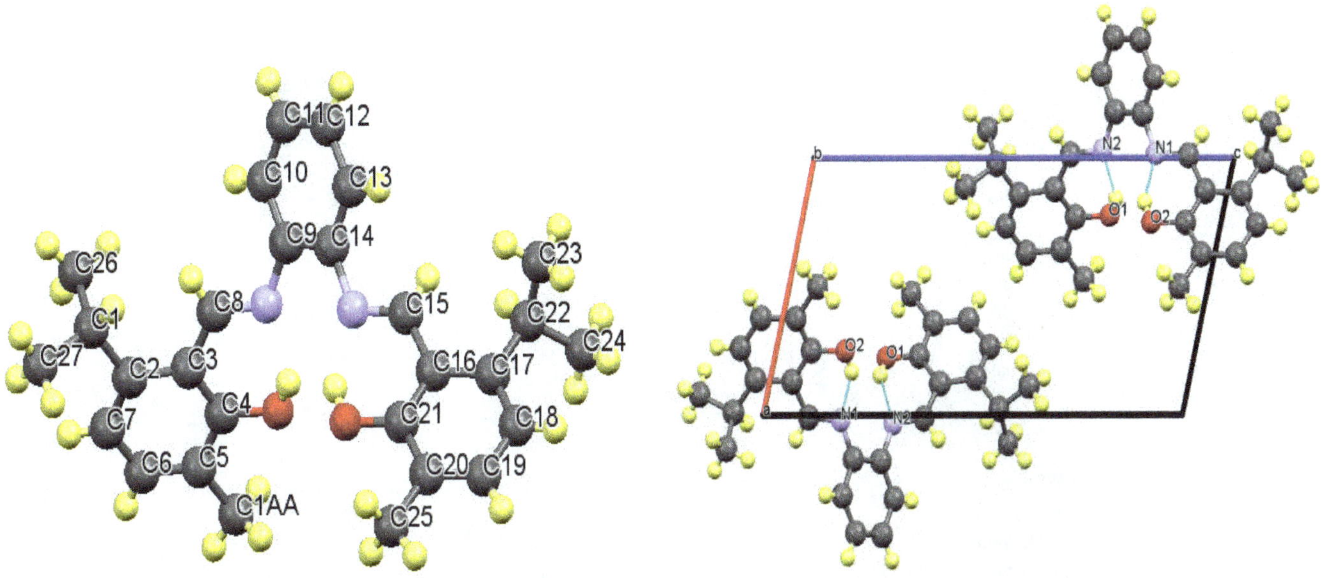

Figure 3: ORTEP diagram with atomic labeling and crystal packing diagram viewed along b with O—H--N intramolecular hydrogen bond is shown as a light blue dashed line.

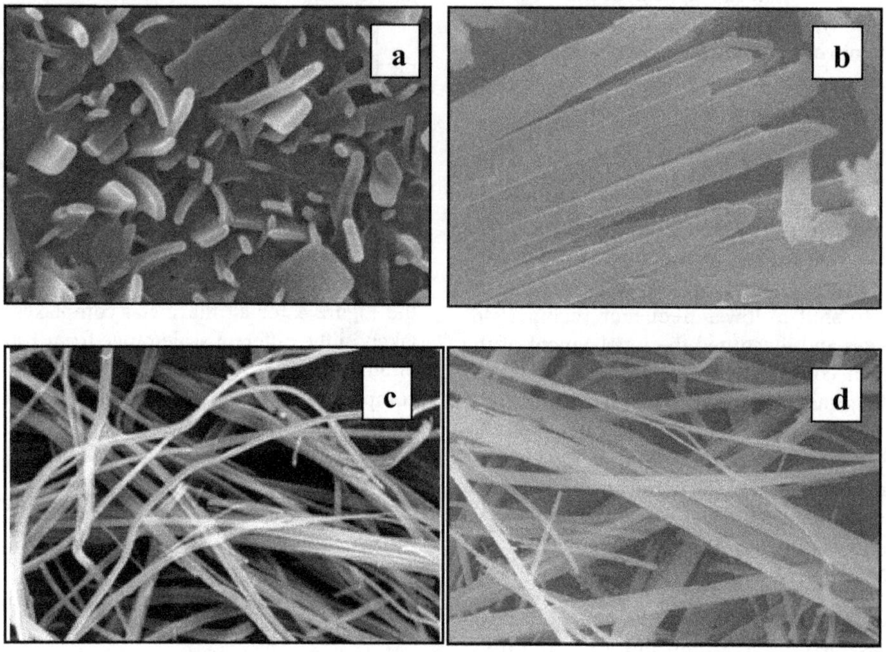

a=Mn(III), b=Co(II), c=Ni(II), d=Cu(II)

Figure 4: SEM photograph of all the mononuclear complexes.

Compound Name	EC_{50}
Carvacrol	0.1137
Carvacrol aldehyde	0.1107
Ligand	0.1142
Mn(II)	0.1124
Co(II)	0.1172
Ni(II)	0.1158
Cu(II)	0.1135
Ascorbic acid	0.1203

Table 3: The antioxidant activity representation.

Empirical formula	$C_{28}H_{32}N_2O_2$
Formula weight	428.55
Temperature/K	296.15
Crystal system	monoclinic
Space group	$P2_1$
a/Å	8.561(2)
b/Å	8.705(2)
c/Å	16.556(4)
α/°	90
β/°	103.736(11)
γ/°	90
Volume/Å^3	1198.6(5)
Z	2
ρ_{calc} g/cm^3	1.187
μ/mm^{-1}	0.074
F(000)	460.0
Crystal size/mm^3	0.23 × 0.22 × 0.21
Radiation	MoKα (λ=0.71073)
2Θ range for data collection/°	2.532 to 50
Index ranges	-10 ≤ h ≤ 10, -10 ≤ k ≤ 10, -19 ≤ l ≤ 19
Reflections collected	16503
Independent reflections	4178 [R_{int}=0.0816, R_{sigma}=0.0720]

Data/restraints/parameters	4178/1/297
Goodness-of-fit on F^2	1.127
Final R indexes [I>=2σ (I)]	R_1=0.1051, wR_2=0.2745
Final R indexes [all data]	R_1=0.1275, wR_2=0.2945
Largest diff. peak/hole / e Å^{-3}	0.92/-0.29
Flack parameter	-1.2(10)

Table 4: Crystallographic parameters, data collection and refinement is given as follows.

Bond Lengths		**Bond Angles**	
C_8-N_2	1.277(8)	O_1-C_4-C_3	123.1(6)
N_1-C_{15}	1.286(8)	O_2-C_{21}-C_{16}	121.5(6)
N_2-C_9	1.422(8)	N_2-C_8-C_3	124.1(6)
N_1-C_{14}	1.422(9)	N_1-C_{15}-C_{16}	122.0(6)
O_1-C_4	1.339(8)	C_8-N_2-C_9	118.6(6)
C_{21}-O_2	1.349(8)	C_{15}-N_1-C_{14}	115.5(6)

Table 5: Selected bond lengths/A° and angles/ ° for Ligand HL.

D-H...A	d(D-H)	d(H...A)	d(D...A)	<(D-H...A)
O_2-H_2...N_1	0.820	1.847	2.580	148.10
C_{27}-H_{27}B... O_1	0.960	2.895	3.819	161.80
C_{24}-H_{24}B...O_2	0.960	2.916	3.841	161.93
C_{12}-H_{12}...O_2	0.930	2.867	3.717	152.52

Table 6: The intramolecular hydrogen bonds and angles.

the copper complex follows the trend and also provides information about the unpaired electron that it is localized in $d_{X^2-Y^2}$ orbital having $^2B_{1g}$ as ground state which is consistent with the square planar geometry [33]. The extent of geometrical distortion is measured by the f=g/A ratio [34]. It has been reported that complex falls between 105–135 cm^{-1} and also ratio for tetragonally distorted complex falls in the range 135-250 cm^{-1} [35]. The current complex has f=118 cm^{-1} indicating that copper complex has square planar geometry. The value of the exchange coupling interaction between two Cu(II) ions in terms of G has been explained by the Hathway expression (0023)/(0023). From the expression, if G>4·0, the exchange interaction is negligible while when the value of G<4.0 significant exchange coupling is present in the solid complex [36]. This result indicates that the exchange coupling effects are not effective in the present complex. The value of in plane sigma bonding parameter α^2 was expected from following expression,

$$\alpha^2=-(A_{\parallel}/0.036)+(g_{\parallel}-2.0023)+3/7(g_{\perp}-2.0023)+0.04$$

The value of α^2=0.5, indicates complete covalent bonding, while the value of α^2=1.0 suggests complete ionic bonding. The observed value of α^2 is less than 1, which indicated that the complex has some covalent character in the ligand environment [37].

Conclusion

In this investigation we are reporting the synthesis and characterization of symmetrical salen based ligand and their Mn(III), Co(II), Ni(II) and Cu(II) complexes. The single crystal structure of Schiff base ligand have been solved by single x-ray crystallography which having two imine nitrogen atoms and two phenolic oxygen atoms act as tetradentate donor. The Schiff base ligand and metal complexes were screened for their biological activities such as, *in vitro* antibacterial, antifungal and antioxidant activities. The results indicated Ni(II) and Cu(II) complexes exhibit better antibacterial activity against *Escherichia coli, Pseudomonas aeruginosa, Bacillus subtilis* and *Staphylococcus aureus* as compared to carvacrol and standard drug (ciprofloxacin). The Co(II) complex show the good antifungal activity against *Candida albicans, Aspergillus flavus, Cryptococcus neoformans* as campared to parent molecule (carvacrol). The Mn(III) and Cu(II) complex shows the good antioxidant activity as comparable to standard drug (ascorbic acid).

Acknowledgements

Samina K Tadavi gratefully thanks UGC, New Delhi for Rajiv Gandhi National Fellowship for ST candidates. We are also thankful to Prof. TN Guru Row, IISC, Bangalore for extending help in solving crystal structure of schiff base ligand.

References

1. Cozzi PG (2004) Metal-Salen Schiff base complexes in catalysis practical aspects. Chemical Society Reviews 337: 410-421.
2. Budige G, Puchakayala MR, Kongara SR, Hu A (2011) Synthesis characterization, and biological evaluation of mononuclear Co (II), Ni (II), Cu (II) and Pd (II) complexes with new N2O2 Schiff base ligands. Chemical and Pharmaceutical Bulletin 59: 166-171.
3. Yimer AM (2014) Chemical preparation, spectro-magnetic and biocidal studies on some divalent transition metal complexes of schiff's base derived from 1-Phenyl-2-(Pyridin-2-yl) ethane-1, 2-dione and ethylenediamine. Modern Chemistry & Applications 2: 1-3.
4. Akila E, Usharani M, Rajavel R (2013) Metal (II) complexes of bioinorganic and medicinal relevance: Antibacterial, Antioxidant and DNA cleavage studies of tetradentate complexes involving O, N-donor environment of 3, 3'-dihydroxybenzidine-based Schiff bases. International Journal of Pharmacy and Pharmaceutical Sciences 5: 573-581.
5. Djebbar SS, Benali BO, Deloume JP (1998) Synthesis characterization electrochemical behaviour and catalytic activity of manganese (II) complexes with linear and tripodal tetradentate ligands derived from Schiff bases. Transition Metal Chemistry 23: 443-447.
6. Klement R, Stock F, Elias H, Paulus H (1999) Copper (II) complexes with derivatives of salen and tetrahydrosalen a spectroscopic, electro chemical and structural study. Polyhedron 18: 3617-3628.
7. Niederhoffer EC, Timmons JH, Martell AE (1984) Thermodynamics of oxygen binding in natural and synthetic dioxygen complexes. Chemical Reviews 84: 137-203.
8. Martell AE, Sawyer DT (1988) Oxygen Complexes and Oxygen Activation by Transition Metals.
9. Holm RH (1987) Metal-centered oxygen atom transfer reactions. Chemical Reviews 87: 1401-1449.
10. Dixit PS, Srinivasan K (1988) The effect of clay-support on the catalytic epoxidation activity of a manganese (III)-Schiff base complex. Inorganic Chemistry 27: 4507-4509.

11. Singh P, Goel RL, Singh BP (1975) Synthesis Characterization and Biological Activity of Schiff Bases. J Indian Chem Soc 52: 958-959.

12. Mohindru A, Fisher JM, Rabinovitz M (1983) Bathocuproine sulphonate a tissue culture-compatible indicator of copper-mediated toxicity. Nature 303: 64-65.

13. Raman N, Raja YP, Kulandaisamy A (2001) Synthesis and characterisation of Cu (II), Ni (II), Mn (II), Zn (II) and VO (II) Schiff base complexes derived fromo-phenylenediamine and acetoacetanilide. Journal of Chemical Sciences 113: 183-189.

14. Greenwood D, Slack RCB, Peutherer JF (1992) Medical Microbiology. 14 edn. ELBS London, UK.

15. Saundane AR, Rudresh K, Satyanarayan ND, Hiremath SR (1998) Pharmacological screening of 6H, 11H-indolo [3, 2-c] isoquinolin-5-ones and their derivatives. Indian Journal of Pharmaceutical Sciences 60: 379-383.

16. Chandrappa CP, Govindappa M, Kumar AN (2013) In vitro antimitotic, antiproliferative and DNA fragmentation assay of ethanol extract of Carmona retusa (Vahl.) Masam. Applied Cell Biology 2: 52-57.

17. Dery R, Davison R, Befus A (2001) International Archives of Allergy and Immunology 124: 201-204.

18. Rajput JD, Bagul SD, Tadavi SK, Karandikar PS (2016) Design, Synthesis, and Biological Evaluation of Novel Class Diindolyl Methanes (DIMs) Derived from Naturally Occurring Phenolic Monoterpenoids. Medicinal Chemistry 6: 123-128.

19. Kumar D, Rawat DS (2013) Synthesis and antioxidant activity of thymol and carvacrol based Schiff bases. Bioorganic and Medicinal Chemistry Letters 23: 641-645.

20. Mohie EM, Shishtawy RM (2013) Synthesis and Antimicrobial Activity of Aluminium (III), Nickel (II) and Zinc (II) Schiff base Complexes Derived from o-Phenylenediamine and Salicylaldehyde. Asian Journal of Chemistry 25: 2719-2721.

21. Bilgicli AT, Tekin Y, Alici EH, Yaraşir MN (2015) α-or β-Substituted functional phthalocyanines bearing thiophen-3-ylmethanol substituents: synthesis, characterization, aggregation behavior and antioxidant activity. Journal of Coordination Chemistry 68: 4102-4116.

22. Harinath Y, Reddy DHK, Kumar BN, Apparao C (2013) Synthesis, spectral characterization and antioxidant activity studies of a bidentate Schiff base, 5-methyl thiophene-2-carboxaldehyde-carbohydrazone and its Cd (II), Cu (II), Ni (II) and Zn (II) complexes. Spectrochimica Acta Part A: Molecular and Biomolecular Spectroscopy 101: 264-272.

23. Rodríguez DMJ, Fernández MI, González NAM, Maneiro M (2006) Novel Manganese (III) Complexes with the Schiff Base N, N' - (1, 2-Phenylene) -bis (3-Hydroxysalicylidenimine). Synthesis and Reactivity in Inorganic, Metal-Organic and Nano-Metal Chemistry 36: 655-662.

24. Mota VZ, Carvalho GS, Corbi PP, Bergamini FR (2012) Crystal structure and theoretical studies of the keto-enol isomerism of N, N'-bis (salicylidene)-o-phenylenediamine (salophen). Spectrochimica Acta Part A: Molecular and Biomolecular Spectroscopy 99: 110-115.

25. Kavitha P, Reddy KL (2016) Synthesis spectral characterisation morphology biological activity and DNA cleavage studies of metal complexes with chromone Schiff base. Arabian Journal of Chemistry 9: 596-605.

26. Lashanizadegan M, Boghaei DM, Shivapour Z, Seraj S (2010) Synthesis and Crystal Structure of Non-symmetric Tetradentate Complexes of Ni (II), Cu (II), Co (II), and Zn (II). Synthesis and Reactivity in Inorganic, Metal-Organic and Nano-Metal Chemistry 40: 373-377.

27. Youssef NS, Zahany EA, Barsoum BN, Seidy AM (2009) Synthesis and characterization of copper (II), cobalt (II), nickel (II), and iron (III) complexes with two diamine Schiff bases and catalytic reactivity of a chiral diamine cobalt (II) complex. Transition metal chemistry 34: 905-914.

28. Dutta RL, Syamal A (1992) Elements of Magnetochemistry. 2nd edn. Elsevier.

29. Ray A, Banerjee S, Rosair GM, Gramlich V (2008) Variation in coordinative property of two different N2O2 donor Schiff base ligands with nickel (II) and cobalt (III) ions: characterisation and single crystal structure elucidation. Structural Chemistry 19: 459-465.

30. Rabindra RP, Shilpa A (2011) Synthesis Characterization, and DNA-Binding and-Cleavage Properties of Dinuclear CuII Salophen/Salen Complexes. Chemistry and Biodiversity 8: 1245-1265.

31. Mohamed GG, Omar MM, Hindy AM (2005) Synthesis, characterization, and biological activity of some transition metals with Schiff base derived from 2-thiophene carboxaldehyde and aminobenzoic acid. Spectrochimica Acta Part A: Molecular and Biomolecular Spectroscopy 62: 1140-1150.

32. Klement R, Stock F, Elias H, Paulus H (1999) Copper (II) complexes with derivatives of salen and tetrahydrosalen: a spectroscopic, electrochemical and structural study. Polyhedron 18: 3617-3628.

33. Raj BB, Kurup MP, Suresh E (2008) Synthesis, spectral characterization and crystal structure of N-2-hydroxy-4-methoxybenzaldehyde-N'-4-nitrobenzoyl hydrazone and its square planar Cu (II) complex. Spectrochimica Acta Part A: Molecular and Biomolecular Spectroscopy 71: 1253-1260.

34. Boraey HA, Emam SM, Tolan DA, Nahas AM (2011) Structural studies and anticancer activity of a novel (N 6 O 4) macrocyclic ligand and its Cu (II) complexes. Spectrochimica Acta Part A: Molecular and Biomolecular Spectroscopy 78: 360-370.

35. Tabl AS, Saied FA, Hakimi AN (2008) Spectroscopic characterization and biological activity of metal complexes with an ONO trifunctionalized hydrazone ligand. Journal of Coordination Chemistry 61: 2380-2401.

36. Hathaway B, Billing DE (1970) The electronic properties and stereochemistry of mono-nuclear complexes of the copper (II) ion. Coordination Chemistry Reviews 5: 143-207.

37. Raman N, Ravichandran S (2005) Synthesis and Characterization of a New Schiff Base and its Metal Complexes Derived from the Mannich Base, N-(1-piperidinobenzyl) acetamide. Synthesis and Reactivity in Inorganic, Metal-Organic, and Nano-Metal Chemistry 35: 439-444.

22

Synthesis, Characterization and Electrochemical Properties of Composite Membrane by an Aqueous Sol-Gel Method

Fakhra Jabeen* **and M Sarfaraz Nawaz**

Department of Chemistry, Jazan University, Jazan, Saudi Arabia

Abstract

The composite membranes were prepared by sol-gel process, and the membrane potential has been measured for characterizing the ion-transport phenomena across a charged membrane using electrolytes (KCl, NaCl and LiCl). The membrane potential offered by the electrolytes is in the order of LiCl>NaCl>KCl. The results have been used to estimate fixed-charge density, distribution coefficient, charge effectiveness and transport properties of electrolytes of this membrane. The fixed-charge density is the most important parameter, governing transport phenomena in membranes. It is estimated by the TMS method; it is dependent on the feed composition due to the preferential adsorption of some ions. The results indicate that the applied pressure is also an important variable to modify the charge density and, in turn, the performance of membrane. The experimental results for membrane potential are quite consistent with the theoretical prediction. The morphology of the membrane surface is studied by Scanning Electron Micrographs (SEM).

Keywords: Composite membrane; Membrane potential; Electrolytes; SEM; Sol-gel method

List of Symbols

Nomenclature

AR: analytical reagent

C_1,C_2: concentrations of electrolyte solution on either side of the membrane (mol/l)

$\overline{C}_{2+}$: cation concentration in membrane phase 1 (mol/l)

$\overline{C}_{2+}$: cation concentration in membrane phase 2 (mol/l)

C_i: ith ion concentration of external solution (mol/l)

$\overline{D}$: ith ion concentration in membrane phase (mol/l)

$\overline{D}$: charge density in membrane (eq/l)

F: Faraday constant (C/mol)

$K_{\pm}$: distribution coefficient of ions

80-160: pressure (MPa)

q: charge effectiveness of the membrane

R: gas constant (j/K/mol)

SCE: saturated calomel electrode

SEM: scanning electron microscopy

TMS: Teorell, Meyer and Sievers

t_+: transport number of cation

t_-: transport number of anion

$\overline{v}$: mobility of cations in the membrane phase (m^2/v/s)

$\bar{U}$: $(\overline{v} - \overline{v})/(\overline{v} + \overline{v})$

$\overline{v}$: Mobility of anions in the membrane phase (m2/v/s)

V_k: Valency of cation

V_x: Valency of fixed-charge group

$\gamma'_{\pm}, \gamma''_{\pm}$: mean ionic activity coefficients

$\overline{\omega}$: Mobility ratio

$\Delta\overline{\Psi}_m$: Observed membrane potential (mV)

$\Delta\overline{\Psi}_m$: Theoretical membrane potential (mV)

$\Delta\overline{\Psi}_{diff}$: Donnan potential (mV)

$\Delta\overline{\Psi}_{diff}$: Diffusion potential (mV)

Introduction

Ion-exchange membranes (IEM) carry the fixed positive or negative charges (called anion-exchange membranes, AEM or cation-exchange membranes, CEM, respectively). They are generally used in the treatment of ionic aqueous solutions, e.g., electrodialytic concentration of seawater, desalination of saline water, demineralization process, acid and alkali recovery and others [1-4]. Ion-exchange charged membranes, which are now extensively utilized in industries, have attracted considerable attentions due to their extraordinary properties and practical demands and thus a large number of researchers have concentrated on these investigations for many years [5]. With the rapid development of industry and population explosion throughout the world, the demand for fresh water has become increasingly urgent due to the scarcity of drinking water resource and the contamination of industry to environment. Thus, the treatment of industrial wastewater is becoming imperative; while innovative technologies, which are used to prepare fresh water such as the desalination of brackish water and to treat the industrial refuses, have attracted numerous researchers. Among these novel methods, ion-exchange membrane-based technologies have been regarded as both effective and economical due

***Corresponding author:** Fakhra Jabeen, Assistant Professor, Department of Chemistry, Jazan University, Jazan, Saudi Arabia
E-mail: fakhrajabeen@gmail.com

to its lower operation expense and secure process, etc. [6-8]. Composite membranes have high thermal and chemical stability, long life and good defouling properties in their application, and they can have catalytic properties [9]. These properties have made these membranes desirable for industrial applications in the food, pharmaceutical and electronic industries. The sol-gel technique is an extremely flexible method to produce inorganic materials with highly homogeneous and controlled morphology [10-12]. Recently, due to the mild reaction conditions that can be used, the great potential of sol-gel processes, both hydrolytic and non-hydrolytic, has been extensively investigated for the synthesis of organic/inorganic materials [13].

A potential difference can be observed and measured, at least partly ionically perm selective, membrane in contact with two solutions at following cases: (1) same electrolyte of different concentration; and (2) same ionic strength but different counter-ions or co-ions. The former is called concentration potential and the latter bi-co-ionic/bi-counter-ionic potential [14,15]. These potentials are of great interest in connection with the analysis of effective charge density, ionic transport number, and selectivity as well as interaction between charged species and membranes in both single charged membrane and bipolar membranes and thus caused great attention for many years [16-18]. For this purpose, a potential model correlating the intrinsic parameters of the membrane and ionic species with transport properties is actually needed and a body of such models has been obtained for single charged membranes and bipolar membranes [19-22]. It is now recognized that the electrical charge on the pore wall of membranes plays an important role in its separation performance and fouling behavior [23-25]. The choice of a membrane with suitable charge or electrical potential property can lead to optimization of existing processes or allow selective separations. For these reasons there is much interest in characterizing the charge or potential property of membranes. The electrical potential difference which is generated when an electrolyte solution flows across a charged membrane under a concentration gradient is among the most convenient experimental techniques for studying such electrical potential properties of porous membranes [26].

In the present investigation, a composite titanium-vanadium phosphate membrane is developed by sol-gel process using polystyrene as a binder. Fixed-charge density, the most effective parameter, has been evaluated and utilized to calculate membrane potentials for different electrolyte concentrations using TMS method [27-31]. In addition to the fixed-charge density, distribution coefficient, transport numbers, mobility, charge effectiveness and other related parameters were calculated for characterizing the composite membrane.

Materials and Methods

Preparation of membrane

Titanium-Vanadium phosphate precipitate was prepared by mixing a 0.2 mol titanium (III) chloride (Otto Kemi, India with 99.989% purity) and vanadium (III) chloride (Merck, Germany with 99.989% purity) with 0.2 mol tri-sodium phosphate (E. Merck, India with 99.90% purity) solutions. The precipitate was washed properly with deionized water to remove free electrolytes and then dried at 100°C. The precipitate was ground into fine powder and was sieved through 200 mesh (granule size <0.07 mm). Pure crystalline polystyrene (Otto Kemi, India, AR) was also ground and sieved through 200 mesh. The titanium-vanadium phosphate along with appropriate amount of polystyrene powder was mixed thoroughly using mortar and pestle. The mixture was then kept into a cast die having a diameter of 2.45 cm and placed in an oven maintained at 300°C for about an hour to equilibrate the reaction mixture [32]. The die containing the mixture was then transferred to a pressure device (SL-89, UK), and various pressures such as 80, 100, 120, 140 and 160 MPa were applied during the fabrication of the membranes. As a result titanium-vanadium phosphate membrane of approximate thicknesses 0.095, 0.090, 0.085, 0.080 and 0.075 cm were obtained, respectively. The membranes prepared by embedding 25% of polystyrene by weight were suitable, and the greater or lesser than this weight did not show reproducible results and appeared to be unstable. Membranes prepared in this way were stable and further subjected to microscopic and electrochemical examinations for cracks and homogeneity of the surface.

Scanning electron microscopy (SEM)

The prepared samples at various pressures was heated in the tabular furnace for 3 hours and then cooled. A very thin transparent polymer glue tape was applied on the sample and then placed on an aluminum stub of 15 mm diameter. Thereafter, the sample was kept in a chamber at a very low pressure where the entire plastic foil containing the sample was coated with gold (60 μm thickness) for 5 minutes. The scanning electron micrograph of gold coated specimen was recorded, operating at an accelerating voltage of 10 kV using the scanning electron microscope (GEOL JSM-840).

Measurement of membrane potential

The freshly prepared charged membrane was installed at the center of the measuring cell, which had two glass containers, one on either side of the membrane. Both collared glass containers are having a hole for introducing the electrolyte solution and Saturated Calomel Electrodes (SCEs). The half-cell contained 40 ml of the electrolyte solutions. Electrochemical cells of the type C_1 SCE Solution and C_2 Membrane Solution SCE were used for measuring membrane potential using Osaw Vernier Potentiometer. In all measurements, the electrolyte concentration ratio across the membrane was taken as $C_2/C_1=10$. All solutions were prepared by using Analytical Reagent (AR) grade chemicals and ultra-pure distilled water. The electrodes used were saturated calomel electrode and were connected to a galvanometer. The solutions in both containers were stirred by a magnetic stirrer to minimize the effects of boundary layers on the membrane potential. The pressure and temperature were kept constant throughout the experiment and the potentials were measured at 25°C.

Results and Discussion

The composite membranes using polystyrene as a binder were prepared by sol-gel process. The membranes were found to have the following properties:

- They were thermally stable up to 500°C.
- They were resistant to compaction.
- They were inert to harsh chemical ($K_2Cr_2O_7$, H_2O_2, HNO_3, H_2SO_4, etc.) as they did not decompose in their presence.
- They did not show any swelling.
- They were stable after long usage, i.e., they were durable.

The characterization of membrane morphology has been studied by using SEM [33]. The information obtained from SEM images have provided guidance in the preparation of well-ordered precipitates, composite pore structure, micro/macro porosity, homogeneity, thickness, surface texture and crack-free membranes [34]. The SEM surface images of the composite membranes were taken at different applied pressure and are presented in Figure 1. Inorganic composite membranes have the ability to generate potential when two electrolyte solutions of unequal concentration are separated by a membrane and driven by different chemical potential acting across the membrane [35]. The electrical character of the membrane regulates the migration of charged species, and diffusion of electrolytes from higher to lower concentration takes place through the charged membrane. The values of membrane potential $\Delta\bar{\Psi}_m$ measured across membranes in contact with various 1:1 electrolytes (KCl, NaCl and LiCl) were dependent on concentration of electrolytes present on both sides of the membrane at 25 ± 1°C are given in Table 1. The observed potential was low (mV, +ve). It was found to increase on decreasing the concentration of electrolytes (KCl, NaCl and LiCl), which is a usual behavior of inorganic membranes. The selectivity character of ion-exchange membranes were reported on the basis of membrane potential values, performed on uni-uni and multi-uni valents electrolytes as 1:1, 2:1 and 3:1. The reversal in sign from positive to negative values of membrane potential occurred with the 2:1 and 3:1 electrolytes. This is evidently due to the adsorption of multivalent ions, which led to a state where the net positive charge left on the membrane surface made the anion selective with 2:1 or 3:1 electrolytes. The membrane potential was also seen to be largely dependent on the pressure applied during the membrane fabrication. Application of higher pressure at composite membranes led to reduction in their thicknesses, contraction in pore volume and consequently offered a progressively higher fixed-charge density [36].

Figure 1: SEM images of composite membranes prepared at different applied pressures (80 MPa, 120 MPa and 160 MPa).

The charge property of the membrane matrix greatly influences the counter-ion than co-ion as well as the transport phenomena in the solutions. The surface charge concept of the TMS model for charged membrane is an appropriate starting point for the investigations of actual mechanisms of ionic or molecular processes which occur in membrane phase [27-31]. The TMS model assumes uniform distribution of surface charge and consists of Donnan potential and diffusion potential. According to the TMS, the membrane potential $\Delta\bar{\Psi}_m$ is applicable to an idealized system and is given by

$$\Delta\bar{\Psi}_m = 59.2\left(\log\frac{C_2}{C_1}\frac{\sqrt{4C_1^2+\bar{D}^2}+\bar{D}}{\sqrt{4C_2^2+\bar{D}^2}+\bar{D}}+\bar{U}\log\frac{\sqrt{4C_2^2+\bar{D}^2}+\bar{D}\bar{U}}{\sqrt{4C_1^2+\bar{D}^2}+\bar{D}\bar{U}}\right),\ \bar{U}=(\bar{u}-\bar{v})/(\bar{u}+\bar{v}) \qquad (1)$$

Where $\bar{v}$ and $\bar{v}$ are the ionic mobilities ($m^2/V/s$), of cation and anion, respectively, in the membrane phase. The charge densities of inorganic membranes were estimated from the membrane potential measurement and can also be estimated from the transport number. From the plots in Figure 2, the charge density parameters can be evaluated for a membrane carrying various charge densities, $\bar{D} \leq 1$ for different 1:1 electrolytes systems. The theoretical and observed potentials were plotted as a function of $-\log C_2$ as shown in Figure 2. Thus, the coinciding curve for various electrolytes system gave the value for the charge density $\bar{D}$ within the membrane phase.

Therefore, the increase in the values of $\bar{D}$ with higher applied pressure is due to successive increase of charge per unit volume as well as the modification in the surface microstructure of the membrane. The plot of charge density $\bar{D}$ of the membrane for 1:1 electrolytes (KCl, NaCl and LiCl) versus pressures is shown in Figure 3. The order of charge density of various electrolytes is found to be KCl>NaCl>LiCl throughout the range of applied pressure at which the membranes were prepared. The surface charge model may work as a tool to improve the performance of the membrane filtration process. Since, the charge density is an important parameter governing transport phenomena and the charge property of the membrane dominates the electrostatics interaction between the membrane and particles in the feed solution due to the prefential adsorption of some ions. Therefore, by controlling the solution physico-chemistry, the optimum charge property of the membrane can be obtained as desired.

The TMS equation (1) can also be expressed by the sum of Donnan potential $\Delta\Psi_{Don}$ between membrane surfaces and external solutions and the diffusion potential $\Delta\bar{\Psi}_{diff}$ within the membrane.

	Applied pressure (MPa)														
	Membrane Potential (O)														
C_2 (mol/l)	80			100			120			140			160		
	KCl	NaCl	LiCl	KCl	NaCl	LiCl	KCl	NaCl	LiCl	KCl	NaCl	LiCl	KCl	NaCl	LiCl
0.001	49.5	50.6	51.7	50.6	52.0	53.2	52.2	53.3	54.4	53.3	54.5	55.8	54.3	55.4	56.4
0.001	45.0	45.9	47.1	46.2	47.3	48.5	47.1	48.2	49.3	48.0	49.2	50.3	49.3	50.6	51.8
0.01	22.5	23.4	24.8	24.0	25.2	26.3	25.0	25.9	27.3	25.5	27.2	28.5	27.0	28.4	30.0
0.1	7.53	8.53	9.62	9.02	10.5	12.2	10.0	11.1	12.3	11.0	12.3	13.5	12.1	13.6	15.1
1.0	5.08	6.09	7.50	6.50	7.60	9.00	7.50	8.54	9.60	8.52	9.57	10.6	9.60	10.3	11.9
	Applied pressure (MPa)														
	Membrane Potential (T)														
C_2 (mol/l)	80			100			120			140			160		
	KCl	NaCl	LiCl	KCl	NaCl	LiCl	KCl	NaCl	LiCl	KCl	NaCl	LiCl	KCl	NaCl	LiCl
0.001	58.94	58.94	58.94	59.02	59.01	59.01	59.12	59.13	59.12	59.14	59.15	59.15	59.16	59.16	59.16
0.001	47.25	47.25	47.25	49.57	49.47	49.57	54.48	54.49	54.49	55.90	55.90	55.90	56.78	56.78	56.78
0.01	11.45	11.46	11.48	13.44	13.45	13.47	20.40	20.41	20.43	24.01	24.02	24.04	27.13	27.15	27.16
0.1	1.32	1.35	1.36	1.55	1.58	1.59	2.47	2.50	2.51	3.05	3.07	3.09	3.62	3.65	3.66
1.0	0.24	0.26	0.28	0.26	0.28	0.30	0.36	0.37	0.39	0.41	0.43	0.45	0.47	0.48	0.51
Charge density (eq/l)	1.48	1.27	0.99	1.74	1.43	1.20	4.14	3.41	2.68	6.91	4.76	3.63	10.2	6.74	4.34

Table 1: Observed and theoretical membrane potentials (mV) across the composite membranes in contact with 1:1 electrolyte solution at different concentrations C_2/C_1=10 at 25 ± 1°C.

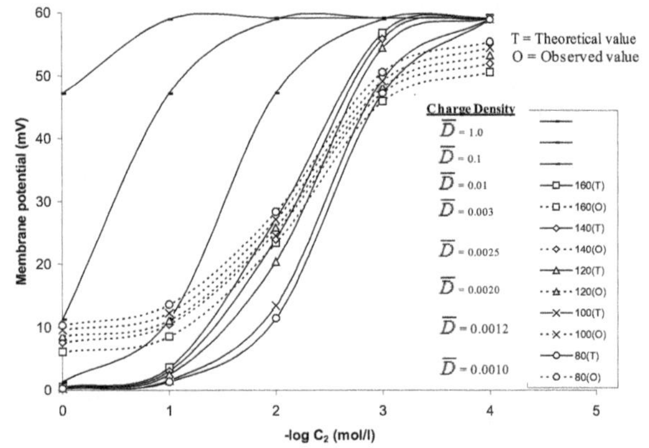

Figure 2: Plots of membrane potentials versus –log C_2 at different concentrations of NaCl electrolyte solution for composite membranes prepared at different pressures 80-160 MPa.

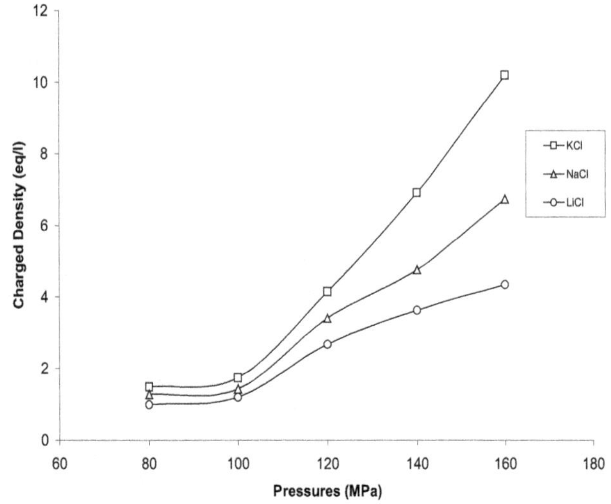

Figure 3: Plots of fixed-charge density $\bar{D}$ versus pressures (80-160 MPa).

$$\Delta\bar{\Psi}_m = \Delta\Psi_{Don} + \Delta\bar{\Psi}_{diff} \tag{2}$$

$$= -\frac{RT}{V_kF}\ln\left(\frac{\gamma''_\pm C_2\bar{C}_{1+}}{\gamma'_\pm C_1\bar{C}_{2+}}\right) - \frac{RT}{V_kF}\frac{\bar{\omega}-1}{\bar{\omega}+1}\times\ln\left(\frac{(\bar{\omega}+1)\bar{C}_{2+}+(V_x/V_k)\bar{D}}{(\bar{\omega}+1)\bar{C}_{1+}+(V_x/V_k)\bar{D}}\right) \tag{3}$$

The R, T and F have their usual significance; $\gamma''_\pm$ and $\gamma''_\pm$ are the mean ionic activity coefficients; $\bar{\omega}=\frac{\bar{u}}{\bar{v}}$ is the mobility ratio of the cation to the anion in the membrane phase and $\bar{C}_{2+}$ and $\bar{C}_{2+}$ are the cation concentrations in the membrane phase first and second, respectively. The cation concentration is given by the equation

$$\bar{C}_+ = \sqrt{\left(\frac{V_x\bar{D}}{2V_k}\right)^2 + \left(\frac{\gamma_\pm C}{q}\right)^2} - \frac{V_x\bar{D}}{2V_k} \tag{4}$$

Here V_k and V_x refer to the valency of cation and fixed-charge group on the membrane matrix, q is the charge effectiveness of the membrane and is defined by the equation

$$K_\pm = \frac{\bar{C}_i}{C_i}, \tag{5}$$

Where $K_\pm$ is the distribution coefficient. It is expressed as

$$K_\pm = \frac{\bar{C}_i}{C_i}, \tag{6}$$

Where $\bar{C}_i$ the ith ion concentration in the membrane is phase and C_i is the ith ion concentration of the external solution. The transport properties of the membrane in various electrolyte solutions are important parameters to further investigate the membrane phenomena as shown in Eq. (7)

$$\Delta\bar{\psi}_m = \frac{RT}{F}(t_+ - t_-)\ln\frac{C_2}{C_1}, \frac{t_+}{t_-} = \frac{\bar{u}}{\bar{v}} \tag{7}$$

Equation (7) was first used to calculate the values of transport numbers t_+, mobility ratio $\bar{\omega}=\frac{\bar{u}}{\bar{v}}$ and finally Ū as given in Table 2. The values of mobility $\bar{\omega}$ of the electrolytes in the membrane phase were found to be high at lower concentration of all the electrolytes (KCl, NaCl and LiCl). Further increase in concentration of the electrolytes led to a sharp drop in the values of $\bar{\omega}$ as given in Table 2. The high mobility is attributed to higher transport number of comparatively free cations of electrolytes and also be similar trend as the mobility in least concentrated solution. The values of the parameters K_+, q and $\overline{C}_+$ derived for the system have also been included in Table 2. Using Eq. (6) it was found that the values of distribution coefficients increased at lower concentration of electrolytes. As the concentration of electrolytes increased, the values of distribution coefficients sharply dropped and, thereafter, a stable trend was observed as shown in Figure 4. The large deviation in the value of $K_\pm$ at the lower concentration of electrolytes was attributed to the high mobility of comparatively free charges of the strong electrolyte and thus, reached into the membrane phase easily compared to higher concentrated electrolytes solution. In order to interpret the variation of the charge effectiveness depending on those values, that the ion-pairing effect causes the difference between the effective charge density and the fixed-charge density in membrane phase. In our membrane, counter ion Cl^- is same for 1-1 electrolytes therefore, the variation in the q values are follow the similar trend and the order is LiCl>NaCl>KCl up to the C_2=0.01 mol/l and then drop in the q values were analyzed from Figure 5. When, the external electrolyte concentration is higher or lower, a number of counter ions go into the membrane due to imbalance in the counter ion concentration of external electrolyte and fixed charged group in the membrane phase. Therefore, the ion association with the fixed charged group and counter ions in the membrane is enhanced as a result the charge effectiveness has a lower value whereas in the moderate concentration region the counter ion concentration in the external electrolyte and the fixed-charge density in the membrane are comparable. Therefore, a less number of ion pair formation and consequently higher values of the charged effectiveness, the optimum value of charge effectiveness are obtained at C_2=0.01 mol/l and then decreased steeply. The order of the charge effectiveness of 1-1 electrolytes may depend on increasing ionic charge density of co-ion adsorption on the charged membranes. The membrane potential derived in this way (theoretical) and the experimentally obtained membrane potentials at different concentrations for various electrolytes systems have been compared and provided in Figure 6. It may be noted that the experimental data follow the theoretical curve quite well. However, some deviations may be due to various non ideal effects, such as swelling effect and osmotic effects. These effects are often simultaneously present in the charged membranes.

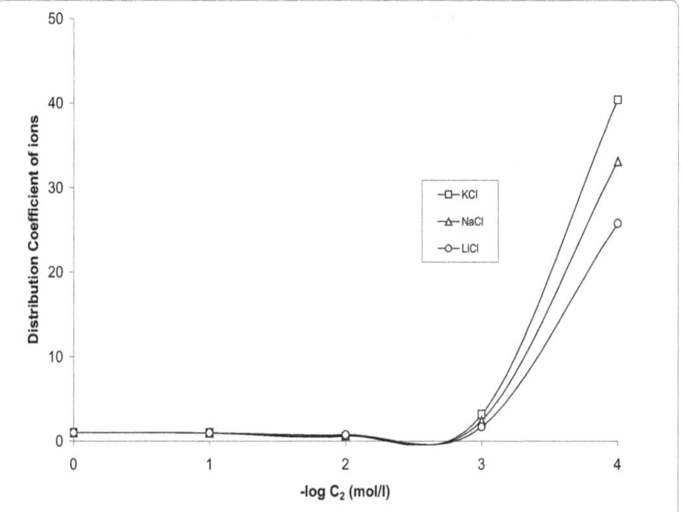

Figure 4: Plots of Distribution coefficient of ions ($K_\pm$) versus –log C_2 (mol/l).

Conclusion

In the present study, the composite membranes were prepared by sol-gel process, and results indicate that the sol-gel approach is appropriate for composite membrane synthesis. The sol-gel technique is an extremely flexible method to produce inorganic materials with highly homogeneous and controlled morphology. The experimental results were analyzed on the basis of the TMS approach, and it was found that the calculated values agree well with the experimental results. The fixed-charge density is the central parameter governing transport phenomena in membranes. The electrical charge on the pore wall of membranes plays an important role in its separation performance and fouling behavior and it depends upon the feed composition and applied

KCl (Electrolyte)						
C_2(mol/l)	t_+	Ū	$\overline{C}_+$	$K_\pm$	q	$\overline{C}_+$
0.0001	0.93	0.86	13.3	40.40	0.157	0.00002
0.0010	0.89	0.78	8.09	3.140	0.554	0.00023
0.0100	0.71	0.42	2.45	0.586	1.239	0.00334
0.1000	0.58	0.16	1.38	0.958	0.895	0.07993
1.000	0.56	0.12	1.27	0.995	0.780	0.77073
NaCl						
0.0001	0.94	0.88	15.7	33.10	0.173	0.00002
0.0010	0.90	0.80	9.00	2.410	0.633	0.00021
0.0100	0.72	0.44	2.57	0.659	1.170	0.00413
0.1000	0.59	0.18	1.44	0.965	0.897	0.08176
1.000	0.57	0.14	1.33	0.996	0.811	0.80377
LiCl						
0.0001	0.95	0.90	19.0	25.80	0.196	0.00001
0.0010	0.91	0.82	10.1	1.680	0.758	0.00018
0.0100	0.73	0.46	2.70	0.732	1.111	0.00505
0.1000	0.60	0.20	1.50	0.973	0.901	0.08375
1.000	0.58	0.16	1.38	0.997	0.881	0.87453

Table 2: The values of t_+, Ū, $\overline{C}_+$, $K_\pm$, q, $\overline{C}_+$ evaluated from using Equation 7 and Equations 4-6 respectively, from observed membrane potentials for various electrolytes at different concentrations for composite membranes prepared at 120 MPa pressure.

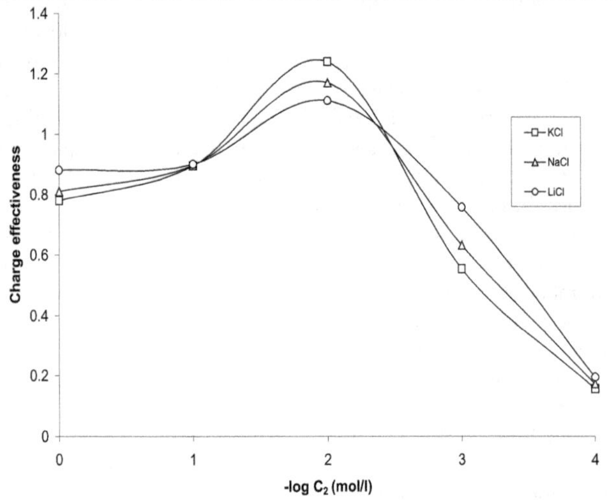

Figure 5: Plots of Charge effectiveness versus –log C_2 (mol/l) for 1:1 electrolyte at different concentrations.

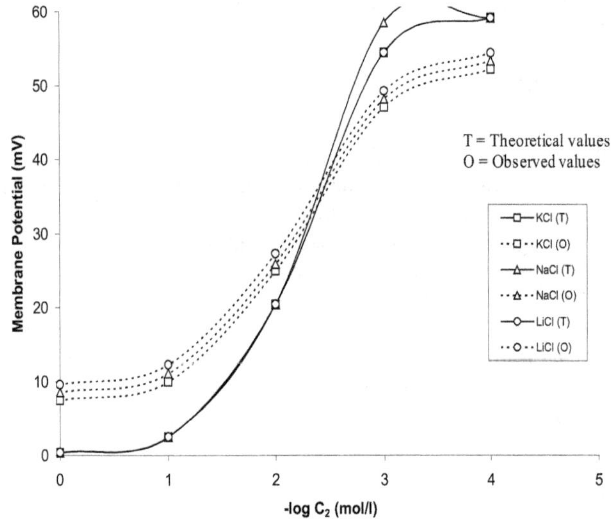

Figure 6: Membrane potentials across composite membranes using various electrolyte (1:1) solutions at different concentrations.

pressure due to the prefential adsorption of some ions. The charge effectiveness of membrane is greatly influenced form applied pressure and increase in adsorption of co-ions on charged membrane, order is KCl<NaCl<LiCl. Thus, this membrane can be suited for commercial application.

Acknowledgements

The authors gratefully acknowledge the Head of Department, Department of Chemistry for providing necessary research facilities.

References

1. Sata T, Sata T, Yang W (2002) Studies on cation-exchange membranes having permselectivity between cations in electrodialysis. J Membr Sci 206: 31-60.
2. Volgin VM, Davydov AD (2005) Ionic transport through ion-exchange and bipolar membranes. J Membr Sci 259: 110-121.
3. Balster J, Krupenko O, Pünt I, Stamatialis DF, Wessling M (2005) Preparation and characterisation of monovalent ion selective cation exchange membranes based on sulphonated poly(ether ether ketone). J Membr Sci 263: 137-145.
4. Xu T (2005) Ion exchange membranes: State of their development and perspective. J Membr Sci 263: 1-29.
5. Xu T, Yang W (2001) Sulfuric acid recovery from titanium white (pigment) waste liquor using diffusion dialysis with a new series of anion exchange membranes-static runs. J Membr Sci 183: 193-200.
6. Linde C, Kedem O (2001) Asymmetric ion exchange mosaic membranes with unique selectivity. J Membr Sci 181: 39-56.
7. Saito K, Ishizuka S, Higa M, Tanioka A (1996) Polyamphoteric membrane study: 2. Piezodialysis in weakly amphoteric polymer membranes. Polymer 37: 2493-2498.
8. Ito H, Toda M, Ohkoshi K, Iwata M, Fujimoto T, et al. (1988) Artificial membranes from multiblock copolymers. 6. Water and salt transports through a charge-mosaic membrane. Ind Eng Chem Res 27: 983-987.
9. Lin YS, Burggraaf AJ (1991) Preparation and Characterization of High-Temperature Thermally Stable Alumina Composite Membrane. J Amer Ceram Soc 74: 219-224.
10. Brinker CJ, Scherer GW (1990) Sol-Gel Science. Academic Press, San Diego, USA.
11. Mann S, Burkett SL, Davis SA, Fowler CE, Mendelson NH, et al. (1997) Sol-Gel Synthesis of Organized Matter. Chem Mater 9: 2300-2310.
12. Licoccia S, Polini R, D'Ottavi C, Fiory FS, Di Vona ML, et al. (2005) A simple and versatile Sol-Gel method for the synthesis of functional nanocrystalline oxides. J Nanosci Nanotechnol 5: 592-595.
13. Vioux A (1997) Nonhydrolytic Sol-Gel Routes to Oxides. Chem Mater 9: 2292-2299.
14. Bockris JO'M, Reddy AKN (1970) Modern Electrochemistry: An Introduction to an Interdisciplinary Area. Plenum Press, New York, USA.
15. Schmid G, Schwarz H (1998) Electrochemistry of capillary systems with narrow pores. IV. Dialysis potentials. J Membr Sci 150: 189-196.
16. Kimura Y, Lim HJ, Iijima T (1984) Membrane potentials of charged cellulosic membranes. J Membr Sci 18: 285-296.
17. Koizumi S, Imato T, Ishibashi N (1997) Bi-ionic membrane potential across a liquid anion-exchange membrane containing triphenyltin chloride. J Membr Sci 132: 149-158.
18. Kudela V, Richau K, Bleha M, Paul D (2001) Orientation effects on bipolar and other asymmetric membranes as observed by concentration potentials. Sep Purif Technol 22: 655-662.
19. Koter S (2001) Transport number of counterions in ion-exchange membranes. Sep Purif Technol 22: 643-654.
20. Suendo V, Minagawa M, Tanioka A (2002) Membrane potential of a bipolar membrane: the effect of concentration perturbation of the intermediate phase around a certain value. J Electroanal Chem 520: 29-39.
21. Tasaka M, Kiyono R, Yoo DS (1999) Membrane Potential across a High Water Content Anion-Exchange Membrane Separating Two Solutions with a Common Counterion but Two Different Co-ions. J Phys Chem B 103: 173-177.
22. Dammak L, Larchet C, Auclair B (1999) Theoretical study of the bi-ionic potential and confrontation with experimental results. J Membr Sci 155: 193-207.
23. Fievet P, Szymczyk A, Labbez C, Aoubiza B, Simon C, et al. (2001) Determining the Zeta Potential of Porous Membranes Using Electrolyte Conductivity inside Pores. J Colloid Interface Sci 235: 383-390.
24. Jimbo T, Tanioka A, Minoura N (1999) Pore-surface characterization of poly(acrylonitrile) membrane having amphoteric charge groups by means of zeta potential measurement. Coll Surfaces A: Physicochem Eng Aspects 159: 459-466.
25. Sung JH, Chun MS, Choi HJ (2003) On the behavior of electrokinetic streaming potential during protein filtration with fully and partially retentive nanopores. J Colloid Interface Sci 264: 195-202.
26. Bowen WR, Cao X (1998) Electrokinetic effects in membrane pores and the determination of zeta-potential. J Membr Sci 140: 267-273.
27. Teorell T (1935) An Attempt to Formulate a Quantitative Theory of Membrane Permeability. Exp Biol Med (Maywood) 33: 282-285.
28. Teorell T (1935) Studies on the "Diffusion Effect" upon Ionic Distribution. Some Theoretical Considerations. Proc Natl Acad Sci USA 21: 152-161.
29. Meyer KH, Sievers JF (1936) La perméabilité des membranes I. Théorie de la perméabilité ionique. Helv Chim Acta 19: 649-664.

30. Meyer KH, Sievers JF (1936) La perméabilité des membranes. II. Essais avec des membranes sélectives artificielles. Helv Chim Acta 19: 665-677.

31. Meyer KH, Sievers JF (1936) La perméabilité des membranes. IV. Analyse de la structure de membranes végétales et animales. Helv Chim Acta 19: 987-995.

32. Beg MN, Siddiqi FA, Singh SP, Prakash P, Gupta V (1979) Studies with inorganic precipitate membrane: evolution of thermodynamically effective fixed charge density and test of the most recently developed theory of membrane potential based on the principles of non-equilibrium thermodynamics. Electrochem Acta 24: 85-88.

33. Santos LRB, Pulcinelli SH, Santilli CV (1997) Formation of SnO_2 Supported Porous Membranes. J Sol Gel Sci Technol 8: 477-481.

34. Huang L, Wang Z, Sun J, Zhili LMQ, Yan Y, et al. (2000) Fabrication of Ordered Porous Structures by Self-Assembly of Zeolite Nanocrystals. J Am Chem Soc 122: 3530-3531.

35. Barragán VM, Rueda C, Ruiz-Baura C (1995) On the Fixed Charge Concentration and the Water Electroosmotic Transport in a Cellulose Acetate Membrane. J Colloid Interface Sci 172: 361-367.

36. Winter R, Czeslik C (2000) Pressure effects on the structure of lyotropic lipid mesophases and model biomembrane systems. Z Kristallogr Cryst Mater 215: 454-474.

On Electrochemical Managing the Properties of Aqueous Coolant

Alexander S*

NRC Kurchatov Institute, Moscow, Russia

Abstract

Liquid water as a chemical compound with the wide band gap is characterized by varying their Fermi level as a linear identifier of water oxidation-reduction potential (ORP). This potential is the management tool for changing chemical properties of the aqueous coolant by forced shifting Fermi level in the band gap at the expense of insignificant deviation (|z|<10) of water composition, H_2O_{1-z}, from the stoichiometric one (z=0). The hypo-stoichiometric state (z>0) with the negative ORP is realized when Fermi level is shifted to the local donor level, ε_{HO}, by electro-reducing the aqueous coolant in the electrochemical cell with the strongly polarized anode and the quasi-equilibrium cathode, occupying εH_2O by electrons, and forming hydroxonium radicals, H_3O, as the strongest reducers. Opposite, the hyper-stoichiometric state (z<0) with the positive ORP is realized in the electrochemical cell with the strongly polarized cathode and the quasi-equilibrium anode when Fermi level is shifted to the local acceptor level, radicals, ε_{OH}, as the strongest oxidizers. ε_{OH}, forming in water hydroxyl.

Keywords: Aqueous coolant; Non-stoichiometric water; Band gap; Fermi level; Oxidation-reduction potential

Introduction

A series of theoretical works on the electronic properties of liquid water have appeared in the present decade aided by the rapid increase in computational power [1]. In particular, a density of states (DOS) and a band gap as an energy difference which separates the occupied molecular orbital and the empty electronic states in the liquid water have merited the attention for fundamental studying. They are not understood quite in comparison with thermodynamics and structure of water, but they are very important for understanding water as participant and medium of chemical reactions. This medium is described as a dielectric with the broad band gap, g=6.9 eV [2] as a difference between electron energies at the top of valence band or a highest occupied molecular orbital and the bottom of conduction band as a lowest unoccupied molecular orbital. At the same time, there are two allowed local electron states in the band gap of liquid water such as an occupied-by-electron level, ε_{OH} , of hydroxide ion, OH , and the vacant one of hydroxonium cation, $\varepsilon_{H3O}{}^{+}$, located symmetrically nearby the band-gap middle with the energy difference between them of 1.75 eV [3]. This theoretical concept allows to eliminate the inconsistencies [4] in reconciling the electrochemical properties of these well known aqueous ions in the frame of electronic band structure [3]. For liquid water as a chemical compound of variable composition, it is useful to define a quantity called Fermi level, F [5,6] as an electrochemical potential which indicates the tendency of liquid water to donate or accept the proton. If ε_F is high, there is a strong tendency for liquid water to donate protons, i.e., it is reducing. Opposite, if Fermi level in aqueous medium is low, there is the strong tendency for it to accept protons when this matter is oxidizing [7]. Now, it is very important to understand what application of this approach can be used in practice. Just the solution of such the question is the subject of the present paper specifically electrochemical keeping the given quality of the aqueous coolant in power water reactors.

The band structure of liquid water

In according to [3] the liquid water contains the allowed energy levels, ε_{HO} and ε_{OH}, in the band gap for inherent constituents of liquid water as ions of hydroxonium, H_3O^+ and hydroxide, OH^- due to the reversible self-dissociation of liquid water by reaction [8].

$$2H_2O \leftrightarrow H_3O^+ + OH^- \quad (1)$$

Their radicals (H_3O, OH) as the hydrated hydrogen, $H{\cdot}H_2O$, and half-oxygen, (1/2) ($O{\cdot}H_2O$), are interpreted as the mentioned levels occupied by electrons and holes respectively and located symmetrically nearby the middle of the band gap (Figure 1) with $\varepsilon_{HO}-\varepsilon_{OH}$=1.75 eV.

In general, the position of Fermi level ε_F in the band gap of dielectric is the threshold of 50% population of the all allowed electronic levels in its band gap. For the aqueous coolant, this level as the electrochemical potential [5–7] is a single-valued characteristic of water oxidation-reduction potential (ORP) [6]:

$$ORP=-(\varepsilon_F-\varepsilon_{SHE})/e \quad (2)$$

where ε_{SHE}=–6.21 eV is the Standard Hydrogen Electrode [3] and *e* is the charge of electron.

ORP becomes negative when Fermi level is shifting to the donor level, ε_{H2O} (Figure 1) which is occupied by electrons and forms hydroxonium radicals, H_3O, as the strongest reducers transforming water to hypo-stoichiometric one, H_2O_{1-z}, with z>0. Opposite in the hyper-stoichiometric water (z<0), Fermi level is shifted to the acceptor level, ε_{OH}, which forms hydroxyl radicals, ε_{OH}, as the strongest oxidizers and ORP is positive.

These states are easily realized in pure water by the following standard reactions [8]

$$2H_3O{+}+2e^- \leftrightarrow H_2\uparrow +2H_2O \quad (3)$$

$$(1/2)\ O_2+H_2O+2e^- \leftrightarrow 2OH^- \quad (4)$$

The corresponding Fermi levels, $\varepsilon_{F(2)}$ and $\varepsilon_{F(3)}$, are shown in Figure 2 by red lines. It is known [9, 10] that the population [H_3O] and [OH] of the energy levels, ε_{HO} and ε_{OH}, by electrons and holes can be defined by the proportions of the species concentrations: [H_3O+]/[H3O] and [OH]/[OH^-], that are given by Maxwell–Boltzmann distribution of electrons and holes in the corresponding energy levels [3]

$$[H_3O+]/[H_3O]=\exp[(\varepsilon_{H2o}-\varepsilon_F)/k_B\,T], \quad (5)$$

***Corresponding author:** Alexander S, NRC Kurchatov Institute, Moscow, Russia
E-mail: shimkevich_al@nrcki.ru

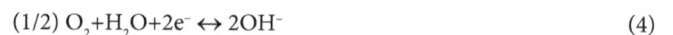

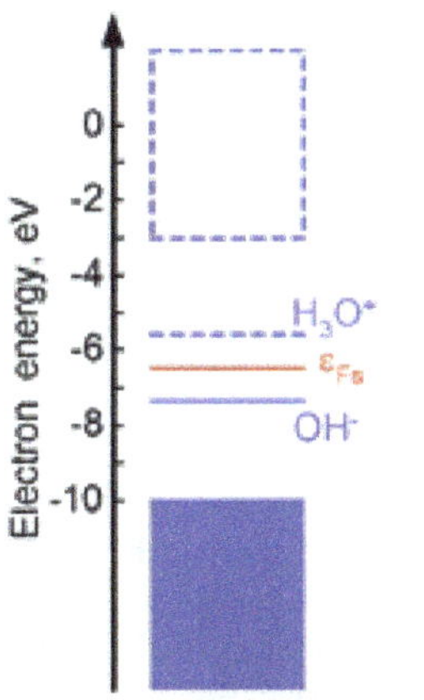

Figure 1: The band structure of stoichiometric liquid water with Fermi level (red line in the middle of band gap); the blue box is the valence band and the dotted one is the conduction bands; the full blue line denotes occupied-by-electrons energy level, ε_{OH}, for hydroxide ions, OH– and dotted blue line denotes the vacant one, for hydroxonium ions, H_3O.+

$[OH]/[OH-]=\exp\left[(\varepsilon_{OH}-\varepsilon_F)/k_BT\right]$ (6)

FB where T is Kelvin temperature, and kB is Boltzmann constant equal to $8.62\cdot10^{-5}$ eV/K.

Then, using the molar concentrations, $[H_3O^+]$ and $[OH^-]$, of hydroxonium and hydroxide ions in the famous dissociation ratio [11].

$[H_3O^+]\cdot[OH^-]=Kw$ (7)

with the constant, Kw=10–14 M2 at T=300 K, we can transform the index z of non-stoichiometric water, H_2O_{1-z}

z=0.018{$[H_3O]$–[OH]} (8) to the form

$z=0.018\{\exp[(\varepsilon_F-\varepsilon_{HO})/kBT-2.3pH]-\exp[(\varepsilon_{OH}-\varepsilon_F)/kBT-32.2+2.3pH]\}$ (9)

where $pH=-\lg[H_3O^+]$.

The composition confines of chemical water stability

As seen in Figure 2 Fermi level, $\varepsilon_{F(3)}$, in hypo-stoichiometric water is controlled by the concentration of hydroxonium radicals, $[H_3O]$, as species of occupied-by-electrons level, ε_{HO}. At the same time, Fermi level, $\varepsilon_{F(4)}$, of hyper-stoichiometric water is controlled by hydroxyl concentration [OH]. These non-stoichiometric states of liquid water are realized at $[H_3O^+]=[OH^-]=10^{-7}$ M. In the case of T=300 K and

$P_{h2}=P_{e2}=1$ atm, $[H_2]=1.6\cdot10^{-3}$ M, $[O_2]=2.7\cdot10^{-4}$ M [12], $[H_3O]\sim2\cdot10^{-11}$ M and $[OH]\sim8\cdot10^{-13}$ M [3]. Then, the Eq. (9) becomes

$z=0.44.10^{-16}\{\exp[38.7\Delta_F-2.3pH]-\exp[2.3pH-38.7\Delta_F-32.2]\}$. (10)

for $\varepsilon_F=\varepsilon_{Fs}+\Delta_P$, $\varepsilon_{Fs}=-6.45$ eV, $\varepsilon_{HO}=-5.58$ eV, $\varepsilon_{OH}=-7.32$ eV [3], and kBT=0.026 eV. This equation allows to plot Fermi level in any aqueous solution as a function of its non-stoichiometry, z, as shown in Figure 3. One can see that the region of chemical water stability is shifted to the hypo-stoichiometric state, $H_2O_{1-}z$ (z>0), of aqueous medium which is achieved easier in an acidic solution than in the basic one by least shifting Fermi level from the band-gap middle to the local donor level, ε_{HO}. Opposite, the hyper-stoichiometric one, H_2O_{1-z} (z<0), with a little variation of z is achieved easier in a basic solution than in the acidic one by shifting Fermi level from the band-gap middle to the local acceptor level, ε_{OH}. Such the consideration can be applied to the aqueous coolant of pressurized water reactors with operational parameters: T=600 K,

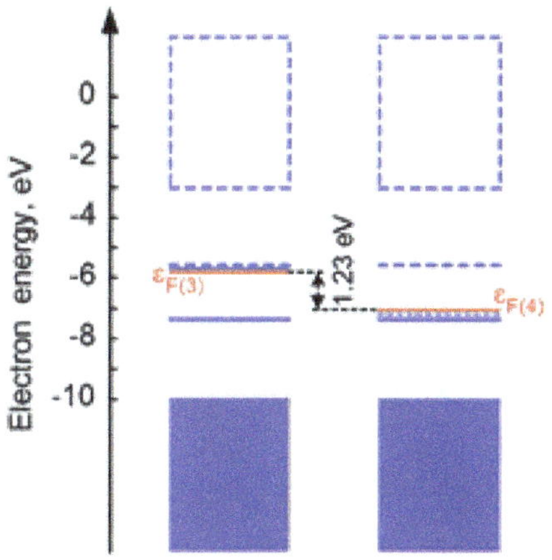

Figure 2: The non-stoichiometric states of liquid water with Fermi levels, $\varepsilon_{F(3)}$ and $\varepsilon_{F(4)}$, are realized in pure water at 300 K by the reactions (3) and (4) at pH=7; the full blue lines denote local occupied-by-electrons levels, ε_{HO} and ε_{OH}, as hydroxonium radicals, H_3O, and hydroxide ions, OH–; dotted ones denote hydroxonium ions, H_3O+, and hydroxyls, OH.

$P_{H2}=P_{o2}=160$ atm. Then, we will have $[H_2]=7.7\cdot10^{-2}$ M,

or $[O_2]=1.8\cdot10^{-3}$ M [12], $[H_3O]=1.4\cdot10^{-10}$ M or $[OH]\sim2\cdot10^{-12}$ M, and Eq. (10) is transformed to

$z=0.98.10^{-9}\{\exp[19.2\Delta_F-2.3pH]-\exp[2.3pH-19.2\Delta_F-26.45]\}$. (11)

for $k_{BT}=0.052$ eV. One can see that here the region of chemical water stability by order of magnitude is greater than in Figure 3. At the same time, Fermi level as a function of water non-stoichiometry, z, at 600 K is more conservative than at 300 K and more sensitive to aqueous-solution type due to its line bundle. At the same time as seen in Figures 3 and 4, the quick response of Fermi level to changing the water non-stoichiometry, z, in the stoichiometric point (z=0) takes place at 600 as well as at 300 K. Therefore, a little additive of any oxidant or antioxidant in pure water changes its ORP (Fermi level) appreciably. It is interesting to note that such the effect can be gotten without chemical additives by electro-oxidizing (or electro-reducing) the pure liquid water in a special electrochemical cell with one polarized electrode and the other in equilibrium with the aqueous medium.

Electro-oxidation of the pure liquid water

As seen in Figure 3, the region of liquid-water chemical stability is defined by the very narrow range $-10^{-14}\le z<4.10^{-13}$ of non-stoichiometric composition, H_2O_{1-z}, and changing Fermi level in the band gap up to 2 eV. The forced variation of ε_F can be carried out by the electrochemical cell with the voltage of ~2 V between the strongly polarized cathode and the anode in quasi-equilibrium [13]. This can change physical and chemical properties of pure liquid water as well as its ORP up to the ones of a strong acid by keeping liquid water in the hypo-stoichiometric state (z>0) when the external potential applied to the strongly polarized cathode intensively generates hydroxide ions, OH–, in the narrow layer of unstable liquid water of near the cathode (Figure 5) by reactions:

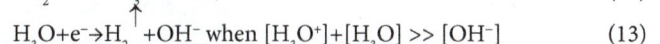

$2H_2O+e^-\rightarrow H_3O+OH^-$, (12)

$H_3O+e^-\rightarrow H_2\uparrow+OH^-$ when $[H_3O^+]+[H_3O]\gg[OH^-]$ (13)

At the same time, Fermi level is shifting to the energy level, ε_{OH}, and the hydroxide ions forcedly migrate in the bulk of water due to the action of electric field. They are discharged to hydroxyls by quasi-equilibrium anodic reaction [13]

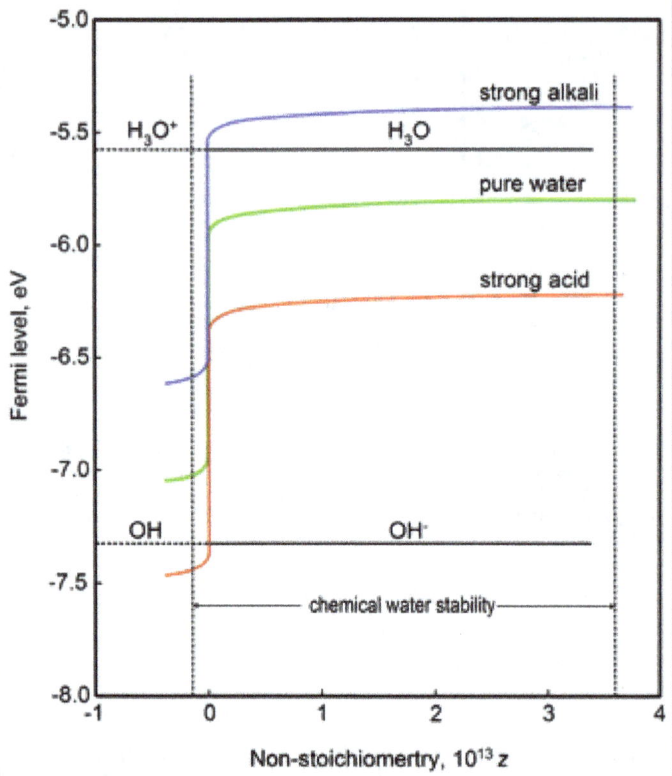

Figure 3: Fermi level as a function of non-stoichiometry, z, in pure liquid water, $H_2O_{1-z,}$ (green line), in the aqueous solution of strong alkali (blue line), and the solution of strong acid (red line) at T=300 K and P=1 atm; dotted vertical lines denote the composition confines of chemical water stability and the horizontal lines denote the local energy levels, εH_3O and εOH, for hydroxonium, H_3O+/H_3O, and hydroxyl, H/OH– respectively.

$$OH^- - e^- \rightarrow OH \quad (14)$$

but hydroxonium radicals, H_3O, diffused out of the cathode layer (Figure 5) of electrochemical cell put electrons to hydroxyl radicals by reaction

$$H_3O + OH \rightarrow H_3O^+ + OH^- \quad (15)$$

All this forms the positive bulk charge in liquid water near the cathode (Figure 5) at the condition

$$[H_3O^+] > [OH] >> [OH^-] > [H_3O] \quad (16)$$

which indicate on the strong acidic reaction of the chemically stable hypo-stoichiometric water with the high mole fraction ($[OH] >> 10^{-7}$ M) of hydroxyls as the strongest oxidizers (Table 1).

The advantage of this approach is the high efficiency of oxidation reaction, the simplicity of the procedure, low cost, and there is no need for special sorbents because water itself becomes the agent for oxidizing pollutants [13,14]. Thus, the electrochemical oxidation with the advantage of environmental compatibility is the promising procedure for removing pollutants from waste water by hydroxyl radicals produced on the quasi-equilibrium anode out of hydroxide ions electro- generated by the strongly polarized cathode. It is very important to generate them intensively since the life-time of hydroxyl radicals is very short (only few nanoseconds) [15] due to quick forming the solvated acceptors of electrons by bulk reaction.

$$H_3O^+ + OH \rightarrow (H_2O)_2^+ \quad (15)$$

Electro-reducing the pure liquid water

Changing the polarity of the electrochemical cell, we will get the strongly polarized anode and the cathode in quasi-equilibrium and by applied potential, convert the hyper-stoichiometric unstable water (z<0) in the anode layer (Figure 6) in the hypo-stoichiometric state (z>0) by reactions [16].

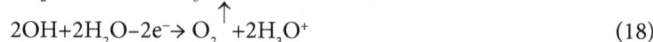

$$2H_2O - e^- \rightarrow OH + H_3O^+ \quad (17)$$

$$2OH + 2H_2O - 2e^- \rightarrow O_2\uparrow + 2H_3O^+ \quad (18)$$

One can see that Fermi level is shifting higher the energy level, εH_3O, and the hydroxonium ions forcedly migrate in the bulk of water due to the action of electric field. This gives the basic properties for processed liquid water which is kept in the stable hypo-stoichiometric state by negative charge formed in the bulk and cathode layer by quasi-equilibrium cathode reactions [16].

$$H_3O^+ + e^- \rightarrow H_3O \quad (19)$$

$$H_3O + e^- \rightarrow H_3O^- \quad (20)$$

$$H_3O^- + OH \rightarrow H_3O + OH^- \quad (21)$$

because a little part of hydroxyl radicals, OH, diffuses in the bulk out of the anode layer of electrochemical cell. All this forms the negative bulk charge in liquid water near the anode (Figure 6) at the condition

$$[H_3O^-] + [H_3O] > [OH^-] >> [H_3O^+] >> [OH] \quad (22)$$

Oxidizer	ORP, V
Hydroxyl radical Oxygen (atomic) Oxygen (molecular)	2.80 2.42 1.23

Table 1: Oxidation potential of chemical oxidants [14].

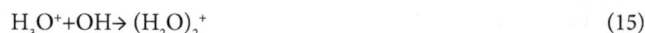

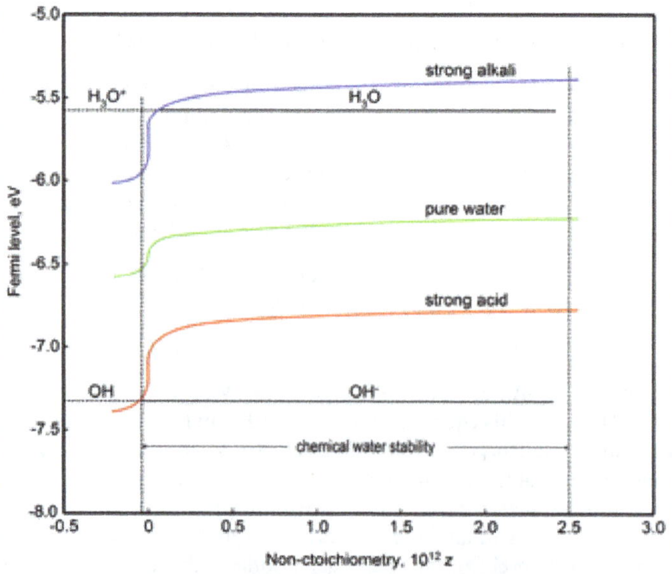

Figure 4: Fermi level as a function of non-stoichiometry, z, in pure liquid water, H_2O_{1-z} (green line), in the aqueous solution of strong alkali (blue line), and the solution of strong acid (red line) at *T*=600 K and *P*=160 atm; dotted vertical lines denote the composition confines of chemical water stability and the horizontal lines denote the local energy levels, εH_3O and εOH, for hydroxonium, H_3O+/H_3O, and hydroxyl, OH/OH– respectively.

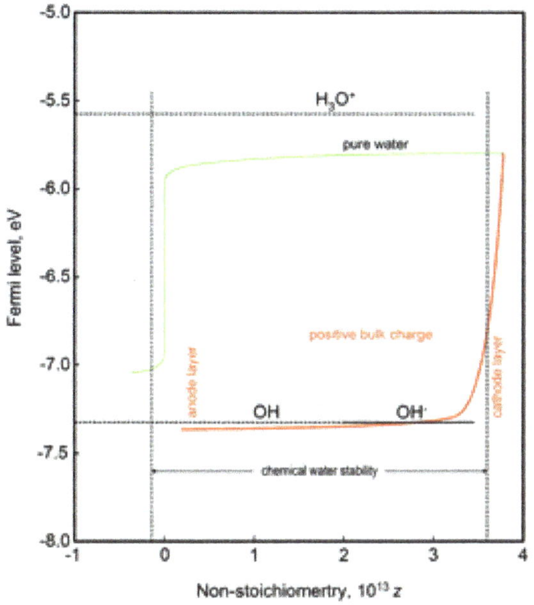

Figure 5: The change of Fermi levels of the initial non-stoichiometric pure liquid water, H_2O_{1-z} (green line in Figure 3) by the electrochemical cell with the voltage of 2V between the strongly polarized cathode and the quasi-equilibrium anode (red line) at *T*=300 K and *P*=1 atm; dotted vertical lines denote the composition confines of chemical water stability and the horizontal lines denote the local energy levels, ε_{HO} and ε_{OH}, for hydroxonium ions, H_3O+, and hydroxyl/hydroxide species, OH/OH– respectively.

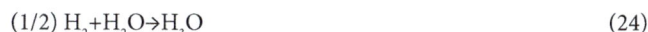

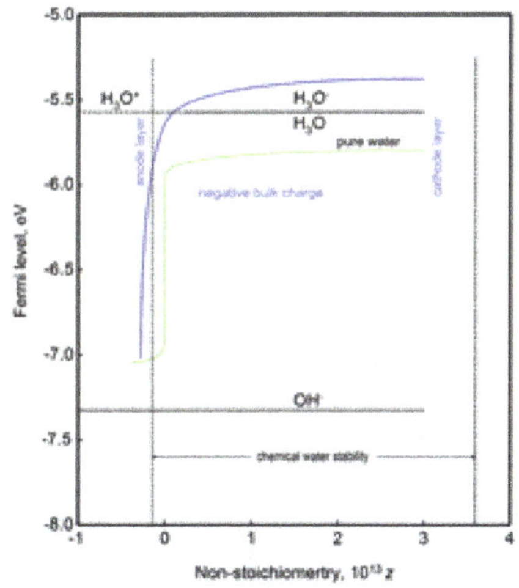

Figure 6: The change of Fermi levels of the initial non-stoichiometric pure liquid water, H_2O_{1-z}, (green line in Figure 3) by the electrochemical cell with the voltage of 2 V between the strongly polarized anode and the quasi-equilibrium cathode (blue line) at *T*=300 K and *P*=1 atm; dotted vertical lines denote the composition confines of chemical water stability and the horizontal lines denote the local energy levels, εH_3O and ε_{OH}, for hydroxonium species, $H_3O_+/H_3O/H_3O_-$ and hydroxide anions OH_- respectively.

which indicate on the strong basic reaction of the chemically stable hypo-stoichiometric water with the high mole fraction ($[H_3O^-]+[H_3O]>>10^{-7}$ M) of the hydride anions as proton acceptors and the hydroxonium radicals as electron donors in the bulk of water [16].

Thus, electrochemical production of these very active antioxidants can be more effective than the gaseous hydrogen can do them in the aqueous coolant ($[H_3O]\sim10^{-10}$ M) for holding the negative ORP of hydrogen water chemistry in PWR.

Discussion of results

The plot of Fermi levels in the non-stoichiometric aqueous medium is differed essentially from the one in the electrochemical cell with one strongly polarized electrode and the other in quasi-equilibrium as seen in the Figures 5 and 6. The mechanisms for forming oxidizers, [OH], or reducers, $[H_3O]$, are also differed: in aqueous emulsion of gases, they are formed by kinetically- limited reactions of dissociation:

$$(1/2)\ O_2+H_2O\rightarrow 2OH \quad (23)$$

$$(1/2)\ H_2+H_2O\rightarrow H_3O \quad (24)$$

but in the electrochemical cell, they are generated without limitation by reactions (14) and (19) that essentially increases their molar portion in liquid water. It can change the current aqueous chemistry of PWR which is the oxidative one in essence due to continuous feed water additives naturally containing oxygen. This impurity is not desirable for the PWR first-loop coolant due to the continuous growth of oxide films on the surface of fuel cladding. Therefore, it is important to organize an effective technological process for removing oxygen from the feed water by converting ORP of the aqueous coolant in the negative region (2) and obtain the corrosion-passive one that can inhibit the growth of oxide films on the surface of fuel cladding and local break-up of them. Thus, pH and ORP mechanism are independent in processing liquid water. The first can be changed in stoichiometric water, H_2O, by adding equivalent amount of anions and cations. The second can be changed without changing pH in ventilating the pure liquid water by bubbles of oxygen (hydrogen) or by processing the initial aqueous medium in the electrochemical cell with one strongly polarized electrode and the other in quasi-equilibrium. Since non-stoichiometry of H_2O_{1-z} varies at very narrow interval of $|z| \leq 10^{-12}$ (Figure 4), ORP is highly sensitive to external conditions and can change without visible varying a composition of the aqueous medium. Therefore online monitoring ORP (Fermi level) by precise sensor is very important [6].

Conclusions

It is shown that the region of chemical water stability is shifted to the hypo-stoichiometric state, H_2O_{1-z} (z>0), of aqueous medium which is achieved easier in an acidic solution than in the basic one by least shifting Fermi level from the band-gap middle to the local donor level, ε_{HO}. Opposite, the hyper-stoichiometric one, H_2O_{1-z} (z<0), with a little variation of z is achieved easier in a basic solution than in the acidic one by shifting Fermi level from the band-gap middle to the local acceptor level, ε_{OH}. It turned out that the interval of chemical stability of the PWR coolant is greater by order of magnitude than the one for water at the room temperature as well as more conservative and more sensitive to aqueous-solution type. At the same time, the quick response of Fermi level to changing the water non-stoichiometry, z, takes place in the stoichiometric state (z=0). Therefore, a little additive of oxidant or antioxidant in pure water changes ORP (Fermi level) appreciably. Such the effect may be gotten also by electro-oxidizing (electro-reducing) liquid water in the electrochemical cell with one polarized electrode and the other in equilibrium with the aqueous medium. Thus, the electrochemical oxidation with the advantage of environmental compatibility can become the effective method for removing pollutants from wastewater by hydroxyl radicals produced on the quasi-equilibrium anode out of hydroxide ions electro-generated by the strongly polarized cathode. In turn, the electrochemical production of very active reducers, H_3O and H_3O^-, can be more effective for maintaining the negative ORP in PWR coolant than the gaseous hydrogen in feed water can do this.

Acknowledgements

Author thanks the Russian Foundation of Basic Research (RFBR) for supporting this work (grant # 16-08-00029a) and appreciates his colleagues at active discussing all the aspects of electrochemical managing the properties of aqueous coolant for PWR.

References

1. Garbuio V, Cascella M, Pulci O (2009) Excited state properties of liquid water. J Phys Condens Matter 21: 1-15.
2. Do Couto PC, Guedes RC, Costa Cabral BJ (2004) The density of states and band gap of liquid water by sequential Monte Carlo and quantum mechanics calculations. Brazilian J Physics 34: 42-47.
3. Alexander S (2014) Electrochemical View of the Band Gap of Liquid Water for Any Solution. World J Condensed Matter Physics 4: 243-249.
4. Do Couto PC (2007) Understanding electronic properties of water: a theoretical approach to the calculation of the adiabatic band gap of liquid water. Thesis for PhD degree, Lisbon University.
5. Alekseev PN, Yu M, Semchenkov AL, Shimkevich (2012) Aqueous nanofluid as a two-phase coolant for PWR. Sci Technol Nuclear Install pp: 1-6.
6. Shimkevich AL, Yu I, Shimkevich (2012) On 2D water chemistry. Nuclear Plant Chem 46: 1-39.
7. Alexander LS (2013) On arising nano hydrides in reduced alkaline solution. Am J Modern Physics 2: 185-189.
8. Bard AJ, Parsons R, Jordan J (1985) Standard Potentials in Aqueous Solutions. New York.
9. Kittel Ch, Kroemer H (1980) Thermal Physics. WH Freeman, San Francisco.
10. Kittel Ch (2004) Introduction to Solid State Physics (eds.) New York.
11. Bandura AV, Lvov SN (2006) The ionization constant of water over wide ranges of temperature and density. J Phys and Chem Ref Data 35: 15-30.
12. Kaye GWC, Laby TH (1986) Tables of physical and chemical constants (eds.) New York.
13. Alexander S (2014) On performance capabilities of alkaline anolyte in wastewater management in Proc. Int Conf on Water Chemistry of Nuclear Reactor Systems 2: 882- 893.
14. Hannmann L, Powers K, Shepherd O, Taylor H (2012) Removal of ciprofloxacin from water with chemical oxidation. Worcester Polytechnic Institute.
15. Gamal AM (2010) Comparative efficiencies of the degradation of CI Mordant Orange 1 using UV/H2O2 Fenton and photo Fenton processes. Life Sci J 7: 51-59.
16. Shimkevich AL (2014) On catholyte application for hydrogen water chemistry in PWR in Proc. European Nuclear Conference, Marseille, France.

24

Assessment of Major and Trace Elements in Aquatic Macrophytes, Soils and Bottom Sediments Collected Along Different Water Objects in the Black Sea Coastal Zone by Using Neutron Activation Analysis

Nekhoroshkov PS[1]*, Kravtsova AV[2], Kamnev AN[3], Duliu O[4], Bunkova OM [3], Frontasyeva MV[1] and Yermakov IP[2]

[1]*Frank Laboratory of Neutron Physics, Joint Institute for Nuclear Research, Russia*
[2]*Faculty of Biology, Lomonosov Moscow State University, Moscow, Russia*
[3]*Faculty of Soil Science, Lomonosov Moscow State University, Moscow, Russia*
[4]*Department of the Structure of Matter, Faculty of Physics, University of Bucharest, Romania*

Abstract

The levels and compartmentalization of Na, Mg, Al, Cl, K, Ca, Sc, Ti, V, Cr, Mn, Fe, Co, Ni, Zn, As, Se, Br, Rb, Sr, Mo, Sb, I, Cs, Ba, La, Ce, Sm, Eu, Tb, Hf, Ta, Au, Th, and U in *Phragmites australis Carex conescens* L and *Cladophora sericea* from the Caucasian coast of the Black Sea Anapa recreational region was investigated by Neutron Activation Analysis. The study touches upon subject of the sediment-to-plant and root-to-leaf elemental transfer as well as of the influence of anthropogenic pollution on wetland ecosystems in zone of resort. The content of the majority of considered elements was found higher in the belowground organs of *P. australis* than in the aboveground tissues while a reverse regularity was evidenced for *C. conescens*. The levels of elements decrease from bottom sediments to aquatic plants with the notable exception of the halogens Cl, Br and I that presented 5 to 100 fold higher content in plants than in sediments. The increased levels of As, Mo, and Sb in some soil and sediment samples most probably indicate the anthropogenic pollution. It recommends them for a continuous monitoring of the same area.

Keywords: Major and trace elements; Neutron activation analysis; Black Sea; *Phragmites australis*; *Carex conescens; Cladophora sericea*

Introduction

The aquatic macrophytes are widely used for assessing the environmental situation in fresh as well as seawater [1-10]. As the accumulation of trace metals in organisms depends on the concentration of pollutants in water and sediments as well as on exposure time, a tissue analysis of aquatic macrophytes may provide cumulative evaluation of exposure [11-13]. The concentrations of chemical elements in aquatic plants can be more than 100,000 times higher than in the associated water [14]. This accumulation ability of certain macrophytes is used for monitoring purposes in relatively clean and recreation zones, where low level of contamination might be difficult to detect [15]. Our previous investigations in the Black Sea [16-18] evidenced for the increased elemental concentrations in marine algae in more polluted waters reflecting their great potential for biomonitoring of water quality. They proved not only the existence of a certain degree of anthropogenic contamination but also the suitability of aquatic plants for biomonitoring of trace elements. To extend our studies regarding the elemental content of more than 35 elements for the territory of an important but poor investigated recreation zone of the Caucasian coast of the Black Sea, the aquatic macrophytes *Phragmites australis, Carex conescens* L as well as the green algae *Cladophora sericea* (Hudson) Kutzing were used. *P australis* is one of the most distributed macrophytes in aquatic ecosystems, and numerous studies showed its capacity of trace element bioaccumulation [19-23]. Thus Duman reported that the roots of *Phragmites australis* from fresh water Lake Sapanca in Turkey were found to be good accumulators of Cu, Mn, Ni, Zn. The studies of in the estuaries of Italian rivers affected by municipal wastewaters and agricultural activities showed a good correlation of Al, As, Cr, Cu, Mn, Ni and Zn in *P. australis* with the elemental content in corresponding sediments and water. Also a strong positive correlation between the concentrations of Al, As, Co, Cr, Cu, Fe, Mn, Ni, Se, Sr and Zn in the sediments and all organs (rhizome, stem and leave) of *P. australis* sampled from the Tisza River in Serbia was found by [23]. The investigations in the constructed wetland in North Italy [24] and in the Hokersar wetland, Ramsar site of Kashmir Himalaya, India [25] showed that *P. australis* is appropriate species for phytoextraction and phytoremediation of the environment. Analysis of the elemental composition of *P. australis*, collected in the Anapa region in 2013-2014, showed that the concentration ratios with the absolute value that is greater than 1 (pointing to the pollution of the area) are determined only for As. Maximal values of biological absorption coefficients were found for the As, Fe, K, Mn, Zn in roots. The data of using the species of *Carex* (sedges) in biomonitoring purposes are scarce in comparison with *Phragmites*. Horovitz [26] reported the content of Ag, Co, Cr, Cs, Fe, Rb, Sc, Th, Zn and in *Carex pendula* sampled in botanical garden in Germany. Pederson and Harper [27] studied the chemical composition of some major forage plants of mountain summer ranges of southeastern Utah, USA, reported the content of K, Ca and Mg in *Carex geyeri*. Ohlson [28] studied the content of Al, Ca, Cu, Fe, K, Na, Mg, Mn, Mo, Zn in eleven plants from the mires of central and north Sweden found that the largest variation in elemental concentration of roots and leaves was observed in *Carex rostra*. He also reported that the concentration of K in tissues of *Carex* species was highly correlated with its concentration in the substrate. The species of green algae of genus *Cladophora* has frequently been suggested as a suitable organism to monitor water contamination and its practical use in monitoring river, lake and sea pollution has been reported from a range of countries [29-34]. Thus Whitton reported that there were highly significant correlation between Cu, Fe, Zn content in *Cladophora glomerata* from rivers and streams in Northern England and water. The similar results

***Corresponding author:** Nekhoroshkov PS, Frank Laboratory of Neutron Physics, Joint Institute for Nuclear Research, Russia
E-mail: p.nekhoroshkov@gmail.com

were reported by for Cr, Ni and V determined in *Cladophora glomerata* from refinery sewage lagoon (Bratislava). Levkov and Krstic found that the levels of Co, Cu, Fe, Mn and Zn in *Cladophora glomerata* reflected their load in the River Vardar, Macedonia, and recommended it as a precise biomonitoring tool for determination and quantification of heavy metal pollution in this river. In the distribution patterns of Ca, Cu, K, Mg, Mn, Na, Ni, Zn and in the green algae *Cladophora* sp. from the Southern Baltic is assessed. The study concluded that *Cladophora* sp. can be used the most successfully as biomonitor of Cu and Zn content in the Baltic Sea because of its ability to accumulate metal contaminants from seawater, tolerance to metals, simple morphology and adequate tissue for analysis. The preliminary study of elemental composition of *Cladophora sericea*, collected in the Anapa region in 2013, showed that the plant to soil ratios greater than one and pointing towards a possible contamination process were detected only for As, and Sr. For our study, we have chosen three types of phototrophic macrophytes as ones of the most convenient organisms-biomonitors. Moreover, they occur in different ecological conditions and are the first ones that take the fall of the coastal pollution runoff accordingly, we have investigated the hydrophyte filamentous marine green alga *Cladophora sericea* (Hudson) Kutzing, helophyte *Phragmites australis* (Cav) Trin. Ex Steud as well as the hygrophyte *Carex conescens Cladophora sericea* lives in shallow sandy areas of the Black Sea; absorb major and trace elements by all surface of their body, do not have root system, *Phragmites australis* lives along the coastal zones of rivers and seas. The well-developed root system makes more than 80% of the total biomass. Plants absorb major and trace elements from soil, sediment and water by additional roots. *Carex conescens* grows on the banks of the rivers. Unlike *Phragmites australis* is plant has a small root system so it absorbs major and trace elements only from the soil. The main goals of the study consist of: quantifying the content of a wide range of major as well as trace elements in *Phragmites australis, Carex conescens, Cladophora sericea* and corresponding soil and bottom sediments samples assessing the elemental content in different parts of plants (leaves and roots) quantifying the element mobility from sediment to organs, as well as within the plant, providing new data on the geochemistry of sediments and soil from the Anapa region quantifying the level of the anthropogenic pollution of the study area. The results of this project will be further presented and discussed.

Materials and Methods

Study area and sampling

A resort city of Anapa (Krasnodar region) located on the Caucasian coast of the Black Sea is characterized by humid subtropical climate and long sandy beach. The Anapka river crossing the territory of the town connects Anapa reed beds with the Black Sea [35]. The investigated area (Figure 1; Table 1) includes the municipal waste dump at the Krasnyi hutor and some reservoirs, i.e., a lake, a river and reed beds at the foot of the hill and below the dump situated on the highland. These water bodies form an indivisible watershed of the river Anapka which estuary occupies the main city beach within the city recreation zone. The solid waste city dump of Anapa is located near the Krasnyi hutor, 4.65 km from the Black Sea. During 10 years, the total area of the dump increased from 9 to 26 hectares in 2013. There is a lake located in the distance of 0.62 mile downhill from the dump at village Krasnyi hutor 7). The next sampling point is Anapa reed beds this marshland is situated at the hollow, where the Kotloma and Kumatyr Rivers get its confluence, not far from the Anapa station. The length of the Anapka reed beds is 1 mile long. The station 2b is situated in old bed of the Anapka river. The mouth of the Anapka river is located at the main city beach. The samples of vegetation (live and dead leaves and roots of *Phragmites* and *Carex*, algae *Cladophora*) (n=35) and the corresponding soil (n=40) and bottom sediment (BS) (n=15) were collected at 7 sites

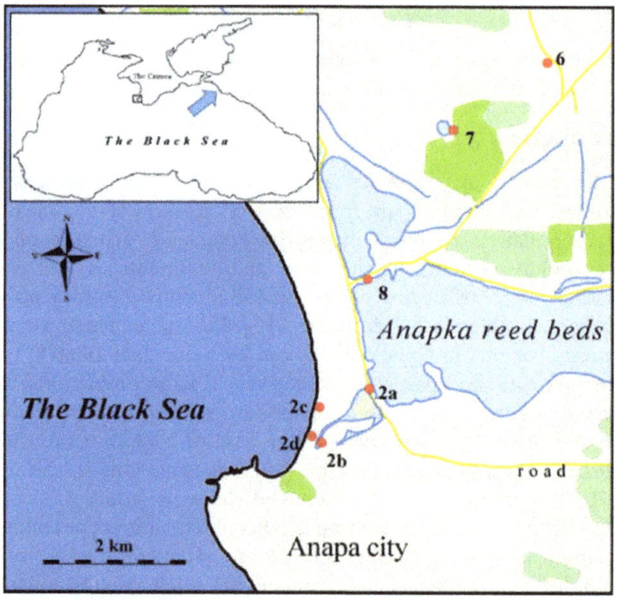

Figure 1: Sampling sites in the studied area near Anapa city (Anapa transect or cross-section.

Sampling point	Latitude (N)	Longitude (E)	Type	Summary description
City dump	44°57'31.76"	37°21'51.01"	Soil	Dump without vegetation
Lake near Krasnyi Hutor (village)	44°56'51.52"	37°20'41.70"	Sediments Plants	Waste liquid disposal, polluted runoff
Anapa reed beds	44°55'35.97"	37°19'47.57"	Soil Sediments Plants	Traffic, Gas station
Anapka river	44°54'35.84"	37°19'45.33"	Soil Sediments Plants	Traffic
Old bed of Anapka river	44°54'10.27"	37°19'10.18"	Soil Sediments Plants	Beach, objects of recreation
. Mouth of Anapka river	44°54'21.59"	37°19'06.89"	Soil Sediments Plants	Beach
Anapa Bay	44°54'11.74"	37°19'06.87"	Sediments Plants	Beach, marine traffic

Table 1: The location of sampling points as well as the type of collected material.

along the transect located near Anapa city in summer of 2013 and 2014. The sampling sites are shown in Figure 1 while Table 2 presents more details regarding the sampling points location as well as a summary description of each category of samples. The samples of soil and bottom sediments were collected according to GOST (state standard) [36,37]. The sampling process ensures compliance with the two requirements

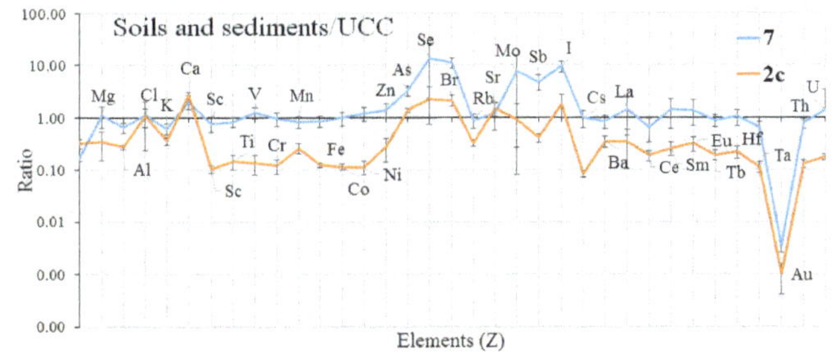

Figure 2: The diagram of the elemental content in soils and sediments normalized on UCC for 7 and 2c stations.

the amount of the sampled material to be enough for the analysis, and in the case of soils or sediments, their texture should be an average one over the studied media. Both soil and bottom sediments samples were air dry in a warm and ventilated room until constant weight, them milled and sieved through 0.04 mm mesh. Special attention was given to the preparation of the plant material samples, which was dried at 105°C until constant weight, ground in an agate mortar. There were no further chemical preparations of the samples, which excluded the errors due to the reagents.

Neutron activation analysis

Elemental analysis of the samples was carried out by INAA at the reactor IBR-2 of the Frank Laboratory of Neutron Physics (FLNP) of the Joint Institute for Nuclear Research (JINR), Dubna, Russia. The analytical procedures and the basic characteristics of the employed experimental facility are described in detail elsewhere [38]. The samples of about 0.3 g were packed in polyethylene bags for short-term irradiation and in aluminum cups for long-term irradiation. To determine the short-lived isotopes of Mg, Al, Cl, Ca, Ti, V, Mn and I the samples were irradiated

El.	SRM 1632c		SRM 433		SRM 667	
	Certified	Determined	Certified	Determined	Certified	Determined
Na	299 ± 5	300 ± 8	13500 ± 4050	13150 ± 160	-	-
Mg	384 ± 32	362 ±15	11500 ± 230	11430 ± 120	-	-
Al	9150 ± 137	9350 ± 187	78200 ± 782	77980 ± 890	-	-
Cl	1139 ± 41	1120 ± 36	-	-	-	-
K	1100 ± 33	1100 ± 201	16600 ± 2224	16300 ± 250	-	-
Ca	1450 ± 290	1430 ±130	-	-	-	-
Sc	2.9 ± 0.03	2.91 ± 0.07	14.6 ± 4.38	15.1 ± 0.15	13.7 ± 0.7	12.3 ± 0.24
Ti	517 ± 32	511 ± 21	-	-	-	-
V	23.7 ± 0.52	25.3 ± 0.78	160 ± 2.08	152 ± 11	-	-
Cr	8.24	13.7	136 ± 1	136 ±4	178 ±16	172 ± 8.5
Mn	13 ± 0.52	13.2 ± 0.46	316 ± 3.16	313 ± 5	920 ± 40	924 ± 18
Fe	7350 ± 110	7350 ± 250	40800 ± 408	40805 ±1673	44800 ± 986	39926 ± 1200
Co	3.48 ± 0.2	3.91 ± 0.24	39.4 ± 0.4	39.4 ± 2.9	23 ± 1.3	19 ± 0.2
Ni	9.32 ± 0.51	10.5 ± 3.2	39.4 ± 0.39	39.4 ± 0.14	128 ± 8.96	23 ± 1
Zn	12.1 ± 1.29	11.2 ± 1.8	101 ± 1	101 ± 3	175 ± 13	148 ± 3
As	6.18 ± 0.27	6.25 ± 0.35	18.9 ± 0.2	18.9 ± 0.4	17.1 ± 5.13	17.5 ± 4
Se	1.33 ± 0.07	1.33 ± 0.09	0.78 ± 0.03	0.72 ± 0.3	1.59 ± 0.08	1.49 ± 0.1
Br	18.7 ± 0.39	17.9 ± 0.5	67 ± 7.97	70 ± 5	99.7 ± 2.5	99.7 ± 2.7
Rb	7.5 ± 0.3	7.5 ± 1.3	99.9 ± 8.49	102 ± 14	-	-
Sr	63.8 ± 1.4	63.4 ± 5.3	302 ± 3	302 ± 20	224.5 ± 67.3	200 ± 10
Mo	0.8 ± 0.24	0.79 ± 0.28	-	-	-	-
Sb	0.46 ± 0.03	0.46 ± 0.04	1.96 ± 0.04	1.96 ± 0.06	0.96 ± 0.05	0.74 ± 0.04
I*	-	-	-	-	-	-
Cs	0.59 ± 0.01	0.59 ± 0.02	6.4 ± 0.26	6.2 ± 0.06	7.8 ± 0.7	6.7 ± 0.08
Ba	41 ± 2	41 ± 3	268 ± 19	268 ± 12	-	-
La	-	-	33.7 ±1.61	31 ± 5	27.8 ± 1	27.8 ± 1.1
Ce	11.9 ± 0.2	17.5 ± 3.6	64.5 ±19.4	73.9 ± 1.5	56.7 ± 2	57 ± 3
Sm	1.08 ± 0.03	1.08 ± 0.04	-	-	4.66 ± 0.2	4.25 ± 0.5
Eu	0.12 ± 0.003	0.32 ± 0.1	1.18 ± 0.35	2.42 ± 0.04	1 ± 0.01	1.0 ± 0.1
Tb	-	-	0.696 ± 0.2	0.7 ± 0.03	0.68 ± 0.02	0.60 ±0.02
Hf	0.59 ± 0.01	0.59 ± 0.05	3.66 ± 1.1	4.41 ± 0.08	-	-
Ta	-	-	1.03 ± 0.31	1.00 ± 0.02	0.88 ± 0.02	0.88 ± 0.02
Au	-	-	-	-	0.017 ± 0.005	0.017 ± 0.005
Th	1.4 ± 0.03	1.4 ± 0.04	9.8 ± 0.3	9.8 ± 0.3	10 ± 0.5	9.14 ± 0.09
U	0.51 ± 0.01	0.51 ± 0.02	2.45 ± 0.2	2.23 ± 0.2	2.26 ± 0.15	2.29 ± 0.3

Table 2: The NAA data and certified values of reference materials (mean ± error, in μg g^{-1} dry weight).

(I concentration was determined using SRM 1515 (apple leaves): certified value 0.3±0.09; determined value 0.26±0.12.)

for 3 min in the reactor channel with a neutron flux density of $1.3 \cdot 10^{12}$ n/(cm^2 s). Gamma spectra of induced activity were measured for 12-15 min after 20 min of decay. The elemental contents of the long-lived isotopes of Na, K, Sc, Cr, Fe, Co, Ni, Zn, As, Se, Br, Rb, Sr, Mo, Sb, Cs, Ba, La, Ce, Sm, Eu, Tb, Hf, Ta, Au, Th, and U were determined using epithermal neutrons in a cadmium-screened irradiation channel with a neutron flux density of $1.6 \cdot 10^{12}$ n/(cm^2 s). Samples were irradiated for 90 h, repacked and then measured twice after 4–5 d of decay during 30 min and after 20 days of decay during 1.5 h. To process gamma spectra of induced activity and to calculate concentrations of elements in the samples, software developed at FLNP, JINR was used [39]. The uncertainties in the determined concentrations were in the range of 5-15%, and of 30% or more for those elements which concentrations in the samples were at the detection limit. Quality control was provided by using National Institute for Standard and Technology (NIS) reference materials (SRM): NIST 1632c (trace elements in coal), NIST 433 (marine sediment), NIST 667 (estuarine sediment) as well as NIST 1515 (apple leaves) irradiated in the same conditions together with the samples under investigation. The NAA data and certified values of reference materials are given in concentration were determined using SRM 1515 (apple leaves): certified value 0.3 ± 0.09; determined value 0.26 ± 0.12.

Data analysis

To unify the major and trace composition of each plant we used the Reference Plant (RP) contents [40] as normalizing factors. In this way, it was possible to compare the distribution of the considered elements in all species of plants chosen for the present study. A similar approach we used in the case of soils and sediments by considering the Upper Continental Crust (UCC) [41] as reference average rock. Therefore, all data regarding the elemental composition of the Anapa soils and sediments samples were normalized to the corresponding content of the UCC. The accurate data on concentrations with the wide number of elements for "average sediment" are presented in UCC. The normalized on UCC data of concentrations in soils and sediments were used for comparison the levels of elements between different stations. The levels

Element	Station 7		Station 2c		UCC[1]	SNC[2]
	Soils	BS	Soils	BS		
Na	4200 ± 700	4300 ± 700	8000 ± 2000	7400 ± 2200	24259	–
Mg	20000 ± 6300	10200 ± 5600	5300 ± 3000	5000 ± 3400	14957	–
Al	60000 ± 7000	38000 ± 600	23000 ± 2000	20000 ± 400	81505	–
Cl	260 ± 160	800 ± 300	430 ± 150	240 ± 70	370	–
K	16000 ± 3000	11200 ± 600	8300 ± 1400	8900 ± 2600	23244	–
Ca	44000 ± 6300	73000 ± 11000	68200 ± 10400	75300 ± 12100	25658	–
Sc	12.1 ± 3.5	7.8 ± 0.2	1.48 ± 0.30	1.46 ± 0.01	14	–
Ti	3500 ± 400	2400 ± 300	600 ± 180	450 ± 70	3897	5030
V	136 ± 18	88 ± 5	14 ± 6	11.5 ± 1.2	97	126
Cr	89 ± 25	86 ± 38	12 ± 4	9 ± 2	92	109
Mn	704 ± 116	463 ± 50	210 ± 50	180 ± 13	774	930
Fe	33500 ± 8600	34800 ± 8800	4900 ± 600	4800 ± 100	39176	–
Co	17.7 ± 5.1	15.0 ± 0.1	2.0 ± 0.3	1.9 ± 0.3	17.3	–
Ni	58.2 ± 19.2	53.1 ±7.8	5.3 ± 1.7	5.5 ± 0.4	47	47
Zn	86.3 ± 11.4	112 ± 51	19 ± 10	14 ± 3	67	106
As	14.8 ± 2.5	17.6 ± 5.1	6.7 ± 0.7	6.5 ± 0.4	4.8	–
Se	1.1 ± 0.9	1.6 ± 1.5	0.20 ± 0.15	0.2 ± 0.1	0.09	–
Br	18.6 ± 4.9	17.3 ± 1.1	3.5 ± 0.7	2.7 ± 0.6	1.6	–
Rb	83.3 ± 21.9	53.8 ± 3.3	27.9 ± 4.5	25.9 ± 0.1	84	–
Sr	370 ± 250	438 ± 49	470 ± 170	500 ± 4	320	216
Mo	6.2 ± 5.3	14.2 ± 14.1	1.0 ± 1.0	1.0 ± 0.1	1.1	–
Sb	1.7 ± 0.4	2.4 ± 0.9	0.20 ± 0.03	0.17 ± 0.01	0.4	–
I	12.3 ± 2.7	16.5 ± 3.2	2.1 ± 0.7	4.5 ± 0.9	1.4	–
Cs	5.5 ± 1.7	3.3 ± 0.04	0.40 ± 0.07	0.40 ± 0.01	4.9	–
Ba	530 ± 170	524 ± 84	225 ± 60	200 ± 41	624	720
La	36.5 ± 18.6	66.9 ± 55.3	10.8 ± 8.2	10.8 ± 6.4	31	–
Ce	41.9 ± 24.5	39.6 ± 6.8	11.5 ± 2.7	13.7 ± 0.4	63	–
Sm	5.6 ± 1.8	9.8 ± 7.4	1.2 ± 0.4	1.3 ± 0.3	4.7	–
Eu	1.7 ± 0.6	0.8 ± 0.9	0.3 ± 0.1	0.30 ± 0.01	1	–
Tb	0.6 ± 0.2	0.6 ± 0.04	0.14 ± 0.03	0.140 ± 0.004	0.7	–
Hf	6.1 ± 1.9	4.3 ± 1.1	1.2 ± 0.3	1.1 ± 0.2	5.3	–
Ta	0.6 ± 0.2	0.5 ± 0.1	0.10 ± 0.03	0.09 ± 0.01	0.9	–
Au	0.01 ± 0.001	0.01 ± 0.001	0.002 ± 0.002	0.002 ± 0.001	1.5	–
Th	8.6 ± 2.1	7.6 ± 0.3	1.4 ± 0.3	1.27 ± 0.02	10.5	–
U	4.3 ± 6.5	2.2 ± 0.8	0.5 ± 0.1	0.440 ± 0.002	2.7	–

1 elements in the average UCC according to [41]
2 elements in soils of the North Caucasus according to [43]

Table 3: The average for 2013-2014 years' elemental content of soils and bottom sediments (BS) for two different stations of Anapa region, upper continental crust (UCC) and average soils of the North Caucasus (SNC) (mean ± standard deviation, µg/g dry weight).

in UCC have the good agreement with the local data for Anapa region (Table 3). For a better description of the local conditions for each sampling site we determined the content of the same elements in soil as well as in sediments. This procedure was used for a more complete analysis of the distribution of elemental content of all considered elements in plants, soil and sediments collected from the Anapa region. Besides the above mentioned statistical analysis techniques, we have also used some graphic analysis procedure such as the ternary diagrams. They allowed to reveal at which extent the content of Cl, Br and I could be used to discriminate the different species of studied plants. All computations were performed using the LibreOffice 5.0.2 and Past 3.0 [42].

Results and Discussion

Accumulation of elements in soils and bottom sediments

The levels of the major and trace elements in soils and sediments from two stations located at 1 km (station 7) and 4 km (station 2c) from the city dump is given in Table 3. The content of the same elements in the average UCC and the levels of some elements in soils of the North Caucasus [43] are listed. Elements in the average UCC according to elements in soils of the North Caucasus according to ref. [43]. The determined concentrations of the majority of elements in soils and BS for each station belonged to close ranges. In that case, we would contemplate these milieus for plants as one. For further analysis, the average values were calculated as arithmetic means for soils (data from surface and from 0-20, 20-40, 40-60 cm layers) and bottom sediments (only from surface). The standard deviation for joint is given on the Figure 3. As follows from Table 3 Cl, Se, Br, and I concentrations in soils and BS from both stations (Figure 2) are higher than in UCC. It can be explained by the location of Anapa region near the sea and the fact that atmospheric supply from the marine environment is the predominant source of these elements in the soil [44,45]. As described in [46] and [47] soil contamination may be considered when concentrations of an element in soils were two- to three times greater than the mean background levels. For our study station 7 (the closest to city dump) hypothetically was the most polluted and the station 2c which situated on the shore was used as background for whole transect. The increasing levels of As, Mo, and Sb in soils and BS from the most polluted station 7 probably indicates the anthropogenic pollution with these elements. Increasing trend of levels of elements from the relatively pristine to polluted area probably ensue from influences of local disposal dump and traffic impacts. The concentrations of all elements (except for V and Ni) reported by for the soils of the North Caucasus are higher than those determined in the soil samples from the most polluted station 7 near city dump of Anapa. Our data were also compared to results of [48] who determined in laboratory conditions the levels of several elements

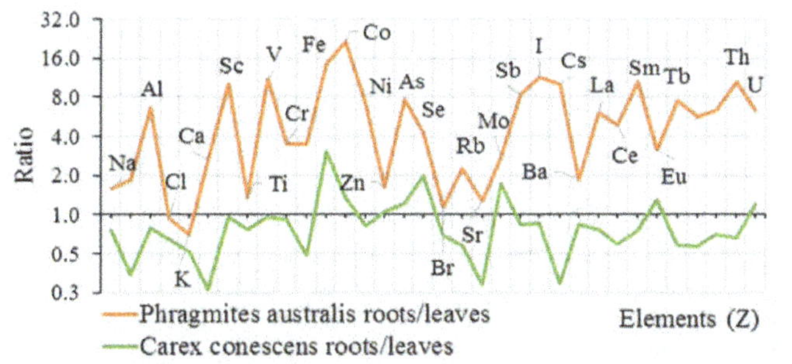

Figure 3: Average differences between leaves/roots ratios for *Phragmites australis* and *Carex conescens*.

for non-polluted, low polluted and moderate polluted soil from the Southern part of Russia using the integral index of biological state of soil (Table 4). It helps to realize the level of local differences in elemental content of soils from the standard levels for whole region, elements

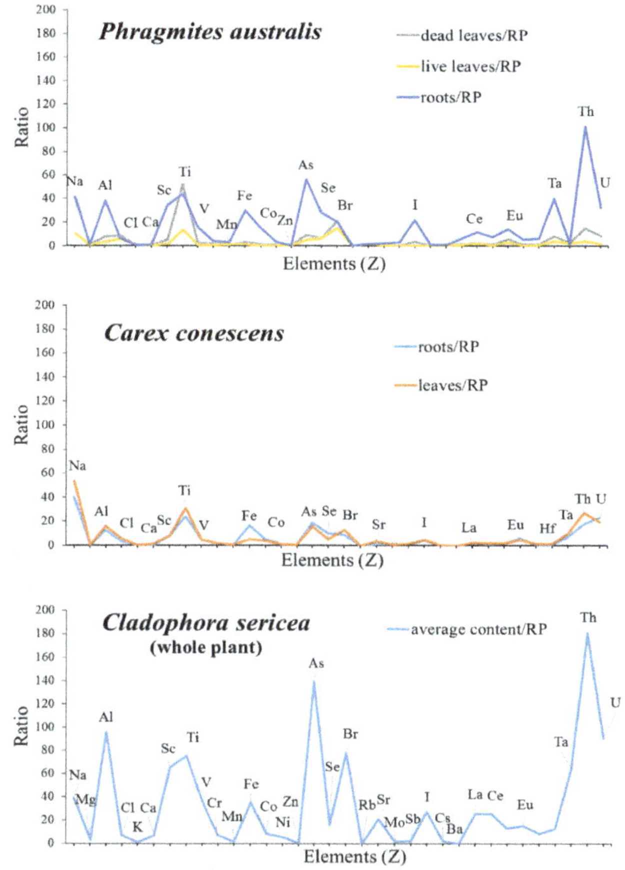

Figure 4: Studied plant/Reference plant ratios for different parts of *Phragmites australis*, *Carex conescens* and *Cladophora sericea*.

Elements	Soils in Anapa region		Soils in the Southern part of Russia		
	Max	Median	Non-polluted	Low polluted	Moderate polluted
V	150	30	<200	200-300	300-850
Cr	**105**	30	<70	70-90	90-170
Mn	900	370	<1000	1000-1600	1600-1800
Co	24	4	<18	18-36	36-250
Ni	80	12	<50	50-100	100-700
Zn	**270**	50	<125	125-200	200-850
As	**36.8**	7.1	<17	17-30	30-160
Se	**2.31**	0.25	<0.7	0.7-1.4	1.4-9
Sr	**840**	510	<250	240-450	450-3200
Mo	15.7	1.1	<8	8-400	>400
Sb	2.1	0.6	<5	5-12	12-200
Ba	690	250	<900	900-1500	1500-4000

Table 4: Maximal and median elemental concentrations (μg/g dry weight) in soils from Anapa region (our data) and values for non-polluted and polluted soils from the Southern part of Russia.

1 elements in soil according to [48].
The concentration of elements which relate to moderate polluted range are given in bold

in soil according to the concentration of elements which relate to moderate polluted range are given in bold. The maximal concentrations of Cr, Zn, As, Se and Sr in soils of Anapa region that were determined at the stations 6 and 7 (the nearest to city dump) are similar with the values reported for moderate polluted soils. Nevertheless, all median values of studied elements Figure 5 in soils of Anapa region are within the range of concentrations determined for non-polluted soils (Table 4) and less than maximum permissible levels of elements established in different countries (Table 5) [49]. Data of maximum permissible levels are widely used for ecological management in assessment of environmental impacts. It was concluded that the soils in study region were in low-polluted state despite the sources of anthropogenic stress.

Accumulation and compartmentalization of elements in water and coastal-aquatic plants

The data about accumulation of elements in different organs of plants were analyzed at the all stations, but after that, the average levels of elemental concentrations for whole Anapa region were calculated as arithmetic mean values obtained from all sampling stations. It helped to realize the ability of different species of plants to reflect the chemical features of environment, including the local pollution influences.

The concentrations of all elements determined (except for K and Cl) are higher in roots of *P. australis* than in leaves. In particular, the leaf/root ratios range from 0.86 for Br to 0.05 for Co. For Sc, V, Fe, Co, I, Cs, Sm and Th the root concentrations are one order of magnitude higher than concentrations in leaf. The obtained results confirm the data that *Phragmites australis* is prevalently a root bioaccumulator species. It is well known that roots are generally the main pathway of trace elements to plants. However, other tissues of *P. australis* in particular, leaves, show the ability readily to translocate such elements as Na, Ti, Zn, Br, and Sr (Figure 3). In contrast to *P. australis*, the concentration of all elements, except for Fe, Se, Mo, Eu and U are higher in leaves of *Carex conescens* than in roots. Our results emphasized the differences between accumulation features of these two species. The Carex conescens could be used as a good bio concentrator of majority of elements from soils and bottom sediments but *P. australis* could be used as a good comparative biomonitor (root type) in clean and polluted areas due to its self-cleaning processes. The obtained results were compared to the available data for *Phragmites, Carex,* and *Cladophora*, reported by other authors (Table 6) to represent the variability of concentrations in different regions. The concentrations of most elements in leaves and roots of *Phragmites australis* sampled in the mountain lake in Italy and in the mouth of the longest Sicilian river are higher compared to our results. The exceptions are Ti, Mn, As, Sb and Ti, V, As, Se, which values in roots and leaves, respectively, are higher in the present study. The values of Co, Zn, Rb, and Th in *Carex pendula* sampled in Germany in botanical garden are higher than our data; the reverse trend is observed for Sc, Cr, Fe and Cs. The elemental content of *Cladophora* reported by different authors varies in a wide range depending on the sampling region and the species. Thus, the levels of Mg, Ca and Mn in *Cladophora glomerata* from the lake Karasevoe in Siberia are one order of magnitude higher compared to our results [50,51]. In contrast, the content of Ca, Co and Ni in *Cladophora* sp. from the Baltic Sea [52] is one order of magnitude lower than those determined in the present study. The levels of Fe and Zn in *Cladophora glomerata* sampled in the Danube river [53] are 2-fold higher than our data. Thus, the exact concentrations of elements in studied species are absent or not widely available. As a result, it is necessary to determine the range of variability in different pollution conditions. According to the wide variability of elemental content of studied plants across the regions we normalized our data on values for so called reference plant for comparative analysis. The results of normalized elemental concentrations against Reference Plant (RP) show that roots and leaves of *P. australis* are good accumulators of Na, Ti, and Br and, in contrast, contain lower levels of Zn, Rb, and Ba than RP. (Figure 4) In *Carex* roots and leaves the levels of Na, Ti, As, Th, and U are one order of magnitude higher than in RP. In contrast, Mg, K, Mn, Zn, Rb, Cs, and Ba show lower levels in comparison to RP concentrations. The concentrations of the majority of elements in algae

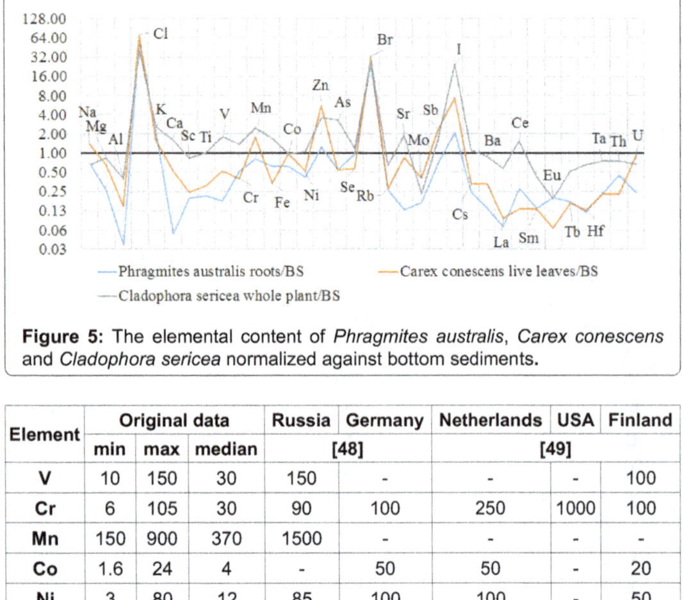

Figure 5: The elemental content of *Phragmites australis, Carex conescens* and *Cladophora sericea* normalized against bottom sediments.

Element	Original data			Russia	Germany	Netherlands	USA	Finland
	min	max	median	[48]		[49]		
V	10	150	30	150	-	-	-	100
Cr	6	105	30	90	100	250	1000	100
Mn	150	900	370	1500	-	-	-	-
Co	1.6	24	4	-	50	50	-	20
Ni	3	80	12	85	100	100	-	50
Zn	6	270	50	100	300	500	2500	200
As	3	36.8	7	2	50	30	30	5
Se	0.06	2.31	0.25	-	10	-	-	-
Mo	0.2	15.7	1.1	-	10	40	-	-
Sb	0.1	2.1	0.6	4.5	-	-	-	2
Ba	150	690	250	-	-	400	-	-

Table 5: Maximum permissible levels of elements in soils established in different countries.

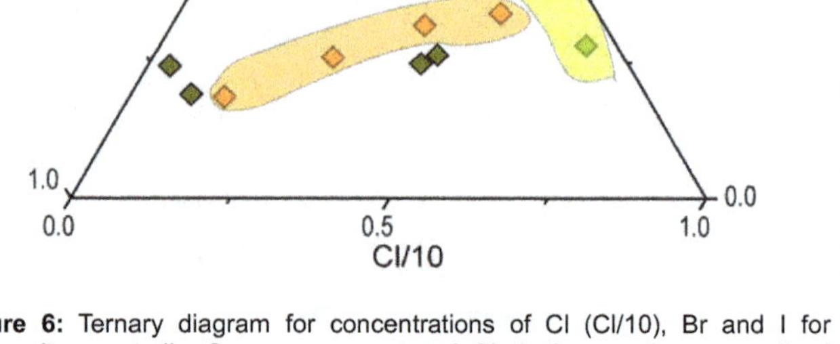

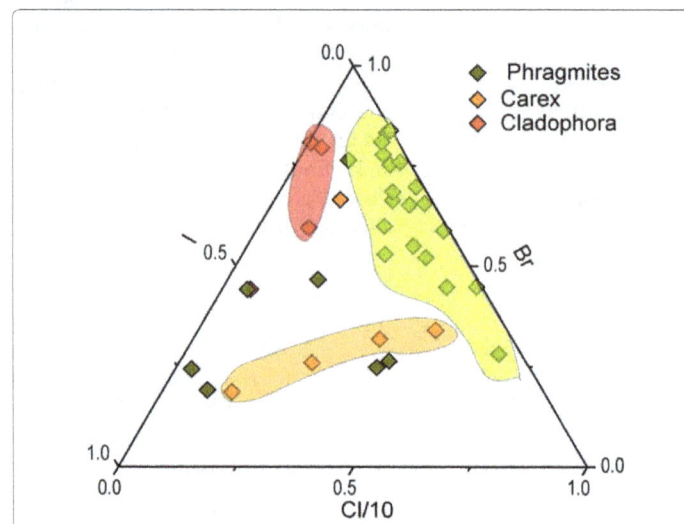

Figure 6: Ternary diagram for concentrations of Cl (Cl/10), Br and I for *Phragmites australis, Carex conescens* and *Cladophora sericea* normalized against content of these elements in sediments (BS). With some exceptions (*Phragmites* samples), all other points form three distinct clusters corresponding to each type of plant.

Element	Phragmites australis			Carex pendula	Cladophora sp.	Cladophora glomerata	
	Roots (1)	Roots (2,3)	Leaves (2,3)	Whole plant (4)	Whole plant	Whole plant (5)	Whole plant (7)
Na	–	–	–	–	17100	–	3000
Mg	1550	–	–	–	7800	–	23000
Al	–	3153	389	–	–	–	–
K	17000	–	–	–	24500	–	11000
Ca	–	–	–	–	8500	–	170000
Sc	–	–	–	0.19	–	–	–
Ti	–	<0.05	<0.05	–	–	–	–
V	14.5	9.2	<0.05	–	–	–	–
Cr	3.06	6.97	0.69	0.12	–	–	–
Mn	300	475.8	308	–	470	500	18000
Fe	2990	5561	453	460	2400	10000	2300
Co	–	8.0	0.22	0.03	0.5	–	–
Ni	6.52	9.12	1.69	–	3.1	–	6.3
Zn	54	104	28.4	595	60	200	–
As	–	<0.05	<0.05	–	–	–	5
Se	–	<0.5	<0.5	–	–	–	–
Rb	–	–	–	3.8	–	–	–
Sr	–	48.5	82.3	–	90	–	–
Mo	–	16.8	8.5	–	–	–	–
Sb	–	<0.05	<0.05	–	–	–	–
Cs	–	–	–	0.01	–	–	–
Ba	–	47.3	20.1	–	–	–	–
Th	–	–	–	0.45	–	–	–

Table 6: Elemental content of different species of *Phragmites*, *Carex* and *Cladophora* (µg/g dry weight).

Cladophora are at least one order of magnitude higher than in RP. The levels of Zn and Rb, that are lower than RP concentrations, become the exception. (Figure 4) The different composition of *Phragmites australis* and *Carex conescens* with *Cladophora sericea* is explainable by fully different uptake mechanisms of elements either by all surface of plant from water (*Cladophora*) or by roots (*Phragmites* and *Carex.*). Also some elements may characterize the different types of plants (for example, algae). Thus, the level of as that is a part of phosphatides in algae and plays an important role in glycometabolism [54] is 140-fold higher in *Cladophora* than its concentration in RP. (Figure 4) The similar patterns of elemental accumulation for all three species are found for several elements. Thus, the levels of Na, Ti and Br are higher than in RP; the reverse trend is revealed for Zn and Rb. It could be explained by abundance or lower concentrations of mentioned elements in the surrounding environment (soils, BS).

Transport of major and trace elements from bottom sediments to plants

The element distributions between the two compartments follow the order: bottom sediment>plant due to differences in concentrations. The differences between species accumulation with taking into account the type of accumulation (roots for *P. australis* and live leaves for *C. conescens*) were represented by normalizing concentrations of elements in plants from the same station on values in bottom sediments. (Figure 5) BS was used as a milieu, which at the same station reflects the local elemental fingerprint of water and other components. It is known that most rooted macrophytes uptake chemicals primarily from sediment pore water but it is also reported that some rooted submersed plants may absorb metals directly from water when they are not readily available in sediments and/or in high concentrations in the surroundings [55]. The one more way of coming the elements to plants is an uptake mechanism of them from air. Plants may absorb Cl, Br, and I directly from the atmosphere; and the marine environment is the main source of these halogens for plants [45-53]. It is found that the levels of Br and I in algae *Cladophora* are higher than in *Phragmites* and *Carex.* (4) Our results are in agreement with the statement that algae are one of the best accumulators of these elements [56]. To reveal the differences of halogens accumulation in *Phragmites*, *Carex* and *Cladophora* the ternary diagram for the levels of Cl (Cl/10), Br and I in plants normalized against content of these elements in sediments is built. (Figure 6) After that for ternary diagram the values was proportionally reduced to relative units (by using Origin™ 8). *Phragmites* is characterized by high levels of Cl at the majority of sampling sites. In *Carex* the content of Br is equal at all stations except one. *Cladophora* is characterized by high levels of Br and I, while the content of Cl is the minimal. This results demonstrate the specific accumulation features of plants. For example, *Cladophora sericea* accumulates Cl in small relative amounts in comparison to Br and I. *Phragmites australis* in the major cases selects I and Cl regardless Br. In that sense the *Carex conescens* demonstrates the most flexible ability for accumulation of these halogens.

Conclusion

The similarity in elemental concentration in soils and sediments at the majority of sampling stations is established. (Table 3) Sediments act as the primary source of elements for water plants. Regarding Cl, Br and I, the atmospheric supply from the marine environment is the predominant source. The concentration of majority of elements in soils of Anapa region are corresponded to values reported for non-polluted zones. The exception are the most polluted stations (6 and 7) near city dump, where elemental levels are several times higher if compered to median values. The study shows that *Phragmites australis* is prevalently a root bioaccumulator species; in contrast, the concentrations of all elements except for Fe, Se, W, and Mo are higher in leaves of *Carex conescens* than in roots. The different composition of *Cladophora sericea* and *P. australis* with *Carex conescens* is explainable by different elemental uptake, either mainly by entire surface of plant from water (*Cladophora*) or by roots from sediments (*Phragmites australis* and *Carex conescens*). Translocation of elements varies depending on the physiological property of elemental uptake and is generally more intense through plant tissues than from sediments to plants. Leaves of *P. australis* show the ability to readily translocate such elements as Na, Ti, Zn, Br, and Sr. Among the elements determined the highest translocation between roots and leaves of *Carex conescens* is found for Sc, V, Cr, and Zn. The results of normalized elemental concentrations against Reference Plant show that roots and leaves of *Phragmites australis* are good accumulators of Na, Ti, and Br and, in contrast, contain lower levels of Zn, Rb, and Ba than RP. In *Carex conescens* roots and leaves the levels of Na, Ti, As, Th, and U are one order of magnitude higher than in RP. In contrast, Mg, K, Mn, Zn, Rb, Cs, and Ba show lower levels in comparison to RP concentrations. The concentrations of the majority of elements in algae *Cladophora sericea* are at least one order of magnitude higher than in RP. *Cladophora sericea* accumulated Cl in small relative amounts in comparison to Br and I. *Phragmites australis* in the major cases selected I and Cl regardless Br. In that sense the *Carex conescens* demonstrated the most flexible ability for accumulation of these halogens. The found ratios BS to plants demonstrated the different ability of this three species to reflect the local elemental fingerprints. The levels of majority of elements in *Phragmites australis*, *Carex conescens*, *Cladophora* could be used in future biomonitoring studies on local and regional scales. These plants are potentially useful for monitoring of pollution in general, and for the most elements examined in particular.

References

1. Burdin KS (1985) The basis of biological monitoring. Moscow State University Press, p: 160.
2. Burdin KS, Gavrilenko EE, Zolotukhina E, Kamnev AN (1988) Study of the macrophytes of the seas of the USSR as biomonitoring objects of heavy metals in the sea water. Ocean Biology, pp: 16-25.
3. Burdin KS, Voskoboinikov GM, Zubov EV, Kamnev AN (1993) Marine macro algae as monitoring objects in Arctic ocean bioindication environmental state bioassay and the waste destruction technology. Apatity: MMBI RAS, pp: 72-81.
4. Kamnev AN (1989) Structure and functions of the brown algae.
5. Vozzinskaya VB, Kamnev AN (1994) Ecological and biological foundations of cultivation and utilization of marine benthic algae.
6. Phillips DJH, Rainbow PS (1993) Biomonitoring of trace aquatic contaminants. 2nd edn. Chapman and Hall, London.
7. Wang W, Gorsuch JW, Hughes JS (1997) Plants for environmental studies. p: 576.
8. Gerhardt A (2000) Biomonitoring of polluted water. Zurich Trans Tech Publ.
9. Markert B, Breure T, Zechmeister H (2003) Bioindicators and biomonitors principles concepts and applications. Amsterdam Elsevier.
10. Marques JC, Salas F, Patrício J, Teixeira H, Neto JM, et al. (2009) Ecological indicators for coastal and estuarine environmental assessment a user guide. Southampton WIT Press.
11. Kamnev AN, Frontasyeva MV, Kravtsova AV, Nekhoroshkov PS (2014) Assessment of elemental composition of macrophytes, soil and bottom sediments along a transect in the coastal zone of Anapa, studied by neutron activation analysis.
12. Burdin KS, Zolotuhina E (1998) Heavy Metals in Aquatic Plants (Accumulation and Toxicity). Moscow: Dialog MSU. Russian.
13. Kamnev AN, Bunkova OM, Bogatyrev LG (2015) Sabinins ideas and their embodiment. Mineral composition of macrophytes is one of the most important indicator of their contribution to a biogeochemical exchange of the World Ocean. Issues of modern Algology.
14. Albers PH, Camardese MB (1993) Effects of acidification on metal accumulation by aquatic plants and invertebrates 1 Constructed wetlands. Environ Toxicol Chem 12: 959-967.
15. Bonanno G (2011) Trace element accumulation and distribution in the organs of Phragmites australis (common reed) and biomonitoring applications. Ecotoxicology and Environmental Safety 74: 1057-1064.
16. Bunkova O, Kamnev A, Yakovlev A, Suhova T (2014) Studying of mineral structure of Cystoseira barbata in a sublittoral zone of the northeast coast of the Black Sea. Proceedings of World Conference on Marine Biodiversity Oct 12-16.
17. Kravtsova AV, Milchakova NA, Frontasyeva MV (2015) Levels spatial variation and compartmentalization of trace elements in brown algae Cystoseira from marine protected areas of Crimea (Black Sea). Marine pollution bulletin 97: 548-554.
18. Pankova ES, Kamnev AN, Golubeva EI (2015) Features of the distribution of heavy metals in brown algae Cystoseira barbata. International popular science journal Europe-Asia. Earth Sciences 5: 25-28.
19. Duman F, Cicek M, Sezen G (2007) Seasonal changes of metal accumulation and distribution in common club rush (Schoenoplectus lacustris) and common reed (Phragmites australis). Ecotoxicology 16: 457-463.
20. Baldantoni D, Ligrone R, Alfani A (2009) Macro- and trace-element concentrations in leaves and roots of Phragmites australis in a volcanic lake in Southern Italy. Journal of Geochemical Exploration 101: 166-174.
21. Maddison M, Soosaar K, Mauring T, Mander U (2009) The biomass and nutrient and heavy metal content of cattails and reeds in wastewater treatment wetlands for the production of construction material in Estonia. Desalination 246: 120-128.
22. Bonanno G (2013) Comparative performance of trace element bioaccumulation and biomonitoring in the plant species Typha domingensis Phragmites australis and Arundo donax. Ecotoxicology and Environmental Safety 97: 124-130.
23. Strbac S, Sajnovic A, Kasanin GM, Vasic N (2014) Metals in sediment and Phragmites australis (common reed) from Tisza river Serbia. Applied Ecology and Environmental Research 12: 105-122.
24. Bragato C, Schiavon M, Polese R, Ertani A, Pittarello M, et al. (2009) Seasonal variations of Cu, Zn, Ni and Cr concentration in Phragmites australis Trin ex Steud in a constructed wetland of North Italy. Desalination 246: 35-44.
25. Ahmad SS, Reshi ZA, Shah MA, Rashid I, Ara R, et al. (2014) Phytoremediation potential of Phragmites australis in Hokersar wetland-A Ramsar site of Kashmir Himalaya. International Journal of Phytoremediation 16: 1183-1191.
26. Horovitz CT, Schock HH, Horovitz-Kisimova LA (1974) The content of scandium, thorium, silver, and other trace elements in different plant species. Plant and Soil 40: 397-403.
27. Pederson JC, Harper KT (1979) Chemical composition of some important plants of southeastern Utah summer ranges related to mule deer reproduction. The Great Basin Naturalist 39: 122-128.
28. Ohlson M (1988) Variation in tissue element concentration in mire plants over a range of sites. Ecography 11: 267-279.
29. Whitton BA, Burrows IG, Kelly MG (1989) Use of Cladophora glomerata to monitor heavy metals in rivers. Journal of Applied Phycology 1: 293-299.
30. Malea P, Haritonidis S, Kevrekidis T (1995) Metal content of some green and brown seaweeds from Antikyra Gulf. Hydrobiologia 310: 19-31.
31. Chmielewska E, Medved J (2001) Bioaccumulation of heavy metals by green algae Cladophora glomerata in a refinery sewage lagoon. Croatica Chemica Acta 74: 135-145.
32. Vershinin A, Kamnev A (2001) Harmful algae in Russian European coastal waters. Proceedings of 9th Int Conf on Harmful algal blooms 2000 Feb 7-11 Hobart Australia UNESCO.
33. Levkov Z, Krstic S (2003) Use of algae for monitoring of heavy metals in the River Vardar Macedonia. Mediterranean Marine Science 3: 99-102.
34. Zbikowski R, Szefer P, Latała A (2007) Comparison of green algae Cladophora sp. and Enteromorpha sp. as potential biomonitors of chemical elements in the southern Baltic. Science of the total environment 387: 320-332.
35. Pogorelov AV, Dulepa SV, Lipilin DA (2013) The space monitoring on the territory of the Krasnodar region. Geomatics 4: 64-71.
36. GOST (State Standard) 17.4.4.02-84 Nature Protection. Soils. Methods of Sampling and Preparation of Material for Chemical, Bacteriological, and Helminthological Analysis.
37. GOST (State Standard) 17.1.5.01-80 Nature protection. Hydrosphere. General requirements for sampling of bottom sediments of water objects for their

pollution analysis.

38. Frontasyeva MV (2011) Neutron activation analysis for the Life Sciences. A review Physics of Elementary Particles and Atomic Nuclei 42: 332-378.

39. Dmitriev AY, Pavlov SS (2013) Automated quantitative determination of elements in samples by neutron activation analysis at the IBR-2 at LNP JINR 10: 58-64.

40. Markert B (1992) Establishing of Reference Plant for inorganic characterization of different plant species by chemical fingerprinting. Water Air and Soil Pollution. 64: 533-538.

41. Rudnick RL, Gao S (2003) Composition of the continental crust. Treatise on geochemistry 3: 1-64.

42. Hammer, Harper DAT, Paul DR (2001) PAST paleontological statistics software package for education and data analysis. Palaeontologia Electronica 4: 1-9.

43. Dyachenko V, Matasova I, Ponomareva O (2014) The trace elements concentrations dynamics in the soil landscapes of the Southern Russia. Universal Journal of Geoscience 2: 28-34.

44. Fuge R (1988) Sources of halogens in the environment, influences on human and animal health. Environmental Geochemistry and Health 10: 51-61.

45. Frontasyeva MV, Steinnes E (2004) Marine gradients of halogens in moss studies by epithermal neutron activation analysis. Journal of Radioanalytical and Nuclear Chemistry 261: 101-106.

46. Logan TJ, Miller RH (1983) Background levels of heavy metals in Ohio farm soils. Research circular 275. AGDEX, pp: 508-530.

47. Chen M, Ma LQ, Harris WG (1999) Baseline concentrations of 15 trace elements in Florida surface soils. Journal of Environmental Quality 28: 1173-1181.

48. Kolesnikov SI, Kazeev KS, Denisova TV, Dadenko EV (2012) Development of a regional ecological regulations of content of contaminants in soils of Southern Russia. Polythematic Online Scientific Journal of Kuban State Agrarian University 82: 1-17.

49. Mynbayeva BN, Imanbekova TG (2013) Assessment of standards of soil's contamination by heavy metals (analytical review). The Bulletin of the National Academy of Sciences of the Republic of Kazakhstan 4: 29-39.

50. Bonanno G, Lo Giudice R (2010) Heavy metal bioaccumulation by the organs of Phragmites australis (common reed) and their potential use as contamination indicators. Ecological Indicators 10: 639-645.

51. Maltsev AE, Leonova GA, Bogush AA, Bulycheva TM (2014) Eco geochemical assessment of anthropogenic pollution of flooded open pits ecosystems in Novosibirsk. Ecology of industrial production 2: 44-53.

52. Bojanowski R (1973) The occurrence of major and minor chemical elements in the more common Baltic seaweed. Oceanologia. 2: 81-152.

53. Ravera O (2001) Monitoring of the aquatic environment by species accumulator of pollutants: a review. Journal of Limnology 60: 63-78.

54. Kabata PA, Pendias H (2001) Trace elements in soils and plants. 3rd edn. Boca Raton London New York: CRC Press.

55. Guilizzoni P (1991) The role of heavy metals and toxic materials in the physiological ecology of submersed macrophytes. Aquatic Botany. 41: 87-109.

56. Saenko GN (1992) Metals and halogens in marine organisms.

Boric Acid Production from a Low-Grade Boron Ore with Kinetic Considerations

Mahdi H[1*], Davood M[2], Mohsen V[3] and Behzad S[3]

[1]*College of Engineering, University of Tehran, Tehran, Iran*
[2]*Faculty of Engineering, University of Zanjan, Zanjan, Iran*
[3]*Research and Engineering Company for Non-Ferrous Metals, Zanjan, Iran*

Abstract

The most important Iranian boron reserves are in the basin of Ghezel Ozan, a river in the West and Northwest area of Zanjan. In the present study, boric acid production from an Iranian low-grade borate ore by hydrometallurgical process was investigated. In order to produce boric acid, boron ore was reacted with sulphuric acid. The influence of four parameters on the course of reaction such as pH, temperature, liquid to solid ratio, and reaction time was examined. Optimum condition for leaching part was obtained in temperature of 90°C, reaction time of 2 hours, L/S ratio of 3, and pH of 1. Under these conditions, the recovery of boron acidic leaching was reported to be 92.21%. Neutralization of pulp was done by lime. Finally, boric acid was obtained by crystallization. The purity of produced boric acid was 99.56%. The data obtained from the acidic leaching kinetics indicated that the dissolution of boron ore is fluid film diffusion controlled reaction and the reaction activation energy equals to 11.6 kJ/mol. Enthalpy of activation (⟦ΔH⟧^‡) and entropy of activation (⟦ΔS⟧^‡) were 11.2 kJ/mol and -246.3 J/(mol.K) respectively.

Keywords: Boron ore; Optimization; Boric acid; Kinetics

Introduction

Boron is one of the most significant industrial elements. The total world boron ore reserves are estimated to be equivalent to 885 billion tons [1]. The major boron producers are Turkey (39% of global production), USA (24%), Argentina (14%), and Chile (10%) [2]. One of the most-used and commercially important boron compounds is boric acid. Due to increasing demands for the boron compounds, especially boric acid which is used in many branches of industry, such as in the medical, pharmaceutical and electronic sectors, the production of these compounds has enhanced recently [3]. Boric acid (H_3BO_3) can be produced from different boron minerals such as colemanite, tincal, kernite, ulexite or from the brines having dissolved boron salts [4]. Boron compounds found in nature in the form chiefly as calcium, magnesium and sodium borate. In Ca Borate ores, concentration is generally carried out by disintegration, washing and classification in the size fractions in large sizes. Concentrate is obtained through attrition tumbling and hand sorting. In Na Borate ores, attrition scrubbing to the ore is followed by classification by the use of screens and cyclone. All the water is kept at near saturation with boron since Na borate is soluble in water [5]. Magnesium borates occur in small quantities in various boron deposits of the world. Commercial Hydroboracite deposits are found in Inder, a place in Siberia. Hydroboracite in Turkey occurs in the Büyük Günevi mine of the Yakal Borasit Ltd [6]. In some boric acid processes, borax is reacted with aqueous hydrochloric acid or nitric acid [7,8]. The production cost is very high in this method due to the use of strong acids, which lead to a short life of the equipment. In Europe, boric acid is industrially produced based on the reaction of colemanite with sulphuric acid at 90°C [9]. Tunc and Kocakerim investigated the leaching of ulexite in sulphuric acid. The parameters were temperature, particle size, solid-to-liquid ratio, acid concentration, and stirring rate. The results showed that temperature and stirring rate had positive effect on the leaching process. Conversion rate increased with the acid concentration up to 1 mol L^{-1} and decreased with the concentrations above 1 mol L^{-1} [10]. The effect of sulphuric acid concentration on the dissolution of impurities from colemanite was studied by Kalafatoğlu et al. Results demonstrated that if the acid concentration was held around 5%, the dissolution of clay and the other impurities would decrease [11]. In another paper, the crystallization kinetics of gypsum during the dissolution of colemanite in a batch reactor at different temperatures were studied; (60-90°C, stirring rate: 150-400 rpm), and the initial concentration of the reactants. It was found that the crystal growth of gypsum on seed crystals follows a second order kinetics from the solution supersaturated in calcium and sulphate ions [12]. Another study examined the effect of particle size of colemanite on gypsum crystallization in batch reactor [13]. Dissolution kinetics of colemanite in sulphuric acid in the absence and presence of ultrasound was studied by Okur et al. [14]. An Avrami-type equation was used successfully to explain kinetic data. Activation energy was 30 kJ/mol in both situations. Iran is one of the countries that have boron mines. There are less than 5 active boron mines in Iran. The most important Iranian boron reserves are in the basin of Ghezel Ozan, a river in the West and North-west area of Zanjan, the North-western region of Iran. Moshampa borate ore containing the minerals, hydroboracite (($CaMgB_6O_{11}.6H_2O$) 90.9% of all valuable mineral), and pandermite ($Ca_4B_{10}O_{19}.7H_2O$) is located in the northwest of Zanjan. The first utilization of this mine started in 2005. This mine being extracted in a non-advanced way. Hence, the objectives of this study were to investigate the main factors involved in the leaching of boron ore by sulphuric acid solution, such as solid/liquid ratio, temperature, pH and time, and also to determine what processes control the rate of the dissolution of boron ore or what kinetic model can be applied.

Materials and Methods

Materials

The sample under the study was obtained from damp of Moshampa boron mine, Zanjan, Iran. Prior to this study, the boron ore sample was dried, grounded, and homogenized. The chemical analysis was carried out with a Spectro XEPOS model XRF, as well as boron content by inductively coupled plasma-atomic emission spectrometry (ICP-AES)

***Corresponding author:** Mahdi H, College of Engineering, University of Tehran, Tehran, Iran, E-mail: haghani.mahdi@ut.ac.ir method. The chemical analysis of the boron ore is given in Table 1. As can be seen from Table 1, studied sample is mostly composed of B and Ca. Furthermore, the characterization of minerals in the boron ore was performed by X-ray powder diffraction (XRD) under the condition of Cu Kα at 40 kV and 30 mA as shown in the Figure 1. The results of XRD analysis revealed that $CaSO_4.2H_2O$ and $CaCO_3$, were the major, $Ca_4B_{10}O_{19}.7H_2O$ (pandermite) and $CaMgB_6O_{11}.6H_2O$ (Hydroboracite) were the minor mineralogical phases in Moshampa boron ore. SEM of the boron ore were performed by Scanning Electronic Microscope model XL-30 which was manufactured by Philips. The SEM image of boron ore (Figure 2) showed that they consist of large sheet form particles, sized less than 100 μm. In the leaching experiments, sulphuric acid was used which made by Merck with a purity of 98% and a density of 1.89 g/mL. Acidic leach of boron was examined in four parameters of pH, temperature, liquid to solid ratio, and reaction time. For each parameter, four levels were chosen as shown in Table 2.

Methods

Boric acid production took place in a glass beaker of 5 Littre volume, equipped with a mechanical stirrer submerged in a thermostatic bath. The mechanical stirrer (Heidolf RZR 2020) had a controller unit and the bath temperature was controlled using thermometer (within ± 0.5°C). Leaching experiments were performed based on our previous experience in related works and preliminary tests. Time, temperature, pH, and L/S amount were chosen as the four variables to be investigated. For each experimental condition, the experiment was repeated twice, and the arithmetic average of the results was used in the plotting of yield curves. The ranges and values of parameters for leaching tests are given in Table 2. After adding acid to the reaction beaker and setting the temperature at the desired value, a known weight of sample was added to the reactor while the content of the reactor was stirred at a certain speed (700 rpm). pH was adjusted by adding drops of concentrated sulphuric acid (98%). Under desired circumstance, B_2O_3 was absorbed in the solution phase. In the next step, impurities of 'Fe' were precipitated using lime as iron hydroxide in pH of 5. Obtained filtrate in this part was crystallized by reducing the temperature and eventually boric acid was produced. Optimization was carried out in classic method, similar to the mineral processing part.

Results and Discussion

Investigation of effective parameters in acidic leaching

Equations 1 and 2 present the dissolution of boron in acid reactions. Thermodynamic data showed that the dissolution of boron in sulfuric acid occurs due to Eq. (1) and 2, which is calculated with HSC 5.1 software. The Eh-pH diagrams for the B-S-water system at the temperature of 90°C, as seen in Figure 3, illustrates at 90°C, the lines related to the specie of H_3BO_3 is the stable type of reaction's yield.

Content [wt. %]							
B_2O_3	CaO	MgO	SiO_2	Fe_2O_3	Al_2O_3	K_2O	Na_2O
8.95	29.59	3.51	5.29	0.92	0.89	0.19	5.03

Table 1: Chemical analysis of the boron ore.

Parameters	Units	Level 1	Level 2	Level 3	Level 4
pH	-	1	1.5	1.75	2
Temperature	°C	50*	70	80	90
L/S	L/kg	1.5*	2	3	5
Time	min	30*	60	90	120

*Base level of each parameter

Table 2: Special parameters in precipitation of Lead with Zinc powder.

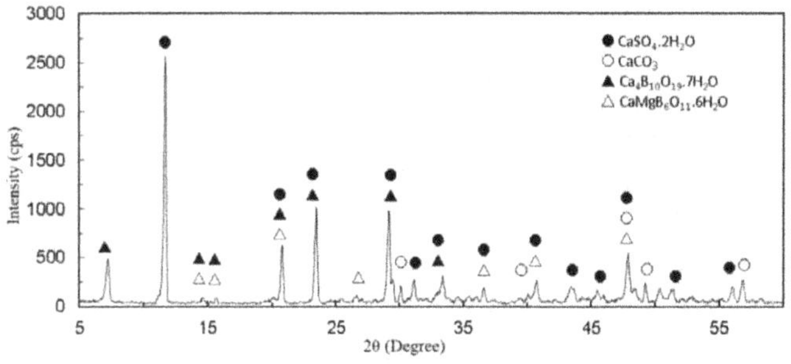

Figure 1: XRD analysis of the boron ore.

Figure 2: SEM analysis of the boron ore.

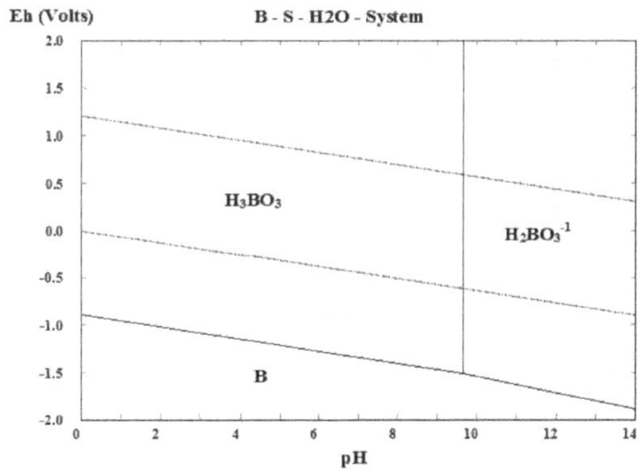

Figure 3: Eh-pH diagrams for the B-S-water system at the temperature of 90°C.

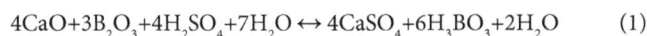

$$4CaO+3B_2O_3+4H_2SO_4+7H_2O \leftrightarrow 4CaSO_4+6H_3BO_3+2H_2O \quad (1)$$

ΔG of -1145.79 kJ/mol at 90°C

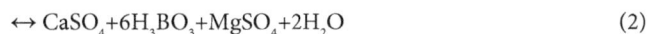

$$CaO+MgO+3B_2O_3+2H_2SO_4+9H_2O$$

$$\leftrightarrow CaSO_4+6H_3BO_3+MgSO_4+2H_2O \quad (2)$$

ΔG of -493.21 kJ/mol at 90°C

Experiments were carried out to investigate the effect of pH on the dissolution of boron under constant conditions of temperature of 50°C, L/S ratio of 1.5, and reaction time of 30 min. Increasing the dissolution of boron resulted in an increase in pH. As it can be seen in Figure 4, maximum recovery is in minimum pH. In this pH, the amount of dissolution was 59.71%. Therefore, to investigate the other parameters, the pH of 1 was chosen as the optimum pH. Figure 5 shows the effect of temperature on the boron leaching. As it was shown in Figure 5, the maximum dissolution of boron was noticed in the temperature of

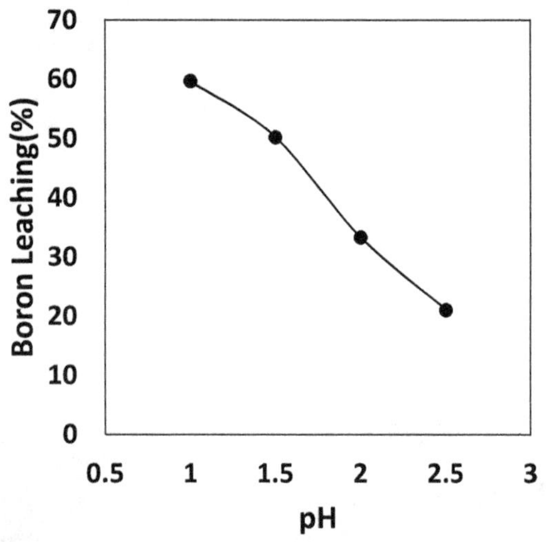

Figure 4: Effect of pH on boron leaching (temperature: 50°C; time: 30 min, L/S ratio: 1.5).

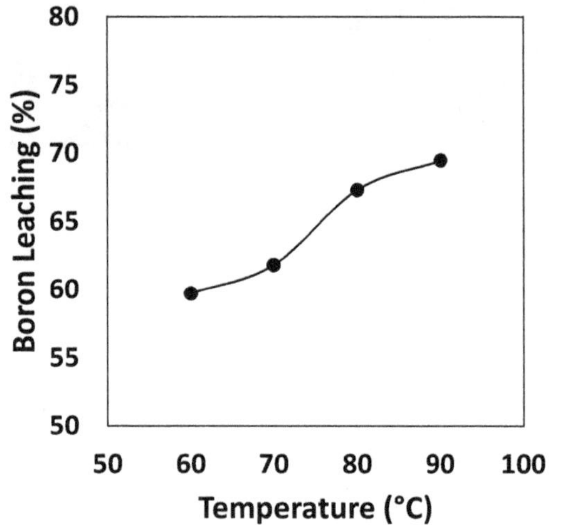

Figure 5: Effect of temperature on boron leaching (pH: 1; time: 30 min, L/S ratio: 3).

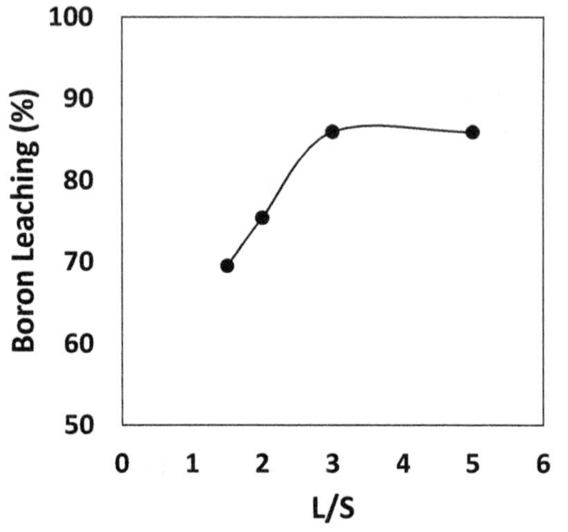

Figure 6: Effect of L/S ratio on boron leaching (pH: 1; time: 30 min, temperature: 90°C).

90°C that equaled 69.48%. Constant factors in this experiment were as follows: pH of 1, L/S ratio of 1.5, and reaction time of 30 min. Figure 6 shows the effect of liquid on solid ratio on the boron leaching. As it is shown in Figure 6, increasing the L/S ratio up to 3 resulted in an increase in boron leaching amount, 85.91%. Increasing of L/S ratio to more than 3 does not show significant change in solubility of boron. Constant factors in this experiment were as follows: pH of 1, temperature of 90°C, and reaction time of 30 min. The reaction time was the last factor which was tested. As it is shown in Figure 7, the optimal mode for dissolution was 92.21%, at the time of 120 min. Constant factors in this experiment were as follows: pH of 1, temperature of 90°C, and L/S ratio of 3.

Kinetic modelling

For the kinetics study, the experiment was conducted in the optimum condition achieved by design of experiment software and central composite method. The sample of the solution was taken between determined time intervals. As it is shown in Figure 8, efficiency of acidic leaching increased sharply during the first 30 minutes. Information on times higher than the 30 minutes was excluded in kinetic study due to low growth rate of efficiency. In chemical and hydrometallurgical processes fluid-solid reaction rate maybe generally controlled by one of the following steps: diffusion through fluid film, diffusion through ash or the chemical reaction at the surface of the core or unreacted materials in the solid liquid systems [15]. The rate of the process is controlled by the slowest of the sequential steps. As the reaction of Boron particles precede, the thickness of outer shell of insoluble product progressively increased while the inner core of unreacted particle decreased. It is clear from Figure 9 that the rate of reaction increases with time. This is due to the growth of the reactant surface and the increase in path length for the diffusion of ions [16,17].

Chemical reaction rate controlling system

The fraction of Boron reacted at any time, t, in a reaction control process can be calculated from the following equation [15]: where XB is the fraction of Boron reacted. The time for complete disappearance of a particle, τ, can be calculated from:

$$t/\tau=1-(1-XB)1/3 \quad (3)$$

where $\tau=(\rho R0)/(b\ KS\ C)$

where ρ is density of Boron ore, R0 is radius of the unreacted particle, b is stoichiometric coefficient of the reaction, C is concentration of sulphuric, and KS is rate constant of the reaction. Based on the experimental data plotted in Figure 10, the right-hand side of Eq. (3) is plotted against reaction time in Figure 8. It is obvious that the data do not fit with a straight line and it is concluded that the chemical reaction could not be rate determining in the leaching system used here.

Solid product diffusion control

Diffusion of the reagent or dissolved species through a solid reaction product at any time, t, can be calculated from the following equation [15]:

$$t/\tau=1-3\ (1-XB)2/3+2(1-XB) \quad (4)$$

In order to test the possibility of diffusion through a solid reaction product, the right-hand side of Eq. (4) is plotted against time and is shown in Figure 11. This is also evident from the high R2 values for film diffusion model as shown in Table 3, which also lists the values of apparent rate constants for Boron with their equivalent correlation coefficients. The kinetic data showed poor fit to the ash diffusion and chemical control model, especially at higher temperatures. Therefore, the data can be correlated with diffusion through fluid film.

Activation energy

The temperature dependence of the reaction rate constant can be calculated by the Arrhenius equation:

$$Ka = A \exp(-Ea / RT) \quad (5)$$

where A is frequency factor, Ea is activation energy of the reaction, R is universal gas constant, and T is absolute temperature. The values of Ka at different temperatures can be calculated from the slope of the lines shown in Figure 9 and Eq. (5). The Arrhenius plot (lnKa vs. 1/T), shown in Figure 12, gives an activation energy of 11.6 kJ/mol (2.77 kcal/mol) for the reaction from the slope of the lines in the temperature range between 333 and 364 °K. The activation energy of a diffusion-controlled process is characterized to be from 1 to 3 kcal/mol, while for a chemically controlled process it usually is greater than 10 kcal/mol.

According to the linear form of the Eyring-Polanyi equation (Eq. (6)) the values for $\Delta H^{\ddagger}$ and $\Delta S^{\ddagger}$ can be determined from kinetic data obtained from a ln k/T vs. 1/T plot (Figure 12.1) The Equation is a straight line with negative slope $-\Delta H^{\ddagger}/R$, and a y-intercept, $\ln k_B/h + \Delta S^{\ddagger}/R$.

$$\ln k/T = -\Delta H^{\ddagger}/R \; 1/T + \ln k_B/h + \Delta S^{\ddagger}/R \quad (6)$$

Where:

k=reaction rate constant, T=absolute temperature, $\Delta H^{\ddagger}$=enthalpy of activation, R=gas constant, kB=Boltzmann constant, h=Planck's constant, $\Delta S^{\ddagger}$=entropy of activation.

Hence, with a simple calculation enthalpy of activation ($\Delta H^{\ddagger}$=11.2 kJ/mol) and entropy of activation ($\Delta S^{\ddagger}$=-246.3 J/(mol.K)) were evaluated.

Characteristic of produced boric acid

The results of XRD analysis for Boric acid from acidic leaching reaction revealed that boric acid was the major mineralogical phase in the product, as shown in the Figure 13.

Scanning electron microscopy (SEM) was used for a morphological study of Boric acid deposits under the optimum conditions. SEM micrographs of the sample are presented in Figure 14 and shows sheet form structures crystallization.

Working diagram

The final process of production of boric acid and boron ore is shown in Figure 15. Firstly, process of grinding the ore is done right on mesh generator. In second step, ore is dissolved in acidic condition with H_2SO_4 to pH=1 in temperature of 90°C. After 120 min of acidic leach in Liquid/Solid rate of 3:1, neutralization process is carried out with lime because of precipitation of iron. In the following steps, solid–liquid separation operation is required. Solid part of alkaline leaching separation section is deposited as a waste. Filtrate of neutralization separation section is used by crystallization. Finally, solid boric acid is obtained by filtration and drying of crystals to scale up experiments. To scale up the process, 10 Kg of boron ore was used from Moshampa mine in Province of Zanjan, Iran. Process properties, acidic leaching, neutralization, and crystallization efficiencies were obtained, as shown in Table 4. It explains that acidic leaching rate of boron can reach 99.74%. After precipitation step, 95% of iron is removed and around 98.5% of boron can be extracted from the ore since composition of iron precipitate was environmentally safe in the iron precipitation step. The contents of leaching residue were lower than hazardous limits of the

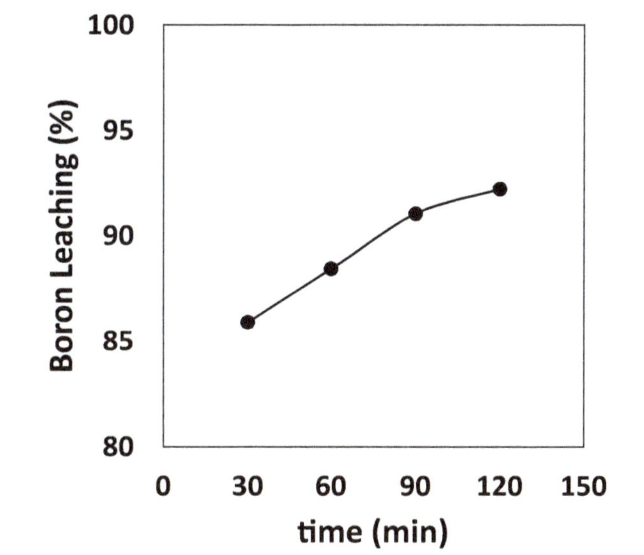

Figure 7: The Effect of reaction time on boron leaching (pH: 1; L/S ratio: 3, temperature: 90°C).

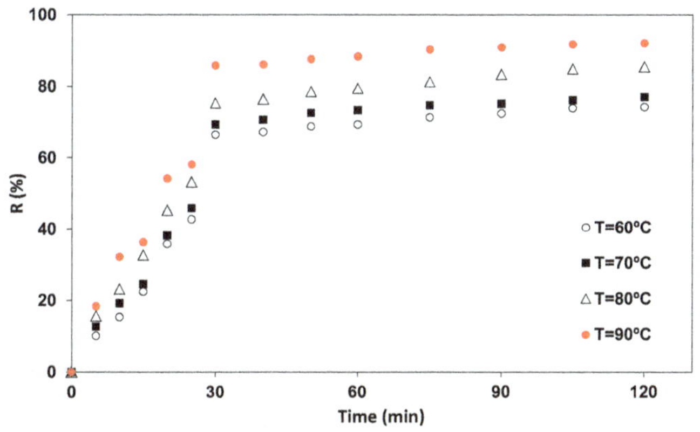

Figure 8: Efficiency of acidic leaching during the time.

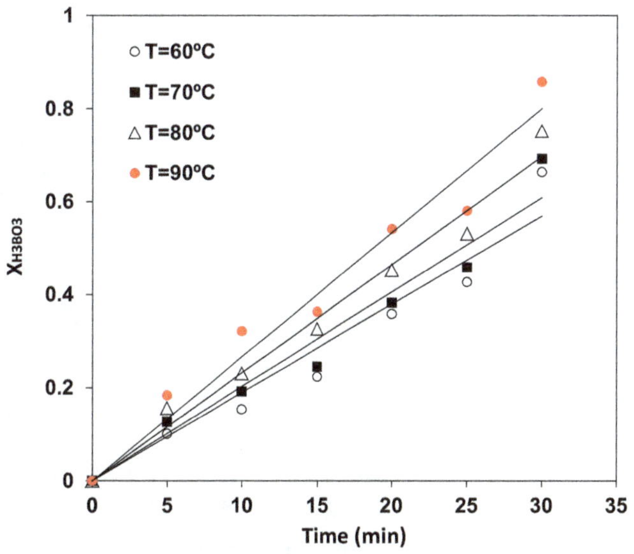

Figure 9: Leaching of Boron with time at various temperatures.

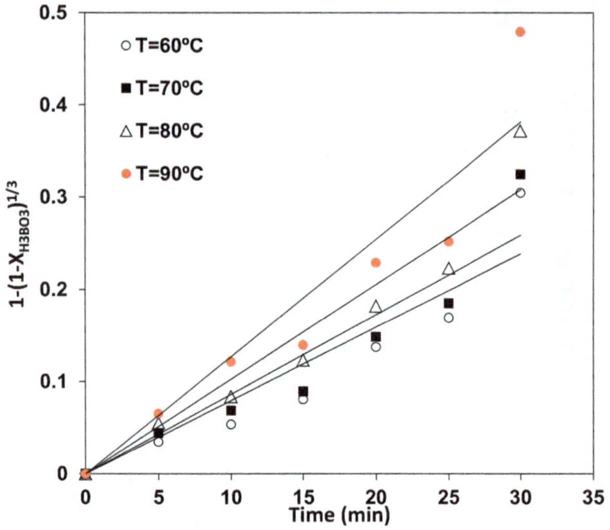

Figure 10: Plot of chemical system during the time at various temperatures.

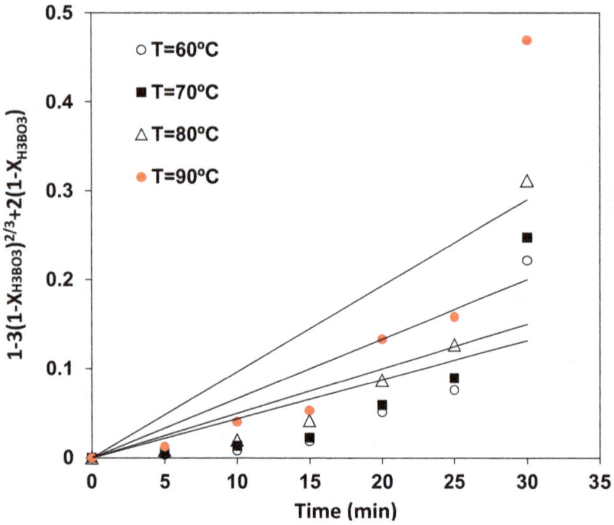

Figure 11: Plot of diffusion through a solid system during the time at various temperatures.

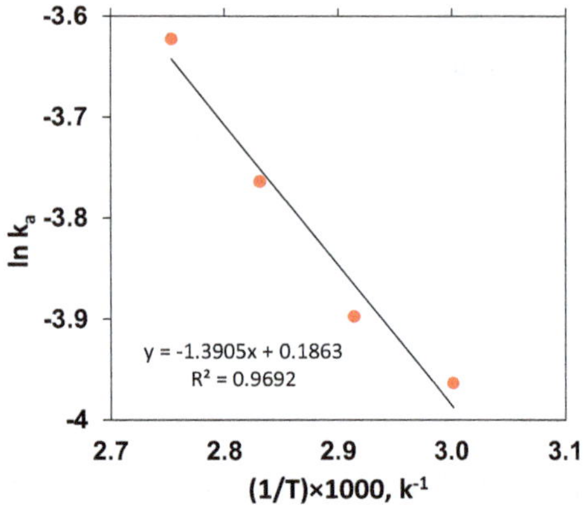

Figure 12: Arrhenius plot for leaching of Boron in the temperature range 333-364 K^{-1}.

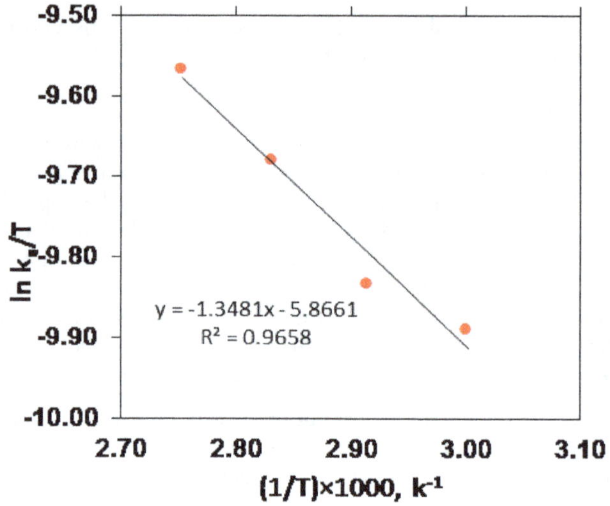

Figure 12.1: Arrhenius plot for leaching of Boron in the temperature range 333-364 K^{-1}.

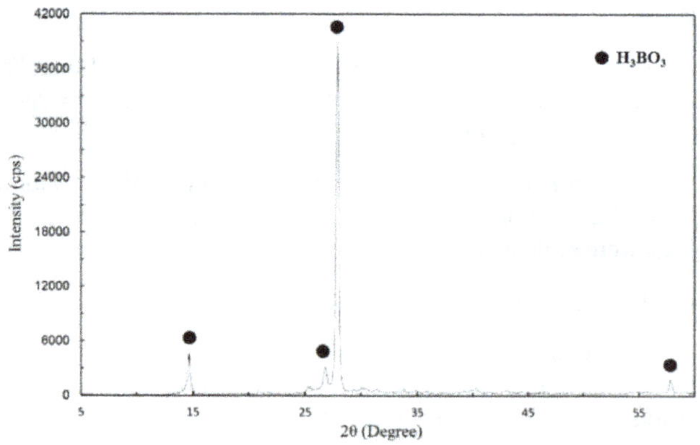

Figure 13: XRD analysis of product of Boric acid.

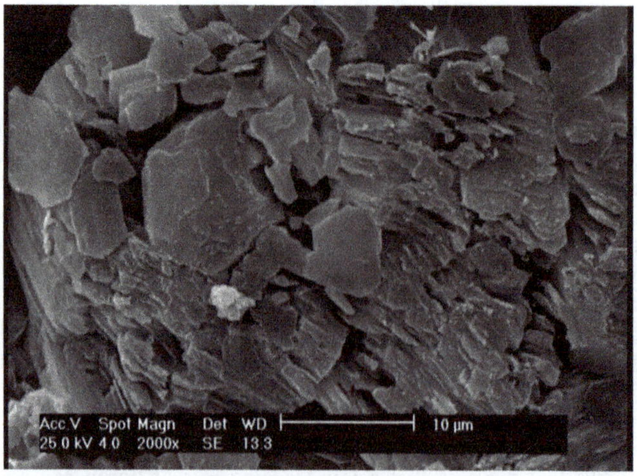

Figure 14: SEM micrograph of precipitated Boric acid sample.

Maximum Concentration of Contaminants for Toxicity Characteristic (United States Environmental Protection Agency and China GB/5085.3-2007). Finally, crystallization rate of boric acid can reach 99.74%.

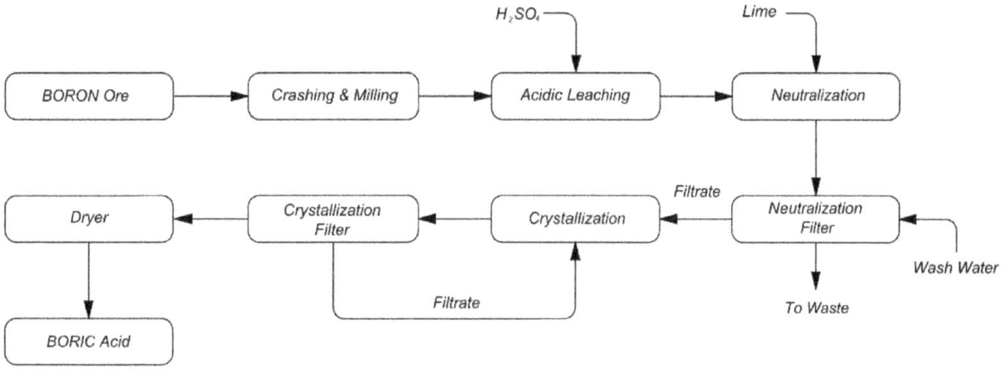

Figure 15: Working Flow Diagram for production of Boric acid.

Temperature (°C)	XB	1-3(1-XB)$^{2/3}$+2(1-XB)	1-(1-XB)$^{1/3}$
60	0.9454	0.6164	0.8776
70	0.9558	0.6348	0.8898
80	0.9796	0.6913	0.9237
90	0.9626	0.6539	0.8835

Table 3: The results of the R^2 values for each model.

Step/Condition	Solid (gr)	Value (L)	B_2O_5 (%)	B (gr/L)	Mass Balance	stage recovery %
1. Feed	10000	-	8.95	-	-	-
2. Acidic Leach	-	-	-	-	99.44	91.71
Solid	9740	-	0.71	-	-	-
Filtrate	-	26.3	-	6.64	-	-
3. Neutralization	-	-	-	-	99.52	99.45
Solid	347	-	0.18	-	-	-
Filtrate	-	26.8	-	6.48	-	-
4. Crystallization	-	-	-	-	99.23	75.28
Solid	745.12	-	81.83	-	-	-
Filtrate	-	26.3	-	1.62	-	-
5. Product	745.12				98.30	90.50

Table 4: Results of Scale-up experiment.

Conclusions

In this study, dissolution of boron ore sample with H_2SO_4 acid at various experimental conditions was examined and research findings obtained in this study are as follows. pH has a significant effect on extraction of boron from ore. As pH in leaching test increases, the extraction recoveries of sample tend to increase. An increase in reaction temperatures caused an increase in sample dissolved rate. Optimum condition for acidic leaching section was obtained in temperature of 90°C, reaction time of 2 hours, L/S ratio of 3, and pH of 1. In the optimum point, the recovery of boron acidic leaching was achieved 92.21%. After precipitation of iron, boric acid was obtained by crystallization. Purity of produced boric acid was 99.56%. Dissolution of boron ore was followed fluid film diffusion controlled reaction and the reaction activation energy was equaled to 11.6 kJ/mol. Enthalpy of activation (⟦ΔH⟧^‡) and entropy of activation (⟦ΔS⟧^‡) were 11.2 kJ/mol and -246.3 J/(mol.K) respectively.

Acknowledgements

The authors are thankful to the Research and Engineering Co. for Non-Ferrous Metals for financial and technical support and the permission to publish this paper.

References

1. Tokatli C, Arslan EKN, Arzu OE (2016) Ecosystem Quality Assessment of an Aquatic Habitat in a Globally Important Boron Reserve: Emet Stream Basin (Turkey). Environment and Pollution 59: 116-141.
2. William HS, Vengosh A (2016) Global boron cycle in the Anthropocene. Global Biogeochemical Cycles 30: 219-230.
3. Sert H, Yildiran H (2011) A study on an alternative method for the production of boric acid from ulexite by using trona. Journal of Ore Dressing 13: 1.
4. Levent S, Pamuko Y, Gonen M (2016) Extraction of Boric Acid from Tincal Mineral by Supercritical Ethanol. The Journal of Supercritical Fluids 109: 67-73.
5. Onal G, Burat F (2008) Boron Mining and Processing in Turkey. Mineral Resources Management 24: 49-60.
6. Demircioglu A (2011) Boron Minerals of Turkey-Hydroboracite. Mineral Research and Exploration Institute of Turkey 543: 1-26.
7. Flores HR, Mattenella LE, Valdez SK (2002) Physical and physicochemical properties of borates from Argentine Puna. Technological Information 132: 103-108.
8. Mergen A, Demirhan MH, Bilen M (2003) Processing of Boric Acid from Borax by a Wet Chemical. Advanced Powder Technology 14: 279-293.
9. Kuskay B, Bulutcu AN (2011) Design Parameters of Boric Acid Production Process from Colemanite Ore in the Presence of Propionic Acid. Chemical Engineering and Processing 50: 377-383.

10. Tunc M, Kocakerim M, Yapici S (1999) Dissolution Mechanism of Ulexite in H_2SO_4 Solution. Hydrometallurgy 51: 359-370.

11. Kalafatoglu IE, Oes N, Ozdemir SS (2000) Dissolution behavior of colemanite with sulfuric acid. Proceedings of IV National Chemical Engineering Conference, Istanbul, Turkey, pp: 263-268.

12. Cetin E, Eroglu I, Ozkar S (2001) Kinetics of Gypsum Formation and Growth during the Dissolution of Colemanite in Sulfuric Acid. Journal of Crystal Growth 231: 559-567.

13. Gaye OC (2004) The Effect of Stirring Rate on Dissolution of Colemanite and Particle Size of Gypsum Crystals during the Boric Acid production in a Batch reactor. Ultaniyat Bor Symposium, Turkey 23: 319-325.

14. Okur H, Tekin T, Ozer K, Bayramoglu M (2002) Effect of Ultrasound on the Dissolution of Colemanite in H_2SO_4. Hydrometallurgy 67: 79-86.

15. Levenspiel O (1999) Chemical reaction engineering. Industrial & engineering chemistry research. ACS Publications 38: 4140-4143.

16. Xuin GH, Yu D, Su Y (1986) Leaching of scheelite by hydrochloric acid in the presence of phosphate. Hydrometallurgy 16: 27-40.

17. Pohlman SL, Olson FA (1974) A kinetic study of acid leaching of chrysocolla using a weight loss technique. Solution Mining Symposium, AIME.

26

Kinetics and Mechanistic Approach to the Chromic Acid Oxidative Degradation of Atropine Drug in Perchlorate Solutions and the Effect of Ruthenium(III) Catalyst

Ahmed Fawzy[1,2,*], Ishaq A. Zaafarany[1], Rabab S. Jassas[1], Rami J. Obaid[1], Saleh A. Ahmed[1,2]
[1]*Chemistry Department, Faculty of Applied Science, Umm Al-Qura University, 21955 Makkah, Saudi Arabia*
[2]*Chemistry Department, Faculty of Science, Assiut University, 71516 Assiut, Egypt*

Abstract

The effect of ruthenium(III) catalyst on the kinetics of oxidation of atropine drug (ATR) by chromic acid in perchlorate solutions was studied spectrophotometrically at a fixed ionic strength of 1.0 mol dm^{-3} and at 25°C. Both uncatalyzed and Ru(III)-catalyzed oxidation reactions showed a first order dependence in [Cr(VI)], and less than unit order dependences with respect to both [ATR] and [H$^+$]. The reaction was first order in [Ru(III)]. The effects of both ionic strength and dielectric constant of the reactions medium were investigated. Addition of Mn(II) was found to decrease the oxidation rate. The rate of Ru(III)-catalyzed oxidation of atropine was found to be about 10-fold higher than that of the uncatalyzed reaction. In both cases, the main oxidation products of atropine were identified as tropine, benzaldehyde, methanol, and carbon dioxide. Plausible mechanisms for both uncatalyzed and Ru(III)-catalyzed oxidations were proposed and the rate-law expressions associated with these mechanisms were derived. The activation parameters related to the second order rate constants were evaluated and discussed.

Keywords: Atropine; Oxidation; Chromic acid; Ruthenium(III); Kinetics; Mechanism

Introduction

Alkaloids have a number of pharmacological activities including antimalarial, antiasthma, anticancer, cholinomimetic, antibacterial, psychotropic and stimulant activities [1-3]. Among the most famous of the alkaloids is tropine alkaloid or atropine (ATR) which is an anticholinergic drug containing two cyclic structures (alicyclic nitrogen-containing alcohol tropine and aromatic tropic acid) joined by an ester linkage [4]. This structure allows for its rapid absorption through the blood-brain barrier. Atropine is structurally similar to cocaine as illustrated below.

Atropine 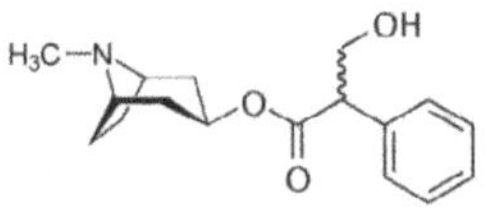

Cocaine 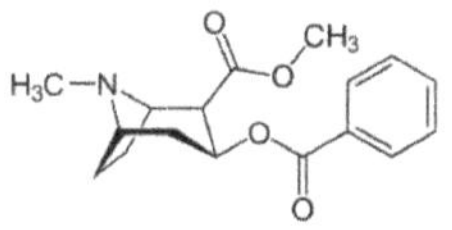

Atropine occurs naturally in plants in the nightshade family including deadly nightshade, Jimson weed and mandrake [5]. It is a secondary metabolite of such plants and serves as a drug with a wide variety of effects. Atropine is considered as a core medicine in the World Health Organization (WHO) for a main health care system. Furthermore, it is the most essential drug in the treatment of nerve agent poisoning. Its degradation by microorganisms has been reported by several groups [6] and in the initial stage, the hydrolysis of the ester linkage to give two separate cyclic components takes place. Alkaloids may play an important role in the chemistry of chromium because of its carcinogenic and mutagenic activities [7] which due to chromium(VI) metabolism by various cellular components. Among the various metabolic pathways, generation of chromium(V) intermediate by a variety of biologically active reductants is a prime suspect. Chromium(V) is a putative DNA-damaging agent and has been shown to be a long-lived intermediate in the redox reaction of chromium(VI) [8-10]. Epidemiological and animal studies, as well as *in vitro* mutagenicity assays [11], indicate that chromium(VI) compounds are dangerous for biological systems but chromium(III) compounds are considered as non-toxic [12]. Furthermore, chromium(VI) is considered as one of the most significant oxidants for oxidation of organic compounds [13,14]. On the other hand, some transition metal ions are widely used as homogeneous catalysts for oxidation of organic substrates [15-17]. Kinetic studies on the homogeneous catalyzed oxidation of organic substrates are considered to be an important field of chemistry because of their roles in the biological systems. Although some work on the oxidation of atropine by various oxidants has been performed [18-22], there is a lack of literature on the kinetics of oxidation of this drug by chromic acid in absence or presence of a catalyst. This leads us to study the present reactions. The objectives of this study were to check the reactivity of atropine drug towards chromic acid in perchlorate solutions, to understand the active species of the reactants in such medium, to check the catalytic activity of Ru(III) and to propose the oxidation mechanisms of the drug.

Experimental

Materials

The stock solution of atropine was prepared by dissolving the sample, atropine sulfate monohydrate $(C_{17}H_{23}NO_3)_2.H_2SO_4.H_2O$ (Aldrich), in doubly distilled water. Chromic acid solution was freshly prepared before each experiment and it was standardized spectrophotometrically. Solution of ruthenium(III) chloride was prepared according to the procedure reported earlier [23]. Other chemicals employed in the present investigation were of reagent grade and their solutions were prepared by dissolving the required amounts of the samples in doubly distilled water.

***Corresponding author:** Ahmed Fawzy, Chemistry Department, Faculty of Applied Science, Umm Al-Qura University, 21955 Makkah, Saudi Arabia
E-mail: afsaad13@yahoo.com

Kinetic measurements

Kinetic runs were followed under pseudo-first order conditions in an excess of atropine over chromic acid. The courses of the uncatalyzed and Ru(III)-catalyzed oxidation reactions were followed by tracing the decay in chromium(VI) absorbance at λ_{max}=350 nm, its absorption maximum. Absorbance measurements were carried out on Shimadzu UV-VIS-NIR-3600 double-beam spectrophotometer with a temperature controlling system. The observed-first order rate constants of uncatalyzed (k_U) and catalysed (k_C) reactions were calculated as the gradients of ln(absorbance) versus time plots. The rate constants were the main values of at least three kinetic measurements. The rate constants were reproducible to within 3-4%.

Results

Spectral changes

Spectral changes during the chromic acid oxidations of atropine in the absence and presence of Ru(III) catalyst are shown in Figure 1a and b respectively. In both cases, the recorder spectra indicate gradual decay of Cr(VI) band due to its reduction by atropine drug.

Stoichiometry and product characterization

Reaction mixtures containing various amounts of Cr(VI) and atropine at constant [H$^+$], ionic strength, and temperature were allowed to react for 24 h in closed vessels for completion of the oxidation reactions. The unconsumed [Cr(VI)] was determined spectrophotometrically at 350 nm. The results indicated that two moles of Cr(VI) are consumed by three mole of atropine drug to yield the oxidation products as shown in the following equation:

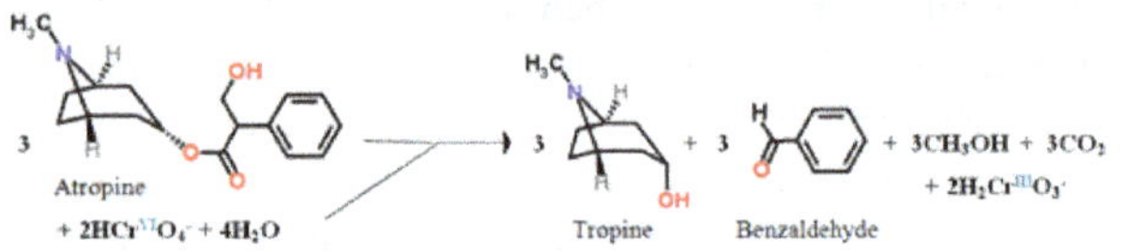

This equation is consistent with the product characterization. Tropine and benzaldehyde as the main reaction products were identified by spectral analysis as described elsewhere [24-26]. Tropine was also identified by its hydrazone derivative [24]. Methyl alcohol was confirmed by sodium test [24] and carbon dioxide was detected by lime water.

Effect of [chromic acid]

The effect of chromic acid on the oxidation rates of both uncatalyzed and Ru(III)-catalyzed reactions was investigated by varying its concentration in the range of (1.0 - 10.0) × 10^{-4} mol dm^{-3}. Plots of ln(absorbance) versus time were linear up to at least 75% of the reactions completion. Furthermore, increasing the initial oxidant concentration did not significantly affect the rates of the reactions. These observations suggest that the order of reactions with respect to the oxidant is unity.

Effect of [atropine]

The observed-first order rate constants for both paths were evaluated at different [ATR] with other variables constant. The results showed that increasing [ATR] increased the oxidation rates (Table 1). Plots of log k_U and log k_C versus log[ATR] were linear with slopes of 0.51 and 0.57 for uncatalyzed and catalyzed reactions, respectively, (Figure 2) suggesting that the orders of the reactions with respect to atropine concentration were less than unity.

Effect of [perchloric acid]

It was found that increasing [H$^+$] increased the oxidation rates as listed in Table 1 which suggested that the oxidation reactions were acid-catalyzed. Plots of log k_U and log k_C versus log [H$^+$] were found to be linear with positive slopes (Figure 3) confirming the less than unit order kinetics in [H$^+$].

Effect of [Ru(III)]

The effect of ruthenium(III) catalyst was examined by measuring the oxidation rate of atropine at various concentration of Ru(III), namely (2.0-18.0) × 10^{-5} mol dm^{-3}. The oxidation rate increased as [Ru(III)] increased. A plot of log k_C versus log [Ru(III)] was linear with unit slope as shown in Figure 4 indicating that the reaction order with respect to the catalyst concentration was unity.

Effect of manganese(II)

The involvement of Cr(IV) as an intermediate species of chromium during these reactions was examined by addition of different concentrations of Mn(II) to the reaction mixtures up to 0.01 mol dm^{-3}. The rate of the uncatalyzed reaction inhibited with increasing the concentration of Mn(II) as illustrated in Figure 5.

Effect of temperature

The rates of both uncatalyzed and Ru(III)-catalyzed oxidations of

10^4 [Cr(VI)] (mol dm^{-3})	10^2 [ATR] (mol dm^{-3})	[H$^+$] (mol dm^{-3})	10^5 [Ru(III)] (mol dm^{-3})	I(mol dm^{-3})	10^3 k_U (s^{-1})	10^3 k_C (s^{-1})
1.0	1.0	0.5	6.0	1.0	9.7	95.7
3.0	1.0	0.5	6.0	1.0	8.9	94.2
5.0	1.0	0.5	6.0	1.0	9.3	94.6
7.0	1.0	0.5	6.0	1.0	8.8	96.2
10.0	1.0	0.5	6.0	1.0	10.1	94.9
5.0	0.2	0.5	6.0	1.0	3.0	36.8
5.0	0.6	0.5	6.0	1.0	6.7	71.0
5.0	1.0	0.5	6.0	1.0	9.3	94.6
5.0	1.4	0.5	6.0	1.0	11.6	127.3
5.0	1.8	0.5	6.0	1.0	13.7	158.1
5.0	1.0	0.1	6.0	1.0	3.5	30.1
5.0	1.0	0.3	6.0	1.0	6.8	60.7
5.0	1.0	0.5	6.0	1.0	9.3	94.6
5.0	1.0	0.7	6.0	1.0	11.1	123.4
5.0	1.0	0.9	6.0	1.0	12.4	148.3
5.0	1.0	0.5	2.0	1.0	9.3	46.7
5.0	1.0	0.5	6.0	1.0	9.3	94.6
5.0	1.0	0.5	10.0	1.0	9.3	138.0
5.0	1.0	0.5	14.0	1.0	9.3	194.2
5.0	1.0	0.5	18.0	1.0	9.3	257.6
5.0	1.0	0.5	6.0	1.0	9.3	94.6
5.0	1.0	0.5	6.0	1.5	8.9	95.5
5.0	1.0	0.5	6.0	2.0	10.2	96.2
5.0	1.0	0.5	6.0	2.5	9.6	95.7
5.0	1.0	0.5	6.0	3.0	10.7	97.1

Table 1: Effects of variations of [Cr(VI)], [ATR], [H$^+$], [Ru(III)] and ionic strength, I, on the pseudo-first order rate constant values in the uncatalyzed and Ru(III)-catalyzed oxidations of atropine by chromic acid in perchlorate solutions at 25°C.

Reaction	$\Delta S^{\neq}$ (Jmol^{-1}K^{-1})	$\Delta H^{\neq}$ (kJ mol^{-1})	$\Delta G^{\neq}_{298}$ (kJ mol^{-1})	$E_a^{\neq}$ (kJ mol^{-1})
Uncatalyzed	-103.32	48.02	78.81	46.52
Catalyzed	-84.53	33.47	58.66	35.33

Table 2: Activation parameters for the second order rate constant in the uncatalyzed and Ru(III)-catalyzed oxidations of atropine by chromic acid in perchlorate solutions

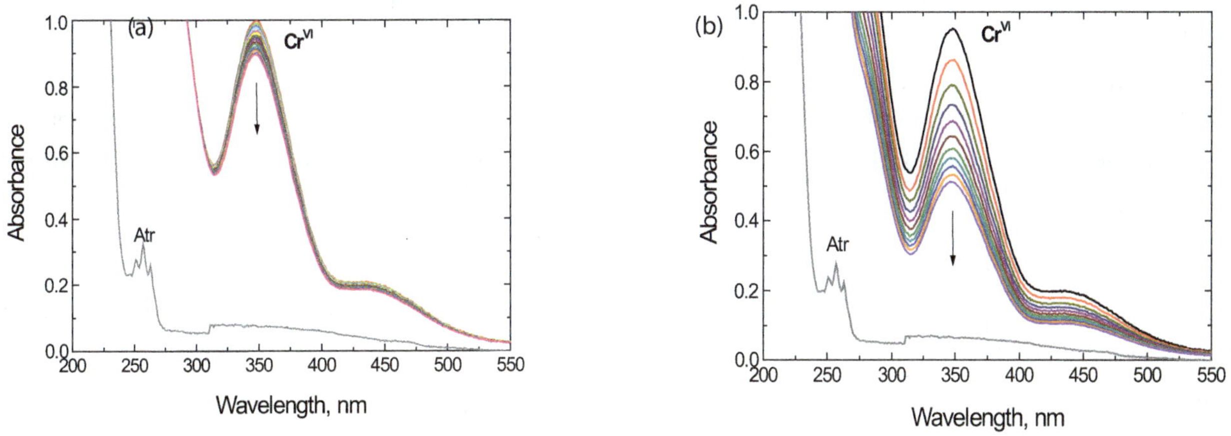

Figure 1: Spectral changes during: (a) uncatalyzed, and (b) Ru(III)-catalyzed oxidations of atropine by chromic acid in perchlorate solutions. [Cr(VI)]=5.0 × 10^{-4}, [ATR]=0.01, [H^+] =0.5 and I=1.0 mol dm^{-3} at 25°C.

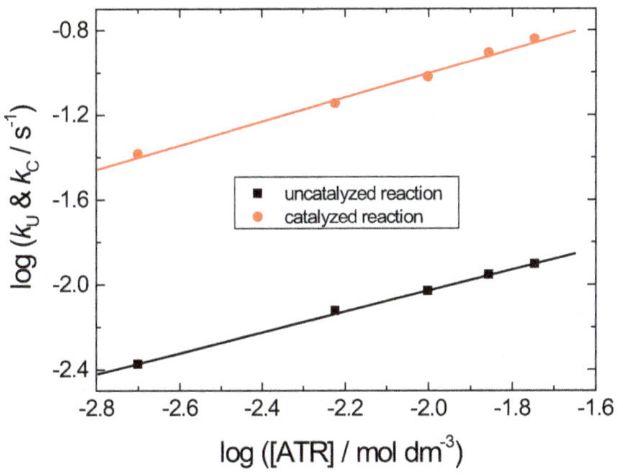

Figure 2: Plots of log k_U and log k_C versus log [ATR] in the: (a) uncatalyzed, and (b) Ru(III)-catalyzed oxidations of atropine by chromic acid in perchlorate solutions. [Cr(VI)]=5.0 × 10^{-4}, [H^+]=0.5 and I=1.0 mol dm^{-3} at 25°C. [Ru(III)]=6.0 × 10^{-5} mol dm^{-3}.

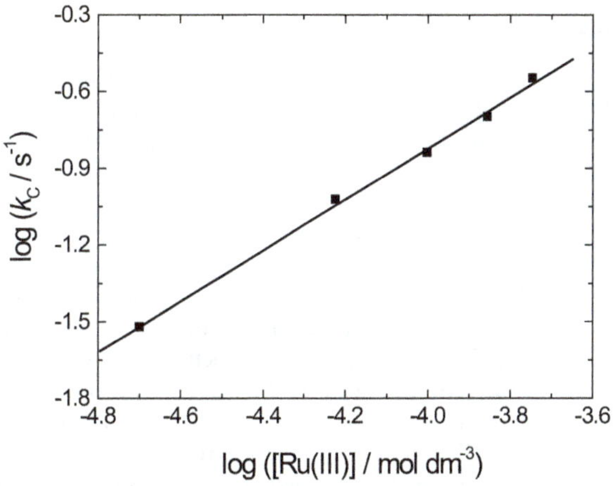

Figure 4: A plot of log k_C versus log [Ru(III)] in the Ru(III)-catalyzed oxidation of atropine by chromic acid in perchlorate solution. [Cr(VI)]=5.0 × 10^{-4}, [ATR]=0.01, [H^+]=0.5 and I=1.0 mol dm^{-3} at 25°C.

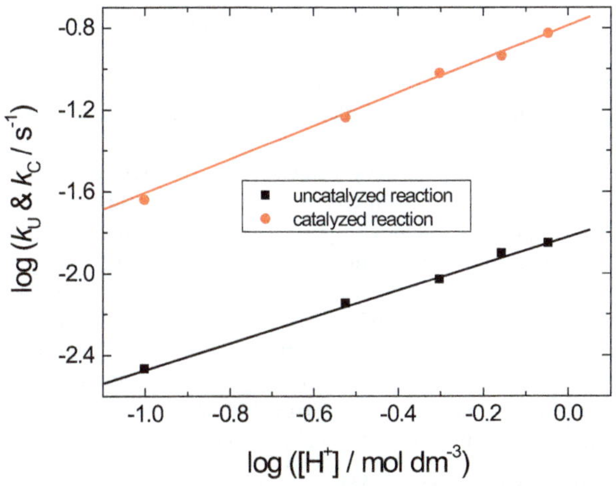

Figure 3: Plots of log k_U and k_C versus log [H^+] in the: (a) uncatalyzed, and (b) Ru(III)-catalyzed oxidations of atropine by chromic acid in perchlorate solutions. [Cr(VI)]=5.0 × 10^{-4}, [ATR] = 0.01 and I=1.0 mol dm^{-3} at 25°C. [Ru(III)]=6.0 × 10^{-5} mol dm^{-3}.

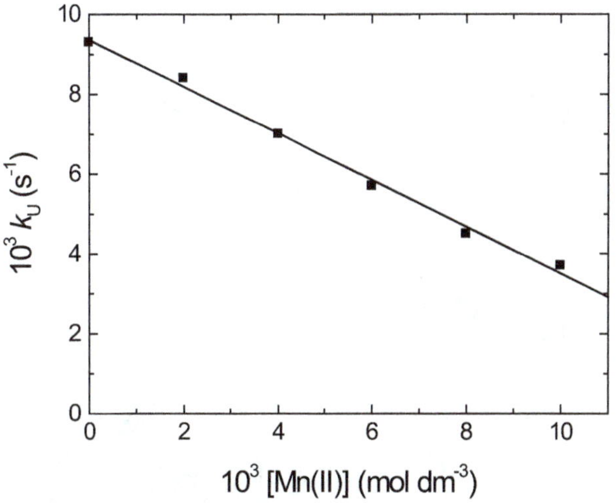

Figure 5: Effect of Mn(II) on the rate of uncatalyzed oxidation of atropine by chromic acid in perchlorate solution. [Cr(VI)]=5.0 × 10^{-4}, [ATR]=0.01, [H^+]=0.5 and I=1.0 moldm^{-3} at 25°C.

atropine were measured at different temperatures between 288 and 308 K, and all other conditions being constant. The values of k_U and k_C were found to increase with raising temperature and the activation parameters of the second order rate constant were determined using Eyring and Arrhenius plots and were listed in Table 2.

Polymerization test for free radical intermediates

The intervention of free radicals in both uncatalyzed and Ru(III)-catalyzed reactions was examined as follows: the mixtures to which known quantities of acrylonitrile had been added, were kept in an inert atmosphere for 6 h at room temperature. On diluting the reaction mixture with methanol, no white precipitates were formed thus confirming the absence of free radicals in the reactions.

Discussion

It is reported [27-31] that aqueous solutions of chromic acid contain ions such as CrO_4^{2-}, $HCrO_4^-$ and $Cr_2O_7^{2-}$, besides other protonated species such as H_2CrO_4, $HCr_2O_7^-$ and $H_2Cr_2O_7$ [32]. Increasing the oxidation rates with increasing $[H^+]$ in the present work suggested [32,33] that the protonated chromate (H_2CrO_4) may be the reactive species of Cr(VI).

Mechanism of uncatalyzed oxidation reaction

The reaction between atropine and chromic acid in perchlorate solutions was found to exhibit a less than unit order dependence on [ATR] suggestion formation of a complex (C_1) between atropine drug and chromic acid which was also proved kinetically by a non-zero intercept of the plot of $1/k_U$ versus 1/[ATR] [34] as shown in Figure 6. The complex formation of chromium(VI) with D-xylose and L-arabinose [35] and with tyrosine [36] in aqueous perchlorate solutions was also reported. The negligible effect of both ionic strength and dielectric constant of the medium is consistent with a reaction between two neutral molecules [37,38], i.e., between ATR and H_2CrO_4. The cleavage of such complex leads to the formation of one of the final oxidation products of atropine (tropine), Cr(IV) intermediate and tropic acid cation. The latter is rapidly hydrolyzed to the remainder oxidation products, benzaldehyde, methyl alcohol and carbon dioxide. This step is followed by a reaction between the second mole of atropine and another mole of chromic acid to give the oxidation products of atropine and another Cr(IV) intermediate. Finally, the third mole of atropine reacts with the formed two Cr(IV) intermediate species leading to the formation of the oxidation products of atropine and Cr(III) as the final oxidation product of chrmioum(VI), satisfying the observed reaction stoichiometry as illustrated in Scheme 1.

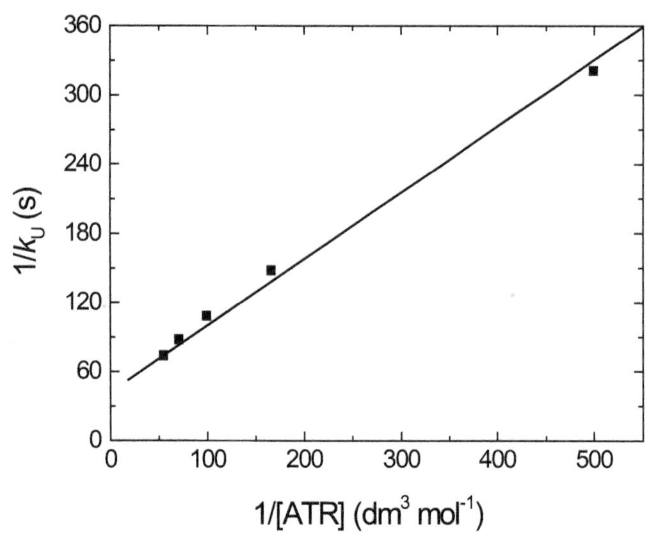

Figure 6: Verification of Eq. (4) in the uncatalyzed oxidation of atropine by chromic acid in perchlorate solutions. [Cr(VI)]=5.0 × 10^{-4}, $[H^+]$=0.5 and I=1.0 mol dm^{-3} at 25°C.

The suggested mechanistic (Scheme 1) leads to the following rate law expression,

$$\text{Rate} = \frac{k_1K_1K_2[\text{HCrO}_4^-][\text{ATR}][\text{H}^+]}{1+K_1[\text{H}^+]+K_1K_2[\text{ATR}][\text{H}^+]} \qquad (1)$$

which explains all of the observed kinetic orders of different species.

Under pseudo-first order conditions,

$$\text{Rate}=\frac{-d[\text{HCrO}_4^-]}{dt} = k_U[\text{HCrO}_4^-] \qquad (2)$$

Comparing Eqs. (1) and (2) gives,

$$k_U = \frac{k_1K_1K_2[\text{ATR}][\text{H}^+]}{1+K_1[\text{H}^+]+K_1K_2[\text{ATR}][\text{H}^+]} \qquad (3)$$

Equation (3) can be rearranged into the following forms,

$$\frac{1}{k_U}=\left(\frac{1+K_1[\text{H}^+]}{k_1K_1K_2[\text{H}^+]}\right)\frac{1}{[\text{ATR}]}+\frac{1}{k_1} \qquad (4)$$

$$\frac{1}{k_U}=\left(\frac{1}{k_1K_1K_2[\text{ATR}]}\right)\frac{1}{[\text{H}^+]}+\left(\frac{1}{k_1K_2[\text{ATR}]}+\frac{1}{k_1}\right) \qquad (5)$$

According to Eqs. (4) and (5), a plot of $1/k_U$ versus 1/[ATR] at constant $[H^+]$ should be linear with a positive intercept on $1/k_U$ axis as is observed experimentally (Figure 6) and a plot of $1/k_U$ against $1/[H^+]$ at constant [ATR] also should be a straight line with a positive intercept on $1/k_U$ axis and was found to be so (Figure 7) confirming the validity of the proposed mechanism.

Mechanism of ruthenium(III)-catalyzed oxidation reaction

In acid media, the reactive species of Ru(III) chloride is suggested [39-41] to be $[RuCl_5(H_2O)]^{2-}$. The experimental results showed that the oxidation of atropine by chromic acid in the presence of small amounts of Ru(III) is similar to the uncatalyzed oxidation with respect to stoichiometry, reaction orders and effects of both ionic strength and dielectric constant of the reaction medium. The reaction was first order with respect to Ru(III). Therefore, the proposed catalyzed oxidation mechanism (Scheme 2) is likely to be similar, except for the participation of the catalyst. Thus, we propose that atropine forms an intermediate complex with the reactive species of Ru(III) catalyst in a pre-equilibrium step. Kinetic evidence for complex formation was obtained from the non-zero intercept of the plot of $[Ru(III)]/k_C$ versus 1/[ATR] (Figure 8). Such complex between atropine and Ru(III) was reported earlier [19] in the oxidation of atropine by copper(III) periodate complex in aqueous alkaline medium. The formed complex then attached by chromic acid in a slow step resulting in decomposition of the complex with regeneration of the catalyst, as well as formation of the final oxidation product tropine, Cr(IV) intermediate and tropic acid cation. The latter is rapidly hydrolyzed to the remainder oxidation products, benzaldehyde, methyl alcohol and carbon dioxide.

The suggested mechanism leads to the following rate law expression,

$$\text{Rate} = \frac{k_2K_1K_3[\text{HCrO}_4^-][\text{ATR}][\text{Ru(III)}][\text{H}^+]}{1+K_1[\text{H}^+]+K_1K_3[\text{ATR}][\text{H}^+]} \qquad (6)$$

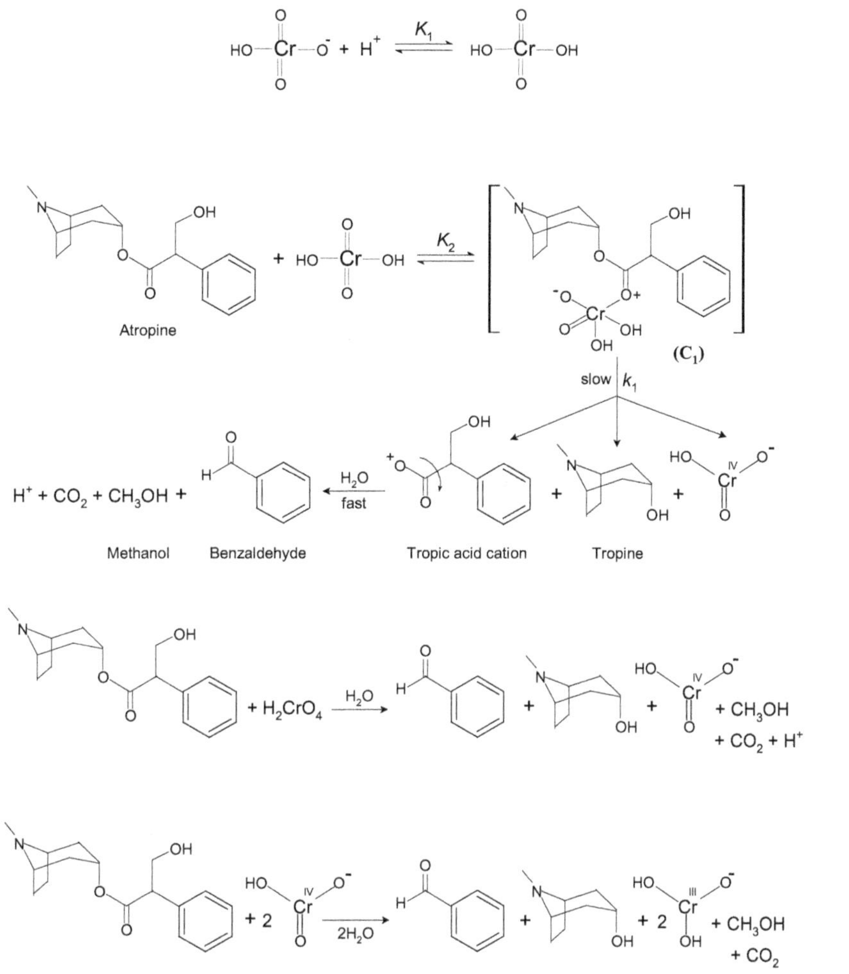

Scheme 1: Mechanism of the uncatalyzed oxidation of atropine by chromic acid in perchlorate solutions.

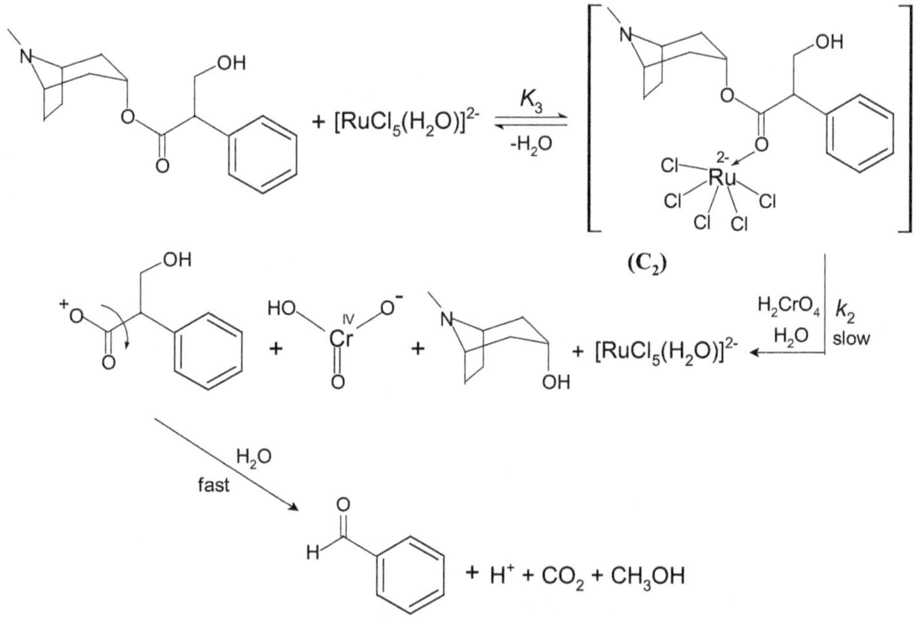

Scheme 2: Mechanism of Ru(III)-catalyzed oxidation of atropine by chromic acid in perchlorate solutions.

The rate law (6) is consistent with all observed orders with respect to different species.

Under pseudo-first order conditions,

$$\text{Rate} = \frac{-d[\text{HCrO}_4^-]}{dt} = k_C[\text{HCrO}_4^-] \quad (7)$$

$$k_C = \frac{k_2K_1K_3[\text{ATR}][\text{Ru(III)}][\text{H}^+]}{1+K_1[\text{H}^+]+K_1K_3[\text{ATR}][\text{H}^+]} \quad (8)$$

and with rearrangement,

$$\frac{[\text{Ru(III)}]}{k_C} = \left(\frac{1+K_1[\text{H}^+]}{k_2K_1K_3[\text{H}^+]}\right)\frac{1}{[\text{ATR}]}+\frac{1}{k_2} \quad (9)$$

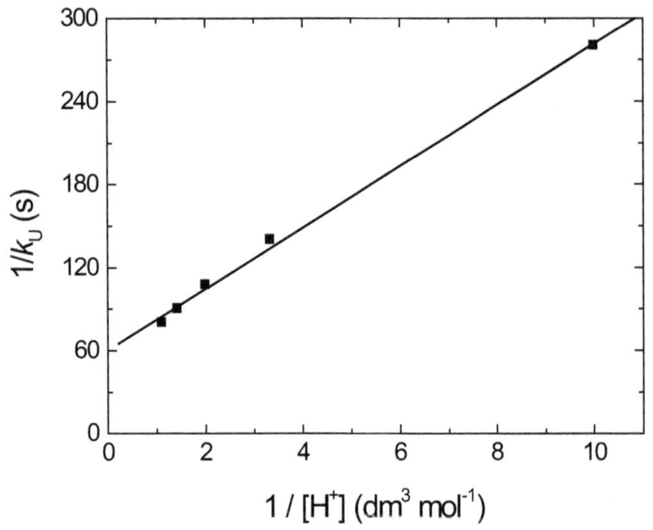

Figure 7: Verification of Eq. (5) in the uncatalyzed oxidation of atropine by chromic acid in perchlorate solutions. [Cr(VI)]=5.0 × 10^{-4}, [ATR]=0.01 and I=1.0 mol dm^{-3} at 25°C.

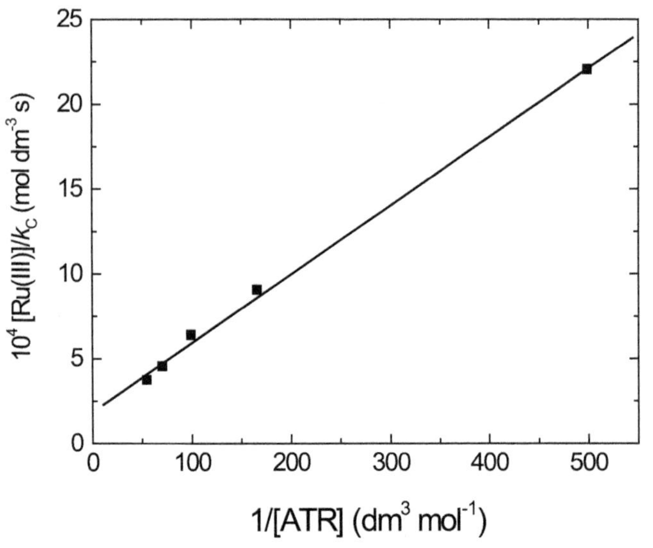

Figure 8: Verification of Eq. (9) in the Ru(III)-catalyzed oxidation of atropine by chromic acid in perchlorate solutions. [Cr(VI)]=5.0 × 10^{-4}, [ATR]=0.01, [Ru(III)]=6.0 × 10^{-5} and I=1.0 mol dm^{-3} at 25°C.

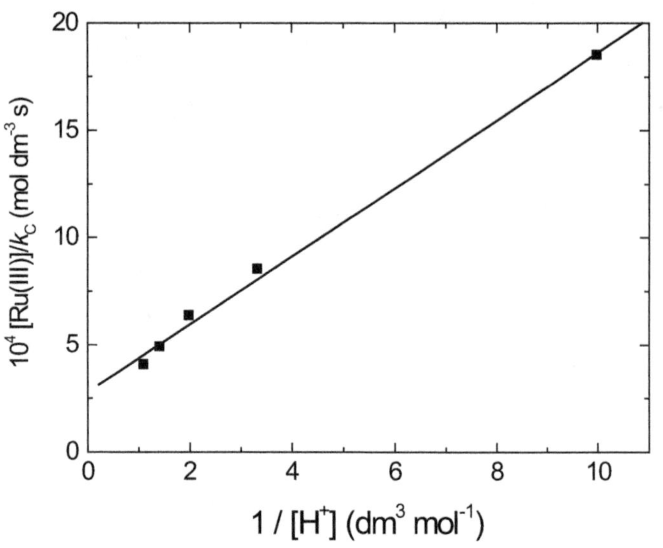

Figure 9: Verification of Eq. (10) in the Ru(III)-catalyzed oxidation of atropine by chromic acid in perchlorate solutions. [Cr(VI)]=5.0 × 10^{-4}, [ATR]=0.01, [Ru(III)]=6.0 × 10^{-5} and I=1.0 moldm^{-3} at 25°C.

$$\frac{\text{Ru(III)}}{k_C} = \left(\frac{1}{k_2K_1K_3[\text{ATR}]}\right)\frac{1}{[\text{H}^+]}+\left(\frac{1}{k_2K_3[\text{ATR}]}+\frac{1}{k_2}\right) \quad (10)$$

Equations (9) and (10) suggests that plots of [Ru(III)]/k_C versus 1/[ATR] at constant [H$^+$] and [Ru(III)]/k_C versus 1/[H$^+$] at constant [ATR] should be linear with positive intercepts. The experimental results satisfied these requirements, as shown in Figures 8 and 9 respectively. The determined activation parameters listed in Table 2 showed that the values of entropy of activation ($\Delta S^{\neq}$) were negative suggesting formation of compacted intermediate complexes of inner-sphere nature [40]. Also, the values of enthalpy of activation ($H^{\neq}$) and $\Delta S^{\neq}$ were both favorable for electron transfer processes.

Conclusions

The kinetics of uncatalyzed and Ru(III)-catalyzed oxidations of atropine drug by chromic acid in perchlorate solutions have been studied. Under comparable experimental conditions, the rate of Ru(III)-catalyzed oxidation of atropine was found to be about 10-fold higher than that of the uncatalyzed reaction. In both cases, the main oxidation products of atropine were identified by spectral and chemical analyses as tropine, benzaldehyde, methanol and carbon dioxide.

References

1. Kittakoop P, Mahidol C, Ruchirawat S (2014) Alkaloids as important scaffolds in therapeutic drugs for the treatments of cancer, tuberculosis, and smoking cessation. Curr Top Med Chem 14: 239-252.
2. Cushnie TP, Cushnie B, Lamb AJ (2014) Alkaloids: an overview of their antibacterial, antibiotic-enhancing and antivirulence activities. Int J Antimicrob Agents 44: 377-386.
3. Qiu S, Sun H, Zhang AH, Xu HY, Yan GL, et al. (2014) Natural alkaloids: basic aspects, biological roles, and future perspectives. Chin J Nat Med 12: 401-406.
4. Bartholomew BA, Smith MJ, Trudgill PW, Hopper DJ (1996) Atropine metabolism by pseudomonas sp. Strain AT3: Evidence for nortropine as an intermediate in tropine breakdown and reactions leading to succinate. Appl Environ Microbiol 62: 3245-3250.
5. Brust JCM (2004) Neurological aspects of substance abuse. 2nd edn. Philadelphia: Elsevier, p: 310.
6. Rorsch A, Berends FA, Bartlema CH, Stevens WF, Winsinck F, et al. (1971)

The isolation and properties of *Pseudomonas* strains growing on atropine and producing an atropine esterase. Proc K Ned Akad Wet Ser C 74: 132-147.

7. Branca M, Micera G, Dessi A (1988) Reduction of chromium(VI) by D-galacturonic acid and formation of stable chromium(V) intermediates. Inorg Chem Acta 153: 61-65.
8. Wetterhahn JK (1982) Microsomal reduction of the carcinogen chromate produces chromium (IV). J Am Chem Soc 104: 874-881.
9. Rossi SC, Gorman N, Wetterhahn KE (1988) Mitochondrial reduction of the carcinogen chromate: formation of chromium(V). Chem Res Toxicol 1: 101-107.
10. Alexander DR (1984) Inhibitory action of hexavalent chromium (Cr(VI)) on the mitochondrial respiration and a possible coupling to the reduction of Cr(VI). J Biochem Pharmacol 33: 2461-2446.
11. Levis AG, Bianchi V, Langard S (1982) Biological and environmental aspects of chromium. Elsevier Biomedical Press, Amsterdam, p: 171.
12. Katz SA, Salem H (1993) The toxicology of chromium with respect to its chemical speciation: a review. J Appl Toxicol 13: 217-224.
13. Barnhart J (1997) Chromium in soil: perspectives in chemistry, health, and environmental regulation. J Soil Contamin 6l: 561-568.
14. Costa M (1997) Toxicity and carcinogenicity of Cr^{VI} in animal models and humans. Crit Rev Toxicol 27: 431-442.
15. Fawzy A (2014) Influence of copper(II) catalyst on the oxidation of L-histidine by platinum(IV) in alkaline medium: a kinetic and mechanistic study. Transition Met Chem 39: 567-576.
16. Fawzy A (2015) Kinetics and mechanistic approach to the oxidative behavior of biological anticancer platinum(IV) complex towards L-asparagine in acid medium and the effect of copper(II) catalyst. Int J Chem Kinet 47: 1-12.
17. Fawzy A, Asghar BH (2015) Kinetics and mechanism of uncatalyzed and silver(I)-catalyzed oxidation of L-histidine by hexachloroplatinate(IV) in acid medium. Transition Met Chem 40: 287-295.
18. Byadagi KS, Hosahalli RV, Nandibewoor ST, Chimatadar SA (2012) Oxidation of a anticholinergic drug atropine sulfate monohydrate by alkaline copper(III) periodate complex: a kinetic and mechanistic study. Z Phys Chem 226: 233-249.
19. Byadagi KS, Nandibewoor ST, Chimatadar SA (2013) Catalytic activity of ruthenium(III) on the oxidation of an anticholinergic drug-atropine sulfate monohydrate by copper(III) periodate complex in aqueous alkaline medium - decarboxylation and free radical mechanism. Acta Chim Slov 60: 617-627.
20. Meti M, Nandibewoor S, Chimatadar S (2014) Spectroscopic investigation and oxidation of the anticholinergic drug atropine sulfate monohydrate by hexacyanoferrate(III) in aqueous alkaline media: a mechanistic approach. Turk J Chem 38: 477-487.
21. Abdullah S, Al-Ghreizat SK, Abdel-Halim HM (2015) Kinetics of oxidation of atropine by alkaline $KMnO_4$ in aqueous solutions. Asian J Chem 27: 3877-3882.
22. Do Pham DD, Kelso GF, Yang Y, Hearn MTW (2014) Studies on the oxidative N-demethylation of atropine, thebaine and oxycodone using a FeIII-TAML catalyst. Green Chem 16: 1399-1405.
23. Radhakrishnamurti PS, Swamy PRK (1979) Kinetics of ruthenium(III)-catalysed oxidation of aromatic aldehydes by alkaline ferricyanide. Proceed Indian Acad Sci - Chem Sci 88: 163-170.
24. Furniss BS, Hannaford AJ, Smith WG, Tatchell AR (2004) In: Vogel's textbook of practical organic chemistry. 5th edn. Pearson Education Ltd.
25. Vogel AI (1973) Text book of practical organic chemistry. 3rd edn. ELBS, London, Longman, p: 332.
26. Feigl F (1975) Spot tests in organic analysis. New York, Elsevier, p: 195.
27. Wiberg KB (1965) Oxidation in organic chemistry. Academic Press, New York, USA.
28. Sen Gupta KK, Chakladar JK (1974) Kinetics of the chromic acid oxidation of arsenic(II). J Chem Soc Dalton Trans 2: 222-225.
29. Sen Gupta KK, Chakladar JK, Chatterjee AK, Chakladar JK (1973) Kinetics of the oxidation of hypophosphorous and phosphorous acids by chromium(VI). J Inorg Nucl Chem 35: 901-908.
30. Espenson JH (1970) Oxidation of transition metal complexes by chromium(VI). Accoun Chem Res 3: 347-351.
31. Khan Z, ud-Din K (2001) Effect of manganese(II) ions on the oxidation of malic and oxalo ethanoic acids by aqueous $HCrO_4$. Transition Met Chem 26: 672-679.
32. Milazzo G, Caroli S, Sharma VK (1978) Tables of standard electrode metal potentials. Wiley & Sons, New York, USA.
33. Bailey N, Carrington A, Lott KAK, Symons MCR (1960) Structure and reactivity of the oxyanions of transition metals. Part VIII. Acidities and spectra of protonated oxyanions. J Chem Soc, pp: 290-297.
34. Sasaki Y (1962) Equilibrium studies on polyanions. 9. The first steps of acidification of chromate Ion in 3 M $Na(ClO_4)$ Medium at 25 degrees C. Acta Chem Scand 16: 719-734.
35. Michaelis L, Menten ML (1913) The kinetics of invertase action. Biochem Z 49: 333-369.
36. Odebunmi EO, Obike AI, Owalude SO (2009) Kinetics and mechanism of oxidation of D-xylose and L-arabinose by chromium(VI) ions in perchloric acid medium. Int J Biolog Chem Sci 3: 178-185.
37. Naik PK, Chimatadar SA, Nandibewoor ST (2008) A kinetic and mechanistic study of the oxidation of tyrosine by chromium(VI) in aqueous perchloric acid medium. Transition Met Chem 33: 405-410.
38. Frost AA, Person RG (1971) Kinetics and Mechanism. Wiley Eastern, New Delhi, India.
39. Amis ES (1966) Solvent effects on reaction rates and mechanism. Academic Press, New York, USA.
40. King EL, Pandow ML (1952) The Spectra of Cerium(IV) in Perchloric Acid Evidence for Polymeric Species. J Am Chem Soc 74: 1966-1969.
41. Weissberger A (1974) In Investigation of rates and mechanism of reactions in techniques of chemistry. John Wiley & Sons, p: 421.

Hirsutism and Health Related Quality of Life

Yahya M Hodeeb[1], Amal M Al Dinary[2], Hassan M Hassan[1]* and Dina A Samy[1]

[1]*Department of Dermatology, Veneriology and Andrology, Faculty of Medicine, Al Azhar University, Cairo, Egypt*
[2]*Department of Community Medicine, Faculty of Medicine, Al Azhar University, Cairo, Egypt*

Abstract

Background: Quality of life (QoL) is an emerging general parameter of patients' wellbeing. It is a multifactorial concept consisting of individual perception of physical, psychological and social functioning. Hirsutism is a common disorder of excess growth of hair in an androgen-dependent male distribution in women. Hirsutism in women results in significant psychological and social problems. It impacts negatively upon the QoL of women and is the cause of stress, anxiety and depression.

Aim of the study: To investigate the effect of hirsutism on QoL of hirsute women.

Patients and methods: One hundred female patients with hirsutism over the age of eighteen years were enrolled in this study. Each woman was asked to fill a self-report questionnaire.

Results: The results of this study showed that not only the QoL, but also the self-related health status are seriously affected by the level of hirsutism in women. The higher the level of hirsutism, the worse the QoL, measured by DLQI. Hirsutism has a great negative impact on QoL in women and causes psycho-logical problems. Psychological or psychiatric treatment has been suggested for this group of patients. However, according to our results the outcome of QoL, anxiety and depression level is significantly associated with the level of hairiness. So it is more appropriate to offer effective medical treatment for hirsutism than to just offer psychotherapy and refer the women back to self-treatment.

Conclusion: We can conclude that hirsutism has a great negative impact on QoL in women as QoL, anxiety and depression level is significantly associated with the level of hairiness. Although hirsutism is not a serious or life threatening disease, it produces social, psychological and emotional disability, it is more appropriate to offer effective medical treatment for hirsutism plus psychotherapy.

Keywords: Hirsutism; Psychotherapy; Androgens; Physician

Introduction

Hirsutism is excessive growth of terminal hair in women in skin areas sensitive to androgens. It is a sign of increased androgen activities in the hair follicles, either as a result of increased circulating level of androgens or increased sensitivity of the hair follicles to normal circulating level of androgens [1]. The areas' most affected are the face and the lower abdomen [2].

Hirsutism is an international issue and approximately 5% to 15% of women have reported to be hirsute [3]. Excess hair is cosmetically concerning for women and can significantly affect self-esteem. Mediterranean women generally have a medium amount of body and facial hair, whereas Asian women have a minimal amount [4].

The concept health is defined as "a state of complete physical, mental and social well-being and not merely the absence of disease or illness" [5].

The definition of Quality of Life (QoL) proposed by the World Health Organization is "the individuals' perceptions of their position in life in the context of the culture and value systems in which they live, and in relation to their goals, expectations, standards, and concerns [6,7].

At the beginning of the new Millennium, Smith [8] concluded in his review of the literature that "QoL is currently underpinning a significant proportion of new social science research".

QoL is the product of the interplay among social, health, economic and environmental conditions, which affect human and social development [9].

Moreover, QoL provides a global evaluation of one's life that can be used to determine the subjective experience of living with a condition, affect planning for the future, and potentially affect acceptance and adherence to treatment. However, every disease has some explicit and implicit effects on the life of the patient; use of medicines may cure the patient from infection but seems to be ineffective to reduce the physical, psychological and mental distortions which he or she faces during infection time period [10].

QoL broadens the definition of health outcomes beyond the traditional clinical endpoint to represents the implication of disease and treatment in terms of what people are able to do and how they feel. QoL assumed to be used to evaluate issues unrelated to the context of health care that usually includes subjective evaluations of both positive and negative aspects of life [11]. In the medical context, QoL is often defined in terms of functional status [12]. Within the public health context, the term corresponds with the WHO definition of health; which is defined as a state of complete physical, mental, and social well-being not a mere absence of disease or infirmity [13].

Researchers have shown that excessive growth of hair in women was the second most serious rated factor after infertility that negatively influenced their QoL, and that these women had higher depression scores and greater body dissatisfaction. Women with excessive hair growth experience it as a theft of womanhood and talk about themselves in masculine terms, such as having male hair or a full beard [14].

***Corresponding author:** Hassan M Hassan, Department of Dermatology, Veneriology and Andrology, Faculty of Medicine, Al Azhar University, Cairo, Egypt, E-mail: hasnasar56@hotmail.com

Health Related QoL is influenced not only by the disease itself but also by socio-demographic, psychological, lifestyle, biomedical factors and gender. Low socio-economic status, immigrant status, single statuses are all related to poorer QoL [15].

The impact of hirsutism symptoms on a woman's QoL may be profound and can result in psychological distress that threatens her feminine identity. The condition may therefore result in altered self-perception, a dysfunctional family dynamic, and problems at work. Many aspects of the disorder can very conceivably cause a significant amount of emotional stress [16].

Also, patients with hirsutism displayed significantly higher social fears than controls. They also showed more anxiety and psychotic symptoms, whereas there were no significant differences in depression, somatization, anger-hostility and cognitive symptoms [17].

Excessive hair growth in women with Polycystic Ovary Syndrome (PCOS) was the second most seriously rated factor that negatively influenced their QoL [1]. The changes that occur in women's physical appearance as a result of PCOS, (such as obesity, acne and hirsutism), might contribute to psychological morbidity and a feeling of being stigmatized [18].

Aim of the Work

The aim of this work is to investigate the effect of hirsutism on Health Related Quality of Life (HRQoL) of hirsute women attending the Dermatology Clinic, Al Hussein Hospital for problem of excessive hair growth.

Patients and methods

Both qualitative and quantitative methods have been used to investigate the impact of hirsutism on the quality of life of the patient.

Study design: A case control study was used to investigate the current research problem.

Patients and controls

A convenience sample method was used. The studied group consisted of 100 female patients and another 100 females as a control group. The patients attending Dermatology Clinic at Al Hussein University Hospital with hirsutism were recruited in this study. The two groups were matched in age (over 18 to 40 years) and socioeconomic status. The controls were free from hirsutism. Also, they were from females attending other clinics or neighbors.

Inclusion criteria

Hirsute females aged from 16 to 40 years. Ferriman and Gallway [19] [F-G] score was used as a visual method of determining the severity of hirsutism in nine androgenic sensitive skin areas (upper lip, chin, chest, upper back, lower back, upper abdomen, lower abdomen, arm and thigh). Each area has 0 to 4 score and F-G score is summation of all 9 area scores. Those patients with F-G score ≥ 8 or one area score=4 were eligible cases and included in the study. In the next step, the F-G score was categorized based on the following thresholds: F-G score <8 with one area score=4 were considered as one-area limited hirsutism, 8-10=mild, 11-14=moderate and scores >15 were considered as severe hirsutism.

Exclusion criteria

Patients with chronic or debilitating diseases like cardiovascular, thyroid and psychological diseases.

Data collection

1. Patients were asked to complete a form including questions about age, marriage, duration of hirsutism (in study group), method of removing unwanted hair and medical problems diagnosed by their physician before, and to complete a socioeconomic questionnaire calculating their social standard. Social scores to calculate social standards were according to Fahmy and El-Sherbeny [20]; the total score summed 28; high social standards=25-30, middle social standards=20-24, low social standards=15-19, and very low social standards ≤ 15.

2. Also, QoL was assessed by a self-administrated Dermatology Life Quality Index and Hospital Anxiety and Depression score questioners.

Hospital Anxiety and Depression Scale (HADS): The HADS is a self-administered measure used to screen for the presence of depression and anxiety. The HADS was developed to provide clinicians with an acceptable, reliable, valid and easy to use practical tool for identifying and quantifying depression and anxiety. The HADS can be used in a variety of settings (e.g., community, primary care, in-hospital, and psychiatry). The HADS is not intended as a complete diagnostic tool, but as a means for identifying general hospital patients who need further psychiatric evaluation and assistance [21]. A score system for questions 2, 4, 6, 8, 11, 12, 14 was used for anxiety, while a score system for questions 1, 3, 5, 7, 9, 10, 13 was used for depression. A grading of 0-7=non-case, 8-10=borderline case, and >11=case [22].

Dermatology Life Quality Index (DLQI): DLQI is the first dermatology specific instrument and it was developed by Finlay and Khan [23]. The aim of this questionnaire was to measure how much your skin problem has affected your life over the last week. The DLQI is calculated by summing the score of each question resulting in a maximum of 30 and a minimum of 0. The higher the score, the more quality of life is impaired [24]. The scoring of each question was as follows; very much scored 3, a lot scored 2, a little scored 1, not at all scored 0, and question 7 (prevented work or studying) scored 3. Interpretation of the meaning of DLQI scores; 0-1=no effect at all on patient's life, 2-5=small effect on patient's life, 6-10=moderate effect on patient's life, 11-20=very large effect on patient's life, and 21-30=extremely large effect on patient's life [25].

Ethical considerations

An approval to conduct this study was obtained. Also, all the female cases and controls were gave a verbal consent before sharing in the study. The consent informs females that all the filled information will be confidential.

Statistical design and data analysis

Data were coded manually and analysis was conducted through SPSS program, version 16. The results were presented in tables and Figures. Descriptive and analytical statistical analysis was done: Quantitative data; mean ± Standard Deviation (SD) was used as they measure central tendency and dispersion of quantitative data. Qualitative data: number and percentage were used. Chi square (χ^2) test was used for comparison of qualitative data, student's t-test for quantitative data of two independent samples. Analysis of Variance (ANOA) was used for comparison of quantitative data of more than two groups. Linear regression analysis was carried out to assess some factors may affect HADS Scale. Correlation was done between two scales mention. The level of significance was taken at P<0.05.

Results

Figure 1 shows that among case about two thirds (63.0%) of females had moderate hirsutism and about one third (29.0%) had mild hirsutism.

Table 1 illustrates assessment of HADS among studied groups. It was found that the mean of total score of HADS (20.1 ± 3.1) and that of each domain (12.1 ± 2.3 and 7.9 ± 1.2) (anxiety and depression, respectively) were higher among cases; the comparable figures in controls were (4.2 ± 1.9 and 6.8 ± 1). These differences were statistically significant ($p<0.05$). Also, it was found that the mean of total score of QoL was higher in cases (17.4 ± 2.3) than in controls (3 ± 1.2). This difference was statistically significant ($p<0.05$).

Figure 2 show that hirsutism had very large effect on quality of life (91.0%) in cases regarding the QoL score, while 93.0% of controls had no effect.

Table 2 illustrates HADS among studied groups according to marital status. It was found that the mean of total score of HADS and that of each domain (anxiety and depression) score were higher among married and single females of cases than those of controls. These differences were statistically significant ($p<0.05$). Also, the table illustrates DLQI among studied groups according to marital status. It was found that the mean of total score of DLQI score was higher among married and single females of cases than those of controls. These difference was statistically significant ($p<0.05$).

Table 3 illustrates Hospital Anxiety and Depression Scale among studied groups according to occupation. It was found that the mean of total score of HADS and that of each domain (anxiety and depression) score were higher among house wives and worked females of cases than those of controls. These differences were statistically significant ($p<0.05$). Also, the table illustrates DLQI among studied groups according to occupation. It was found that the mean of total score of DLQI score was higher among house wives and worked females of cases than those of controls. This difference was statistically significant ($p<0.05$).

Table 4 illustrates HADS among studied groups according to social class. It was found that the mean of total score of HADS and that of each domain (anxiety and depression) score were higher among different social class females of cases than those of controls. These differences were statistically significant ($p<0.05$). Also, the table illustrates DLQI among studied groups according to social class. It was found that the mean of total score of DLQI score was higher among different social class of cases than those of controls. These differences were statistically significant ($p<0.05$).

In Table 5, it was found that the highest mean of total score of HADS and that of each domain (anxiety and depression) score were recorded among females with severe hirsutism, while the lowest mean of anxiety and total HADS were recorded among those with moderate hirsutism, however the lowest mean of depression score were recorded among females with mild hirsutism. These differences were statistically insignificant ($p>0.05$). Also, it was found that the highest mean of total score of DLQI score was recorded among females with severe hirsutism, while the lowest mean was recorded among females with mild hirsutism. These differences were statistically significant ($p<0.05$).

In Table 6, it was found that the lowest means of total score of HADS and that of each domain (anxiety and depression) were recorded among females who had university education, while highest means were recorded among females who had preparatory education. These differences were statistically significant ($p<0.05$). Also, it was found that the highest mean of DLQI was among females who had university education, while the lowest mean was among females who had primary education. This difference was statistically insignificant.

In Table 7, it was shown that most effective factors that significantly affect DLQI were age of patients, occupation and grade of hirsutism ($P<0.05$) all were positively affect DLQI.(if age of female increase, or if she was house wife and if she had severe grade of hirsutism then DLQI will impaired). Also, it was shown that most effective factor that significantly affect HADS social class of patients ($P<0.05$) as it positively affect HADS (if social class of female is decrease then HADS will impaired).

Table 8 show that daily activity of females with hirsutism had the highest mean among DLQI sections (4 ± 0.8), while work and school activity had the lowest mean (1.4 ± 0.6) among them. This difference was statistically significant ($p<0.05$).

Discussion

In medical practice, it is often impossible to separate the disease from the individual's personal and social context, especially in chronic and progressive diseases [26]. Also, it is known that patients with chronic diseases place a high value on their mental and social well-being as well as pure physical health [27]. Focusing on HRQoL as a national health standard can bridge boundaries between disciplines and between social, mental, and medical services [28].

Hirsutism is a common disorder of excess growth of terminal hair in an androgen-dependent male distribution in women, including the chin, upper lip, breasts, upper back, and abdomen. It affects 5% to 10% of women of reproductive age. Hirsutism is more than a cosmetic problem. It may be linked to significant underlying diseases, often associated with a decreased quality of life, impaired self-image of the patient feminine identity [29].

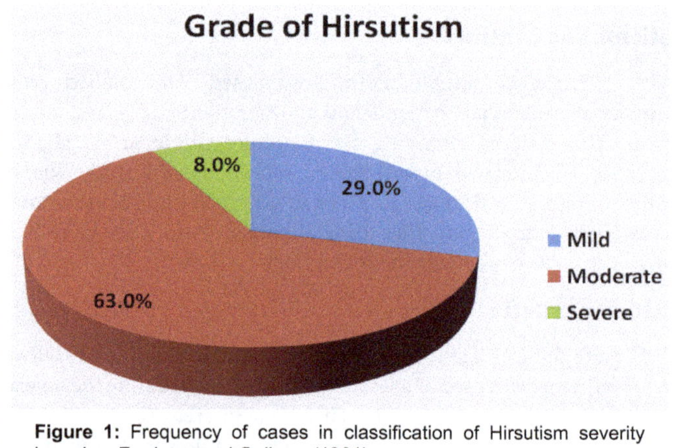

Figure 1: Frequency of cases in classification of Hirsutism severity based on Ferriman and Gallway (1961).

Variables	Cases No=100	Controls No=100	t-test	P value
HADS				
Anxiety scale (Mean ± SD)	12.1 ± 2.3	4.2 ± 1.9	t=27	p<0.05
Depression scale (Mean ± SD)	7.9 ± 1.2	6.8 ± 1	t=7.4	p<0.05
Total scale (Mean ± SD)	20.1 ± 3.1	11 ± 2.3	t=24	p<0.05
DLQI				
Total score (Mean ± SD)	17.4 ± 2.3	3 ± 1.2	t=58	p<0.05

Table 1: Mean ± SD of Hospital Anxiety and Depression Scale (HADS) and Dermatology Life Quality Index (DLQI) among the studied groups.

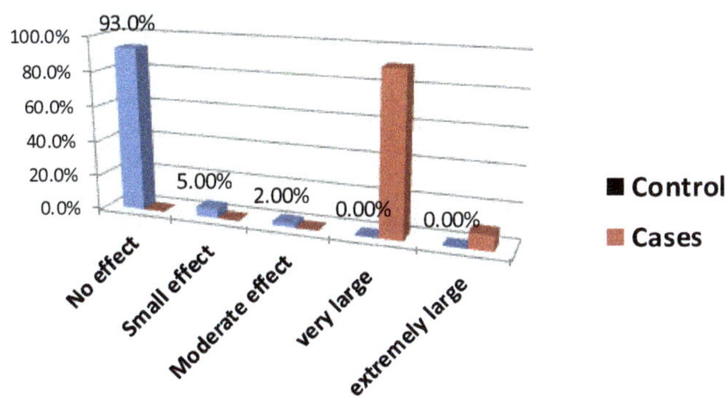

Figure 2: Frequency in meaning of dermatology life quality index (DLQI) scores among studied groups.

Variables	Marital status		ANOVA	P value
	Married	Single		
HADS				
Anxiety scale (Mean ± SD) Case Control	12.2 ± 2.1 4.1 ± 1.8	12 ± 2.7 4.5 ± 2	F=37	p<0.05
Depression scale (Mean ± SD) Case Control	7.9 ± 1.3 6.8 ± 2	8 ± 1 6.7 ± 1.1	F=54	p<0.05
Total scale (Mean ± SD) Case Control	20.1 ± 3 10.9 ± 2.3	20 ± 3.3 11.2 ± 2.1	F=56	p<0.05
DLQI				
Total score (Mean ± SD) Case Control	17.7 ± 2.2 1.1 ± 0.3	16.7 ± 2.2 1.1 ± 0.2	F=46	p<0.05

Table 2: Mean ± SD of Hospital Anxiety and Depression Scale (HADS) and Dermatology Life Quality Index (DLQI) among studied groups according to marital status.

Variables	Occupation		ANOVA	p-value
	Work	House wife		
HADS				
Anxiety scale (Mean ± SD) Case Control	11.7 ± 2.1 4.3 ± 2	12.5 ± 2.3 4 ± 1.7	F=77	p<0.05
Depression scale (Mean ± SD) Case Control	7.8 ± 1.2 6.8 ± 1	8 ± 1.2 6.8 ± 1	F=57	p<0.05
Total scale (Mean ± SD) Case Control	20.1 ± 3 10.9 ± 2.3	20.5 ± 3.1 10.8 ± 2.3	F=56	p<0.05
DLQI				
Total score (Mean ± SD) Case Control	17 ± 2 1 ± 0.2	17.7 ± 2.4 1.2 ± 0.3	F=46	p<0.05

Table 3: Mean ± SD of Hospital Anxiety and Depression Scale (HADS) and Dermatology Life Quality Index (DLQI) among studied groups according to occupation.

Hirsutism has a great impact on the patient's quality of life. The knowledge of the effect of a dermatologic problem on the patient's quality of life is of great importance in the management of that condition and can even change the therapeutic approach. Useful methods to evaluate this impact are the DLQI and HADS [30].

The aim of this study was to evaluate the effects of hirsutism and quality of life, anxiety and depression in hirsute women.

One hundred female patients with hirsutism over the age of sixteen years were enrolled in this study. Each woman was asked to fill the self-administered questionnaire including sociodemographic questions (age, employment, education, income, etc.), DLQI, HADS, and Ferriman-Gallwey scale (F-G scale).

QoL is defined as "an individual's perception of their position in life in the context of the culture and value systems in which they live and in relation to their goals, expectations, standards and concerns [31].

In the present study, the items most affected measured by the DLQI were embarrassment/self-consciousness (item 2), influence on clothing (item 4), affected social or leisure activities (item 5) and problems with partner/close friends/relatives (item 8). The area less affected was work and school. It is noted that the patients with severe level of hirsutism had higher scores on all items more than the patients with mild level (not present data).

These results further strengthen what has been found by Ekback et al. [32], where women were working, but did not participate in activities with work mates outside their work, in case they risked revealing their hairiness because of the embarrassment and guilt they felt over their situation. Because hair removal can be time-consuming, expensive, and frustrating, many hirsute women feel unable to manage their hair removal effectively, and some experience depression to a similar or greater degree than women with breast cancer or psoriasis. This is also in line with the higher levels of 'social fears' that reported among women with hirsutism compared to women without hirsutism [33].

Soliman and Wardle reported that some hirsute women become reclusive and only venture out after dark, while in young people hirsutism can be a cause of bullying, social isolation, and poor educational performance. These problems of low self-esteem and self-consciousness may have an impact on their ability to work because there is evidence to suggest that there are fewer hirsute women in employment than non-hirsute women despite having similar educational experiences and levels of attainment.

Maziar et al. [34] revealed that HRQoL in PCOS patients showed excessive facial hair was significantly impacted and there was improvement in QoL after laser treatment of unwanted facial hair.

Many factors can explain this result. Hirsutism commonly affects young people at a time when they are undergoing maximum psychological, social and physical changes and they are least capable of coping with additional stress. In addition the highly visibility of the disease severely impact QoL. As the social norm of femininity today include hair free body [35].

Variables	Social class			ANOVA	P value
	High	Middle	Low		
HADS					
Anxiety scale (Mean ± SD) Case Control	11.8 ± 2.3 4.2 ± 2.1	12.7 ± 2.2 4 ± 1.3	13.4 ± 1.6 4.4 ± 2.3	F=79	p<0.05
Depression scale (Mean ± SD) Case Control	7.7 ± 1.2 6.7 ± 1	8.1 ± 1 6.9 ± 1	8.8 ± 1 6.6 ± 0.9	F=55	p<0.05
Total scale (Mean ± SD) Case Control	19.5 ± 3 10.9 ± 2.5	20.7 ± 3 10.9 ± 1.7	22.2 ± 2.4 11 ± 3	F=53	p<0.05
DLQI					
Total score (Mean ± SD) Case Control	17.5 ± 2.5 1.2 ± 0.4	17.2 ± 2 0.4 ± 0.06	17.3 ± 1.4 0.3 ± 0.01	F=49	p<0.05

Table 4: Mean ± SD of Hospital Anxiety and Depression Scale (HADS) and Dermatology Life Quality Index (DLQI) among studied groups according to social class.

Variables	Disease severity			ANOVA	P value
	Mild	Moderate	Severe		
HADS					
Anxiety scale (Mean ± SD)	12.5 ± 2.3	11.8 ± 2.1	13.1 ± 3.7	F=1.3	p>0.05
Depression scale (Mean ± SD)	7.7 ± 1.2	7.9 ± 1.1	8.8 ± 1.7	F=2	p>0.05
Total scale (Mean ± SD)	20.1 ± 3.1	19.7 ± 2.9	21.9 ± 4.1	F=1.5	p>0.05
DLQI					
Total score (Mean ± SD)	17.0 ± 2.7	17.2 ± 1.6	20.5 ± 3.1	F=9.8	p<0.05

Table 5: Mean ± SD of Hospital Anxiety and Depression Scales (HADS) and Dermatology Life Quality Index (DLQI) among cases according to severity of hirsutism.

Variables	Education				ANOVA	P value
	Primary	Preparatory	Secondary	University		
HADS						
Anxiety scale (Mean ± SD)	13.4 ± 3.5	13.5 ± 1.8	13 ± 2.5	11.5 ± 2	F=3.8	p<0.05
Depression scale (Mean ± SD)	7.6 ± 0.7	8.5 ± 1.2	8.4 ± 1.1	7.5 ± 1.2	F=3.7	p<0.05
Total scale (Mean ± SD)	21 ± 4	22 ± 3	21.3 ± 3	19.2 ± 2.7	F=4.8	p<0.05
DLQI						
Total score (Mean ± SD)	17.0 ± 1.4	17.2 ± 1.2	17.3 ± 2.4	17.5 ± 2.3	F=0.8	P >0.05

Table 6: Mean ± SD of Hospital Anxiety and Depression Scale (HADS) and Dermatology Life Quality Index (DLQI) among cases according to level of education.

Factors	B	P value
DLQI		
Increase age of patients	0.3	0.009
Marital status	0.0	0.9
Occupation (house wife)	0.3	0.009
Educational level	0.1	0.1
Social classes	-0.09	0.3
Severe grade of hirsutism	0.3	0.009
HADS		
Age of patients	0.03	0.7
Marital status	0.07	0.6
Occupation	0.1	0.3
Educational level	-0.02	0.8
Low social classes	0.3	0.02
Grade of hirsutism	0.05	0.6

Table 7: Linear regression analysis for assessment of some factors may affect Dermatology Life Quality Index (DLQI) Scale and Hospital Anxiety and Depression Scale (HADS).

Sections of DLQI	Mean ± SD	*ANOVA*
Symptoms and feelings	2.4 ± 0.5	
Daily activities	4 ± 0.8	
Leisure	3.2 ± 0.5	
Work and school	1.4 ± 0.6	F=44
Personal relationships	3.7 ± 1.1	
Treatment	2.4 ± 0.3	p value=0.00

Table 8: Dermatology Life Quality Index (DLQI) section scores in women with hirsutism.

In the present study, the mean value for DLQI was high (DLQI=12.6) which is similar to what has been found by Berg and Lindberg [35], in previous study evaluating quality of life in women with hirsutism and in parity with other severe forms of skin diseases, for example psoriasis or atopic dermatitis.

In the present study, women with severe levels of hirsutism (F-G ≥ 15) scored significantly higher both on item and dimension levels of DLQI compared to women with mild (F-G=8-10) levels of hirsutism indicating that there was a highly significant association between the clinical severity and QoL in hirsute patients.

The findings of present study were in concordance with Zhuang et al. [36], who revealed that there was a significant correlation between the clinical severity of hirsutism and DLQI index. The more severe degree of disease, the greater the impact on the DLQI scores. To a certain extent, these results revealed that the clinical severity of hirsutism affect the QoL of patients. Also our findings are in concordance with Reid et al. [37], who said that the more severe form of disease, the more bothered the patient QoL.

Drosdzol et al. [38], in their study on hirsutism found a negative effect of hirsutism on QoL in these patients. A study by Basra et al. [39], considered not only the quality of life of hirsute women in comparison to non-hirsute women, but also the impact on the partners of hirsute women. The study found that a statistically significant difference exists between hirsute and non-hirsute women in relation to health related quality of life. On a scale between 0 (dead) and 100 (perfect health) the mean score for hirsute women was 68.4 while for non-hirsute women it was 87.7. The mean score of the partners of hirsute women was 75.4 indicating that their quality of life was likely to be somewhat affected by the hirsutism of their partner.

The effect of laser treatment was investigated in 45 women with facial hirsutism. A modified DLQI before and one to two months, two to four months and six months after laser treatment in fifteen out of 45 women. The mean DLQI score before treatment was 12.8. The mean DLQI score at one to two months was 7.0, at two to four months it was 9.2 and at four to six months it was 11.5. There was a major improvement in DLQI score at 1-2 months but longer-term benefit was not observed when the hair growth has returned back to pre-treatment levels. In spite of that 70% of the women reported a high level of patient satisfaction and nearly 80% was willing to have further treatment [40].

In the present study a large proportion of the women reported anxiety and depression on the HADS. The levels of anxiety and depression were higher among women with severe F-G score than mild F-G score.

Lipton et al. [41] reported that nearly three-quarters of hirsute women had anxiety, one-third had clinical levels of depression and one-third felt uncomfortable in social situations and tried to prevent others from coming near them.

A study by Clayton et al. [42], suggested that the levels of anxiety and depression of hirsute women were higher than those of women attending outpatient departments with newly diagnosed gynecological or breast cancer.

Hirsutism has a significant negative impact on psychosocial development. It can have serious psychological consequences and undermines a woman's confidence and self-esteem [3].

Our result is in line with results reported by Keegan et al. [18], who said that hirsutism is possibly linked with problems concerning female identity and sexual self-worth; it has a significant negative impact on psychosocial development. It can have serious psychological consequences and undermines a woman's confidence and self-esteem.

However, Hahn et al. [43], found in their study no association between these problems and hirsutism but they found that hirsutism can lead to feelings of decreased sexual self-worth and sexual satisfaction. This difference due to low number of patients (20) and the reason for the disparity between the two studies could lie in ethnical and cultural differences of the two populations studied.

The range of the F-G score in our study was 8-29. However, the mean score was high (12.6) and there were also significant differences in quality of life, anxiety, depression and perceived health status between women with severe (F-G $\geq$ 15) and mild (F-G=8-10) hirsutism, which further point to the fact that the level of hirsutism was the most important factor for women's quality of life.

Hirsutism has a great negative impact on HRQoL in women and causes psychological problems. Psychological or psychiatric treatment has been suggested for this group of patients. However, according to the outcome of HRQoL, anxiety and depression level is significantly correlated with the level of hairiness. So, it is more appropriate to offer effective medical treatment for hirsutism than to just offer psychotherapy and refer the women back to self-treatment. Patients with a suspected distress disorder or signs of severe depression should of course also be referred for psychiatric evaluation [38,44].

Overall, dermatologists should be alert to the following potentially significant factors: sufficient time should be spent on consultation and at the first visit; patients should take a few minutes to complete a simple questionnaire regarding their motivation and expectations of the treatment in order to promote communication between the doctor and patient. Clinicians should answer any questions the patient may have to ease their concerns regarding hirsutism and to correct any impractical expectations, particularly regarding the improvement of disease and time taken for treatment [36].

Conclusion

We can conclude that hirsutism has a great negative impact on QoL in women as QoL, anxiety and depression level is significantly correlated with the level of hairiness. Although hirsutism is not a serious or life threatening disease, it produces social, psychological and emotional disabilities which may cause social and marital problems, it is more appropriate to offer effective medical treatment for hirsutism plus psychotherapy.

References

1. Guyatt G, Weaver B, Cronin L, Dooley JA, Azziz R (2004) Health-related quality of life in women with polycystic ovary syndrome, a self-administered questionnaire, was validated. J Clin Epidemiol 57: 1279-1287.
2. Hunter MH, Carek PJ (2003) Evaluation and treatment of women with hirsutism. Am Fam Physician 67: 2565-2572.
3. Azziz R (2003) The evaluation and management of hirsutism. Obstet Gynecol 101: 995-1007.
4. Himelein MJ, Thatcher SS (2006) Polycystic ovary syndrome and mental health: A review. Obstet Gynecol Surv 61: 723-732.
5. Garvin P, Nilsson E, Ernerudh J, Kristenson M (2015) The joint subclinical elevation of CRP and IL-6 is associated with lower health-related quality of life in comparison with no elevation or elevation of only one of the biomarkers. Qual Life Res.
6. (1998) The World Health Organization Quality of Life Assessment (WHOQOL): development and general psychometric properties. Soc Sci Med 46: 1569-1585.
7. Bonomi AE, Patrick DL, Bushnell DM, Martin M (2000) Validation of the United States' version of the World Health Organization Quality of Life (WHOQOL) instrument. J Clin Epidemiol 53: 1-12.
8. Smith AE (2000) Quality of life: A review. Education and ageing. Triangle J 15: 419-435.
9. Jones CI, Klenow PJ (2010) Beyond GDP? Welfare across countries and over time. NBER Working Paper 16352.
10. Topolski TD, Edwards TC, Patrick DL (2005) Quality of life: how do adolescents with facial differences compare with other adolescents? Cleft Palate Craniofac J 42: 25-32.
11. Gandek B, Sinclair SJ, Kosinski M, Ware JE Jr (2004) Psychometric evaluation of the SF-36 health survey in Medicare managed care. Health Care Financ Rev 25: 5-25.
12. Guyatt GH, Naylor CD, Juniper E, Heyland DK, Jaeschke R, et al. (1997) Users' guides to the medical literature. XII. How to use articles about health-related quality of life. Evidence-Based Medicine Working Group. JAMA 277: 1232-1237.
13. Park K (2011) Concepts of health and disease. In: Park K, (ed.) Park's Textbook of Preventive and Social Medicine 16th edn. Jabalpur: M/S. Banarsidas Bhanot.
14. Kitzinger C, Willmott J (2002) 'The thief of womanhood': women's experience of polycystic ovarian syndrome. Soc Sci Med 54: 349-361.
15. Burström K, Johannesson M, Diderichsen F (2001) Swedish population health-related quality of life results using the EQ-5D. Qual Life Res 10: 621-635.
16. Hajheydari Z, Jamshidi M, Masoudzadeh A (2007) Association between hirsutism and mental health. Neurosciences (Riyadh) 12: 242-244.
17. Sonino N, Fava GA, Mani E, Belluardo P, Boscaro M (1993) Quality of life of hirsute women. Postgrad Med J 69: 186-189.
18. Keegan A, Liao LM, Boyle M (2003) 'Hirsutism': a psychological analysis. J Health Psychol 8: 327-345.
19. Ferriman D, Gallwey JD (1961) Clinical assessment of body hair growth in women. J Clin Endocrinol Metab 21: 1440-1447.
20. Fahmy S, El-Sherbeny A (1983) Determining simple parameters for social classification for health research. Bull High Instet Public Health 13: 95-108.
21. Herrmann C (1997) International experiences with the Hospital Anxiety and Depression Scale--a review of validation data and clinical results. J Psychosom Res 42: 17-41.
22. Zigmond AS, Snaith RP (1983) The hospital anxiety and depression scale. Acta Psychiatr Scand 67: 361-370.
23. Finlay AY, Khan GK (1994) Dermatology Life Quality Index (DLQI)--a simple practical measure for routine clinical use. Clin Exp Dermatol 19: 210-216.
24. Basra MK, Fenech R, Gatt RM, Salek MS, Finlay AY (2008) The Dermatology Life Quality Index 1994-2007: a comprehensive review of validation data and clinical results. Br J Dermatol 159: 997-1035.
25. Hongbo Y, Thomas CL, Harrison MA, Salek MS, Finlay AY (2005) Translating the science of quality of life into practice: What do dermatology life quality index scores mean? J Invest Dermatol 125: 659-664.
26. Chamla D (2004) The assessment of patients' health-related quality of life during tuberculosis treatment in Wuhan, China. Int J Tuberc Lung Dis 8: 1100-1106.
27. Guo N, Marra F, Marra CA (2009) Measuring health-related quality of life in tuberculosis: a systematic review. Health Qual Life Outcomes 7: 14.
28. Selim AJ, Rogers W, Fleishman JA, Qian SX, Fincke BG, et al. (2009) Updated U.S. population standard for the Veterans RAND 12-item Health Survey (VR-12). Qual Life Res 18: 43-52.
29. Pate C (2013) The story plot of living the embarrassment of hirsutism. Arch Psychiatr Nurs 27: 156-157.

30. Lapidoth M, Dierickx C, Lanigan S, Paasch U, Campo-Voegeli A, et al. (2010) Best practice options for hair removal in patients with unwanted facial hair using combination therapy with laser: guidelines drawn up by an expert working group. Dermatology 221: 34-42.

31. Oort FJ (2005) Using structural equation modeling to detect response shifts and true change. Qual Life Res 14: 587-598.

32. Ekback M, Wijma K, Benzein E (2009) "It is always on my mind": women's experiences of their bodies when living with hirsutism. Health Care Women Int 30: 358-372.

33. Somani N, Harrison S, Bergfeld WF (2008) The clinical evaluation of hirsutism. Dermatol Ther 21: 376-391.

34. Maziar A, Farsi N, Mandegarfard M, Babakoohi S, Gorouhi F, et al. (2010) Unwanted facial hair removal with laser treatment improves quality of life of patients. J Cosmet Laser Ther 12: 7-9.

35. Berg M, Lindberg M (2011) Possible gender differences in the quality of life and choice of therapy in acne. J Eur Acad Dermatol Venereol 25: 969-972.

36. Zhuang XS, Zheng YY, Xu JJ, Fan WX (2013) Quality of life in women with female pattern hair loss and the impact of topical minoxidil treatment on quality of life in these patients. Exp Ther Med 6: 542-546.

37. Reid EE, Haley AC, Borovicka JH, Rademaker A, West DP, et al. (2012) Clinical severity does not reliably predict quality of life in women with alopecia areata, telogen effluvium, or androgenic alopecia. J Am Acad Dermatol 66: e97-102.

38. Drosdzol A, Skrzypulec V, Plinta R (2010) Quality of life, mental health and self-esteem in hirsute adolescent females. J Psychosom Obstet Gynaecol 31: 168-175.

39. Basra MK, Finlay AY (2007) The family impact of skin diseases: the Greater Patient concept. Br J Dermatol 156: 929-937.

40. Loo WJ, Lanigan SW (2002) Laser treatment improves quality of life of hirsute females. Clin Exp Dermatol 27: 439-441.

41. Lipton MG, Sherr L, Elford J, Rustin MH, Clayton WJ (2006) Women living with facial hair: the psychological and behavioral burden. J Psychosom Res 61: 161-168.

42. Clayton WJ, Lipton M, Elford J, Rustin M, Sherr L (2005) A randomized controlled trial of laser treatment among hirsute women with polycystic ovary syndrome. Br J Dermatol 152: 986-992.

43. Hahn S, Janssen OE, Tan S, Pleger K, Mann K, et al. (2005) Clinical and psychological correlates of quality-of-life in polycystic ovary syndrome. Eur J Endocrinol 153: 853-860.

44. Swiglo BA, Cosma M, Flynn DN, Kurtz DM, Labella ML, et al. (2008) Clinical review: Antiandrogens for the treatment of hirsutism: a systematic review and metaanalyses of randomized controlled trials. J Clin Endocrinol Metab 93: 1153-1160.

UV-Metric, pH-Metric and RP-HPLC Methods to Evaluate the Multiple pKa Values of a Polyprotic Basic Novel Antimalarial Drug Lead, Cyclen Bisquinoline

Mohammad Faisal Hossain, Cassandra Obi, Anjuli Shrestha and MO Faruk Khan*

SCRiPS, College of Pharmacy, Southwestern Oklahoma State University, USA

Abstract

The purpose of this experiment was to evaluate and compare the pKa values of the poorly water soluble, weakly basic, novel antimalarial drug lead, 4,10-bis (7-chloroquinoline)-1,4,7,10-tetraazacyclododecane (CNBQ). Three separate methods, pH-metric, UV-metric, and Reverse Phase-High Performance Liquid Chromatography (RP-HPLC), were employed to determine the pKa values between 2.0-12.0 pH range. The acetate and phosphate buffers, in addition to methanol and acetonitrile as co-solvents and potassium chloride to maintain the ionic strength, were used as appropriate. In UV-metric method, the drug substance is dissolved in aqueous media eliminating any interference of a co-solvent for measuring the pKa. Consequently, the pKa values obtained by the UV-metric method are considered accurate, as opposed to potentiometric and RP-HPLC methods that require the use of co-solvents. Thus, through the utilization of UV-metric method three pKa values, 5.9, 6.6, and 8.7, were obtained for CNBQ. These studies would be useful to determine the pKa values of the related drug leads under development.

Keywords: pKa; pH-metric; UV-metric; RP-HPLC methods; Drug lead; Drug development

Introduction

The acid dissociation constant (pKa) is the pH at which concentrations of ionized and unionized forms of drugs are equal. It is an essential parameter in drug discovery, particularly in physiological systems where ionization state will affect the rate at which the compound is able to diffuse across membranes including blood-brain barrier [1]. Unless the drug is given intravenously, the drug must pass through several different semipermeable membranes before it reaches the systemic circulation and later in the site of action. The semipermeable cell membranes, due to their inherent hydrophobicity, selectively inhibit the passage of drug molecules. The simplest route for a drug to enter the systemic circulation is by passive diffusion through the semipermeable cell membranes from an area of high concentration to an area of low concentration. This requires the ingenuity from drug developers to design drug substances that will overcome these obstacles and allow the proper passage through the semipermeable membranes.

Most drugs are weak organic acids or bases existing in ionized and unionized forms in aqueous solutions. The unionized forms of drugs are usually more lipophilic and thus readily diffuse across cell membranes. It is in this instance that pKa and bioavailability correlate. When the pH is lower than the pKa, the unionized form of a weak acid predominates, and vice versa for a weak base [2]. Proceeding with this concept, different pH in the body allow for the alteration of solubility, dissociation, and coincident absorption of the drug substance [3] by shifting the concentration of the unionized and ionized forms of drug substances, therefore, validating the importance of determining the pKa of a drug.

The antimalarial drug lead cyclen bisquinoline (CNBQ; Figure 1) demonstrated a potent in vitro anti-malarial activity against chloroquine-sensitive and chloroquine-resistant as well as mefloquine-resistant strains of Plasmodium falciparum. The compound was also found to be a potent antimalarial agent in vivo [4]. Moreover, the drug lead was found to be metabolically stable in vitro in the presence of HLM and cDNA expressing CYP2C8 enzymes [5]. As malaria is caused by invasion of malaria parasites in the blood, it is important that the drug substance intended to treat it is able to enter the systemic circulation. Therefore, evaluation of pKa values is imperative in the early stage of drug discovery and development, because, as previously mentioned, pKa values of the drug substance and its bioavailability are correlated. There are several different methods that can be employed to determine the pKa values of the drugs, such as: pH-metric, UV-metric, NMR, solubility, capillary electrophoresis, HPLC, conductometry, voltammetry, calorimetry, fluorimetry, polarimetry, kinetic, and computational methods [6]. It is challenging to find a single method that will not only measure any and all pKa values, but also be reliable in its findings in just one attempt due to the vast amount of variability in drug structures. This paper will focus on the following three methods: pH-metric, UV-metric, and the RP-HPLC methods. Many of other methods previously mentioned require a substantial amount of instrumentation, test material and/or time, making them unsuitable or not feasible for implementation in this research lab (Figure 1).

Materials and Methods

Materials

CNBQ (Figure 1) was synthesized in our laboratory and the purity checked by HPLC [7]. Chloroquine diphosphate was purchased from Pfaltz & Bauer. The solvents and reagents used were as follows: acetonitrile, methanol, sodium hydroxide, potassium hydroxide, hydrochloric acid, potassium chloride, ammonium acetate, and sodium phosphate dibasic anhydrous. Each solvent and reagent used was HPLC and analytical grade and was purchased from Fisher Scientific. Deionized water used to prepare the solutions and mobile phase was further purified by filtration and degassing.

***Corresponding author:** Faruk Khan MO, SCRiPS, College of Pharmacy, Southwestern Oklahoma State University, 100 Campus Drive, Weatherford, OK 73096, USA, E-mail: faruk.khan@swosu.edu

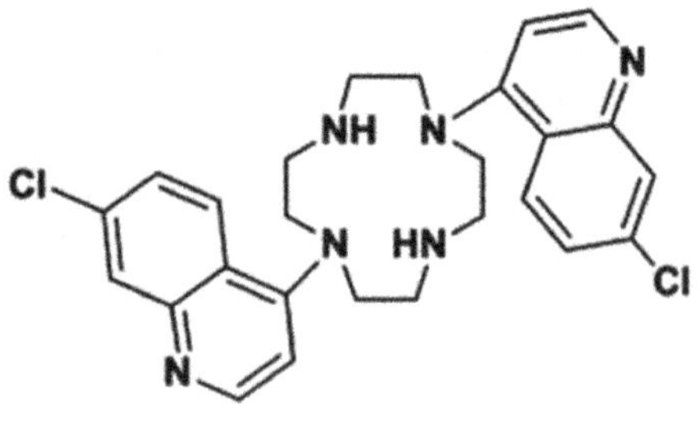

Figure 1: 4,10-bis (7-chloroquinoline)-1,4,7,10-tetraazacyclododecane (Cyclen isquinoline, CNBQ)

Methods

pH-metric method (Potentiometric titration): The pH-metric titration was performed using two different instruments, highly sensitive fully automated Sirius T3 in Sirius laboratory and semi-automated Metrohm in our laboratory. In Sirius laboratory, 0.01 M phosphate buffer was used to prepare the drug solution, and pH of the buffer were accurately adjusted from pH 2.0 to 12.0 with 0.2 interval using 0.5 M HCl and 0.5 M KOH titrants as appropriate. The pH electrode of the potentiometer was calibrated using standard buffers at pHs of 4.0, 7.0 and 10.0. 0.1 M hydrochloric acid was used to prepare the drug solution, and pHs of the buffer was adjusted from pH 2.0 to 12.0 using 0.1 M NaOH/0.1 M KOH titrants as appropriate. In both cases, the ionic strength of the solution was maintained using 0.15 M potassium chloride solution and methanol was used as a co-solvent. The solution temperature was set at 25°C and nitrogen purging was performed to displace the dissolved gases from titrating solutions in both experiments.

In the pH-metric titration, a known volume of reagent is added in a step wise mode to the analyte. The change in the measure of potential is determined by the use of two electrodes, an indicator and a reference electrode [8]. The changes in potential vs. pH are graphed; subsequently producing a sigmoidal curve allowing the determination of the pKa of the compound. The pKa value is the pH at half-neutralization point, which represents the center point on the ascending portion of the sigmoidal curve, and was integrated by automated software.

UV-metric method (Spectrophotometric determination): Spectrophotometric determination of pKa was performed using two different instruments, highly sensitive fully automated Sirius T3 in Sirius laboratory and semi-automated NanoDrop 2000c Spectrophotometer in our laboratory. With the implementation of Sirius T3 instrument, the sample was subsequently titrated in a UV-metric triple titration from pH 2.0 - 12.0 at concentrations of 11 -16 μM under aqueous conditions (0.01 M Phosphate Buffer). The ionic strength of the solution was maintained using 0.15 M potassium chloride solution. The buffer allowed for controlled pH as it was adjusted from pH 2.0 to 12.0 using 0.5 M HCl and 0.5 M KOH titrants. The pKa values were determined using Dip-Probe Absorption Spectroscopy (D-PAS) technique. In the D-PAS technique, a fiber optics dip-probe, a UV-light source (Deuterium Light), and a photodiode array detector were used in conjunction with a titrator to capture the spectral changes which arise during the course of titration. Software set up all experimental data in an absorbance matrix based on Beer's law (Absorbance=Concentration × Extinction Coefficient), and Target Factor Analysis (TFA, at the rate of change at which the compound's UV absorbance was the strongest) detected the corresponding pKa values from the absorbance matrix [9].

The second experiment utilized the NanoDrop Spectrophotometer instrument and 1 cm^2 UV-cell. UV spectra were taken throughout the course of titration with an approximate interval of 0.2 pH unit, between pH 3.0 - 11.0 at concentrations of 0.05 mM under aqueous conditions (0.01M Phosphate Buffer). The ionic strength of the solution was maintained using 0.15 M potassium chloride solution. The pH of the buffer was adjusted by addition of 0.5 M HCl and 0.5 M KOH solution using ACCUMET pH meter. The corresponding pKa values were determined by visual evaluation based on changes in the compound's UV spectra pattern with respect to pH. The pKa value corresponds to the pH at which the rate of change of the UV spectra of the compound is the strongest.

RP-HPLC method: To determine the pKa using RP-HPLC, 0.084 mg/ml solution of CNBQ as a test sample, and 0.032 mg/ml solution of chloroquine diphosphate as a reference sample were prepared in a 50:50 solution of acetonitrile and water. The solutions were then injected using the following chromatographic conditions to determine the retention time in different pH values of the mobile phase. The pKa values were determined from the first derivative curve of the retention time vs. pH values.

Chromatographic separation of CNBQ and chloroquine were successfully achieved on a Waters X-Bridge C-18 column (4.6 mm × 250 mm, 5.0 μm particle sizes, part no. 186003117) purchased from Waters Corporation in an isocratic separation mode with a mobile phase consisting of 50% of acetonitrile and 50% of 0.002 M ammonium acetate. The pH of the mobile phase was set within the range of 7.0-12.0 with an interval of 0.2 with 0.1 N HCl and 0.1 N NaOH. The flow rate was maintained at 1.0 ml/min, the column oven temperature was maintained at 25°C, the injection volume was set at 1 μL, and the effluent was monitored at 254 nm.

Results and Discussion

As mentioned previously, the pKa values of CNBQ were determined using the following three methods: pH-metric, UV-metric and RP-HPLC. Considering the variability in the results obtained from the three aforementioned methods, three pKa values: 5.9, 6.6, and 8.7 were estimated for the compound. Table 1 represents the pKa values observed and calculated from these methods.

The potentiometric determination of pKa is a relatively simple, which can be used for any ionizable compound that does not require presence of chromophore groups for pKa determination [10]. pKa determination by this method is the most economical method, in regard to the short duration of experiment and the ease of reproducibility if carried out correctly. Limitations of this method has been the need for a higher amount of drug substance in order to achieve an accurate result and necessity for the use of co-solvents for poorly water soluble basic compound to prevent precipitation of the drug at higher pH values. CNBQ, being insoluble at higher pH, requires extrapolation using methanol as a co-solvent. Use of a co-solvent impacts the pKa values [11], resulting in a higher level of variability than normally observed.

Table 2 represents the data obtained by pH-metric method. pKa values of CNBQ was determined using Yasuda-Shedlovsky standard extrapolation method [12] as shown in Figure 2, in which X-axis plots the inverse of the dielectric constant of the water-solvent mixture at the experimental percentage of solvent and Y-axis plots the psKa + $\log[H_2O]$; where, psKa is cosolvent dissociation constants, and $[H_2O]$ is the molar water concentration of the given solvent mixture. The extrapolated value represents the pKa value at 100% water concentration (0% co-solvent). Sirus T3 software could conveniently calculate the extrapolated pKa values in water using this method. The pKa values of CNBQ were determined by this method to be 5.8, 6.4 and

pKa Values	pK_{a1}	pK_{a2}	pK_{a3}
UV-metric method (Automated)	5.87 ± 0.01	6.60 ± 0.01	8.73 ± 0.01
pH-metric method (Automated)	5.80 ± 0.01	6.35 ± 0.11	8.43 ± 0.03
UV-metric method (Manual)	5.99	6.81	8.67
RP-HPLC method	---	---	8.8

Table 1: pKa values of CNBQ by different methods

pKa Values	%Methanol	Dielectric Constant	$[H_2O]$	psK_{a1}	psK_{a2}	psK_{a3}
Experiment 1	29.2	65.9	36.4	5.4	6.0	8.3
Experiment 2	38.9	61.5	30.6	5.2	5.9	8.2
Experiment 3	49.5	56.6	24.7	4.9	5.6	8.2

Table 2: pKa values of CNBQ by pH-metric method using Sirius T3

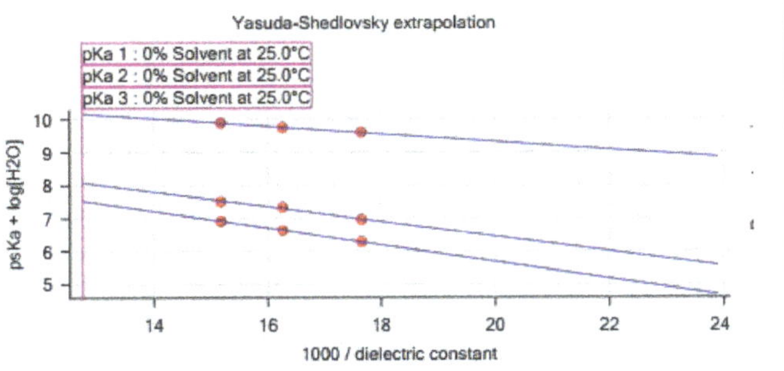

Figure 2: Yasuda-Shedlovsky standard extrapolation in pH-metric method by Sirius T3

8.4 with a R^2 value of 0.99. Another attempt was made to determine the pKa value of CNBQ using a Metrohm potentiometer in our laboratory. In this experiment, methanol and acetonitrile were used as co-solvents. Although the experiment was performed in a controlled setting, it did not produce any reproducible pKa value due to poor solubility of the basic compound causing precipitation of the samples in the solution as pH is increased, and lack of sensitivity of the instrument. In general, a higher quantity of the sample is required to determine the pKa using an ordinary potentiometer that causes precipitation of the sample and thus inaccurate results. However, due to recent advances in technology (Sirius T3, as discussed above), pKa values can now be determined in a more economical approach using smaller amounts of sample, which is very important in the early stage of drug development (Figure 2 and Table 2) [13].

RP-HPLC method was developed to overcome the challenges posed by the properties of current drugs with low solubility. This method is attractive due to its simplicity and its ability to use a variety of isocratic HPLC systems with the use of less drug material [14]. pKa is determined by analyzing the change in retention time of an analyte vs. the pH of the mobile phase of the respective retention time. Utilization of RP-HPLC method has proven to be successful in determining pKa values that resemble literature pKa values [15,16]. The fundamental principle of this method is the variation of retention time based on the pH of the mobile phase. However, obtaining a sharp peak shape and accurate retention time at each point using an isocratic mobile phase in the pH range of 2.0 to 12.0 is quite difficult when dealing with polyprotic basic compound having multiple pKa values. Improvement of peak shape can be accomplished by addition of an organic modifier and an ion pairing reagent to the mobile phase, but addition of these reagents will result in inaccurate pKa values [14]. CNBQ is a polyprotic basic drug and shows a broader peak shape in the lower pH ranges, which are not reproducible. It was revealed in this experiment that good peak shape for CNBQ can be obtained within the pH range of 7.0 to 12.0 as shown in Figure 3. Thus, the pKa determination was performed in the pH range of 7.0 to 12.0, using acetonitrile as an organic phase and acetate buffer as the aqueous phase. Because of the structural similarities between chloroquine and CNBQ, chloroquine was used as a reference throughout the entire experiment. The pKa values for both compounds were determined from the first derivative curve of the retention time vs. pH values (Table 3 and Figure 4). The pKa values for chloroquine obtained by RP-HPLC method were 8.4 and 10.2 (compared to the literature values of 8.37 and 10.49, respectively) [12]. In regards to CNBQ, RP-HPLC method only produced one pKa value of 8.8, which proves the difficultly in determining multiple pKa values using RP-HPLC method beyond the range of 7-12 as mentioned before figures 3 and 4.

The UV-metric method measures absorption, emission, or scattering of electromagnetic radiation with respect to changes in pH, leading to changes in the UV spectra of the molecule [17]. These changes in the multi-wavelength UV spectra can be observed, if the disposition of electrons in chromophore containing conjugated double bonds, carbonyl groups, and other UV absorbing groups change with the molecule's ionization state. There are several methods to calculate the pKa values from UV absorbance data vs. pH [13,17-19]. There are two disadvantages of the UV-metric method. First, if the compound is too basic it may precipitate out; however, this can be overcome if the experiment is performed in the presence of a co-solvent, but would require extrapolation to obtain the final results. Second, if the compound possesses no pH-active chromophore, then the UV-metric method cannot be applied making this the major disadvantage. If the sample has a UV response and exhibits no/less precipitation, then it is possible to conduct the experiment under aqueous conditions using only small amount of the compound [13]. This results in pKa values which are considered more accurate, because there is no interference of a co-solvent and no extrapolation is needed to obtain the final results. In this experiment using automated software of Sirius T3 instrument, TFA detected the corresponding pKa values at which the rate of change in the compound's UV absorbance is the strongest. CNBQ exists in four different species; BH3, BH2, BH, and B at depending on the pH of the medium. Figure 5 represents the spectra for four different species of CNBQ obtained by Sirius T3 instrument. Utilizing this method, three pKa values of 5.87, 6.60, and 8.73, were observed from the spectroscopic data. Table 4 represents the data obtained from automated UV-metric method.

Figure 6 represents the spectra of different species of CNBQ obtained from NanoDrop 2000c Spectrophotometer at different pHs. In this figure, pKa_1, pKa_2 and pKa_3 represent the strongest changes of spectrum from one species to another species. The corresponding pKa values of 5.95, 6.83, and 8.67, were determined by visual evaluation based on changes in the compound's UV spectra pattern with respect to pH (Table 5 and Figure 6). When the pKa values obtained by this manual UV-metric method, conducted in our lab, plotted against those obtained by fully automated Sirius T3 instrument, it exhibited a straight line with R^2 value of 0.993 (Figure 7), signifying the accuracy and validity of the results. Figure 8 represents the four different spectra for the species of CNBQ obtained by NanoDrop 2000c Spectrophotometer (Table 5).

Figure 9 represents the distribution of species of CNBQ in different pH conditions by Henderson-Hasselbalch equation. From the species distribution it is clearly shown that the lone pairs of electrons of nitrogen atoms in the aliphatic and aromatic ring of CNBQ will fully accept three protons (BH3 Species) at lower pH (approximately pH 4.0). CNBQ will be partially ionized at an approximate pH of 6.0 (BH2

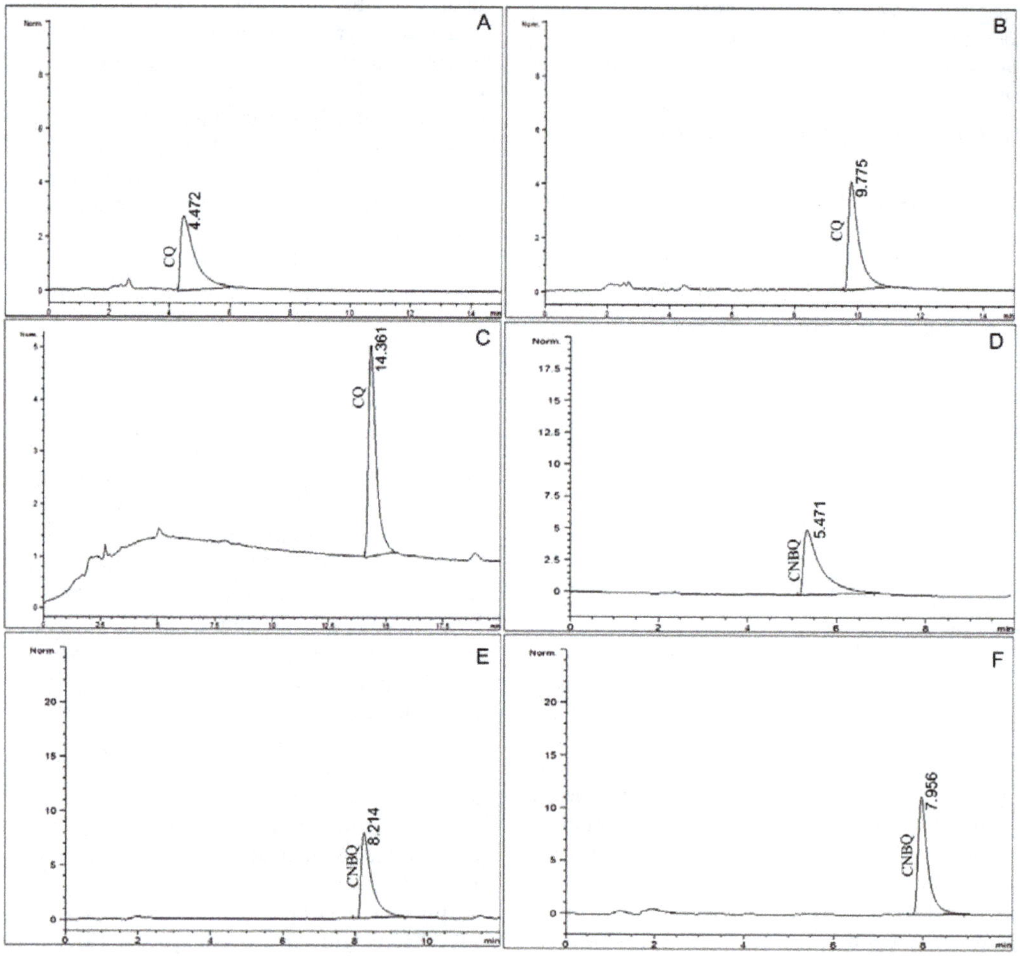

Figure 3: HPLC Chromatograms; A: CQ at pH 7.3, B: CQ at pH 9.5, C: CQ at pH 11.1, D: CNBQ at pH 7.5, E: CNBQ at pH 9.4, F: CNBQ at pH 11.1

Chloroquine			CNBQ		
RT	pH	ΔRT/ΔpH	RT	pH	ΔRT/ΔpH
4.472	7.31	---	5.338	6.97	---
4.460	7.50	0.179	5.471	7.49	0.256
4.528	7.88	0.311	5.647	7.64	1.173
4.584	8.06	0.926	6.000	7.88	1.471
4.797	8.29	1.931	6.341	8.13	1.364
5.048	**8.42**	**3.700**	6.774	8.41	1.546
5.455	8.53	3.231	7.094	8.60	1.684
5.875	8.66	3.604	7.414	8.75	2.133
7.533	9.12	5.468	**7.734**	**8.84**	**3.556**
9.775	9.53	5.715	8.023	9.14	0.963
12.518	10.01	3.894	8.214	9.42	0.682
13.180	**10.18**	**6.417**	8.278	9.64	0.291
13.950	10.30	1.477	8.286	9.74	0.080
14.334	10.56	0.055	8.284	9.97	-0.009
14.361	11.05	-0.178	7.956	11.07	-0.298

Table 3: Retention time (RT) and pH of the mobile phase

Species) and at an approximate pH of 8.0 (BH Species), accepting two and one proton(s), respectively. CNBQ will be completely unionized (B Species) at higher pH (approximately pH 10.0).

To validate the pKa values obtained using in-house RP-HPLC method, chloroquine was used as a reference. The pH-metric method

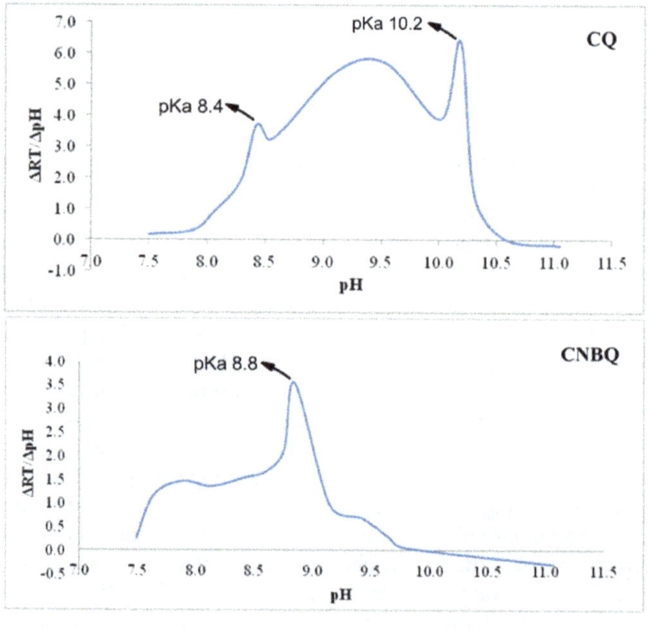

Figure 4: First derivative curves for pKa determination; Top, chloroquine, bottom, CNBQ

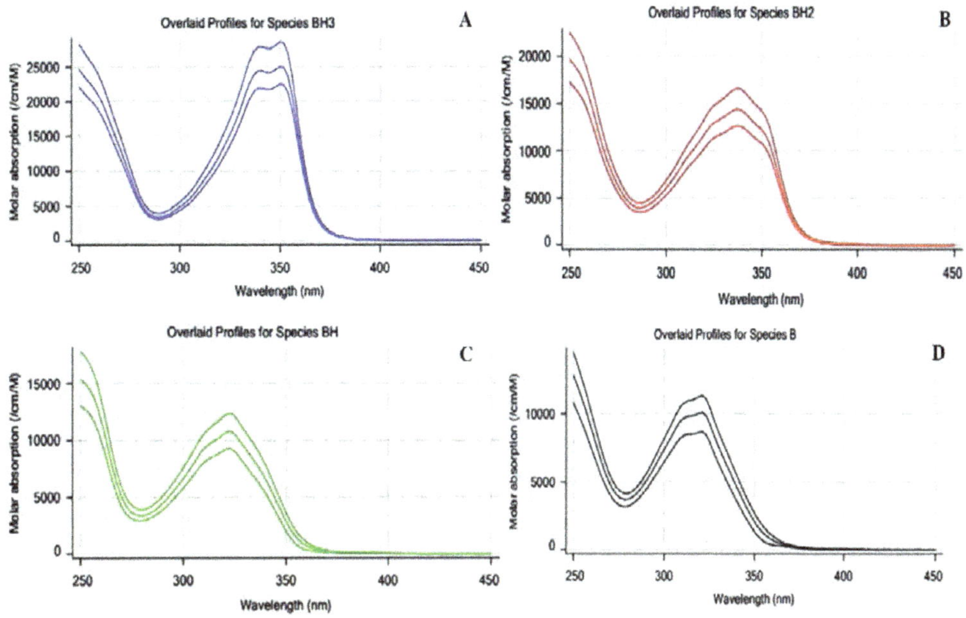

Figure 5: Spectra of different Species of CNBQ obtained from Sirius T3; A: BH3, B: BH2, C: BH and D: B

pKa Values	pK_{a1}	pK_{a2}	pK_{a3}
Experiment 1	5.87	6.59	8.73
Experiment 2	5.89	6.61	8.73
Experiment 3	5.86	6.60	8.73
Average	5.87	6.60	8.73
Std Dev	0.012	0.007	0.004

Table 4: pKa values of CNBQ by UV-metric method using Sirius T3

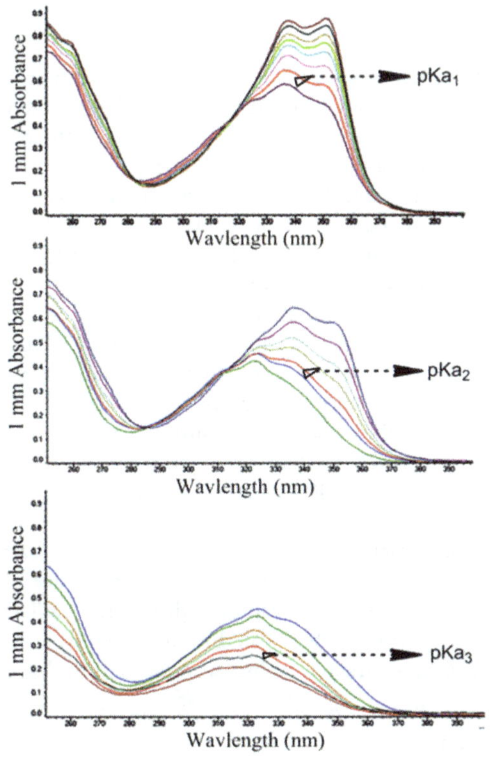

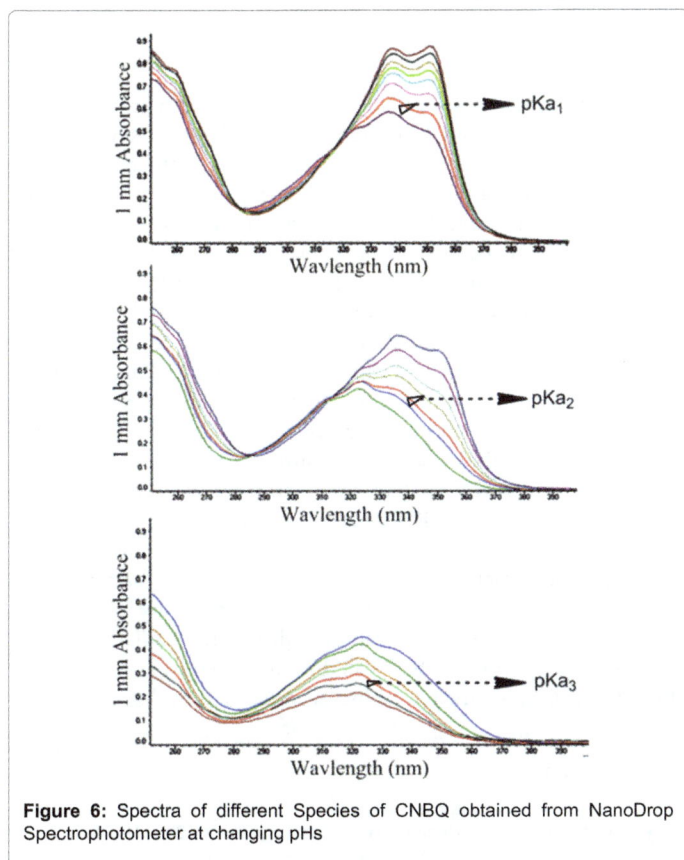

Figure 6: Spectra of different Species of CNBQ obtained from NanoDrop Spectrophotometer at changing pHs

pKa Values	pK_{a1}	pK_{a2}	pK_{a3}
pH of the two consecutive Spectra	5.85	6.74	8.56
	6.05	6.91	8.77
Average	5.95	6.83	8.67

Table 5: pKa values of CNBQ by UV-metric method using NanoDrop spectrophotometer

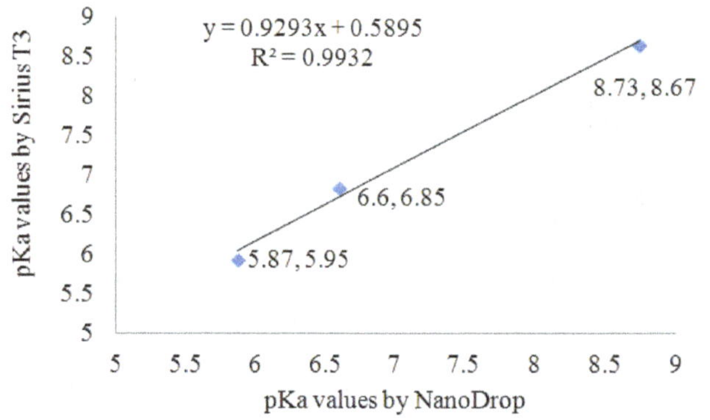

Figure 7: A plot of the pKa values obtained using NanoDrop spectrophotometer against those measured by Sirius T3

and the D-PAS technique for UV-metric method has already been validated using several compounds by Sirius Analytical Ltd. [9,13]. Therefore, UV-metric method using visual inspection was validated by comparing the results obtained at both laboratories (Figure 7). The pKa values obtained from both methods were within ± 0.2 with a R^2 value of 0.99. These minor variations are due to changes in instruments, differences in analytical reagents, and different analyst. Since the compound has excellent UV signal and the analysis was performed under aqueous conditions, it did not require an extrapolation. Hence, UV-metric method stands out as an effective method for the determination of multiple pKa values of a poorly water-soluble, polyprotic basic compound having good UV signal.

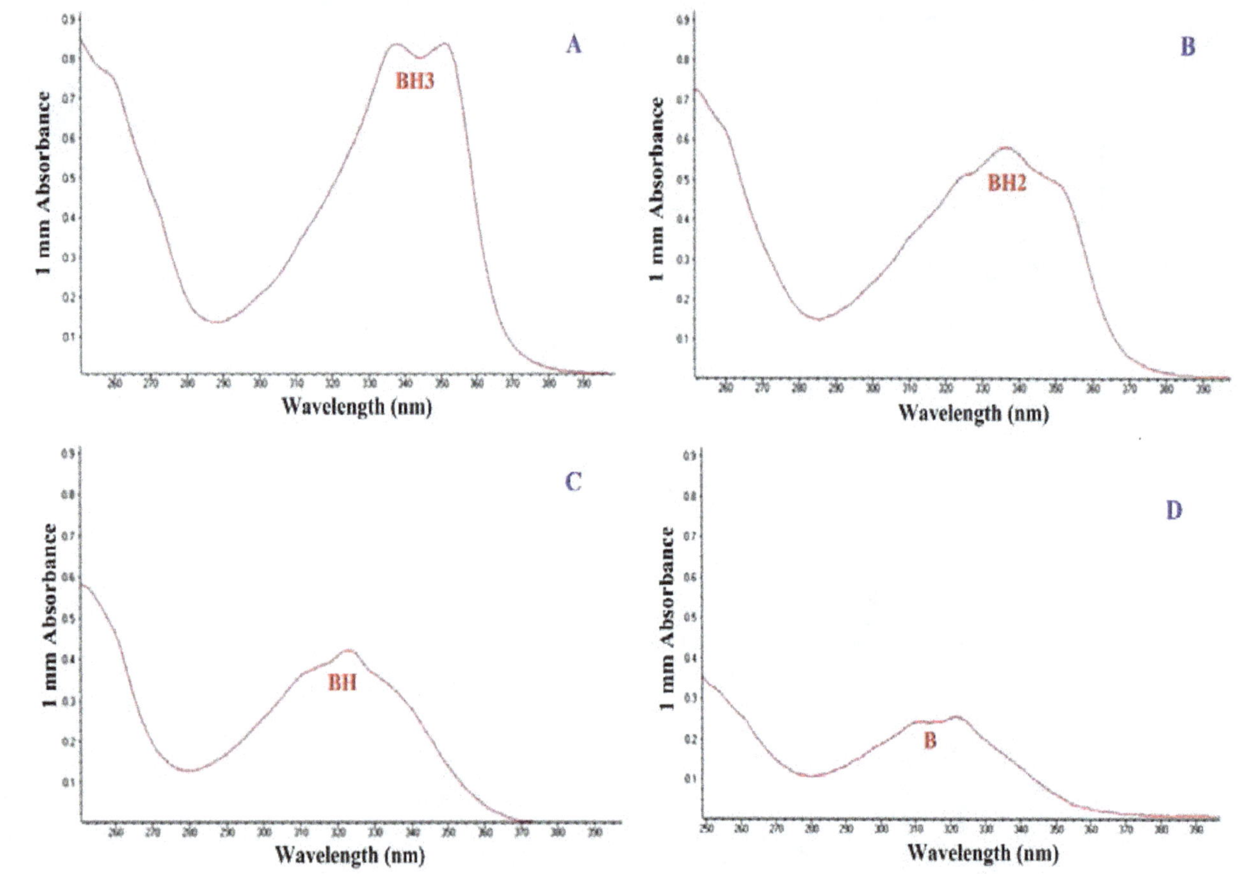

Figure 8: Spectra of different Species of CNBQ obtained from NanoDrop Spectrophotometer; A: BH3, B: BH2, C: BH and D: B

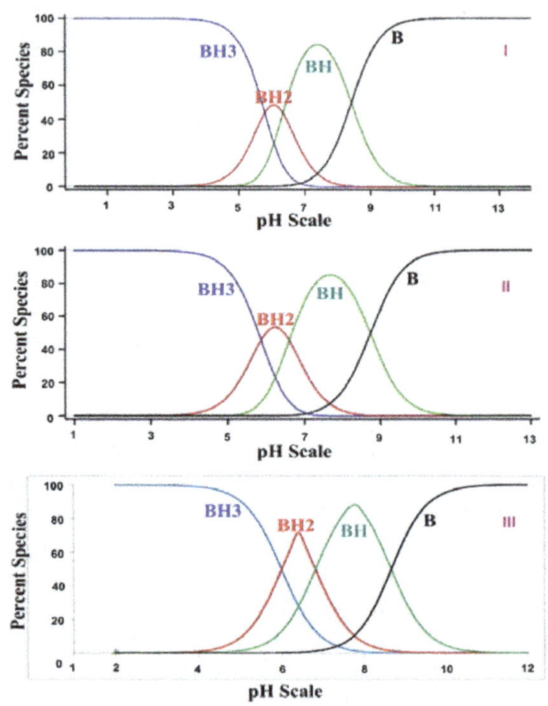

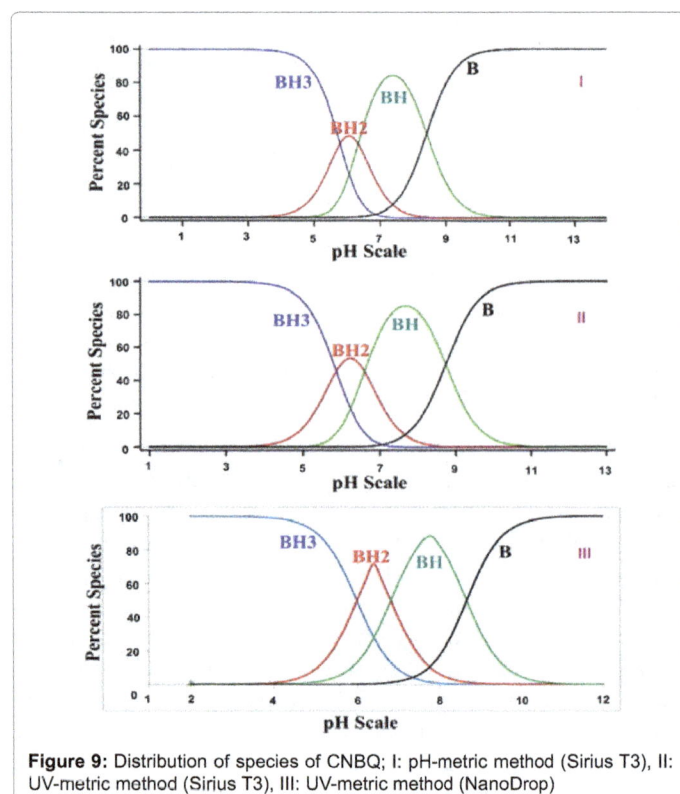

Figure 9: Distribution of species of CNBQ; I: pH-metric method (Sirius T3), II: UV-metric method (Sirius T3), III: UV-metric method (NanoDrop)

Conclusion

From the pKa values obtained, it is apparent that CNBQ is a polyprotic basic drug in which the lone pairs of electrons on nitrogen atoms in aliphatic and aromatic ring will accept three protons in different physiological pH conditions. It is imperative that the method chosen is the most accurate method in determining these pKa values. Considering all the methods discussed, UV-metric method is the most suitable in determining the pKa values for the drug lead CNBQ due to its basicity, inherent insolubility, and multiple pKa values. The RP-HPLC method possesses limitations, in regards to the pH range (pH 7.0 to 12.0 as observed in our experiment), of the mobile phase, making it incapable of determining pKa values that fall outside the range. Use of co-solvents complicates results obtained from both RP-HPLC and pH-metric methods, as the co-solvent will produce a pKa value that is not accurate compared to the actual value. Furthermore, results obtained from pH-metric and RP-HPLC methods require extrapolation making them variable and less accurate than those obtained from the UV-metric method, which does not require co-solvents or extrapolation. In conclusion, RP-HPLC and pH-metric methods are not suitable methods for determining pKa values of drug leads similar to CNBQ. However, UV-metric method can be conveniently and accurately utilized to determine the pKa values of drug leads of this class in our laboratory.

Acknowledgment

This project was supported by the National Institute of General Medical Sciences of the National Institutes of Health through Grant Number 8P20GM103447.

References

1. Manallack DT (2007) The pKa distribution of drugs: Application to drug discovery. Perspect Medicin Chem 1: 25-38.
2. http://www.merckmanuals.com/
3. Chen XQ, Antman MD, Gesenberg C, Gudmundsson OS (2006) Discovery pharmaceutics-challenges and opportunities. AAPS J 8: E402-E408.
4. Khan MOF, Levi MS, Tekwani BL, Khan SI, Kimura E, et al. (2009) Synthesis and antimalarial activities of cyclen 4-aminoquinoline analogs. Antimicrob Agents Chemother 53: 1320-1324.
5. Rudraraju AV, Hossain MF, Shrestha A, Amoyaw PNA, Tekwani BL, et al. (2014) In vitro metabolic stability study of new cyclen based antimalarial drug leads using RP-HPLC and LC-MS/MS. Mod Chem appl 2: 129.
6. Reijenga J, Van hoof A (2013) Development of methods for the determination of pKa values. Anal Chem Insights 8: 53-71.
7. Amoyaw PNA, Pham K, Cain AN, McClain JM, Khan MOF, et al. (2014) Synthesis of novel tetraazamacrocyclic bisquinoline derivatives as potential antimalarial agents. Curr Org Synth 11: 916-921.
8. Qiang Z, Adams C (2004) Potentiometric determination of acid dissociation constants (pK_a) for human and veterinary antibiotics. Water Res 38: 2874-2890.
9. Tam KY, Takács-Novak K (2001) Multi-wavelength spectrophotometric determination of acid dissociation constants: a validation study. Analytica Chimica Acta 434: 157-167.
10. Demiralay EC, Yılmaz H (2012) Potentiometric pK_a determination of piroxicam and tenoxicam in acetonitrile-water binary mixtures. SDU J Sci 7: 34-44.
11. Kamble AD, Barhate, VD, Salunke MH, Narasimham L (2011) Automated potentiometric titration method for determination of pKa values: An application to hydroxy chloroquine sulphate. J Pharm Res 4: 3794-3797.
12. http://www.sirius-analytical.com
13. Box KJ, Comer JEA (2008) Using measured pKa, logP and solubility to investigate supersaturation and predict BCS Class. Curr Drug Metab 9: 869-878.
14. Manderscheid M, Eichinger T (2003) Determination of pKa values by liquid chromatography. J Chromatogr Sci 41: 323-326.
15. Wiczling P, Markuszewski MJ, Kaliszan R (2004) Determination of pKa by pH gradient reversed-phase HPLC. Anal Chem 76: 3069-3077.
16. Huo H, Li T, Zhang L (2013) pKa determination of oxysophocarpine by reversed - phase high performance liquid chromatography. Springerplus 2: 270.
17. Jorge Ramos (2014) UV-Vis Spectrometry, pKa of a dye. 1-7.
18. Chemagination (2009) How to measure pKa by UV-vis spectrophotometry.
19. Martínez CHR, Dardonville C (2013) Rapid determination of ionization constants (pKa) by UV spectroscopy using 96-well microtiter plates. ACS Med Chem Lett 4: 142-145.

Synthesis Characterization and Investigation of Photocatalytic Activity of $H_3PMO_{12}O_{40}/TiO_2/HY$ Nanocomposite for Degradation of Methyl Orange in Aqueous Media

Moosavifar M* and Jafarbahmani M

Department of Chemistry, Faculty of Science, University of Maragheh, Iran

Abstract

Molybdophosphoric acid encapsulated into dealuminated zeolite Y (DAZY) was prepared by template synthesis method. Incorporation of TiO_2 into nanocage of DAZY was performed by impregnation method. The obtained photo catalyst (HPA/TiO_2/DAZY) was characterized by FT-IR, UV–Vis, FESEM, XRD, EDS and ICP technique. This catalytic system was investigated in the photodegradation of methyl orange. The obtained results reveal that the photo catalyst performance depends to photo catalyst loading and TiO_2/(HPA/HY) ratio. The photocatalytic activity of molybdophosphoric acid encapsulated into zeolite cage enhanced with impregnation of TiO_2 into nanocage of dealuminated Y zeolite so that complete removal of methyl orange was occurred.

Keywords: Heterogeneous catalyst; Methyl orange; Photodegradation; Molybdophosphoric acid; Dealuminated zeolite

Introduction

The photocatalytic degradation of organic compounds present in wastewater has attracted a great deal of attention from the viewpoint of green chemistry [1-3]. There are several methods for the removal of organic pollutants from wastewater including chemical and biological oxidation [4] adsorption [5] electrochemical [6] ion exchange [7] and membrane separation [8]. However, all these methods have many drawbacks such as expensive, commercially unattractive, and generate secondary contamination [9]. Meanwhile, photocatalytic degradation is another alternative method for removal of inorganic and organic contaminations in both water and air. TiO_2 is one of the best candidates for this purpose. However, photocatalytic activity is restricted because of high band-gap energy, difficulty in filtering and costly. Therefore, immobilization on inorganic support is good manner to overcome these advantages. On the other hand, the combination of TiO_2 and various supports such as silica, carbon materials and zeolites gives the unique properties extremely small size semiconducting particles lead to catalytic reactions which very different from photo electrochemical reactions observed on bulk TiO_2 semiconducting powders [10-19]. However, in order to improve the efficiency, the usage of photosensitive molecules is a good manner. Heteropoly acids are known to be active, exhibit the unique property of structural stability and redox activity which facilitate photocatalytic activity because of suitable HOMO-LUMO gap [20-24]. Meanwhile, by encapsulating of hetero poly acid, the specific surface area increased which in turn, affected on catalytic activity [20]. observed enhanced photocatalytic activity in the conversion of organic molecules in the presence of hetero polyacid $(PW_{12}O_{40})^{3-}$ (HPA) encaged into SBA-15 [25]. Anandan et al. also investigated increased photocatalytic activity heteropolytungstic acid-encapsulated TiSBA-15 in the degradation of methyl orange [24]. Motivated by the above work, we first modified Y zeolite by post-synthesis method with EDTA treatment in order to improve the surface hydrophilic–hydrophobic properties of support materials. This, in turn, increases the ability of absorption of dye contaminant on zeolite surface and improved the photocatalytic activity. Then molybdophosphoric acid was synthesized into nanocage of dealuminated Y zeolite by ship-in-a-bottle method. Finally, TiO_2 incorporate into zeolite cage by impregnation method. The photocatalytic activity of prepared catalyst was investigated in the degradation of methyl orange.

Experimental

Materials and reagents

All materials were of the commercial reagent grade and were used without any purification. NaY zeolite was purchased from Sigma Aldrich and was dealuminated by post-synthesis method with EDTA treatment. In this case, the controlling of the rate of EDTA addition, EDTA amount and reaction temperature is critical to prevention amorphization [26].

Preparation of zeolite-encapsulated HPA/TiO_2/DAZY

Synthesis of HPA/DAZY: In a typical procedure, NaY zeolite was dealuminated by chemical operation with EDTA treatment. This not only increased the zeolite structure stability in the presence of acidic HPA, but also modified surface hydrophilic–hydrophobic properties and therefore, absorption of organic dye on support is facilitated. 12-molybdophosphoric acid was prepared by ship-in-a-bottle method to the reported procedure [27].

Preparation of HPA/TiO_2/ DAZY: This catalytic system was prepared by impregnation method. In a typical procedure, 0.1 g HPA/DAZY was added to the suspension containing 0.01 g TiO_2 in 50 mL distilled water and was stirred for 24 h. Then the solvent was evaporated under vacuum and dried at 100 °C for 12 h. Then the samples were calcinated at 473 K for 3 h.

General procedure for the photodegradation of methyl orange

In a typical run, the suspension containing 0.02 g catalyst and 50 mL aqueous solution of methyl orange (5×10^{-5} g/L) was stirred

***Corresponding author:** Moosavifar M, Department of Chemistry, Faculty of Science, University of Maragheh, Iran, E-mail: m.moosavifar90@gmail.com first in the dark to establish adsorption/desorption equilibrium to eliminate the error due to any initial adsorption effect. Then irradiation experiments were carried out in a homemade reactor. The suspension was magnetically stirred throughout irradiation. To monitor the organic pollutants methyl orange degradation process, about 3 mL of the suspension was withdrawn and the photo catalysts were separated from the suspensions by 0.45 μm disc. The photodegradation process was monitored by its characteristic absorption band at 465 nm using a double-beam UV-Visible spectrometer.

Results and Discussion

Preparation and characterization of HPA/TiO_2/DAZY

NaY was dealuminated by post-synthesis method with EDTA treatment. In this case, not only mesoporous structure was formed, but also hydrophobicity of surface increased and therefore, the absorption of organic dye on zeolite surface enhanced [13]. Synthesis of HPA/ DAZY was carried out by template synthesis method. In this method, HPA act as a template and zeolite was formed around it. TiO_2 inserted into nanocage of zeolite by impregnation method. The prepared catalyst was characterized by infrared spectroscopy (FT-IR), X-ray diffraction (XRD), field emission scanning electron microscopy (FESEM), energy dispersive spectroscopy (EDS) and ion couple plasma (ICP) technique. The content of Mo and Ti in the synthesized samples was obtained using ICP analysis which equal with 0.0056 g and 0.0055 g /g of the supported catalyst for Mo and Ti respectively. The FT-IR spectra confirm the existence of MPA into nanocage of dealuminated Y zeolite. In the FT-IR spectra of HPA six characteristic peaks are observed at 1617, 1065, 966, 870, 787, 599 cm^{-1}. When the HPA encapsulated into DAZY, these peaks are observed at 1631, 1084 (P–O), 946 (Mo-O_t), 795 (Mo–O_e–Mo) (O_t: terminal oxygen, O_e: edge-sharing oxygen) 710, 559, 465 cm^{-1} (Figure 1). Meanwhile, the presence of absorption bands at 920, 895 and 860 cm^{-1} is related to Ti-O, Ti=O stretching vibration and Ti–O–Ti linkage vibration, respectively (Figure 1) These peaks were disappeared in the spectrum because of overlapping with to that of zeolite and HPA peaks (Figure 1). The XRD pattern of catalyst is brought in Figure 2. In this case, appearance of peaks in 2θ=28 is related to TiO_2 in anatase form which proved TiO_2 inserted on zeolite support. In addition, the XRD pattern of catalyst is identical to that of Y zeolite except with appearance peaks in the region which related to the HPA compound [28,29]. This proved Y zeolite act as support and can be involved HPA and TiO_2. The EDS results proved this claims. The field emission scanning electron micrograph (FESEM) of the HPA/TiO_2/ DAZY indicated the cubo-octahedral units that proved zeolite structure was preserved after the insertion of HPA and TiO_2 in the nanocage of zeolite. In addition the presence of species in nanometer size proved the insertion of TiO_2 into zeolite nanocage (Figure 3). FESEM-EDX spectrum of HPA/TiO_2/ DAZY showed the presence of component of photocatalyst. It proved the formation of metallo complexes into cages of zeolite (Figure 4). In order to detect the photocatalytic activity of HPA/TiO_2/ DAZY, degradation of methyl orange was tested as a function of different experimental parameters which are presented in Figures 5-7.

Effect of catalyst amount: The effect of catalyst amount on removal of methyl orange was investigated by varying the catalyst amount from 0.02 to 0.07 g and the result is brought in Figure 3. It can be seen that with increasing in amount of the catalyst, degradation of methyl orange increased because of increasing in absorbed of dye molecule. Optimum degradation was obtained with 0.03 g catalyst because of high removal of dye contaminant. At high catalyst loading the removal of dye is not significantly great. In higher loading catalyst, light absorption has been increased that can be related to light scattering which was occurred by the catalyst particle and the absorption phenomena was happened with catalyst instead organic dye. Therefore, the photocatalytic activity is reduces and the absorption band increased.

Effect of TiO_2/(HPA/DAZY) ratio: For this purpose variety of TiO_2/(HPA/DAZY) ratio including 1:5, 1:10 and 1:16 was tested in the

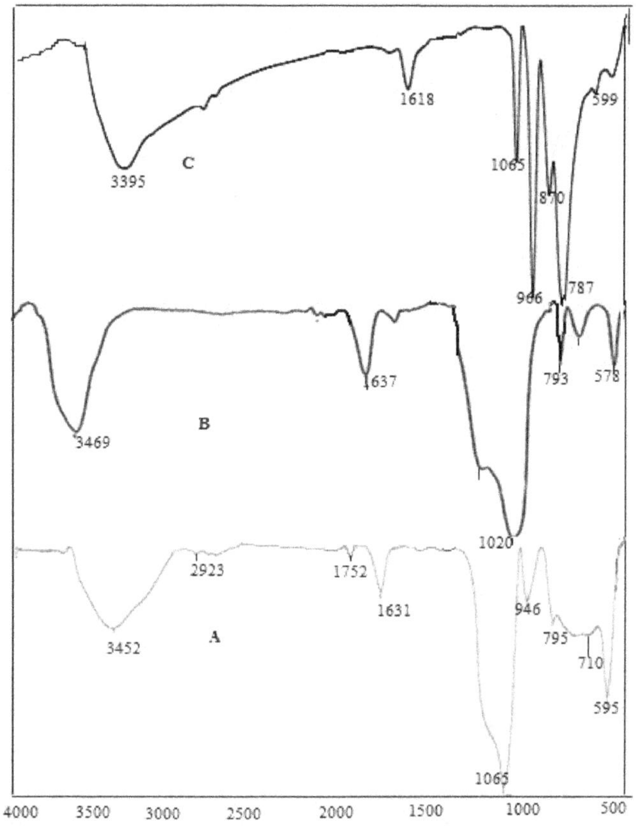

Figure 1: FT-IR spectra of A) HPA/TiO_2/DAZY B) HY C) $H_3PMo_{12}O_{40}$.

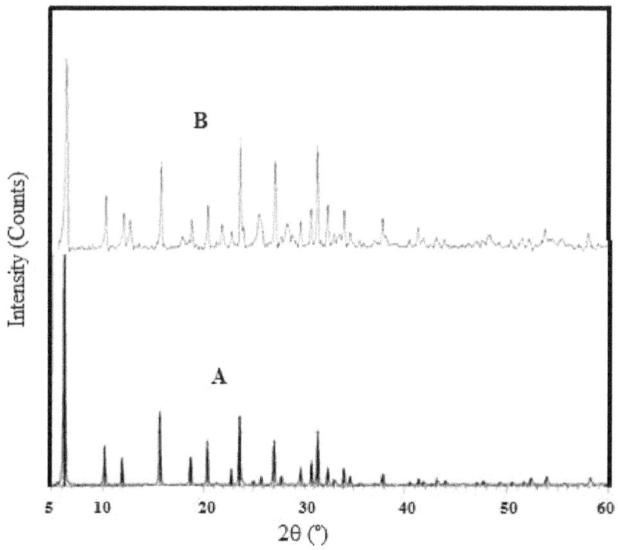

Figure 2: XRD patterns of a) NaY b) HPA/TiO_2/DAZY.

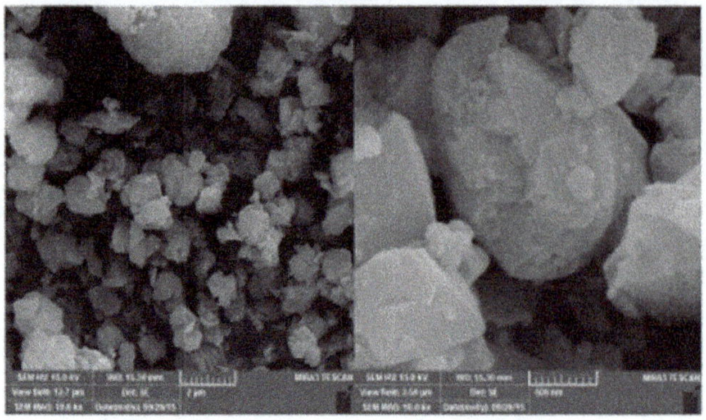

Figure 3: FESEM micrograph of HPA/TiO_2/DAZY.

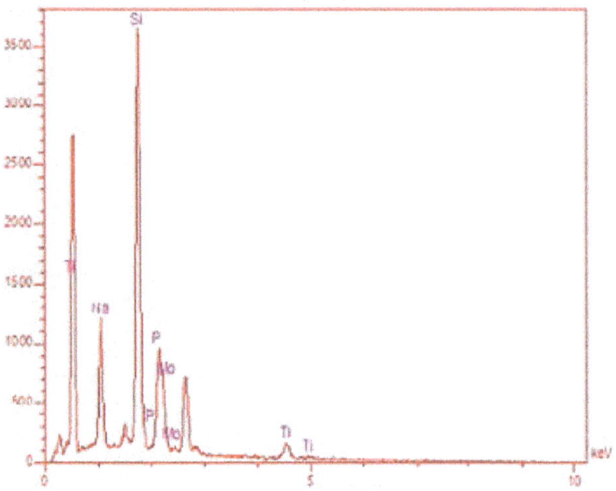

Figure 4: EDS analysis of HPA/TiO_2/DAZY.

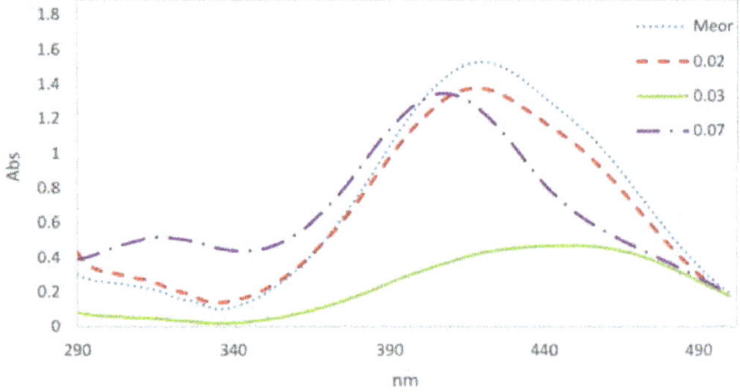

Figure 5: Effect of TiO_2/(HPA/DAZY) ratio on the photodegradation of MO. Concentration of methyl orange was 5.0 × 10^{-5} M.

photodegradation of dye molecules. The results was showed maximum photocatalytic activity was obtained with 1:10 ratio due to high removal of dye molecules. It seems at high TiO_2/(HPA/DAZY) ratio, the contact between photosensitizer and dye molecules reduces which related to decreasing in the interfacial area between the reaction solution and the photo catalyst. Therefore, the removal of dye is reducing [30]. Under the optimum conditions, decolorization of MO was investigated with HPA/TiO_2/DAZY as a function of irradiation time. Figure 4 shows the UV–Vis spectra of methyl orange before and after irradiation in the presence of HPA/TiO_2/DAZY. Methyl orange itself in the absence of HPA/TiO_2/DAZY was photochemically inert as observed by no change in the absorption spectrum. However, in the presence of photo catalyst, the absorption spectrum of methyl orange significantly reduces and after 1 hour, the decolorization of MO is completed.

Mechanism of reaction: The photodegradation of Methyl Orange was initiated by the illumination from a photosensitizer with radiation of energy which it provides the band gap energy of the semiconductor and therefore, separated the photo generated electrons (e-) and holes (h+) in the conduction band (CB) and valence band (VB), respectively. The photoinduced interfacial electron leads to photoreduction of MeOr and converted to photo decoloration product (Figure 8).

Conclusion

In this study, the preparation of HPA/TiO_2/DAZY catalyst was performed using template synthesis and impregnation methods respectively. The physicochemical properties of composites characterized by XRD, FT-IR, FESEM, ICP and EDS technique. XRD studies showed that the appearance of peaks in the regions of TiO_2 and HPA which indicated incorporation of TiO_2 and HPA in nanopores of zeolite. The catalyst displays an efficient photoactivity for the degradation methyl orange under UV illumination. Results indicated HPA/ DAZY hasn't high efficiency in the discoloration of MO and the presence of TiO_2 is necessary for removal of dye molecules. This catalytic system could be successfully recovered by filtration and reused for several times. Therefore, this system would be useful for the cleaning of wastewater containing organic dye using UV-light.

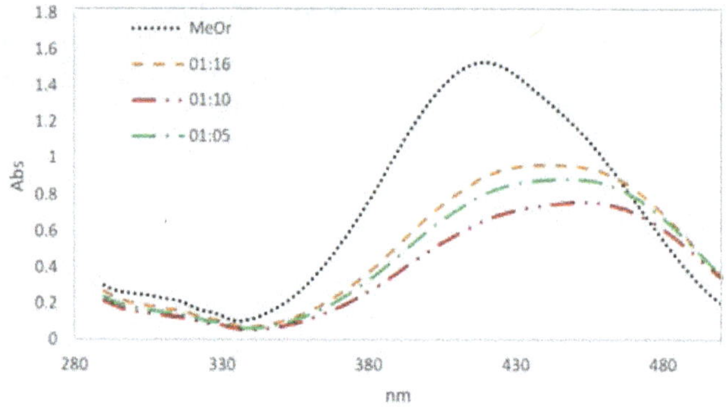

Figure 6: Effect of catalyst amount on the photodegradation of MO. Concentration of methyl orange was 5.0 × 10^{-5} M.

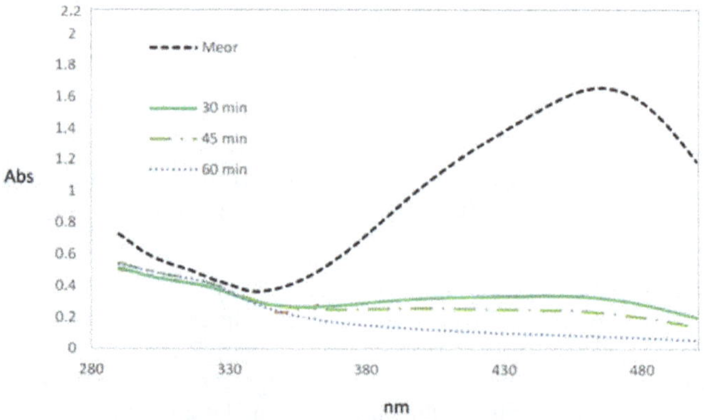

Figure 7: Absorption spectral changes of the filtered solution of methyl orange after irradiation in the presence of HPA/TiO_2/HY. Concentration of methyl orange was 5.0 × 10^{-5} M.

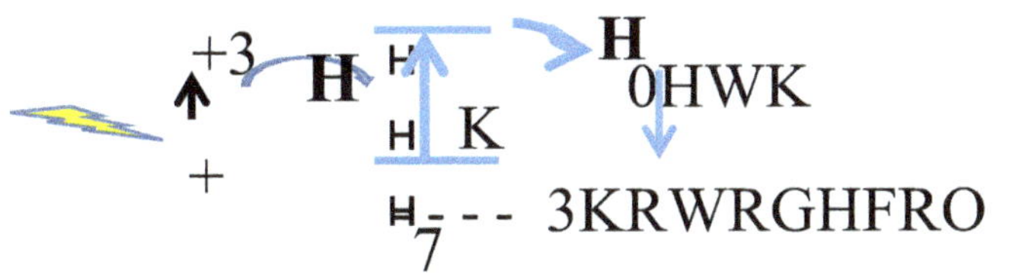

Figure 8: The proposed mechanism for photo decoloration of methyl orange.

References

1. Ollis DF (1993) Photocatalytic Purification and Treatment of Water and Air. p: 820.
2. Greem KJ, Rudham R (1993) Photocatalytic oxidation of propan-2-ol by semiconductor–zeolite composites. J Chem Soc Faraday Trans 89: 1867.
3. Mattews RW (1988) An adsorption water purifier within situ photocatalytic regeneration. Catal J 113: 549.
4. Sarria V, Deront M, Peringer P, Pulgarin C (2003) Degradation of a bio recalcitrant dye precursor present in industrial wastewaters by a new integrated iron (III) photo assisted biological treatment. Applied Catalysis B: Environmental 40: 231-246.
5. Abidin MAZ, Jalil AA, Triwahyono S, Adam SH (2011) Recovery of gold(III) from an aqueous solution on to a durio zibethinus husk. Biochem Eng J 54: 124-131.
6. Jalil AA, Triwahyono S, Razali NAM, Hairom NHH (2010) Adsorption of methyl orange from aqueous solution onto calcined Lapindo volcanic mud. J Hazard Mater 174: 581.
7. Wu JS, Liu LH, Chu KH, Suen SY (2008) Removal of cationic dye methyl violet 2B from water by cation exchange membranes. J Membrane Sci 309: 239-245.
8. Mozia S, Tomaszewska M, Morawski AW (2005) A new photocatalytic membrane reactor (PMR) for removal of azo-dye Acid Red 18 from water. Appl Catal B Environ 59: 131-137.
9. Sapawe N, Jalil AA, Triwahyono S (2013) One-pot electro-synthesis of ZrO2–ZnO/HY nanocomposite for photocatalytic decolorization of various dye-contaminants. Chem Eng J 225: 254-265.
10. Takeuchi M, Hidaka M, Anpo M (2012) Efficient removal of toluene and benzene in gas phase by the tio2/y-zeolite hybrid photocatalyst. J Hazard Mat 133: 237-238.
11. Chen H, Matsumoto A, Nishimiya N, Tsutsumi K (1999) Preparation and characterization of TiO2 incorporated Y-zeolite. Colloids Surf A 157: 295-305.
12. Anpo M, Takeuchi M, Ikeue K, Dohshi S (2002) Design and development of titanium oxide photo catalyst operated under visible and uv irradiation the application of metal ion implantation techniques to semiconducting tio2 and ti/zeolite catalyst. Curr Opin Solid State Mater Sci 6: 381-388.
13. Kamegawa T, Kido R, Yamahana D, Yamashita H (2013) Design of TiO2-zeolite composites with enhanced photocatalytic performances under irradiation of UV and visible light. Microporous Mesoporous Mater 165: 142-147.
14. Bai B, Zhao J, Feng X (2003) Preparation and characterization of supported photocatalysts: HPAs/TiO_2/SiO_2 composite. Mater Lett 57: 3914-3918.
15. Kim Y, Yoon M (2001) TiO2/Y-Zeolite encapsulating intramolecular charge transfer molecules: a new photo catalyst for photoreduction of methyl orange in aqueous medium. J Mol Catal A Chem 168: 257-263.
16. Kuwahara Y, Aoyama J, Miyakubo K, Eguchi T (2012) Tio2 photo catalyst for degradation of organic compounds in water and air supported on highly hydrophobic fau zeolite: structural sorptive and photocatalytic studies. J Catal 285: 223-234.
17. Kamegawa T, Yamahana D, Yamashita H (2010) Graphene Coating of TiO2 Nanoparticles Loaded on Mesoporous Silica for Enhancement of Photocatalytic Activity. J Phys Chem C 114: 15049-15053.
18. Yamashita H, Nose H, Kuwahara Y, Nishida Y (2008) TiO2 photo catalyst loaded on hydrophobic Si_3N_4 support for efficient degradation of organics diluted in water. Appl Catal A Gen 350: 164-168.
19. Kuwahara Y, Maki K, Matsumura Y, Kamegawa T (2009) Hydrophobic Modification of a Mesoporous Silica Surface Using a Fluorine-Containing Silylation Agent and Its Application as an Advantageous Host Material for the TiO2 Photocatalyst. J Phys Chem C 113: 1552-1559.
20. Tiejun C, Yuchao L, Zhenshan P, Yunfei L (2009) Photocatalytic performance of TiO2 catalysts modified by H3PW12O40, ZrO_2 and CeO_2. J Environ Sci 21: 997-1004.
21. Farhadi S, Afshari M, Maleki M, Babazadeh Z (2005) Photocatalytic oxidation of primary and secondary benzylic alcohols to carbonyl compounds catalyzed by $H_3PW_{12}O_{40}$/SiO_2 under an O_2 atmosphere. Tetrahedron Lett 46: 8483-8486.
22. Hu MY (2004) Photocatalytic degradation of textile dye x3b by hetero polyoxometalate acids. Chemosphere 54: 431-434.
23. Anandan S, Yoon M (2007) Photocatalytic degradation of methyl orange using heteropolytungstic acid-encapsulated TiSBA-15. Sol Energy Mater Sol Cells 91: 143-147.
24. Zhang XZ, Yue YH, Gao Z (2001) Studies on 12-Tungstophosphoric Heteropolyacid Supported SBA-15 Catalysts. Chem J Chin Univ 22: 1169-1172.
25. Sulikowski B (1993) The fractal dimension in molecular sieves: synthetic faujasite and related solids. J Phys Chem 97: 1420-1425.
26. Mukai SR, Masuno T, Ogina I, Hashimoto K (1997) Preparation of encaged heteropoly acid catalyst by synthesizing 12-molybdophosphoric acid in the supercages of Y-type zeolite. Appl Catal A Gen 165: 219-226.
27. Marosi L, Arean O (2003) Catalytic performance of Csx(NH4)yHzPMo12O40 and related heteropolyacids in the methacrolein to methacrylic acid conversion: in situ structural study of the formation and stability of the catalytically active species. J Catal 213: 235-240.
28. Ahmad M, Zaidi SMJ, Rahman SU, Ahmad S (2006) Synthesis and proton conductivity of heteropolyacids loaded Y-zeolite as solid proton conductors for fuel cell applications. Micropor Mesopor Mater 91: 296-304.
29. San N, Hatipo A, Glu G, Koçtürk Z (2002) Photocatalytic degradation of 4-nitrophenol in aqueous TiO2 suspensions: Theoretical prediction of the intermediates. J Photochem Photobio A Chemistry 146: 189-197.
30. Aravindhan R, Fathima NN, Rao JR, Nair BU (2006) Wet oxidation of acid brown dye by hydrogen peroxide using heterogeneous catalyst mn-salen-y zeolite: a potential catalyst. J Hazard Mat B 138: 152-159.

Photocatalytic Degradation of Methylene Blue by Using Al_2O_3/Fe_2O_3 Nano Composite under Visible Light

Haile Hassena*

Department of Chemistry, Hawassa College of Teacher Education, Hawassa, Ethiopia

Abstract

Photocatalytic degradation of methylene blue from aqueous solution has been carried out using Al_2O_3/Fe_2O_3 photo catalyst under visible radiation. Effect of various parameters like pH, concentration of dyes, amount of semiconductor and light intensity has been studied on the rate of reaction. Various control experiments were carried out, which indicated that semiconductor Al_2O_3/Fe_2O_3 played a key role in photocatalytic degradation of dye. A suitable tentative mechanism has been proposed for photocatalytic degradation of dye.

Keywords: Methylene blue; Nano composite; Photo catalyst; Sol-gel

Introduction

Waste water from textile, paper and some other industrial processes are usually highly colored, toxic, carcinogenic or mutagenic [1]. These colored compounds are not only aesthetically displeasing but also inhibiting sunlight enetration into the stream and affecting aquatic ecosystem. Dyes usually have complex aromatic molecular structures which make them more stable and difficult to biodegrade. Some dyes are reported to cause allergy, dermatitis, skin irritation, cancer and mutations in humans [2].

Among many organic pollutants, methylene blue (MB) is one of pollutant color for environment undesirable which effects on aesthetic of environment [3]. Thus, environmental contamination by these toxic chemicals has emerged as a serious global problem. On the contrary, bleached dye after degradation of solution is relatively less toxic and almost harmless. Secondly, dye containing coloured water is almost no practical use, but if this coloured solution is bleached to give colorless water, then it may be used for some useful purposes like washing, cooling, irrigation and cleaning. Recently, photocatalytic reactions induced by illumination of semiconductors in suspension have been shown to be one of the most promising processes for the wastewater treatment [4]. Nano sized semiconductors such as TiO_2, ZnO, ZnS, WO_3 and Fe_2O_3 are often used as catalytic agents because of their high stability, low costs, high efficiency and no toxicity [5]. Among various semiconductor photo catalysts, iron oxide (Fe_2O_3) nanomaterials exhibit promising photocatalytic activities due to their environmental friendly behavior, low catalyst cost, high specific surface area, high crystallinity and solar energy application [6-8] and thus, could be an alternative material for environmental application and wastewater treatment [9-11]. Moreover, hematite is the most stable iron oxide under ambient condition and has significant scientific and technological importance, due to stability and interesting band gap of 2.2 eV, for absorption under visible light irradiation. However, the photocatalytic activity of the iron oxide is depending on the particle size, which is difficult to synthesize nano-sized iron oxide by conventional method and to control its crystal size in the photo catalyst [12]. This is due to the agglomeration of nano-particles in the aqueous solution, which causes the reduction of photocatalytic efficiency. One certain way to overcome this drawback is to apply innovative synthetic method of iron oxide nanoparticles of the catalysts that can be easily dispersed in organic medium and homogenously loaded on to the supported materials. Many studies have continuously tried to improve the photocatalytic activity of iron oxide by coupling of different semiconductor oxide nanoparticles [13-16]. Therefore, in this work, binary Al_2O_3/Fe_2O_3 nanocomposite was used as photo catalyst under visible radiation for degradation of MB dye in aqueous solution and investigates the effects of catalyst loading, initial concentration, induced light and pH.

Experimental

Synthesis of photo catalyst

The Al_2O_3/Fe_2O_3 binary mixed nanoxide powder was prepared by sol-gel method. The sol was corresponded to total volume ratio of metal, Butanol, deionized water and nitric acid ratio of 1:20:4:0.1. In each case, ferric nitrate nanohydrate and aluminum nitrate nanohydrate were dissolved in stoichiometric amounts of water, 69% HNO_3 and Butanol then mixed with vigorous stirring at room temperature (25°C). The prepared sol was left to stand for the formation of gel. After the gelation was completed, the gel was aged for 5 days at room temperature and sample was dried at 75°C for 36 h. After grinding the dried samples, they were calcined at 400°C for 3 h. Nano sized materials of the catalyst were analyzed and the procedure was indicated in my previous work [17].

Photocatalytic degradation studies

Photocatalytic activities of the as-synthesized powder were evaluated by decolorization of Methylene blue dye in aqueous solution. The experiments were carried out in the presence of visible light irradiation without any catalyst (blank), with catalyst in dark and in the presence of Al_2O_3/Fe_2O_3 photo catalyst. The photocatalytic reactor consists of a Pyrex glass beaker with an inlet tube for provision of air purging during photo catalysis and outlet tube for the collection of samples from the beaker. Reaction was set up by adding 0.2 g of the as-synthesized powder into 100 mL of MB solution (25 mg/L) in the Pyrex glass beaker of 250 mL volume and the suspension was magnetically stirred in dark for 30 min to obtain adsorption/desorption equilibrium before irradiating the light in the beaker. Before illumination of the samples by visible radiations, air/oxygen was purged into the solution with the help of a porous tube at hand purging in order to keep the suspension of the reaction homogenous.

***Corresponding author:** Haile Hassena, Department of Chemistry, Hawassa College of Teacher Education, Hawassa, Ethiopia, E-mail: hailehassena@ yahoo.com

During the reaction, the solution was maintained at room temperature and the distance of the lamp from the solution was 9 cm. Then the light source was activated, 10 mL of the sample was withdrawn at 15 min time interval over irradiation time for 90 min. The suspension was centrifuged at 3000 rpm for 10 min and filtered to remove the catalyst particles before measuring absorbance. The absorbance of the clear solution was measured at a λ_{max} of 665 nm for quantitative analysis. The incandescent bulb was used as visible light source with a definite power of 40 W, 220 V and 60 Hz frequency. Percentage degradation of MB dye was calculated using the following relation:

$$\% \text{ degradation} = \frac{A_o - A_t}{A_o} \times 100$$

Where A_0 is absorbance of dye at initial stage A_t is absorbance of dye at time t.

Results and Discussion

Photocatalytic degradation study

The photocatalytic activity of as-synthesized nanomaterial was evaluated by the degradation of MB dye in aqueous solution. The decolorization of the MB dye was examined under three different conditions (treatments): visible light irradiation without any catalyst (blank solution), in the presence of catalyst without light irradiation (in dark) and in the presence of Al_2O_3/Fe_2O_3 photo catalyst under visible light irradiation, respectively. For the blank experiment (in the absence of the catalyst) under visible light irradiation, almost insignificant degradation of the dye was observed with only 5.95%. The corresponding plots of percent degradation as a function of time under visible light irradiation without catalyst are shown in Figure 1.

In the presence of photo catalyst (Al_2O_3/Fe_2O_3), but without irradiation, only 14.81% decolorization efficiency was observed throughout in the 90 min. This result confirms that degradation of the MB in the presence of the photo catalyst, but without light irradiation is insignificant. The fact is that no electron-hole pair could be generated in the semiconducting material without assistance of light irradiation. The formation of electrons-holes pairs are responsible for enhancing the oxidation and reduction reactions with the MB dye, which might be adsorbed on the surface of the semiconductor to give the necessary products. The corresponding plots of percent degradation as a function of time in the presence of catalyst without light irradiation are shown in Figure 2.

Actually, the experimental results show that when the dye solution is exposed to visible light irradiation for 90 min in the presence of Al_2O_3/Fe_2O_3 photo catalyst, about 75.10% of the MB dye could be degraded under visible light irradiations. The corresponding plots of percentage degradation of MB dye as function of time under visible irradiation in the presence of Al_2O_3/Fe_2O_3 photo catalyst are shown in Figure 3. Accordingly, the degradation efficiency of MB dye under the visible light was found to be much larger than the degradation efficiency as compare to blank and dark treatment. This enhancement under visible light in the presence of Al_2O_3/Fe_2O_3 photo catalyst could be explained from two reasons. The first one could be the fact that the Al_2O_3/Fe_2O_3 photo catalyst prepared by the sol gel method has a high specific surface area, that could give more active surface sites to absorb water molecules and to form active •OH and HOO• radicals by trapping the photo generated holes. This free active radical drive the photo degradation reactions and eventually leads to the decomposition of organic pollutants in aqueous solution [8]. The higher surface area also facilitates the absorption of dye molecules on the surfaces of Al_2O_3/Fe_2O_3 photo catalyst. Under visible light irradiation, MB molecules are absorbed on the surfaces of nano composite and produced electrons. These electrons are captured by the surface adsorbed O_2 molecules to yield $O_2^{\bullet-}$ and $HO_2\bullet$ radicals, which makes more chance to touch with dye molecules and giving a faster reaction speed then, the MB molecules could be mineralized in time by the super oxide radical ions. Therefore, it can be concluded that the smaller crystalline size of nano composite are favorable for the reduction of O_2 and oxidation of H_2O molecules by trapping electrons and holes, which improves the photocatalytic activity of the nano composites photo catalyst under visible light region. However, the photocatalytic activity of Al_2O_3/Fe_2O_3 nano composite is still not satisfactory for degradation of organic pollutants in wastewater

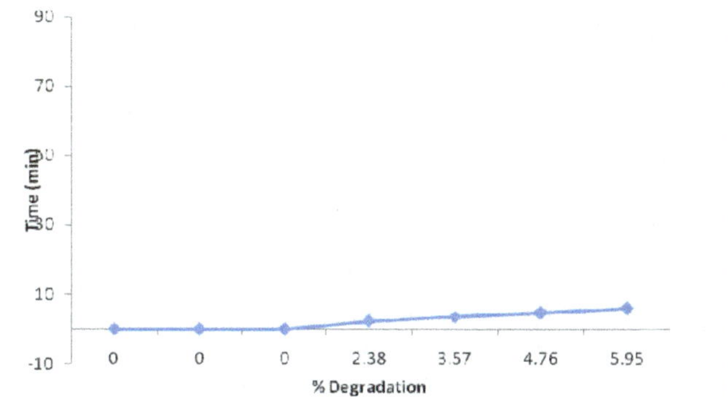

Figure 1: Plots of percentage degradation of MB dye as function of time under visible light irradiations in blank solution (without catalyst).

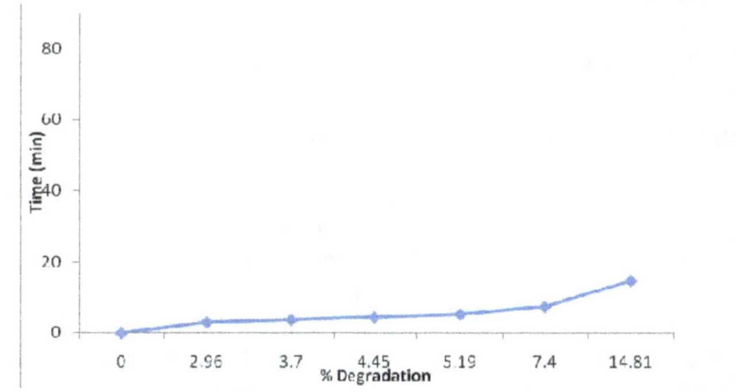

Figure 2: Plots of percentage degradation of MB dye as function of time in the presence of Al_2O_3/Fe_2O_3 photo catalyst under without irradiation.

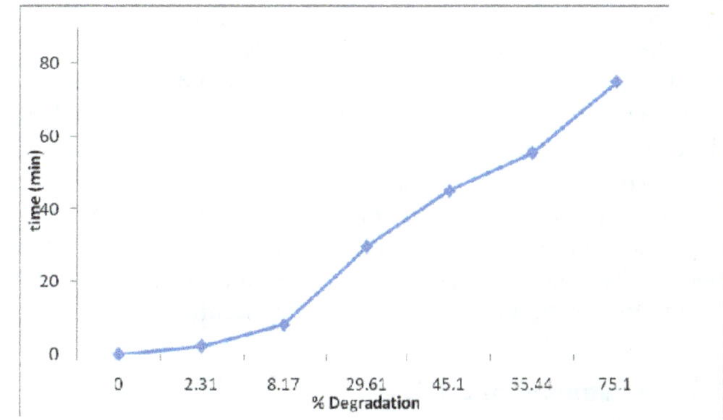

Figure 3: Plots of percentage degradation of MB dye as function of time under visible irradiation in the presence of Al_2O_3/Fe_2O_3 photo catalyst.

as compare to other binary nano composite. This may be due to that, the similar ionic radii and same charge of the materials of iron (III) and aluminum (III) oxide. Subsequently complete substitutions occur between aluminum (III) oxide and iron (III) oxide nanoparticles and no structural defect could be expected [18].

Effect of pH of solution

The pH of the solution is likely to affect the degradation of MB dye in aqueous solution and therefore, the rate of degradation of the dye was studied in the pH range of 4.0 to 9.0. It has been observed that the rate of photo catalytic degradation of MB dye was increased as pH was increased from 4.0 to 5.0 and it got an optimum value at pH 5.0. On further increasing the pH, the rate of the reaction was decreased. This behavior may be explained on the basis that when pH was increased, attraction between cationic dye molecules and hydroxyl ions increases and accordingly the rate of photocatalytic degradation of the dye was increases. But above pH 5.0 value rate of photocatalytic degradation of the dye was decrease, this may be due to the fact that amphoteric nature of catalyst and the surface becomes negatively charged for higher pH values. This causes the electrostatic repulsion between the catalyst and negatively charged MB dye in aqueous solution. On other hand, the zero point charge (pHzpc) of α-Fe_2O_3 is the main parameters that controlling the degradation of MB on composite Fe_2O_3, which are about 4.4. As the solution pH increases from an acidic region to an alkaline region, MB ions in solution exist mainly as acidic form. When the pH is below the isoelectric point of composite Fe_2O_3 catalyst, the surface of Al_2O_3/Fe_2O_3 will be positively charged. Methylene blue dye solutions are easily adsorbed on the Fe_2O_3 hierarchical microspheres surface in the low pH range due to strong electrostatic attraction between dye and Al_2O_3/Fe_2O_3. As pH increases, the Fe_2O_3 surface becomes less positively charged, and the interaction between Al_2O_3/Fe_2O_3 and dyes becomes less and changes to a repulsive force at pH>pHzpc, resulting in a significant decrease of degradation efficiency [19] (Figure 4).

Effect of initial MB concentrations

The effect of initial dye concentration on the degradation efficiency was investigated by varying the initial dye concentration. Different initial concentrations of MB with rang from 0.5×10^{-5} M to 1.4×10^{-5} M were used to evaluate the photo catalytic activity as shown in Figure 5. As a result the photo degradation efficiency was decreased with concentration of MB dye more than 1.1×10^{-5} M after 90 min irradiation time. This fact explain as, the adsorption capacity is higher at lower concentration because more active site is available for MB molecules to be adsorbed on the surface Al_2O_3/Fe_2O_3 photo catalyst. At higher concentration which the more and more molecules of MB adsorbed on the surface Al_2O_3/Fe_2O_3 photo catalyst, which have hinder the photo generation of hydroxyl radicals of OH^- ions ($H^+ + OH^- \rightarrow {}^{\bullet}OH$), which causes the decreasing rate of photo degradation reaction of the catalyst [20,21]. Therefore, much longer time is required to reach the complete degradation of higher concentrations of MB as compared to low concentration of the dye. These results confirm that the Beer-Lambert law: as the initial dye concentration increases, the path length of photons entering the solution was decreases. This results in the lower photon absorption on the catalyst particles, and consequently decreases the photocatalytic reaction rate at higher concentration of MB dye [22].

Effect of amount of catalyst

The amount of semiconductor also affects photo degradation efficiency of the catalyst. Different amounts of photo catalyst were used (0.02 g to 0.11 g) for degradation of MB dyes under visible light irradiation and the results are given as shown in Figure 6. It has been observed that as the amount of photo catalyst of Al_2O_3/Fe_2O_3 was increased, the rate of photo degradation of dye increases but ultimately, the reaction rate become virtually constant after a certain amount (0.05 g) of the semiconductor. This may be due to the fact that as the amount of photo catalyst was increases the number of active sites on the photo catalyst surface and the exposed surface area also increases, but when the concentration of the catalyst increases above the optimum value, the degradation rate decreases due to the interception of the light by the suspension [22,23]. And also excess catalyst prevent the illumination, •OH radical, a primary oxidant in the photocatalytic system decreased

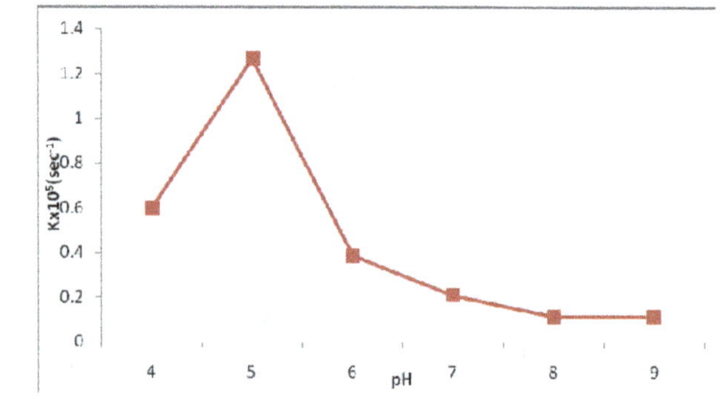

Figure 4: Effect of initial pH on photocatalytic degradation of MB on the presence of catalyst.

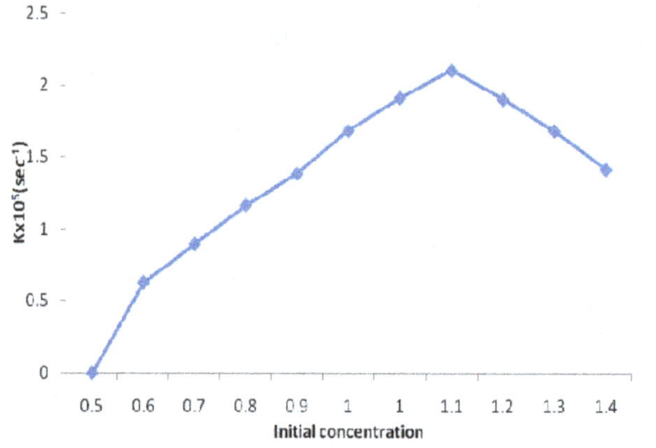

Figure 5: Effect of initial concentration of MB on degradation efficiency.

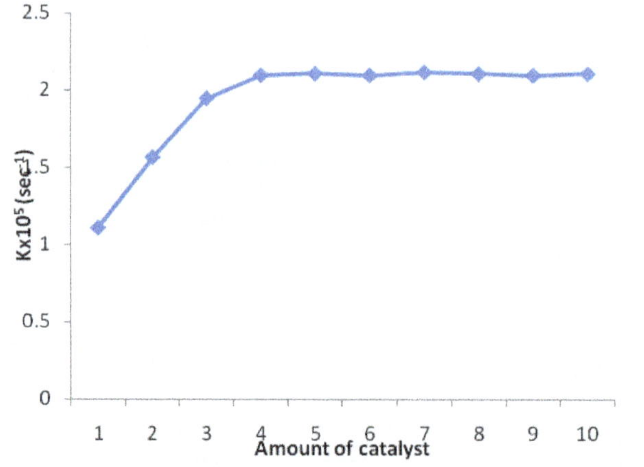

Figure 6: Effect of amount of photo catalyst.

and the efficiency of the degradation reduced accordingly [24]. Furthermore, the increase of catalyst concentration beyond the optimum may result in the agglomeration of catalyst particles, hence more catalyst surface become unavailable for photo absorption, and degradation rate of catalyst became decrease [25].

Effect of light intensity

The effect of light intensity on the photocatalytic degradation of MB was also investigated. The light intensity was varied by changing the distance between the light source and the exposed surface area of semiconductor and observation results were reported in Figure 7. These data indicate that photocatalytic degradation of methylene blue was enhanced with the increase in intensity of light, because an increase in the light intensity will increase the number of photons striking per unit area per unit time of photo catalyst surface. There was a slight decrease in the rate of reaction as the intensity of light was increased beyond 50.0 $mWcm^{-2}$. Therefore, light intensity of medium order was used throughout the experiments.

Degradation mechanism of methylene blue

Methylene blue is a heterocyclic aromatic compound with the molecular formula $C_{16}H_{18}N_3SCl$ [26]. The primary photocatalytic oxidation mechanism is believed to proceed as follows:

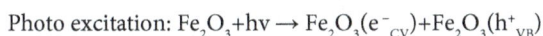

Photo excitation: $Fe_2O_3 + h\nu \rightarrow Fe_2O_3(e^-_{CV}) + Fe_2O_3(h^+_{VB})$

$Fe_2O_3(h^+_{VB}) + H_2O \rightarrow Fe_2O_3 + H^+ + OH^-$

$Fe_2O_3(h^+_{VB}) + OH^- \rightarrow Fe_2O_3 + OH^\bullet$

$Fe_2O_3(e^-_{CB}) + O_2 \rightarrow Fe_2O_3 + O_2^{\bullet -}$

$O_2^{\bullet -} + H^+ \rightarrow HO_2^\bullet$

Degradation pathway of methylene blue provided by -OH radical makes the main contribution to degradation of methylene blue and its intermediates [27] (Figure 8). Since the MB is cationic and not electron donor, the initial step of MB degradation can be ascribed to the cleavage of the bonds of C–S⁺==C functional group in MB:

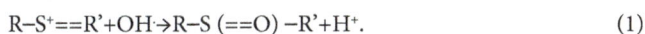

$$R{-}S^+{=}{=}R' + OH^\cdot \rightarrow R{-}S\,({=}{=}O) - R' + H^+. \quad (1)$$

The sulfoxide group can undergo a second attack by an $OH^\cdot$ Radical and producing the sulfone, which causes the definitive dissociation of two benzenic rings:

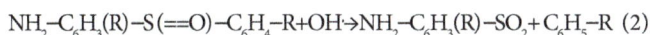

$$NH_2{-}C_6H_3(R){-}S({=}{=}O){-}C_6H_4{-}R + OH^\cdot \rightarrow NH_2{-}C_6H_3(R){-}SO_2 + C_6H_5{-}R \quad (2)$$

and/or

$$NH_2{-}C_6H_3(R){-}S({=}{=}O){-}C_6H_4{-}R + OH^\cdot \rightarrow NH_2{-}C_6H_4{-}R + SO_2{-}C_6H_4{-}R \quad (3)$$

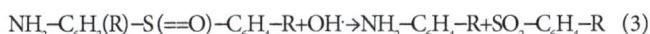

Subsequently, the sulfone can be attacked itself by a third $OH^\cdot$ radical for giving a sulfonic acid:

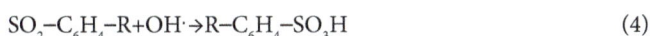

$$SO_2{-}C_6H_4{-}R + OH^\cdot \rightarrow R{-}C_6H_4{-}SO_3H \quad (4)$$

Finally release of SO_4^{2-} ions can be attributed to a fourth attack by $OH^\cdot$

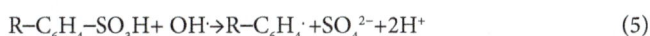

$$R{-}C_6H_4{-}SO_3H + OH^\cdot \rightarrow R{-}C_6H_4^\cdot + SO_4^{2-} + 2H^+ \quad (5)$$

The amino group in MB can be substituted by an $OH^\cdot$ radical, forming the corresponding phenol and releasing a $NH_2^\cdot$ radical which generates ammonia and ammonium ions, estimated to be primary products?

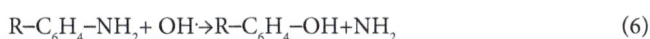

$$R{-}C_6H_4{-}NH_2 + OH^\cdot \rightarrow R{-}C_6H_4{-}OH + NH_2 \quad (6)$$

$$NH_2^\cdot + H^\cdot \rightarrow NH_3 \quad (7)$$

$$NH_3 + H^+ \rightarrow NH_4^+ \quad (8)$$

The other two symmetrical dimethyl-phenyl-amino groups in MB undergo a progressive degrading oxidation of one methyl group by an attack from $OH^\cdot$ radical, producing an alcohol, then an aldehyde, which is spontaneously oxidized into acid, decarboxylates into CO_2 by photo-Kolbe reaction.

Conclusion

Heterogeneous photo catalysis on semiconductor nanoparticles has been shown to be effective methods to removing organic pollutants from wastewater as a complete mineralization of dyes in to CO_2, H_2O and other oxides. The identification of reaction intermediates allowed us to propose a reaction pathway of the degradation which involves breakdown of the cleavage bonds of C–S⁺==C functional group in MB bond by hydroxylation of the aromatic ring, leading at the end to ring opening and formation of carboxylic acids and aldehyde which undergoes oxidation to form CO_2 and water. The photocatalytic degradation of MB dye using Al_2O_3/Fe_2O_3 photo catalyst depends on the initial concentration of MB dye pH, amount of catalyst, and light intensity. Nevertheless, sol-gel method is widely used because the method facilitates the synthesis of nano sized crystallized binary Al_2O_3/Fe_2O_3 oxide catalysts for the degradation of organic pollutants powder of high purity at relatively low temperature.

Acknowledgements

The author would like to thank Hawassa College of Teacher Education for providing necessary laboratory facilities.

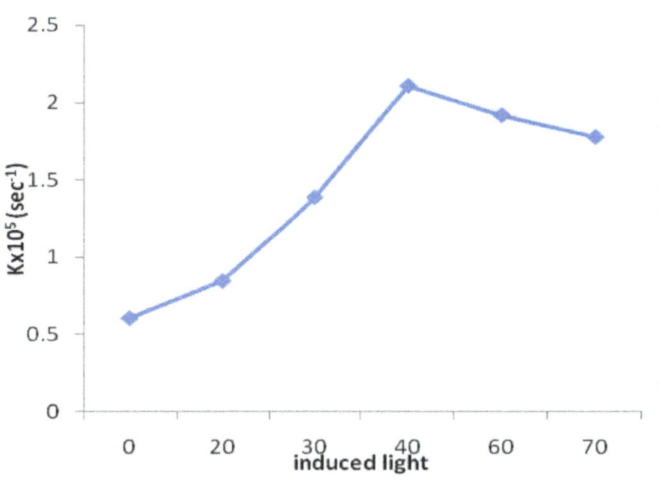

Figure 7: Effect of light intensity.

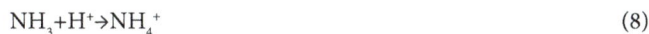

Figure 8: Chemical structure of methylene blue.

References

1. Sima J, Hasal P (2013) Photocatalytic Degradation of Textile Dyes in aTiO_2/UV System. Che Eng Trans 32: 79-84.
2. Ansari R, Mosayebzadeh Z (2010) Removal of basic dye methylene blue from aqueous solutions using sawdust and sawdust coated with polypyrrole. J Iran Chem Soc 7: 339-350.
3. Chen ZJ, Lin BZ, Xu BH, Li XL, Wang QQ, et al. (2010) Preparation and characterization of mesoporous TiO_2-pillared titanate photocatalyst. J Porous Mater 18: 185-193.
4. Dai K, Chen H, Peng T, Ke D, Yi H (2007) Photocatalytic degradation of methyl orange in aqueous suspension of mesoporous titania nanoparticles. Chemosphere 69: 1361-1367.
5. Vaya D, Sharma VK (2010) Study of synthesis and photocatalytic activities of Mo doped ZnO. J Chem Pharm Res 2: 269-273.
6. Leland JK, Bard AJ (1987) Photochemistry of colloidal semiconducting iron oxide polymorphs. J Phys Chem 91: 5076-5083.
7. Zhao S, Wu HY, Song L, Tegus O, Asuha S (2009) Preparation of γ-Fe_2O_3 nanopowders by direct thermal decomposition of Fe-urea complex: reaction mechanism and magnetic properties. J Mater Sci 44: 926-930.
8. Bharathi S, Nataraj D, Mangalaraj D, Masuda Y, Senthil K, et al. (2010) Highly mesoporous α-Fe_2O_3 nanostructures: preparation, characterization and improved photocatalytic performance towards Rhodamine B (RhB). J Phys D Appl Phys 43: 015501.
9. Rhoton FE, Bigham JM, Lindbo DL (2002) Properties of iron oxides in streams draining the Loess Uplands of Mississippi. Appl Geochem 17: 409-419.
10. Lei J, Liu C, Li F, Li X, Zhou S, et al. (2006) Photodegradation of orange I in the heterogeneous iron oxide-oxalate complex system under UVA irradiation. J Hazard Mater 137: 1016-1024.
11. Wang Y, Liu CS, Li FB, Liu CP, Liang JB (2009) Photodegradation of polycyclic aromatic hydrocarbon pyrene by iron oxide in solid phase. J Hazard Mater 162: 716-723.
12. Park JY, Lee YJ, Khanna PK, Jun KW, Bae JW, et al. (2010) Alumina-supported iron oxide nanoparticles as Fischer-Tropsch catalysts: Effect of particle size of iron oxide. J Mol Catal A Chem 323: 84-90.
13. Baldrian P, Merhautova V, Gabriel J, Nerud F, Stopka P, et al. (2006) Decolorization of synthetic dyes by hydrogen peroxide with heterogeneous catalysis by mixed iron oxides. Appl Catal B 66: 258-264.
14. Laat JD, Le TG (2006) Effects of chloride ions on the iron(III)-catalyzed decomposition of hydrogen peroxide and on the efficiency of the Fenton-like oxidation process. Appl Catal B 66: 137-146.
15. Kitis M, Kaplan SS (2007) Advanced oxidation of natural organic matter using hydrogen peroxide and iron-coated pumice particles. Chemosphere 68: 1846-1853.
16. Zelmanov G, Semiat R (2008) Iron(3) oxide-based nanoparticles as catalysts in advanced organic aqueous oxidation. Water Res 42: 492-498.
17. Logita HH, Tadesse A, Kebede T (2015) Synthesis, characterization and photocatalytic activity of $MnO_2/Al_2O_3/Fe_2O_3$ nanocomposite for degradation of malachite green. AJPAC 9: 211-222.
18. Neiva LS, Andrade MC, Costa CFM, Gama L (2009) Synthesis Gas (Syngas) production over Ni/Al_2O_3 catalysts modified with Fe_2O_3. Brazilian Journal of Petroleum and Gas 3: 083-091.
19. Xie K, Wang XX, Liu ZJ, Alsaedi A, Hayat T, et al. (2014) Synthesis of flower-like a-Fe_2O_3 and its application in wastewater treatment. Journal of Zhejiang University Science A 15: 671-680.
20. Al-Shamali SS (2013) Photocatalytic Degradation of Methylene Blue in the Presence of TiO_2 Catalyst Assisted Solar Radiation. Aust J Basic Appl Sci 7: 172-176.
21. Ameta R, Sharma S, Sharma S, Gorana Y (2015) Visible Light Induced Photocatalytic Degradation of Toluidine Blue-O by using Molybdenum Doped Titanium Dioxide. European Journal of Advances in Engineering and Technology 2: 95-99.
22. Byrappa K, Subramani AK, Ananda S, Lokanatha RKM, Dinesh R, et al. (2006) Photocatalytic degradation of Rhodamine B dye using hydrothermally synthesized ZnO. Bull Mater Sci 29: 433-438.
23. Chakrabarti S, Dutta BK (2004) Photocatalytic degradation of model textile dyes in wastewater using ZnO as semiconductor catalyst. J Hazard Mater 112: 269-278.
24. Sun J, Qiao L, Sun S, Wang G (2008) Photocatalytic degradation of Orange G on nitrogen-doped TiO_2 catalysts under visible light and sunlight irradiation. J Hazard Mater 155: 312-319.
25. Huang M, Xu C, Wu Z, Huang Y, Lin J, et al. (2008) Photocatalytic discolorization of methyl orange solution by Pt modified TiO_2 loaded on natural zeolite. Dyes Pigm 77: 327-334.
26. Jian-xiao LV, Ying C, Guo-hong X, Ling-yun Z, Su-fen W (2011) Decoloration of methylene blue simulated wastewater using a UV-H_2O_2 combined system. Journal of Water Reuse and Desalination 1: 45-51.
27. Houas A, Lachheb H, Ksibi M, Elaloui E, Guillard C, et al. (2001) Photocatalytic degradation pathway of methylene blue in water. Appl Catal B 31: 145-157.

Effect of Chemotherapy on Zn, Fe, Mg, Pb, Ca and Se in the Serum

Hasan A*

Department of Chemistry, Faculty of Science, Gazi University, Ankara, Turkey

Abstract

Trace element concentrations in blood serum are very important in various carcinomas. However, the literature lacks in studies of the relationship between the trace element concentrations and cancer types. The present study analyzes blood serums from 40 patients with throat, stomach and lung cancers by inductively coupled plasma optical emission spectrometry (ICP-OES) after low-volume microwave digestion. 7 elements were investigated: Se, Ca, Fe, K, Mg, Pb, and Zn. The elements were categorized into macro elements (Ca, Mg, K) and trace elements (Fe, Zn, Pb, Se) for human blood. The mixture of $HClO_4$:HNO_3:H_2SO_4 was selected for analyzing the trace elements in serums taken from cancer patients. The digested blood serums of pre-chemotherapy and post-chemotherapy were analyzed to investigate the effect of chemotherapy on amounts of trace elements. Decreases were observed in amounts of the essential trace elements except Pb for post-chemotherapy by time, as biochemical reactions were catalyzed by many enzymes and proteins. Certain changes must be made in the scientific approach in cancer therapy. Such changes in treatment of the patient with cancer should include continuous treatment of each patient individually according to the results obtained.

Keywords: Chemotherapy; Trace elements; ICP-OES; Throat cancer; Stomach cancer; Lung cancer

Introduction

Recently, cancer cases have increased considerably in the last decades. The reasons for this increase is connected multi-dimensional and complex relations such as economic development and its effects on human life: the increase in the intake of unnatural foods, breathing polluted air, increased quantity of trace elements in drinking water. The trace elements are play important role in this dramatic increase in cancer cases and more and more factors are affecting the quantities of trace elements in human body. This caused a great deal of attention on the determination of trace elements [1-11]. The change in the balance of trace elements is the most important factor in this situation and any deviation from the optimum levels of them may have a drastic effect on biological processes. In addition, trace elements like Cu, Zn, Ca, Mn, Fe, etc have decisive functions in maintaining human health and are of great importance for the synthesis and structural stabilization of both proteins and nucleic acids. In this context the role of Se in preventing and ratio of Cu/Zn as an indicator for various cancers or tumors are widely investigated. However it is highly difficult to explain the relationship between trace elements and some diseases. Further studies are needed for the clarification of these relations. Blood is the most commonly used sample to identify the trace elements due to ease of sampling. In addition it is the medium of transport of trace elements to tumors and healthy organs. Therefore, whole blood, plasma, and serum are commonly used for determinations. Cancer is known as a genetic disease, but diet and environmental conditions have a significant effect on cancer incidents. For instance, 65-70% of all cancers are associated with the environmental conditions and only 2% is related to genetic factors. Moreover, some of the trace elements are important for human health and some are highly toxic. However an excessive intake any elements causes serious complications including cancer [1-4]. Trace elements in human body are inessential and essential for the growth of the organism due to their electrochemical and catalytic effects. They activate enzymatic reactions or inhibit *in vitro* reactions. Zn, Ca, Mg, Fe and Se are known as essential trace elements due to their catalytic functions for the formation of various essential compounds with proteins. Fe also catalyzes many redox reactions. However some of the trace elements catalyze the oxidative damage of biological macromolecules and generate free radicals [5]. Zn and Pb have been used to induce cancer. The metal ions can interact with nucleic acids to influence base pairing and conformation. Se has anticarcinogenic properties, but its quantity in serum and whole blood or tissue is important. For instance, toxic level of the element causes chronic toxic hepatitis that looks like hyperplasia. It has anticarcinogenic effects by several mechanisms and protective action against chromosomal damage. A part from their role in antioxidant defense system, seleno proteins break downs in hydrogen peroxide and lipid hydroperoxides, which can damage cell membranes and disrupt cellular functions. Pb is a toxic metal and causes a serious hematological damage, brain damage, anemia, and kidney malfunctioning [12]. Zn concentration affects growth retardation, anorexia, delayed sexual maturation, anemia, mental retardation, impaired visual and immunological function, etc. Low levels of zinc cause some dysfunctions of the immune system [13]. Serum concentrations of Zn are modified in some cancers; variations of Zn concentrations have been observed in leukemia. There is an important difference in the role of Se and Zn in a number of cancer types. There is an inverse relationship with cancer and quantity of Se and Zn has a protective effect when it is present in sufficient quantities [14-16]. Pb has an adverse effect on Fe deficiency, and Fe deficiency increases the absorption of Pb [17]. The absorption of Pb is positively correlated by the lower presence of Ca [18]. Fe is an important part of hemoglobin and myoglobin. Mg^{+2} ions are used for activating all enzymes used for transferring phosphate from ATP to acceptor molecule or ADP. Therefore, 1/3 of total quantity of Mg is observed to be bounded to proteins. Chemotherapy used in cancer treatment often causes many side effects depending on the type of cancer, location, drugs and dose and general health. This is because it has an important function on active cell growth and propagation. The most effected cells are the ones in our blood, mouth, digestive system, and hair follicles which cause various side effects. The elimination of these side effects involves the use different medications. Chemotherapy also affects the quantity of trace elements which play important roles in a number

***Corresponding author:** Hasan A, Department of Chemistry, Faculty of Science, Gazi University, Ankara, Turkey, E-mail: haydin@gazi.edu.tr of processes occurring in human body by activating or inhibiting enzymes. There is a little research on the quantity of trace elements in the blood pre-and post-chemotherapy periods. The dissolution of the blood samples in a suitable media is necessary to determine trace elements. Some dry and wet ashing procedures are available. Different acid mixtures such as $HCl:HNO_3$, $HClO_4:HNO_3$, $HClO_4:HNO_3:HCl$ and $HClO_4:HNO_3:H_2SO_4$ for Pb, Cu and Se determination can be used [19]. Especially, there may be substantial loss trace elements during thee decomposition of organic substances. Se is much more volatile than trace metals like Cu and Pb. When acidic mixture does not contain oxidizing substance such as HNO_3, partial losses of selenium as SeO. Cl_2, $SeO_2.2HCl$, etc. has been observed [19-23]. Despite the existence of various analytical methods for the determination of trace elements, most of them are limited with a few trace elements including the ones in blood during the pre-chemotherapy and post-chemotherapy, and the last stage of patient life with cancer. Consequently, this paper investigates the correlation between the quantities of trace elements in serum and chemotherapy for different types of cancers. This study is expected to contribute to medicine as regards to chemotherapy process which is the most common mode of cancer treatment.

Experimental

Reagents and standard solutions

All acids used were of Arist and other reagents were of analytical grade purity. Standard working solutions were obtained from a multielement standard stock solution ICP Multi Element Standard Certipur VI (Merck, Darmstadt, Germany) after suitable dilution. Homemade triply distilled water was used for the preparation and dilution of solutions. Calibration curves were plotted from three replicates measurements and, in all cases, regression coefficients were higher than 0.999. Platinum was chosen because it contained negligible amounts of the metals of interest. The internal standard solution was added to all solutions automatically using an on-line addition (265.945 nm).

Instrumentation and analysis of trace elements

Concentrations of Zn, Fe, Mg, Pb, Ca, and Se in serum of patients were analyzed by ICP-OES using a Perkin–Elmer Optima 4300 DV spectrometer (Shelton, CT, USA), equipped with an AS-90 auto sampler, axial system, a high dynamic range detector and a cross-flow type nebulizer for pneumatic nebulization. The ICP-OES measurement conditions for these elements were optimized to achieve the maximum signal-to-background ratio. A line selection was performed to determine the elements analyzed in the study and detection limits were determined by using an appropriate blank solutions and aqueous standard solutions three times, and thus the LOD was calculated. The instrument was operated under suitable conditions. Different concentrations (0.5, 1.0, 2.0, 5.0, 10.0 and 20.0 mg/L) of trace elements were used for calibration of standard graphs. To verify the assay accuracy and to maintain quality, the standard solutions were run for every 10-test sample.

Blood sample collection preparation and sample analysis

The blood samples collected from 40 volunteer patients in Ankara Oncology Training and Research Hospital to control the results of chemotherapy 0.5 ml serum of the bloods was used for the determination of the trace elements. 5 mL of concentrated HNO_3 was added and waited for 2 minutes. Then, each of the samples waited for 12 h at room temperature by adding 5 ml of concentrated HNO_3 plus approximately 5 drops of H_2O_2. Afterwards, the resulting solutions were heated to approximately 120°C. The digested residues were redissolved and filled in 5 mL flasks. To prepare blank solutions, the same amounts of chemicals used for sample digestion and performing the same dissolution stages were used. For the determination of recovery, a known amount of single element standards was added to the samples with known concentrations prior to wet digestion. One of the most difficult parts of this analysis is getting the sample introduction system clean enough to determine low levels of Zn, Fe, Mg, Pb, Ca, and Se. With the demountable torch cassette design, it is very easy to switch to a different sample introduction system. In addition, this design incorporates a short transfer line between the spray chamber and torch to minimize sample carry-over and contamination. The most difficult part of the analysis is sample introduction system to determine low concentrations of the trace elements. Demountable torch cassette design was used to minimize sample carry-over and contamination. Before the analysis, the nebulizer, spray chamber, injector adapter, injector, and torch were cleaned by soaking for 30 minutes in warm diluted nitric acid. This action provided a clean sample introduction system to measure these low-level analyses.

Results and Discussion

The analysis of heavy metals requires the elimination of the organic compounds. However the digestion of organic compound may result in los of trace elements [24]. Since the volatility of Se is higher than Zn, Fe, Mg, Pb, and Ca it is necessary to use an oxidizing agent in order to obviate partial losses of selenium. The use of oxidizing acids in digestion may cause the conversion of Se(IV) to Se (VI). Therefore one must use high concentrations of HCl (4-7 mol/L) to avoid the back oxidation from Se(IV) to Se (VI) [24-29], since the only compound formed from Se(IV) is selenium hydride. In contrast, HCl concentration has to be higher than 2 mol/L for formation of Se(IV) Aydın observed good results from colorless clear serum solution for trace quantities of Se and the other elements. The researchers proposed that mixture of $HClO_4$: $HNO_3:H_2SO_4$ was suitable for the oxidation of Se and the other metals except high concentration of Pb in organic medium due to precipitation of $PbSO_4$. Serum Trace Elements Profile in Lung, Throat, and Stomach Cancers. There is a vast number of papers on the role of trace elements in various cancer cases and the relation between types of cancer cases and amount of trace elements in human serums serum in the literature. Collecting the samples without contamination and analyzing them for trace elements are highly cumbersome. Trace elements in serums of lung cancer patients have been analyzed pre-and-post-chemotherapy periods. Tables 1-3 compare concentrations of essential trace elements in serums taken from patients with lung, throat and stomach cancers and show the effect of chemotherapy on amounts of trace elements. The elements are categorized into macro elements (Ca, Mg, K) and trace elements (Fe, Zn, Pb, Se). Differences in amount of trace elements were important between pre-chemotherapy (week 1) and post-chemotherapy (week 2-4), but quantities of the elements were low than that of patients without cancers except selenium (Tables 1-3). The source of the observed differences in four weeks for each patient and among patients may be the differences in eaten foods, previous treatments before chemotherapy, genetics of patients, and habit of tobacco, alcohol, dietary supplements consumption frequency of meat consumption. However, there were significant differences between levels of the elements compared to the elements in serum samples in cancer patients as compared to those without lung and throat cancers except selenium. Decreases in amounts of the elements except Pb were observed in the overall amount of some of essential trace elements for post-chemotherapy by time. The decreases in the amounts show a decrease in biochemical reactions for many enzymes and proteins

Patient No	Week	Se (μg/L) Mean ± SD	Ca (mg/L) Mean ± SD	Fe (mg/L) Mean ± SD	K (mg/L) Mean ± SD	Mg (mg/L) Mean ± SD	Pb (μg/L) Mean ± SD	Zn (μg/L) Mean ± SD
1	1	6.387 ± 0.415	2.604 ± 0.023	4.737 ± 0.233	5.342 ± 0.365	0.780 ± 0.045	0.458 ± 0.015	0.124 ± 0.015
	2	6.257 ± 0.265	2.196 ± 0.037	2.120 ± 0.167	6.569 ± 0.387	0.757 ± 0.078	0.590 ± 0.035	0.248 ± 0.019
	3	6.260 ± 0.562	2.485 ± 0.048	4.384 ± 0.197	4.723 ± 0.560	0.921 ± 0.065	0.695 ± 0.037	0.168 ± 0.024
	4	6.905 ± 0.367	3.474 ± 0.054	No observed	1.425 ± 0.433	0.367 ± 0.054	0.938 ± 0.045	0.078 ± 0.033
2	1	6.910 ± 0.612	2.373 ± 0.026	13.42 ± 0.250	6.612 ± 0.423	0.920 ± 0.085	0.017 ± 0.016	0.226 ± 0.056
	2	6.636 ± 0.455	2.502 ± 0.140	6.315 ± 0.167	4.971 ± 0.388	0.737 ± 0.034	0.013 ± 0.011	0.131 ± 0.060
	3	6.576 ± 0.543	2.489 ± 0.079	9.460 ± 0.188	5.412 ± 0.548	1.224 ± 0.145	0.501 ± 0.019	0.140 ± 0.056
	4	6.43 ± 0.397	2.328 ± 0.063	8.042 ± 0.245	5.485 ± 0.275	1.432 ± 0.187	0.621 ± 0.032	0.151 ± 0.047
3	1	7.067 ± 0.451	3.632 ± 0.075	3.597 ± 0.195	3.427 ± 0.197	0.545 ± 0.054	0.033 ± 0.013	0.200 ± 0.054
	2	6.841 ± 0.432	2.491 ± 0.042	4.722 ± 0.345	3.624 ± 0.376	0.739 ± 0.076	0.022 ± 0.009	0.110 ± 0.089
	3	6.093 ± 0.765	2.761 ± 0.049	2.874 ± 0.155	3.829 ± 0.675	0.638 ± 0.098	0.023 ± 0.006	0.070 ± 0.088
	4	5.456 ± 0.227	2.901 ± 0.065	1.687 ± 0.164	3.623 ± 0.377	0.589 ± 0.048	0.027 ± 0.010	0.030 ± 0.045
Ref. 1		3.650 ± 0.17	13.48 ± 0.38	3.61 ± 0.15	15.39 ± 0.36	14.88 ± 0.10	1.74 ± 0.19	4.44 ± 0.15
Ref. 2		4.17 ± 0.06	13.70 ± 0.30	4.78 ± 0.10	15.10 ± 0.14	18.53 ± 0.39	0.01 ± 0.01	5.09 ± 0.25

Table 1: Comparison of trace elements in serums taken from different patients with lung cancer.

Patient No	Week	Se (μg/L) Mean ± SD	Ca (mg/L) Mean ± SD	Fe (mg/L) Mean ± SD	K (mg/L) Mean ± SD	Mg (mg/L) Mean ± SD	Pb (μg/L) Mean ± SD	Zn (μg/L) Mean ± SD
1	1	7.026 ± 0.467	2.991 ± 0.023	4.590 ± 0.033	8.085 ± 0.690	0.697 ± 0.017	0.022 ± 0.019	0.067 ± 0.015
	2	6.977 ± 0.643	3.206 ± 0.056	3.507 ± 0.025	7.214 ± 0.587	0.5190.067	0.012 ± 0.013	0.019 ± 0.012
	3	5.714 ± 0.389	4.046 ± 0.089	3.056 ± 0.047	6.914 ± 0.450	0.487 ± 0.099	0.026 ± 0.013	0.025 ± 0.008
	4	4.618 ± 0.564	4.965 ± 0.054	2.367 ± 0.76	6.523 ± 0.632	0.387 ± 0.087	0.024 ± 0.015	0.038 ± 0.011
2	1	6.835 ± 0786	2.626 ± 0.076	5.650 ± 0.037	4.662 ± 0.941	0.729 ± 0.064	0.015 ± 0.009	0.140 ± 0.007
	2	6.201 ± 0.654	2.777 ± 0.123	4.867 ± 0.024	5.148 ± 0.754	0.987 ± 0.078	0.012 ± 0.007	0.032 ± 0.005
	3	5.872 ± 0.433	2.8920.079	3.690 ± 0.065	6.130 ± 0.367	1.176 ± 0.086	0.010 ± 0.008	0.013 ± 0.015
	4	4.768 ± 0.586	2.987 ± 0.073	2.564 ± 0.041	7.450 ± 0.578	1.289 ± 0.068	0.085 ± 0.011	0.012 ± 0.004
3	1	6.847 ± 0.387	3.467 ± 0.067	3.623 ± 0.069	6.540 ± 0.654	0.457 ± 0.098	0.015 ± 0.005	0.079 ± 0.013
	2	4.849 ± 0.761	4.394 ± 0.084	1.269 ± 0.058	6.858 ± 0.438	0.420 ± 0.054	0.026 ± 0.013	0.081 ± 0.017
	3	6.814 ± 0.659	2.726 ± 0.099	0.954 ± 0.066	6.022 ± 0.854	0.409 ± 0.075	0.016 ± 0.001	No detected
	4	5.643 ± 0.588	3.824 ± 0.058	0.780 ± 0.053	5.908 ± 0.347	0.380 ± 0.063	0.012 ± 0.006	No detected

Table 2: Comparison of trace elements in serums taken from different patients with throat cancer.

Patient No	Week	Se (μg/L) Mean ± SD	Ca (mg/L) Mean ± SD	Fe (mg/L) Mean ± SD	K (mg/L) Mean ± SD	Mg (mg/L) Mean ± SD	Pb (μg/L) Mean ± SD	Zn (μg/L) Mean ± SD
1	1	7.436 ± 0.667	4.474±0.067	5.659 ± 0.036	62.92 ± 0.65	0.686 ± 0.056	0.018 ± 0.012	0.247 ± 0.012
	2	7.054 ± 0.357	3.609±0.087	4.876 ± 0.045	101.05 ± 1.99	0.569 ± 0.064	0.019 ± 0.014	0.370 ± 0.017
	3	6.681 ± 0.653	3.633±0.067	4.056 ± 0.054	78.95 ± 3.87	0.435 ± 0.98	0.043 ± 0.024	0.182 ± 0.011
	4	6.428 ± 0.873	3.657±0.088	2.568 ± 0.076	72.63 ± 6.78	0.356 ± 0.075	0.044 ± 0.011	0.196 ± 0.090
2	1	6.869 ± 1.087	3.042 ± 0.046	5.786 ± 0.067	76.66 ± 0.65	0.608 ± 0.68	0.026 ± 0.015	0.036 ± 0.015
	2	7.390 ± 0.786	3.259 ± 0.064	4.467 ± 0.043	78.13 ± 4.89	0.55 ± 0.84	0.021 ± 0.009	0.059 ± 0.012
	3	6.155 ± 0.679	4.349 ± 0.058	3.285 ± 0.088	68.01 ± 3.87	0.338 ± 0.063	0.019 ± 0.011	0.039 ± 0.017
	4	5.453 ± 0.975	5.231 ± 0.078	2.154 ± 0.051	66.94 ± 7.45	0.276 ± 0.045	0.016 ± 0.007	0.029 ± 0.012
3	1	6.9930.678	3.956 ± 0.059	4.242 ± 0.064	85.32 ± 3.58	0.565 ± 0.079	0.023 ± 0.012	0.054 ± 0.019
	2	6.744 ± 0.543	3.123 ± 0.047	2.369 ± 0.055	62.39 ± 5.66	0.511 ± 0.054	0.021 ± 0.009	0.033 ± 0.013
	3	5.956 ± 0.865	3.166 ± 0.076	1.843 ± 0.092	69.03 ± 4.78	0.315 ± 0.013	0.020 ± 0.011	0.028 ± 0.015
	4	5.354 ± 0.673	3.043 ± 0.054	1.567 ± 0.043	72.012.89	0.310 ± 0.068	0.019 ± 0.006	0.025 ± 0.013

Table 3: Comparison of trace elements in serums taken from different patients with stomach cancer.

used for controlling tumor growth Smoking may affect the amount of trace element levels in serums because of the variation in the amount of trace elements among cigarette brands. Variation in the levels of trace elements including Pb may depend on the inflammatory response attributed to brands of cigarettes. The number of cigarettes smoked and brands of cigarette increase the cigarette content in the serum of smokers, affecting the amounts of trace elements in serum and responding to medical treatment. In preventing cancer, one needs to choose nutrients which augment any of the cellular and organismic defense mechanisms reviewed. This is the foundation for the claim that many enzymes and proteins containing the trace elements used to control tumor growth and/or preventing cancer. In this study, the data which suggest that decreasing the amounts of Se, Ca, Fe, K, Mg, and Zn is associated with increased cancer are reviewed. Several hypotheses have been proposed to explain the effects of various chemotherapy based on its different biochemical activities, but not on the amount of trace elements in serums in pre-and-post chemotherapy periods. The variations in the amounts of elements for post-chemotherapy are because of either the cancer itself and/or previous treatments patients had taken before chemotherapy. Cancer leads to low trace element status, which supports the results presented in this study. In conclusion, the results of the present study confirm the decrease in most of the trace elements in cancer patients after chemotherapy. The decreases in the amounts show that chemotherapy destroys and/or damages body resistance and digestive system. The unwanted cases cause a decrease in biochemical reactions for many enzymes and proteins used to control tumor growth. The present study confirms that anticancer treatment is not appropriate for biological enzyme systems. However, further studies are still necessary to evaluate confirmation of lowering the amounts of trace elements by cancer and anticancer drug toxicity on digestive system, liver, and the other organs. The quantities of trace elements in serum decreases with their absorptions by body. The values given in the tables are the mean values of the data obtained from different patients with the same cancer types. Since the result obtained from different patients varies from each other due to fact that each patient's response to treatment was different. This is because each patient has a different degree and resistance to the disease, different body postures and periods of treatments. The ratios of the elements given in Tables 4 and 5 also decreased. However, this study clearly shows the fact that chemotherapy does not augment the rates of macro and trace elements. As the time passes the anti-cancer activities, synthesis of nucleic acids and proteins, and the structural stability diminish day by day. Consequently, the scientific approach in cancer therapy must be organized according to the stage, period of treatment and the physical state of the patients. Such changes in treatment of the cancer patients should be individually evaluated with continuing treatment according to their test results. Briefly, the characteristics of each patient are different from others. Considering the genetic impact on cancer, the question arises as to why other members of the family sharing the same environment do not have cancer. This is one of the major question to be answered. If genetic factors and lifestyle has a certain influence on the diseases, one would expect that all of the family members would suffer from the same diseases.

Patient No	Week	Ca/Mg	Ca/K	Se/Fe (× 10^{-3})	Se/Zn	Se/Pb
	1	3,338	0.487	1.348	51.508	13,945
	2	2,928	0.334	2.951	25.229	10,587
1	3	2,698	0.526	1.428	37.261	9,007
	4	9,466	2	3.815	44.538	7,361
	1	2,579	0.359	0.515	30.575	40,600
	2	3,394	0.503	1.05	50.656	51,000
2	3	2,033	0.46	0.695	46.97	13,127
	4	1,625	0.424	0.799	42.596	10,357
	1	6,664	1	1.964	35.335	214,000
	2	3,938	0.687	1.448	62.191	309,000
3	3	4,327	0.721	2.12	87.043	264,000
	4	4,925	0.8	3.234	181.186	202,000

Table 4: Variation ratio of elements with time for lung cancer.

Pati ent No	Week	Ca/Mg	Ca/K	Se/Fe (× 10^{-3})	Se/Zn
	1	4.291	0.370	1.531	104
	2	11.904	0.444	1.989	367
1	3	8.141	0.585	1.869	229
	4	12.829	0.761	1.951	122
	1	3.602	0.563	1.209	49
	2	2.814	2.814	1.274	193
2	3	2.459	0.471	1.591	451
	4	2.317	0.401	1.859	397
	1	7.586	0.530	1.890	Uncalculable
	2	10.462	0.641	3.821	Uncalculable
3	3	6.665	0.453	7.142	Uncalculable
	4	10.063	0.647	7.234	Uncalculable

Table 5: Variation ratio of elements with time for throat cancer.

References

1. Fukuda H, Ebara M, Yamada H, Arimoto M (2004) Trace Elements and Cancer. JMAJ 47: 391-395.
2. Ebrahim AM (2003) Study of Selected Trace Elements in Cancerous and Non-Cancerous Human Breast Tissues Using Neutron Activation Analysis.
3. Karimi G, Shahar S, Homayouni N, Rajikan R (2012)Association between Trace Element and Heavy Metal Levels in Hair and Nail with Prostate Cancer. Asian Pac J Cancer Prev 13: 4249-4253.
4. Kwiatek WM, Banas A, Banas K (2005) Iron and other elements studies in cancerous and non-cancerous prostate tissues. J Alloys Comp 401: 178-183.
5. Kaba M, Pirincci N, Yuksel MB, Gecit I (2014) Serum Levels of Trace Elements in Patients with Prostate Cancer. Asian Pac J Cancer Prev 15: 2625-2629.
6. Gecit I, Kavak S, Demir H (2011) Serum trace element levels in patients with bladder cancer. Asian Pac J Cancer Prev 12: 3409-3413.
7. Chan S, Gerson B, Subramaniam S (1998) The role of copper molybdenum selenium and zinc in nutrition and health. Clin Lab Med 18: 673-685.
8. Hebbrecht G, Maenhaut W, Reuck J (1999) Brain trace elements and aging. J Nucl Instrum Methods Phys Res B 150: 208-213.
9. Kok FJ, Van D, Hofman CM, VanDer A (1988) Serum copper and zinc and the risk of death from cancer and cardiovascular disease. Am J Epidemiol 128: 352-359.
10. Miura Y, Nakai K, Sera K, Sato M (1999) Trace elements in sera from patients with renal disease. J Nucl Instrum Methods Phys Res B 150: 218-221.
11. Adeoti ML , Oguntola AS , Akanni EO (2015)Trace elements; copper zinc and selenium in breast cancer afflicted female patients in LAUTECH Osogbo Nigeria. Indian Journal of Cancer 52: 106-109.
12. Liang P, Sang H (2008) Determination of trace lead in biological and water samples with dispersive liquid-liquid micro extraction pre-concentration. Analytical Biochemistry 380: 21-25.
13. Lukasewycz OA, Prohaska JR (1982) Immunization against transplantable leukemia impaired in copper-deficient mice. J Nati Cancer Inst 69: 489-493.
14. Peter RH, Berry MJ (2008) The influence of selenium on immune responses. Mol Nutr Food Res 52: 1273-1280.
15. Prasad AS, Kucuk O (2002) Zinc in cancer prevention. Cancer Metastasis Rev 21: 291-295.
16. Sattar N, Scott HR, Millan DC, Talwar D (1997) Acute-phase reactants and plasma trace element concentrations in non-small cell lung cancer patients and controls. Nutr Cancer 28: 308-312.

17. Kwong WT (2004) Interactions between iron deficiency and lead poisoning: epidemiology and pathogenesis. Science of the Total Environment 330: 21-37.

18. Kerper LE, Hinkle PM (1997) Cellular uptake if lead is activated by depletion of intracellular calcium stores. J Biol Chem 272: 8346-8352.

19. Aydın H, Oruç O (1997) Anodic stripping voltammetric determination of total lead, copper and selenium in whole blood and blood serum. Fresenius J Anal Chem 358: 859-860.

20. Aydin H, Tan GH (1991) Differential-pulse polarographic behaviour of selenium in the presence of copper, cadmium, and lead. Analyst 116: 941-945.

21. Inam R, Aydın H (1996) Determination of Lead Copper and Selenium in Turkish and American Cigarette Tobaccos by Anodic Stripping Voltammetry. Analytical Science 12: 911-915.

22. Aydın H, Somer G (1995) Elemental Determination of Selenium. Encyclopaedia of Analytical Science, Academic Press Limited, London, UK.

23. Somer G, Aydın H (1995) Determination of Selenium Compounds. Encyclopaedia of Analytical Science, Academic Press Limited, London, UK.

24. Watkinson JH (1966) Fluorometric determination of selenium in biological material with 2,3-diaminonaphthalene. Anal Chem 38: 92-97.

25. Nygaard D, Lowry J (1982) Determination of selenium in blood serum by ICP-OES including an on-line wet digestion and Se-hydride formation procedure Anal Chem 54: 803-807.

26. Nakahara T, Kikui N (1985) Introduction to Inductively Coupled Plasma Atomic Emission Spectrometry Spectrochimica Acta 40: 21-28.

27. Massadeh A (2010) Simultaneous determination of Cd, Pb, Cu, Zn, and Se in human blood of jordanian smokers by ICP-OES. Biol Trace Elem Res 133: 1-11.

28. Lavilla I (2009) Elemental fingerprinting of tumorous and adjacent non-tumorous tissues from patients with colorectal cancer using ICP-MS, ICP-OES and chemometric analysis. Bio metals 22: 863-875.

29. Cavusoglu K (2008) Radiotherapy Determination of Change in Plasma Trace Levels of Goren Akciger Cancer Patients 22: 211-222.

Ni and Co Substituted Zinc Ferri-chromite: A Study of their Influence in Photocatalytic Performance

VT Vader*

Walchand College of Arts and Science, Solapur, Maharashtra, India

Abstract

Enhanced photocatalytic activity was demonstrated in the present study for two systems i.e., $Zn_{1-x}Ni_xFeCrO_4$ and $Zn_{1-x}Co_xFeCrO_4$ ($0.0 \le x \le 1.0$) in photo degradation of methyl orange, methyl red and congo red dyes. The basic aim of this study is to find a suitable compound of doped zinc ferrichromite as photocatalyst for degradation of organic dyes. A sol-gel process used for synthesis of crystalline nanopowders gave better photocatalytic activity under the UV light radiations. The result showed that, dyes undergo fast degradation with UV light in presence of doped compounds. The rate of degradation of dyes was estimated spectro-photometrically from residual concentration in dyes.

Keywords: Sol-gel synthesis; Surface area; Pore volume; Photocatalytic activity

Introduction

In the compounds of mixed metal oxides, spinel ferrites find wide applications in both technological and catalytic field. Spinels of AB_2O_4 type, where 'A' and 'B' are tetrahedral and octahedral symmetric oxygen sites containing divalent or trivalent cations have been studied for their innumerable technological applications such as transformer cores, micro-devices in radio frequency coils due to excellent magnetic properties, low magnetic coercivity, high resistivity, small eddy current losses, high curie temperature and chemical stability [1-4]. Their properties like crystal structure, electronic conduction, magnetism and catalytic activities are depends upon the type of magnetic ions residing in A- or B-sites and their sublattice interactions. In the potential applications of Ni and Co based soft ferrites, especially where these elements are partially substituted for zinc in spinel lattice strongly affect the properties of Ferro spinel. $ZnFeCrO_4$ a normal spinel and its degree of inversion depends upon nature of substituted cations. However if A-site get occupied by diamagnetic ion, then B-B interaction become quit predominant. Ni-Zn ferrite or Co–Zn ferrite are inverse spinel where Zn^{2+} ions are present in tetrahedral site while octahedral site is occupied by Ni^{2+}, Co^{2+}, Fe^{2+} and Fe^{3+} ions. The catalytic activity is because of the migration of metallic ions between sub lattices which do not alter a basic structure and makes a catalyst more effective for reactions [5,6].

During recent years, photo catalytic oxidation of dyes using ferrites has gained enormous interest. Most of the dyes used in the pigmentation of leather, textiles, paper, ceramics, cosmetics, inks and food-processing products are derived from azo dyes, which are characterized by the presence of one or more azo groups (-N=N-) in their structure. But during synthesis and processing of materials, much of the dyes get waste within water. Because of toxic nature of azo dyes, this waste produces a great hazard to human and environmental health. Therefore, it is a challenge to researchers for effective removal of dyes of wastewater from industries. Hence for degradation of polluted waste water containing both toxic and non-biodegradable compounds, the development of suitable process is needed [7,8]. The study on ferrospinels based Ni and Co substituted zinc ferrichromites for their structural characterization and electronic properties was recently reported [9,10]. This work is again extended to utilize the synthesized compounds as photocatalysts for degradation of some organic dyes and to determine the detailed degradation kinetics.

Experimental Details

Nanocrystalline Ni and Co doped ferrichromites were prepared separately by sol-gel auto-combustion method using A.R. grade nitrate salts of Fe, Zn, Cr, Ni and Co. For the preparation of system $Zn_{1-x}Ni_xFeCrO_4$ ($0.0 \le x \le 1.0$), calculated amount of zinc nitrate ($Zn(NO_3)_2 \cdot 6H_2O$), chromium nitrate ($Cr(NO_3)_3 \cdot 9H_2O$), iron nitrate ($Fe(NO_3)_3 \cdot 9H_2O$) and nickel nitrate ($Ni(NO_3)_2 \cdot 6H_2O$) were weighed according to required stoichiometry and dissolved separately in deionized water to obtain a clear solution. An equimolar solution of citric acid was then added to all solution and heated at 80°C to about 30 minute under constant stirring. After the solutions to clear and transparent, a small quantity of ammonia was added into the solutions to adjust pH-7-8. Then all solutions were poured into a tray and heated slowly to 110°C which turned into porous dry gel. This mixture was continuously evaporated to almost dryness to obtain a homogeneous uniform material. Here precursors undergo a strong auto-combustion process during ignition of dried gel [11]. The same method was followed for preparation of system $Zn_{1-x}Co_xFeCrO_4$ ($0.0 \le x \le 1.0$) where cobalt nitrate ($Co(NO_3)_2 \cdot 9H_2O$) was used instead of nickel nitrate. All the dried citrate compounds were calcinated separately at 800°C for 6 h (first monitored from TGA-DTA) to breakdown the citrate gel complexes into spinel oxides. Surface area by Brunauer–Emmett–Teller (BET) technique and pore volume distribution for the samples of good activity were obtained by employing nitrogen as adsorbing gas. Initially, samples were regenerated for degassing at 150°C for 1 h and the data were measured using Smart instrument: Model Smart-Sorb 92/93 surface area and pore volume analyzer with the help of liquid nitrogen (77K) and water (300K) atmospheres for both adsorption and desorption of N_2. The photocatalytic activities of synthesized compounds were tested by using anionic dyes such as 10 ppm methyl orange (MO), 10 ppm methyl red (MR) and 5 ppm congo red (CR) solutions for their decolorisation and degradation process. These dye solutions were used as a test contaminant since it has been extensively used as an indicator for the photocatalytic activities owing to its absorption peaks in the visible range. Hence its degradation can be easily monitored by optical absorption spectroscopy. A 100 ml volume of dye in aqueous solution with 100 mg of catalyst was used in all

***Corresponding author:** VT. Vader, Walchand College of Arts and Science, Solapur-413 006, Maharashtra, India, E-mail: vtv_chem@rediffmail.com experiments. The photoreactor consist of a jacketed quartz tube in which a high pressure mercury vapor lamp of 125 W power was placed. At the first, photolysis was also carried for dyes solution using UV radiations without photocatalyst. Prior to irradiation, the suspensions were continuously stirred in the dark for 60 min to ensure the establishment of an adsorption/desorption equilibrium. Then the reactor was irradiated with UV light corresponding to their absorption wavelength. The distance between a lamp and center of the beaker was kept at 8-10 cm. About 2-3 ml of sample was taken at 15 minute intervals of time up to 90 minute. Before measurement, all the samples were centrifuged for 5 minute. The course of degradation reaction and concentration of dye in samples were monitored at regular intervals of time using a UV-Visible spectrophotometer.

Results and Discussion

Photocatalytic activity

The photocatalytic activity of synthesized samples under UV light was evaluated by performing experiments on the degradation of MO, MR and CR in aqueous solution. Figure 1 illustrates the degradation of these dyes by photocatalysis in the presence of different Ni and Co substituted zinc ferri-chromites for one hour. The results showed that, the concentration of dye solution barely changed after a solution had been directly illuminated. The absorption peaks of dyes became weaker along with the irradiation time and hence azo groups as well as aromatic part of the dyes molecule were destructed under UV light. This indicates the zinc substitution by nickel and cobalt in ferri-chromite increase the photocatalytic degradation of dyes. It was found that MO degraded to 30% by catalyst $Zn_{0.4}Ni_{0.6}FeCrO_4$, MR degraded to 20% by same catalyst while CR degraded to 45% by catalyst $Zn_{0.2}Ni_{0.8}FeCrO_4$ after one hour. For Co doped zinc ferri-chromites, MO degraded to 70% by catalyst $Zn_{0.4}Co_{0.6}FeCrO_4$, MR degraded to 40% by same catalyst while CR degraded to 15% by catalyst $Zn_{0.6}Co_{0.4}FeCrO_4$ after one and half hour. This could be explained on the basis of smaller crystallite size and optimal concentration of substituted ions in zinc ferri-chromite for more efficient separation of photo-induced electron-hole pairs. If substitution increases, the surface barrier becomes higher and a space charge region becomes narrower [12-14]. Both Ni^{2+} and Co^{2+} have more partially filled d-orbitals and provide an effective recombination center for electrons and holes. There is more catalysts surface area for absorption of photon and interaction of molecules of reactants with catalyst, due to which number of holes, hydroxyl radicals and supra oxide ions get increased. But with increase in content of doped ions over optimal point i.e., when the substitution content exceeds ($x \geq 0.8$), the space charge region becomes nar-rower and photo-generated electron–hole pair recombination take place, leading to decreased photocatalytic activity of system. In addition, it affects the optical and electronic properties and presumably shift the optical absorption towards the visible region [15-18]. Other reasons of activity of compounds are the surface area, crystallite size, phase purity and surface defects on surface layer. The measured BET surface areas and pore volumes of the active photocatalysts are given in Table 1. Due to different degrees of agglomeration of grains and number of fine pores, an average surface area is found to be 70 m^2g^{-1}. The substituted Ni and Co ions have partially filled d-orbitals. Hence their doping in $ZnFeCrO_4$ produces the surface defects. At higher concentration of substitution, there is a remarkable change in surface area and pore size as the substituted ions occupies the pores. This may be attributed to sintering process and increasing crystallite size. By measuring an absorbance with irradiation time, the concentrations of degraded dyes were determined for which the catalysts show high degree of decolorisation and degradation. Figure 2 shows the plots of $\ln C/C_0$ against irradiation time from which kinetic rate constant (k) of degradation of dyes was calculated by

$$\ln C/C_0 = -k \times t$$

C_0: dye concentration at t=0

C: concentration at time t.

These time varying degradation rates for used photocatalyst are shown in Table 1.

Reusability of photocatalyst

The spent catalysts were tested for its reusability and stability after separating the catalysts from the reactant solution by centrifugation, washing and drying at 120°C in an oven. Their structure was further examined by XRD and SEM image. It was found that there is no much more change in the structure and morphology of used catalyst. The peaks of XRD patterns were again well indexed by Bragg's law showing fundamental peaks of single phase face centered cubic spinel structure. The used catalysts were found to a well-defined crystalline phase as they have before the catalysis. The morphological images visualized by SEM also exhibit an intergranular porosity and the absence of internal pores. The intergranular pores are linked through large pores and this porous structure favors the adsorption and condensation of dye solutions [9,10]. Figures 3 and 4 shows the X-ray pattern and SEM images of used catalysts in photodegradation of dyes.

Conclusions

The progress of photocatalytic study of dyes shows that the facile sol-gel synthesized Ni and Co substituted zinc ferrichromites are the effective catalysts in decolorisation and degradation of methyl orange, methyl red and congo red under UV irradiation. The activity of substituted Ni and Co ferrichromites is due to increased optical absorption and increase life time of charge carriers. These recyclable photocatalyst are reusable as there is no more major change in their structure after photocatalysis.

Acknowledgements

Author is very much thankful to Indian Academy of Sciences, Bangalore, Indian National Academy of Sciences, New Delhi and The National Academy of Sciences India, Allahabad for awarded 'Summer Research Fellow-2015' to pursue research at Indian Institute of Science, Bangalore.

Compound	Surface area $m^2 \cdot g^{-1}$	Pore volume $cc \cdot g^{-1}$	Kinetic rate constant (k) min^{-1}
$Zn_{0.4}Ni_{0.6}FeCrO_4$	68	0.11	0.0049 for MO and 0.004 for MR
$Zn_{0.2}Ni_{0.8}FeCrO_4$	76	0.15	0.00861 for CR
$Zn_{0.4}Co_{0.6}FeCrO_4$	72	0.13	0.00298 for MO and 0.011 for MR
$Zn_{0.6}Co_{0.4}FeCrO_4$	66	0.09	0.015 for CR

Table 1: Surface area, pore volume and kinetic rate constant for degradation of dyes of compounds which works as effective photocatalyst.

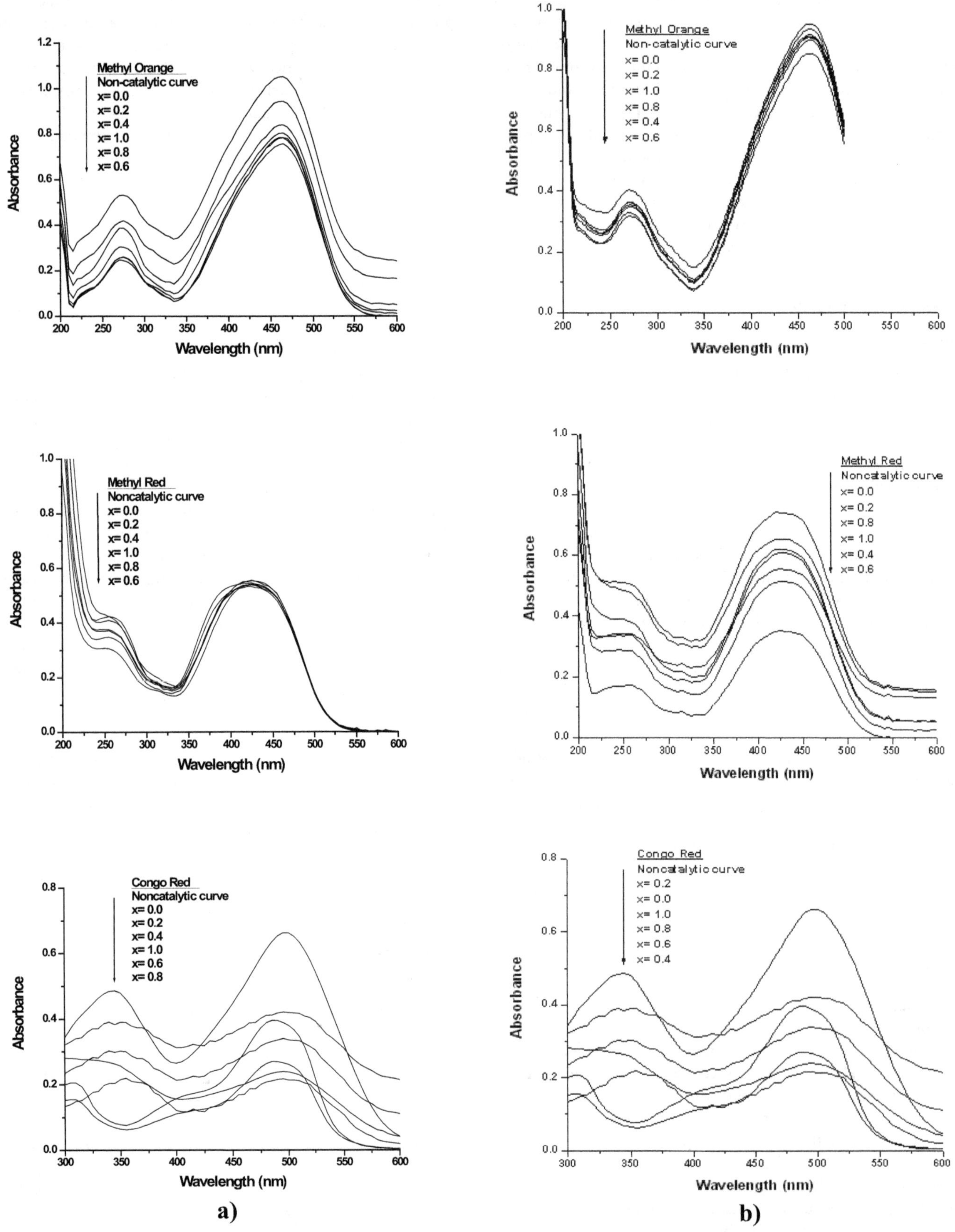

Figure 1: Photodegradation of the MO, MR and CR dyes using a) Ni b) Co doped photocatalyst ($0.0 \leq x \leq 1.0$) after one hour UV irradiation.

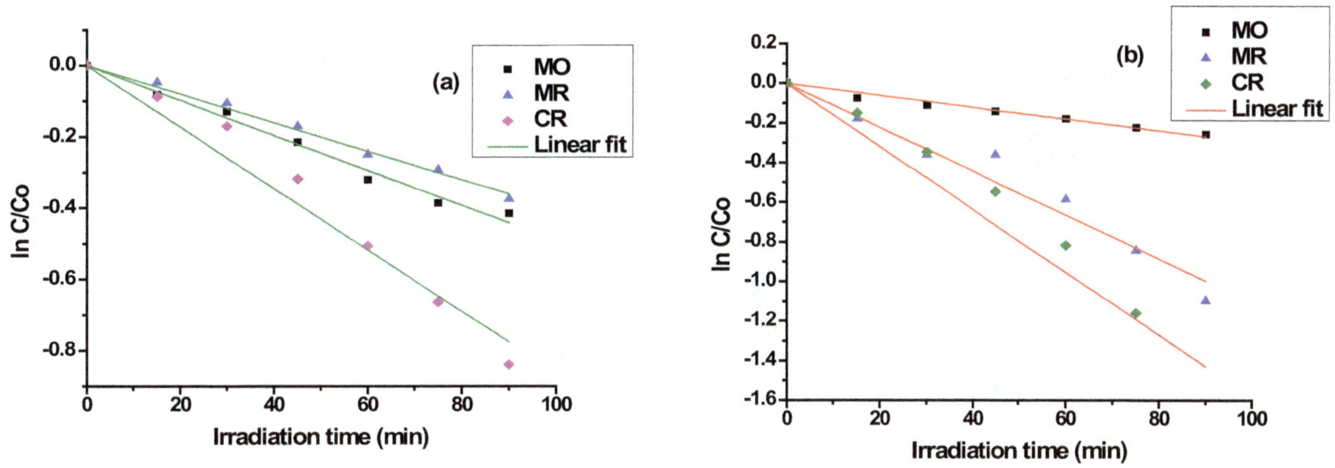

Figure 2: Graph of ln C/Co Vs irradiation time for a) Ni b) Co doped used photocatalyst.

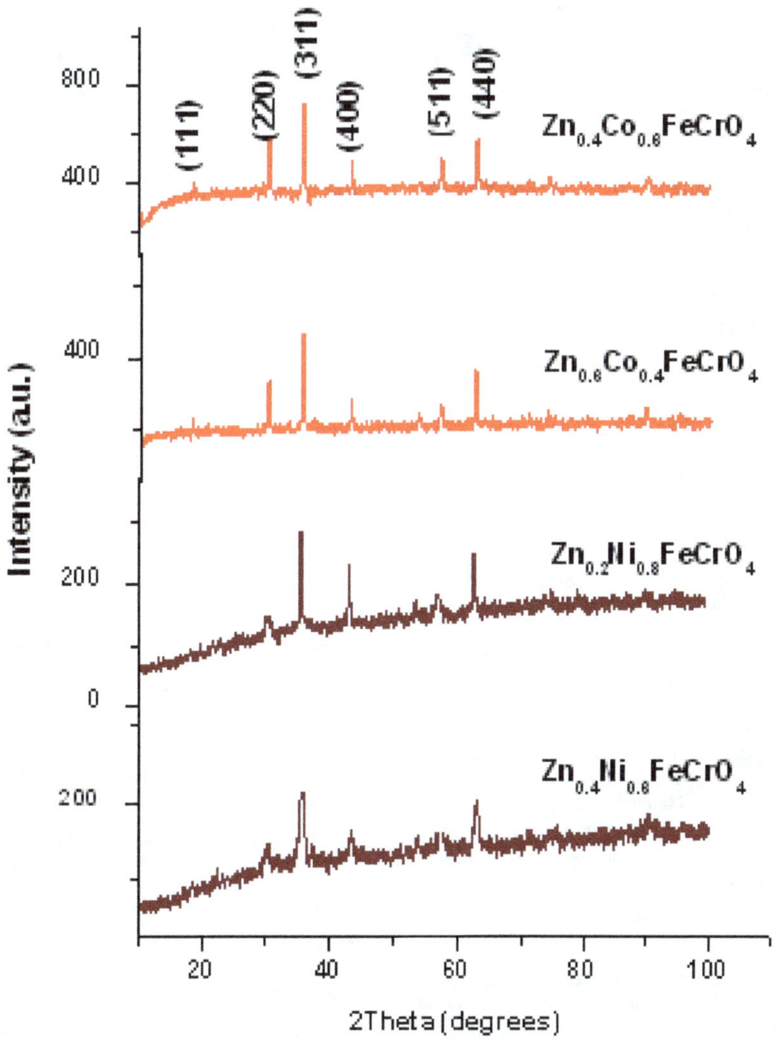

Figure 3: X-ray diffraction patterns of compounds after photocatalysis.

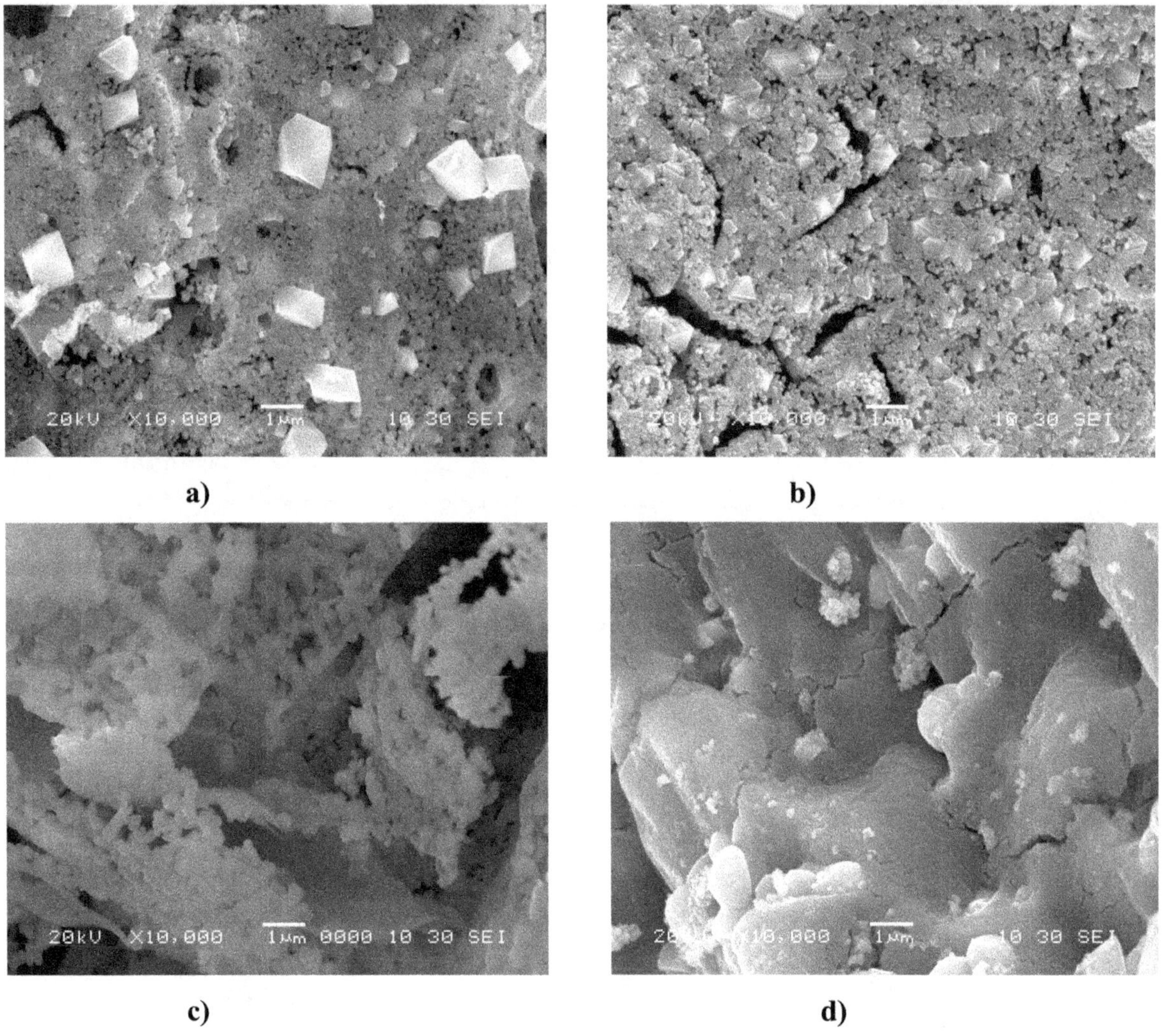

a) $Zn_{0.4}Ni_{0.6}FeCrO_4$ b) $Zn_{0.2}Ni_{0.8}FeCrO_4$
c) $Zn_{0.6}Co_{0.4}FeCrO_4$ d) $Zn_{0.4}Co_{0.6}FeCrO_4$.

Figure 4: Surface morphology by SEM of used compounds after photocatalysis.

References

1. Köseoğlu Y (2015) Structural and magnetic properties of Cr doped NiZn-ferrite nanoparticles prepared by surfactant assisted hydrothermal technique. Ceramics International 41: 6417-6423.
2. Hajalilou AM, Hashim R, Kahrizsangi E, Halimah Kamari MJ (2015) Influence of evolving microstructure on electrical and magnetic characteristics in mechanically synthesized polycrystalline Ni-ferrite nanoparticles. Journal of Alloys and Compounds 633: 306-316.
3. Sawant VS, Rajpure KY (2015) The effect of Co substitution on the structural and magnetic properties of lithium ferrite synthesized by an autocombustion method. Journal of Magnetism and Magnetic Materials 382: 152-157.
4. Reddy V, Byon C, Narendra B, Baskar D, Srinivas G, et al. (2015) Investigation of structural, thermal and magnetic properties of cadmium substituted cobalt ferrite nanoparticles. Superlattices and Microstructures 82: 165-173.
5. Cullity BD (1972) Introduction to Magnetic Materials. Addison-Wesley, New York, USA.
6. Varshney D, Verma K, Kumar A (2011) Substitutional effect on structural and magnetic properties of A_xCo 1-xFe_2O_4 (A= Zn, Mg and x= 0.0, 0.5) ferrites. Journal of Molecular Structure 1006: 447-452.
7. Wang L, Li J, Wang Y, Zhao Z, Jiang Q, et al. (2012) Adsorption capability for Congo red on nanocrystalline MFe_2O_4 (M= Mn, Fe, Co, Ni) spinel ferrites. Chemical Engineering Journal 181: 72-79.
8. Sathishkumar P, Mangalaraja RV, Anandan S, Ashokkumar M (2013) Chem Engin J 220: 302-310.
9. Vader VT, Hankare PP (2012) Solid State Sci 14: 885-889.
10. Hankare PP, Vader VT, Sankpal UB, Patil RP, Jadhav AV, et al. (2011) Synthesis and characterization of cobalt substituted zinc ferri-chromites prepared by sol–gel auto-combustion method. Journal of Materials Science: Materials in Electronics 22: 1109-1115.
11. Vader VT, Achary SN, Meena SS (2014) A facile gel-combustion route for fine particle synthesis of spinel ferrichromite: X-ray and Mössbauer study on effect of Mg and Ni content. Materials Research Bulletin 50: 172-177.
12. Valenzuela MA, Bosch P, Becerrill JJ, Quiroz O, Paez AI, et al. (2002) Preparation, characterization and photocatalytic activity of ZnO, Fe_2O_3 and $ZnFe_2O_4$. Journal of Photochemistry and photobiology A: Chemistry 148 : 177-182.
13. Yuan W, Liu X, Li L (2014) Synthesis, characterization and photocatalytic activity of cubic-like $CuCr_2O_4$ for dye degradation under visible light irradiation. Applied Surface Science 319: 350-357.
14. Borhan AI, Samoila P, Hulea V, Iordan AR, Palamaru MN, et al. (2014) Photocatalytic activity of spinel $ZnFe_{2-x}Cr_xO_4$ nanoparticles on removal Orange I azo dye from aqueous solution. Journal of the Taiwan Institute of Chemical Engineers 45: 1655-1660.

15. Sharma R, Singhal S (2013) Phys B Conden Matt 414: 83-90.

16. Sasikalaa R, Shirole AR, Sudarsan V, Kamble VS, Sudakar CR, et al. (2010) Role of support on the photocatalytic activity of titanium oxide. Applied Catalysis A: General 390: 245-252.

17. Bhukal S, Shivali S, Singhal (2014) Magnetically separable copper substituted cobalt–zinc nano-ferrite photocatalyst with enhanced photocatalytic activity. Materials Science in Semiconductor Processing 26: 467-476.

18. Mahmoodi NM, Bashiri M, Moeen SJ (2012) Synthesis of nickel–zinc ferrite magnetic nanoparticle and dye degradation using photocatalytic ozonation. Materials Research Bulletin 47: 4403-4408.

33

Annual Variation in Population of the Bean Fly Leaf miner, *Melanagromyza phaseoli* (Tryon), Generation Numbers and Related Injury in Common Bean Fields

Hamdy A Salem*, Shadia E Abdel-Aziz and Noeman B Aref

Department of pests and plant protection, Division of Agriculture and Biological Research, National Research Center, Dokki, Cairo, Egypt

Abstract

Four experiments were conducted at Embaba, Giza Governorate by planting beans four times during 2011, 2012 and 2013 to assess the annual variation in population of *Melanagromyza phaseoli* attacking bean plants *Phaseolus vulgaris* (L.) in relation to certain climatic factors, synchronization of plant age with insect counts, generation numbers, and injury on plants. Obtained results cleared that adults and larvae of this fly attack plants as soon as seedlings emergence in all growing seasons. In spring 2011, adult females reached highest counts 1.17, 1.17 and 1.1 individual/sweep at 12th April, 10th May and 7th June, respectively. Their larvae reached highest 2.2 and 15.33 larvae/leaf at 26th April and 31st May. The highest tunnels number during season was 26.87 tunnel/leaf. In summer 2011, larvae reached highest counts 1.4, 29 and 2.3 larvae/leaf at 5th July, 23rd August and 20th September at temperature 27.17, 29.84 and 27.97°C and relative humidity 54.14, 54.57 and 57.71%, respectively. During summer 2012, both adults and larvae reached its highest three times, larvae were 1.9, 4.87 and 7.2 larvae/leaf. Tunnels number was 11.07/leaf in highest case. Leaf miner population was rather low during winter 2012/2013.

Capabilities of insect to produce progeny and to injure plants were discussed. Insect produced three generations/growing season. Temperature and relative humidity revealed their presence inside the optimum range for the insect activity. Plant age had positive relationship with larvae counts in spring and summer seasons. Management program must be directed against insect during period 39-74 day of plant age.

Keywords: Annual variation; *Melanagromyza phaseoli*; Generation numbers; Related injury; *Phaseolus vulgaris*

Introduction

Common bean, *Phaseolus vulgaris* (L.) regard among the most important legume crops in Egypt. It is infesting by bean fly *Melanagromyza Phaseoli* (Tryon) (Agromyzidae); this insect is devastating for its feeding on young plants or young leaves as soon as seedlings emergence [1-3]. Adult females puncture both upper and lower surface of leaves to lay eggs or to feed [4]. Larvae feed as leaf miner and also as stem borer [1,3]. Damage by leaf miner represented in mining, wilting and followed by drying of leaves [5]. In field studies, it is important to follow up the population growth of insect, define the insect counts necessary to apply control measures, timing of insect counts with plant phenology; and select the appropriate sampling technique [6].

In Egypt, Giza governorate this insect was ignored and did not receive any attention from researchers until became dangerous on bean plants. So the objective of this study was to assess the annual variation in larvae and adults population of *M. phaseoli* among different growing seasons in relation to certain climatic factors, synchronization of plant age with insect counts, generation numbers and the related injury on plant due to insect activity.

Material and Methods

Experiments preparation

Four experiments were conducted at Embaba district, Giza Governorate, Egypt. Two experiments were done during 2011 by planting beans at spring and summer, the third one in summer 2012; and fourth experiment at winter 2013. An area about 3 Karats (1/8 feddan) (feddan=4200 m^2) was prepared in each experiment and planted by common bean Giza 6 at dates 24th March 2011, 23rd June 2011, 18th July 2012 and 13th December 2013. Experimental area was divided into three equal parts to use as replicates. Conventional agricultural practices were followed and chemical control was entirely avoided.

Sampling methods

Standard sampling methods were used to estimate leaf miner population according to Pohronezny and Waddil [7]; Gangrade and Kogan [2]; Johnson et al. [8]; Pohronezny et al. [9]; adult flies are very active, cannot estimated accurately on leaves or foliage, therefore sweep net method was used to make 30 double strokes above plants, captured adults were killed in cyanided jar and counted. As well as larvae population was estimated on foliar samples of 30 leaves/experiment every week (10 leaves/replicate), samples were picked up randomly from leaves of the middle level on plants, which harbored the highest density of larvae [10]. Alive larvae were counted using binocular microscope. Tunnels numbers were recorded.

Calculations and statistical analysis

Daily records of temperature and relative humidity during the experimental periods were obtained from the Central Laboratory for Agricultural Climate. Weekly means of two climatic factors used to calculate the simple correlation and regression values according to Gomez and Gomez [11]. Captured adults/sweep; larvae and tunnels numbers/leaf; and percent of infested leaves/sample were calculated.

***Corresponding author:** Hamdy A Salem, Department of pests and plant protection, Division of Agriculture and Biological Research, National Research Center, Dokki, Cairo, Egypt, E-mail: hamdyabdelnaby@ymail.com

Number and duration of insect generations were calculated according to method of Audemard and Miliare [12] and Jacob [13].

Results and Discussion

Assessment the insect population and related injury

Spring experiment: Data in Table 1 cleared that adult of *Melanagromyza phaseoli* had three peaks of activity and reached its highest counts 1.17, 1.17 and 1.1 individuals/sweep at 12th April, 10th May and 7th June, respectively. Larvae reached its highest counts twice 2.2 and 15.33 larvae/leaf at 26th April and 31st May when temperature was 19.47 and 27.91°C; and relative humidity was 53.43 and 43.86% respectively. The highest number of tunnels was 26.8 tunnel/leaf appeared timing with the presence of highest counts of larvae (15.33 larvae/leaf) at 31st May. Highest percent of infested leaves during the season growth was 100% at 31st May.

It is worth to mention that adult fly was catched during the season growth by grand mean 0.745 individuals/sweep; and produced larvae estimated by means 2.7 larvae/leaf; these larvae tunneled and injured leaves by 5.3 tunnels/leaf. This mean that each sampled leaf harbored actual infestation estimated by 2.71 larvae and harbored supposed infestation on leaves=5.35 tunnels (larvae)/leaf wherever each larva mine one tunnel; this tunnel still presented on leaves and did not disappear after larvae developed into another stages and emerged out. Percent of infested leaves was 59.72%.

Summer experiments: Data in Table 2 cleared that adult fly reached its highest counts four times in summer 2011 (2, 0.93, 1.13 and 0.8 individuals/sweep). Larvae had three peaks of activity, reached its highest with more abundant and dominant counts in this season, threating the crop (1.4, 29 and 2.3 larvae/leaf) at 5th July, 23rd August and 20th September, when temperature was 27.17, 29.84 and 27.97°C and relative humidity was 54.14, 54.57 and 57.71%, respectively. The highest number of tunnels (56 tunnels/leaf) appeared timing with the presence of highest counts of larvae (29 larvae/leaf) at 23rd August.

Highest percent of infested leaves/sample during growth season was 93.33% at 6th and 20th of September. Finally, adults captured by mean 0.81 individual/sweep during summer 2011, produced larvae (actual infestation) by mean 4.44 larvae/leaf; these larvae fed, injured and mined leaves by 7.5 tunnels/leaf; so supposed infestation=7.5 larvae (tunnel) leaf; this actual or supposed infestation resulted in 81.96% infestation on sample leaves.

Data in Table 3 cleared that adults reached its highest counts three times in summer 2012, with counts 1.5, 1.23 and 1 individual/sweep. Larvae revealed a trend of activity similar with their mothers and reached its highest counts 1.9, 4.87 and 7.2 larvae/leaf at 7th August, 28th August and 2nd October, when temperature was 30.38, 29.78 and 27.39°C and relative humidity was 57.43, 55.0 and 55.57%, respectively. The highest number of tunnels was 11.07 tunnels/leaf related to the activity of highest counts of larvae (7.2 larvae/leaf). The highest percent of infested leaves/sample was 100% appeared twice during season growth.

Captured adults was estimated by mean 0.79 individuals/sweep during summer 2012, produced progeny=2.38 larvae/leaf; injured and mined leaves by mean 4.58 tunnels/leaf (supposed infestation) Resulted infestation was 82.36% on sampled leaves.

Winter experiment: Data in Table 4 cleared that adult fly reached its highest counts three times (1.3, 1.0 and 0.9 individual/sweep at 28th January, 25th February and 18th March of 2013. But larvae counts in this season was lower than its counts in three previous seasons, it is reached its highest (1.17 larva/leaf) at 21st January in conditions of 14.39°C and 56.86%. Highest number of tunnels during winter 2013 was 2.7 tunnel/leaf. The highest percent of infested leaves/sample was 70%.

During the experimental seasons, it was observed that the captured adults had near values 0.745, 0.81, 0.79 and 0.76 individual/sweep at spring, summer 2011, summer 2012 and winter 2013 (Tables 1-4), respectively. In spite of this, produced larvae differed in its counts, it was 2.71 (larvae/leaf in spring 2011 and duplicated in summer of the same year (4.44 larvae/leaf); and back to 2.38 larvae/leaf in summer 2012; and was minimized (0.46 larvae/leaf) in winter 2013. So that, larvae counts regard the limited parameter for insect in causing its injury on plants. Tunnels number showed similar trend as the larvae counts; but it was dominated and higher than larvae counts, because the larvae completed its development and emerged out tunnels in spite of the still presence of these tunnels; tunnels number changed from 5.301 to 7.5 tunnel/leaf; and from 4.58 to 1.06 tunnel/leaf in spring, summer 2011; and in summer 2012, winter 2013 respectively. Such results are in agreement with results of Assem [1] who found that lowest activity of *M. phaseoli* was recorded in winter. El Gendi et al. [10], El-Khouly et al. [14] and Abou-El-Haggag and Salman [15] who found that larvae of *Liriomyza trifolii* has three peaks of activity per season. They also agree with results of Assem [1], El-Bessomy [16]; Omar and Faris [17] who revealed that larvae of *M. phaseoli* primary mine between leaf

Sampling Date	Plant age (days)	Insect Counts		Related Injury		Climatic factors	
		Adults/sweep	Larvae/leaf	Tunnels no./leaf	%infested leaves	Temperature (°C)	R.H%
April 5	11	0.57	0.13	1	20	19.89	56.29
12	18	1.17	0.33	1.5	23.33	20.06	52.43
19	25	0.9	1.8	1.6	46.67	23.71	37.43
26	32	1.03	2.2	1.9	66.67	19.47	53.43
May 3	39	0.5	0.2	2	73.33	24.3	47.43
10	46	1.17	0.2	2.47	80	23.26	51.14
17	53	0.63	0.73	1.27	60	22.16	50.57
24	60	0.37	1.13	6.4	40	25.41	46.43
31	67	0.7	15.33	26.87	100	27.91	43.86
June 7	74	1.1	6.87	14.4	93.33	26.61	54.0
14	81	0.5	2.33	2.8	73.33	28.37	46.86
21	88	0.3	1.23	1.4	40	27.34	54.43
Total Mean		8.94 0.745	32.48 2.71	63.61 5.301		716.66 59.72	

Table 1: Mean counts of bean fly *Melanagromyza phaseoli* related injury on bean plants and climatic factors during spring 2011 at Giza Governorate.

Sampling date	Plant age (days)	Insect counts		Related injury		Climatic factors	
		Adults/sweep	Larvae/leaf	Tunnels no./leaf	%infested leaves	Temperature (°C)	R.H%
April 5	11	2	1.4	1.62	77	27.17	54.14
12	18	0.97	1.27	3.13	86.67	29.06	55.0
19	25	0.7	1.2	2.33	86.67	29.99	55.57
26	32	0.43	1.13	1.53	86.67	30.03	57.43
August 2	39	0.93	0.8	1.9	73.33	29.57	61.57
9	46	0.4	0.73	1.83	60	29.34	60.14
16	53	1.13	1.53	4.3	80	28.17	60.71
23	60	0.7	29	56	80	29.84	54.57
30	67	0.7	15	15	70	28.93	56.43
September 6	74	0.5	1.53	2.3	93.33	28.53	59.57
13	81	0.6	2.2	4.0	87	28.0	61.57
20	88	0.8	2.3	3.43	93.33	27.97	57.71
27	95	0.7	2.13	4	86.67	27.27	55.43
October 3 2011	102	0.77	1.93	3.67	86.67	22.64	65.0
Total Mean		11.33 0.81	62.15 4.44	105.04 7.5		1147.38 81.96	

Table 2: Mean counts of bean fly *M. phaseoli* related injury on bean plants and climatic factors during summer 2011 at Giza Governorate.

Sampling date	Plant age (days)	Insect counts		Related injury		Climatic factors	
		Adults/sweep	Larvae/leaf	Tunnels no./leaf	%infested leaves	Temperature (°C)	R.H%
July 31	11	1.5	0.9	1.0	27	30.51	58.71
August 7	18	1.3	1.9	2.2	80	30.38	57.43
14	25	0.8	1.75	2.3	90	32.14	47.29
21	32	1.23	1.25	1.7	76.66	29.90	53.71
28	39	0.8	4.87	6.87	87	29.78	55.0
September 4	46	0.6	1.5	2.0	80	28.96	59.0
11	53	0.37	0.45	2.07	60	28.24	59.29
18	60	0.57	1.2	3.47	86.66	27.88	51.71
25	67	1.0	5.4	8.1	100	26.73	58.0
October 2	74	0.7	7.2	11.07	90	27.39	55.57
9	81	0.67	2.2	7.9	96.66	26.17	59.0
16	88	0.47	2.0	4.9	96.66	26.41	57.71
23	95	0.3	0.3	5.9	100	26.06	63.43
Total Mean		10.31 0.79	30.92 2.38	59.48 4.58		1070.64 82.36	

Table 3: Mean counts of bean fly *M. Phaseoli* related injury on bean plants and climatic factors during summer 2012 at Giza Governorate.

Sampling date	Plant age (days)	Insect counts		Related injury		Climatic factors	
		Adults/sweep	Larvae/leaf	Tunnels no./leaf	%infested leaves	Temperature (°C)	R.H%
December 24	11	0.1	0	0	0	15.23	51.29
31	18	0.5	0	0	0	14.75	73.0
January 7 2013	25	0.8	0.3	0.6	30	14.58	64.14
14	32	0.8	0.8	2.8	60	12.39	69.29
21	39	1.13	1.17	1.1	63.64	14.39	56.86
28	46	1.3	0.5	1.0	60	15.88	54.14
February 4	53	0.7	0.9	1.6	70	14.28	60.43
11	60	0.7	0.5	2.7	70	16.18	54.43
18	67	0.8	0.6	2.1	70	15.22	33.0
25	74	1	0.7	1.5	70	16.94	54.57
March 4	81	0.7	0.3	1.1	70	18.45	56.86
11	88	0.5	0.2	0.8	60	18.21	53.29
18	95	0.9	0.3	0.8	60	21.13	45.43
25	102	0.7	0.2	0.9	50	19.09	51.43
Total Mean		10.63 0.76	6.47 0.46	14.77 1.06		733.64 52.4	

Table 4: Mean counts of bean fly *M. Phaseoli* related injury on bean plants and climatic factors during winter 2013 at Giza Governorate.

Experimental Season	1st generation period	2nd generation period	3rd generation period
Spring 2011	From April 1st to 19th April (3 weeks)	From 19th April to 24th May (5 weeks)	From 24th May to 21st June (4 weeks)
Summer 2011	From 5th July to 19th July (3 weeks)	From 19th July to 16th August (4 weeks)	From 16th August to 31st October (7 weeks)
Summer 2012	From July 31st to 21st August (4 weeks)	From 21st August to 25th September (5 weeks)	From 25th September to October 31st (4 weeks)
Winter 2013	From December 31st to January 31st (6 weeks)	From February 1st to 18th February (3 weeks)	From 18th February to 25th April (5 weeks)

Table 5: Number of generations and duration periods of *Melanagromyza phaseoli* on bean plants during 2011, 2012 and 2013.

Experimental Season	Plantage		Temperature		Relative humidity	
	r	b	r	b	r	b
Spring 2011	0.39	0.07	0.51	0.37	-0.27	-0.22
Summer 2011	0.125	0.034	0.22	0.93	-0.36	-0.88
Summer 2012	0.18	0.014	-0.021	-0.23	-0.13	-0.07
Winter 2013	-0.03	-0.0003	-0.43	-0.063	-0.08	-0.007

Table 6: Simple correlation (r) and regression (b) for leaf miner larvae with certain studied factors.

layers, consuming chloroplasts and reducing photosynthesis followed by decreasing yield.

Number of insect generations

Accumulated weekly counts of larvae showed three generations for *M. Phaseoli.* Table 5 revealed that insect had 3 generations/growing season. In spring season, they occupied 3,5 and 4 weeks, respectively, in summer season 2011, occupied 3,4 and 7 weeks, in the following summer they occupied 4,5 and 7 weeks; in the following summer they occupied 4,5 and 4 weeks; in winter season, occupied 6,3 and 5 weeks, respectively. This agrees with results of El-Khouly et al. [14] and Omar et al. [18] who found that larvae of *Liriomyza* spp passed by three generations each season [19].

Synchronization of larvae population with studied factors

The relations between plant age, temperature; relative humidity and annual variation in larvae population of *M. phaseoli* on bean plants, was shown in Table 6. Plant age had positive relationship (correlation) with larvae counts in spring 2011, summer 2011 and summer 2012 where r=0.39, 0.125 and 0.18, respectively; but negative correlation was obtained (r=-0.03) in winter 2013. This agree with results of Salem [20] who found that infestation by *Melanagromyza cunctans* on soybean plants began as the seedlings emergence and continued until the harvest time, captured adults, larvae counts and tunnels numbers were increased with the plant age until 13 weeks (91 days) of plant age. Abd El-Salam et al. [19] mentioned that leaf miner cause damage to faba bean plants during all life stages.

Statistical analysis revealed that weekly mean temperature had insignificant effect on larval population in four experimental seasons (Table 6); also relative humidity had negative and insignificant effect this refer that both temperature and relative humidity was inside the optimum range for the larval activity. These findings agree with results of El-Khouly [14] who found that temperature and relative humidity had insignificant effect on larvae population of *L. congesta.*

On the basis of aforementioned results, larvae of this insect reached its maximum counts during the experimental seasons at range from 1.13 to 29 larvae/leaf; when temperature ranged from 25.41 to 29.84°C; and relative humidity from 43.86 to 60.71% and plant age was between 39-74 days. This refers that this period (39-74 day of plant age) was the most favorable and attractance for (leaf miner infestation; with the range of temperature 25.41-29.84°C and relative humidity 43.86-60.71%, so management program must be directed into insect during this period of plant age or during this ranges of climatic factors [20].

References

1. Assem AA (1961) External morphology, biology and chemical control of the bean fly Melanagromyza phaseoli (Tryon). M.Sc. Thesis, Faculty of Agriculture, Cairo University, Egypt.
2. Gangrade GA, Kogan M (1980) Sampling stem flies in soybean. In: Sampling methods in soybean entomology, Springer. pp. 394-403.
3. Swaminathan R, Singh K, Nepalia V (2012) Insect pests of green gram Vigna raddiata (L.) ilezek and their management. Agricultural Science 197-222.
4. Nagata RT, Wilkinson LM, Nuessly GS (1998) Longevity, fecundity and leaf stippling of Liriomyza trifolii (Diptera: Agromyzidae) as affected by lettuce cultivars and supplemental feeding. J Econ Entomol 91: 999-1004.
5. Nuessly GS, Hentz MG, Beiriger R, Scully BT (2004) Insects associated with faba bean, Vicia faba (Fabaceae) in Southern Florida. Florida Entomol 87: 204-211.
6. Dent D (1991) Insect Pest Management. CABI International Publishing. London, UK.
7. Pohronezny K, Waddil V (1978) Integrated pest management-development of an alternative approach to control of tomato pests in Florida, IFAS Ext. Plant Path Rep. 22, University of Florida, Gainesville.
8. Johnson MW, Oatman ER, Toscano NC (1982) Potential sampling plan for liriomyza sativae on pole tomatoes. Proceedings of 3rd Annual Indus Conference on the leaf miner, Society for American Floriculturists, Alexandria VA50.
9. Pohronezny K, Waddil VH, Schuster DJ, Sonoda RM (1986) Integrated pest management for Florida tomatoes. Plant Dis 70: 96. In: Pedigo LP, Buntin GD (1994) Handbook of sampling methods for arthropods in Agriculture. 603-626 pages.
10. El-Gendy SS, Hanna MA, Mostafa FF (1995) Population dynamics of the leafminer Liriomyza trifolii (Burgess) (Diptera: Agromyzidae) on faba bean and the relative susceptibility of three varieties Fayom. J Agric Res & Develop 9: 288-302.
11. Gomez KA, Gomez AA (1984) Statistical procedures for agricultural research. John Wiley and sons, New York. p. 680.
12. Audemard HO, Miliare G (1975) Le piegeaago du carpocace sexualde sythnesses: Primers results utlilsables pour L.estimation des populations conduite de la lute. Ann Zool Ecol Anim 7: 61-80.
13. Jacob N (1977) A mathematical mode setting limitel Pentra economic or tolerance of attacks in battles integrated fruit Annals moliior. I C p. 15.
14. El-Khouly AS, Khalafalla EME, Metwally MM, Helal HA, El-Mezaien AB (1997) Larval population of Liriomyza congesta (Beak) (Diptera : Agromyzidae) on feba bean and its relation to certain weather factors at Kafr El-Sheikh governorate, Egypt. J Agric Resp 75: 961-967.
15. Abou El-Hagag GH, Salman MA (2001) Seasonal abundance of certain faba bean pests and their associated predators in southern Egypt. Assuit J Agric Sci 32: 49-63.

16. El-Bessomy MAE (1998) Effect of the natural insecticide (Bancol) on Population density of the bean leafminer Melanagromyza phaseoli. J Agric Sci Mansoura Univ 23: 3369-3373.

17. Omar BA, Faris FS (2000) Bio-residual activity of different insecticides on the leafminers and yield components of snap bean (Phaseolus vulgaris L). Egyptian J Agric Res 78: 1485-1497.

18. Omar HIH, Hanafy ABI, El-Roby AM, Yenes WA (2010) Ecological studies of faba bean leafminer Liriomyza trifolii (Burgess) (Diptera : Agromyzidae) on common bean plants Phaselous vulgaris (L) in El-Behera agroecosystem. Egypt J Agric Res 88: 731-737.

19. Abd El-Salam AME, Salem HA, Salem SA (2013) Biocontrol agents against the leafminer, liriomyza trifolii in faba bean fields. Arch Phytopathol Plant Protection 46: 1054-1060.

20. Salem HA (1999) Levels of infestation and damage by certain insect pests of soybean and sunflower crops. Ph. D. Thesis, Faculty of Agriculture, Cairo University, Egypt.

34

Fourier Transform Infrared and Ultraviolet-Visible Spectroscopic Characterization of Ammonium Acetate and Ammonium Chloride: An Impact of Biofield Treatment

Mahendra Kumar Trivedi[1], Alice Branton[1], Dahryn Trivedi[1], Gopal Nayak[1], Khemraj Bairwa[2] and Snehasis Jana[2*]

[1]*Trivedi Global Inc., 10624 S Eastern Avenue Suite A-969, Henderson, NV 89052, USA*

[2]*Trivedi Science Research Laboratory Pvt. Ltd., Hall-A, Chinar Mega Mall, Chinar Fortune City, Hoshangabad Rd., Bhopal, Madhya Pradesh, India*

Abstract

Ammonium acetate and ammonium chloride are the white crystalline solid inorganic compounds having wide application in synthesis and analytical chemistry. The aim of present study was to evaluate the impact of biofield treatment on spectral properties of inorganic salt like ammonium acetate and ammonium chloride. The study was performed in two groups of each compound i.e., control and treatment. Treatment groups were received Mr. Trivedi's biofield treatment. Subsequently, control and treated groups were evaluated using Fourier Transform Infrared (FT-IR) and Ultraviolet-Visible (UV-Vis) spectroscopy. FT-IR spectrum of treated ammonium acetate showed the shifting in wavenumber of vibrational peaks with respect to control. Like, the N-H stretching was shifted from 3024-3586 cm^{-1} to 3033-3606 cm^{-1}, C-H stretching from 2826-2893 cm^{-1} to 2817-2881 cm^{-1}, C=O asymmetrical stretching from 1660-1702 cm^{-1} to 1680-1714 cm^{-1}, N-H bending from 1533-1563 cm^{-1} to 1506-1556 cm^{-1} etc. Treated ammonium chloride showed the shifting in IR frequency of three distinct oscillation modes in NH_4 ion i.e., at v_1, 3010 cm^{-1} to 3029 cm^{-1}; v_2, 1724 cm^{-1} to 1741 cm^{-1}; and v_3, 3156 cm^{-1} to 3124 cm^{-1}. The N-Cl stretching was also shifted to downstream region i.e., from 710 cm^{-1} to 665 cm^{-1} in treated ammonium chloride. UV spectrum of treated ammonium acetate showed the absorbance maxima (λ_{max}) at 258.0 nm that was shifted to 221.4 nm in treated sample. UV spectrum of control ammonium chloride exhibited two absorbance maxima (λ_{max}) i.e., at 234.6 and 292.6 nm, which were shifted to 224.1 and 302.8 nm, respectively in treated sample.

Overall, FT-IR and UV data of both compounds suggest an impact of biofield treatment on atomic level i.e., at force constant, bond strength, dipole moments and electron transition energy between two orbitals of treated compounds as compared to respective control.

Keywords: Ammonium acetate; Ammonium chloride; Biofield treatment; Fourier transform infrared spectroscopy; Ultraviolet spectroscopy

Introduction

Ammonium acetate (CH_3COONH_4) is a white crystalline solid, water soluble compound derived from the chemical reaction ammonia and acetic acid. Being a salt of weak base and weak acid, it possesses several distinct applications like, it is used as an aqueous buffer for High-Performance Liquid Chromatography (HPLC) with Evaporative Light Scattering Detector (ELSD) and Electrospray Ionization Mass Spectrometry (ESI-MS) of proteins [1,2]. It is also used as a food additive to regulate the acidity. Therapeutically, it is reported as an antidiuretic and antipyretic and also as a nutrient [1,3]. Ammonium acetate is also used as an intermediate and catalyst in numerous chemical reactions [1,4]. On the contrary, ammonium acetate also associated with its toxicities like flaccidity of facial muscles, generalized discomfort, tremor, anxiety, and impairment of motor performance [3].

Ammonium chloride (NH_4Cl) is also a white crystalline inorganic salt, having high solubility in water. The natural and mineralogical form of ammonium chloride is known as sal ammoniac. The ammonium chloride has wide application in the field of medicine, agriculture and in food. In medicine, it is used as an expectorant in cough syrup due to irritative effect on the bronchial mucosa. Ammonium chloride causes the nausea and vomiting effects owing to irritative effect on gastric mucosa [5]. It is also used as a systemic acidifying agent for the treatment of severe metabolic alkalosis, and to maintain the urine at acidic pH in the treatment of urinary-tract disorders [6]. In food products, ammonium chloride is used as an additive or feed supplement for cattle and as a nutrient for yeast and other microbes [7,8]. It is also used to improve the crispness of cookies and snacks items. In agriculture, the ammonium chloride is used as an important source of nitrogen in fertilizers [9]. The chemical and physical stability of any chemical compound are most desired qualities that determine its shelf life and effectiveness [10]. Hence, it is advantageous to find out an alternate approach, which could enhance the stability of compounds by altering the structural properties of these compounds. Recently, biofield treatment is reported to alter the physical, and structural properties of various living and non-living substances [11,12]. The relation between mass-energy was described by Einstein through a well-known equation $E=mc^2$ [13]. Planck M gave a hypothesis that energy is a property of matter or substances that neither can be created nor destroyed but can be transmitted to other substances by changing into different forms [14]. According to Maxwell JC, every dynamic process in the human body had an electrical significance [15]. Researchers have experimentally demonstrated the presence of electromagnetic field around the human body using medical technologies such as electromyography, electrocardiography and electroencephalogram [16]. This electromagnetic field of the human body is known as biofield and energy associated with this field is known as biofield energy [17]. Mr. Trivedi has the ability to harness the energy from environment or

***Corresponding author:** Snehasis Jana, Trivedi Science Research Laboratory Pvt. Ltd., Hall-A, Chinar Mega Mall, Chinar Fortune City, Hoshangabad Rd., Bhopal-462026, Madhya Pradesh, India, E-mail: publication@trivedisrl.com universe and can transmit into any object (living or nonliving) around this Globe. The object(s) always receive the energy and responding into useful way, this process is known as biofield treatment [11,12]. Mr. Trivedi's unique biofield treatment is also called as The Trivedi Effect®, and known to alter the characteristics of many things in the verities of research fields including microbiology [11,18], agriculture [19,20], and biotechnology [21,22]. Recently, impact of biofield treatment on atomic, crystalline and powder characteristics as well as spectroscopic characters of different materials were studied and alteration in physical, thermal and chemical properties were reported [12,23,24].

Considering the effects of biofield treatment on various living and nonliving things, the study was aimed to evaluate the impact of biofield treatment on spectral properties of ammonium acetate and ammonium chloride. The effects were analyzed using Fourier Transform Infrared (FT-IR) and Ultraviolet-Visible (UV-Vis) spectroscopic techniques.

Materials and Methods

Study design

The ammonium acetate and ammonium chloride were procured from Sigma-Aldrich, India. Each compound was divided into two parts and coded as control and treatment. The control samples were remained as untreated, and treatment samples were handed over in sealed pack to Mr. Trivedi for biofield treatment under laboratory conditions. Mr. Trivedi provided this treatment through his energy transmission process to the treatment groups without touching the samples. The control and treated samples of ammonium acetate and ammonium chloride were evaluated using FT-IR and UV-Vis spectroscopic techniques.

FT-IR spectroscopic characterization

For FT-IR analysis of control and treated samples of ammonium acetate and ammonium chloride, the samples were crushed into fine powder. Consequently, the crushed powder was mixed in spectroscopic grade KBr in an agate mortar and pressed into pellets with a hydraulic press. FT-IR spectra of were acquired on Shimadzu's Fourier transform infrared spectrometer (Japan) with frequency range of 500-4000 cm^{-1} and a maximum resolution of 0.5 cm^{-1}. The analysis were carried out to evaluate the impact of biofield treatment at atomic level such as force constant, dipole moment, and bond strength in chemical structure [25].

UV-Vis spectroscopic analysis

UV spectra of control and treated ammonium acetate and ammonium chloride were acquired on Shimadzu UV-2400 PC series spectrophotometer with 1 cm quartz cell and a slit width of 2.0 nm. The study was carried out using wavelength in the range of 200-400 nm. The UV spectral analysis was performed to determine the effect of biofield treatment on the energy gap between bonding (π-π^*) and nonbonding (n-π^*) electrons transition [25].

Results and Discussion

FT-IR spectroscopic analysis

The FT-IR spectra of control and treated ammonium acetate are shown in Figure 1 and the IR spectral interpretation results are reported in Table 1. The FT-IR spectrum of control ammonium acetate (Figure 1a) showed the IR peaks at 3024-3586 cm^{-1} for N-H stretching of NH_4 group. These peaks were shifted to higher frequency region i.e., at 3033-3606 cm^{-1} in treated sample (Figure 1b), which indicated an enhanced force constant of N-H bond as compared to control. IR frequency (ν) of stretching vibrational peak depends on two factors i.e., force constant (k) and reduced mass (μ) which can be explained by following equation [26].

$\nu = 1/2\pi c \sqrt{(k/\mu)}$, Here, c is speed of light.

If μ is constant, then the frequency is directly proportional to the force constant; hence, alteration (increase or decrease) in frequency of any bond indicates a respective change in force constant [25].

The C-H stretching's were appeared at 2826-2893 cm^{-1} in control sample that were shifted to lower wavenumber in treated sample i.e., at 2817-2881 cm^{-1}. The C=O asymmetrical stretchings were appeared at 1660-1702 cm^{-1} in control sample, which were shifted to higher wavenumber in treated sample i.e., at 1680-1714 cm^{-1}. This could be due to increased bond strength of C=O bond in treated sample as compared to control. N-H bending was assigned to peaks at 1533-1563 cm^{-1} in control sample of ammonium acetate that were observed at 1506-1556 cm^{-1} in treated sample. It depicted a reduced torsion force of N-H bending after biofield treatment as compared to control. The C=O symmetrical stretching was assigned to peak at 1404 cm^{-1} in control sample, which was observed at higher wavenumber i.e., at 1422 cm^{-1} in treated sample as compared to control. This shifting of C=O bond to higher frequency region was occurred possibly due to increased force constant of C=O bond. The C-H deformation bends were assigned to the peaks at 1281-1342 cm^{-1} in control and 1292-1340 cm^{-1} in treated sample of ammonium acetate. Likewise, the C-O stretching peaks were observed at 1016-1050 cm^{-1} in control sample, which were slightly shifted to lower frequency i.e., at 1006-1043 cm^{-1} in treated sample. This could be due to reduced force constant of C-O bond after biofield treatment as compared to control. Overall, the FT-IR results of ammonium acetate suggest a significant impact of biofield treatment at the atomic level i.e., at dipole moment and force constant of respective bonds. The FT-IR data of control ammonium acetate was well supported by the literature [27].

The FT-IR spectra of control and treated ammonium chloride are shown in Figure 2 and the IR spectral interpretation results are reported in Table 2. Krishnan RS reported that NH_4 ion has tetrahedral symmetry therefore it showed four distinct mode of oscillations i.e., ν_1 and ν_2 due to single and double degenerate, and ν_3 and ν_4 are triply degenerate N-H vibrations, respectively [28,29]. The ν_1 and ν_3 peaks were observed at 3010 cm^{-1} and 3156 cm^{-1}, respectively in control (Figure 2a). Whereas, these were observed at 3029 cm^{-1} (ν_1) and 3124 cm^{-1} (ν_3) in treated sample (Figure 2b). The result showed an upstream shifting of peak ν_1 and downstream shifting of peak ν_3 in threated sample with respect of control. This could be due to biofield induced alteration in force constant of N-H stretching in treated sample as compared to control. Likewise, the vibrational peaks ν_2 and ν_4 were appeared at 1724 cm^{-1} and 1402 cm^{-1}, respectively in control sample and 1441 cm^{-1} (ν_2) and 1401 cm^{-1} (ν_4) in treated sample of ammonium chloride. The result showed an upstream shifting of peak ν_2 i.e., from 1724 cm^{-1} to 1741 cm^{-1} in treated sample, which depicted a corresponding increase in torsional force of ν_2 oscillation. Additionally, the N-Cl stretching was assigned to peak at 710 cm^{-1} in control sample that was shifted to lower frequency at 665 cm^{-1} in treated sample. This similarly suggests a possible decrease in force constant of N-Cl stretching after biofield treatment as compared to control. Overall, the FT-IR spectral data of control and treated ammonium chloride showed an impact of biofield treatment on the internal oscillation of NH_4 group and N-Cl stretching. This could be due to alteration in force constant and dipole moment of ammonium chloride molecules after biofield treatment as compared to control. Because of alteration in force constant and bond strength, the

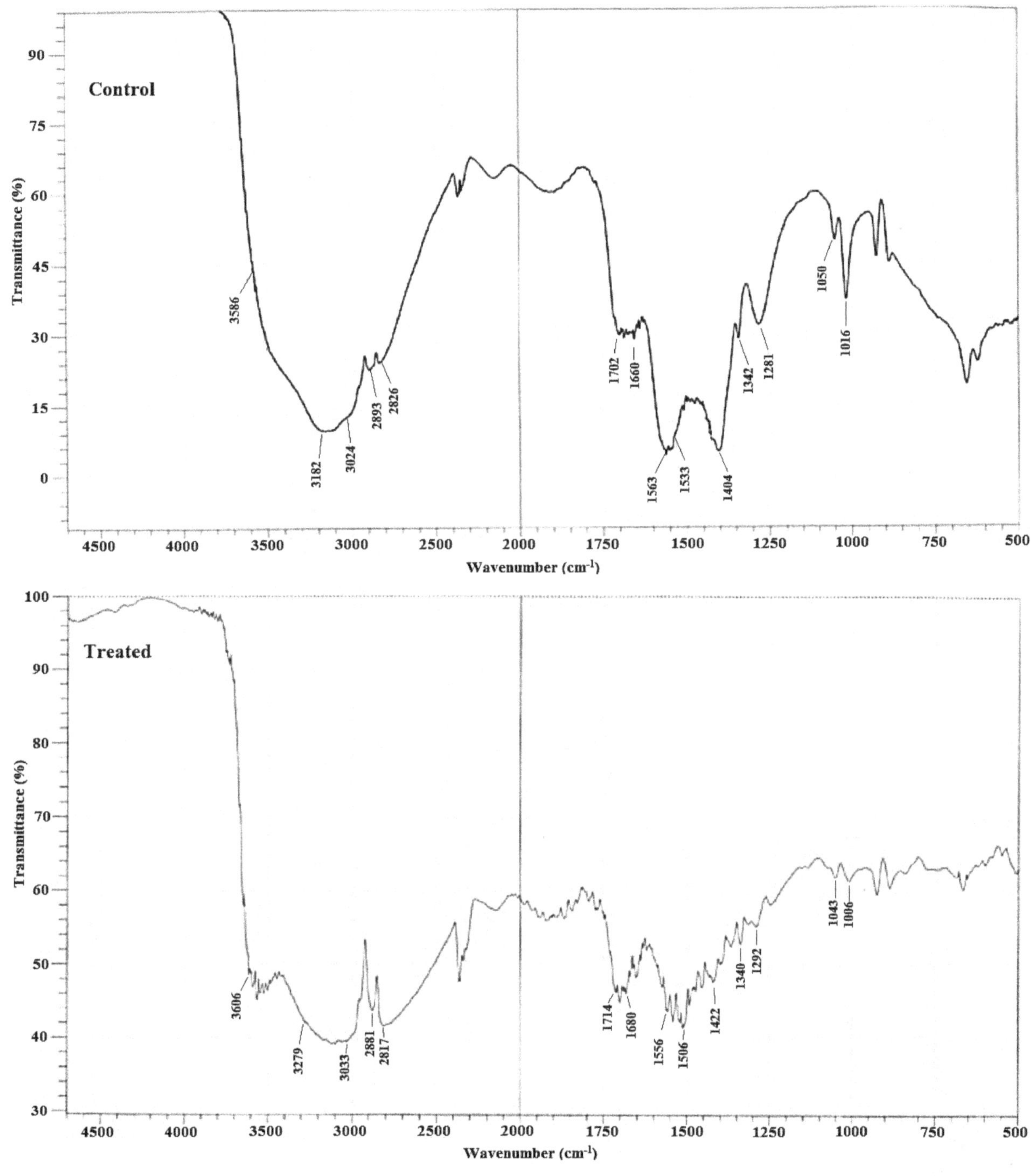

Figure 1: FT-IR spectra of control and treated ammonium acetate.

chemical stability of treated compounds might also be alter. Based on this, it is speculated that biofield treatment could be used to increase the chemical stability of any compound, which might be more useful than the untreated compound.

UV-Vis spectroscopy

UV spectra of control and biofield treated ammonium acetate are shown in Figure 3. The UV spectrum of control ammonium acetate (Figure 3a) showed the absorption maxima (λ_{max}) at 258.0 nm. Whereas, in biofield treated sample of ammonium acetate, this absorption maxima (λ_{max}) was appeared at 221.4 nm and 204.6 nm (Figure 3b). As per existing literature on principle of UV spectrophotometer, the compound can absorbs UV light due to the presence of conjugated pi (π) bonding systems (π- π^* transition) and nonbonding electron system (n-π^* transition). There are certain energy gape between π-π^* and n-π^* orbitals. When this energy gap altered, the wavelength (λ_{max}) was also altered respectively [25]. Based on this, it is speculated that, due to influence of biofield treatment, the energy gap between σ-σ^*, π-π^* or n-π^* transition in ammonium acetate molecules might be altered,

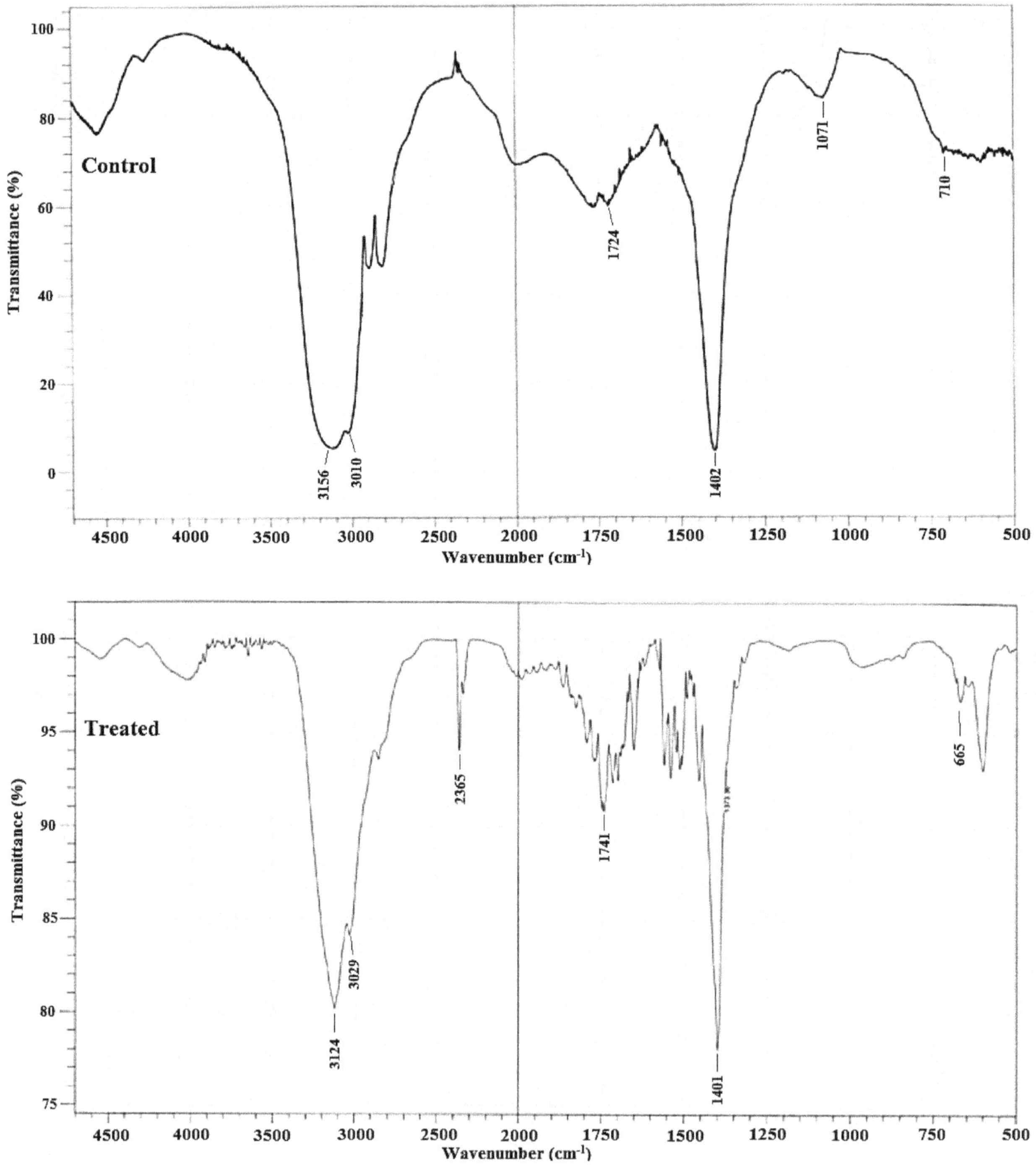

Figure 2: FT-IR spectra of control and treated ammonium chloride.

which causes the shifting of wavelength (λ_{max}) in treated sample with respect to control.

The UV spectra of control and treated ammonium chloride are shown in Figure 4. The control sample (Figure 4a) exhibited the absorbance maxima (λ_{max}) at 234.6 nm and 292.6 nm. Whereas, the biofield treated ammonium chloride exhibited the absorbance maxima (λ_{max}) at 224.1 and 302.8 nm (Figure 4b). This slight shifting of λ_{max} after biofield treatment also suggest a possible alteration in energy gap between σ-σ^*, π-π^* or n-π^* transition in ammonium chloride molecule with respect to control. Altogether, the UV spectral data of both the ammonium acetate and ammonium chloride (control and treated) revealed a considerable impact of biofield treatment on the atomic level of respective compound.

Conclusion

FT-IR spectrum of biofield treated ammonium acetate showed the alteration in wavenumber of IR peaks assigned to N-H, C-H, C=O and C-O stretching as compared to control. Likewise, the biofield treated

Wave number (cm^{-1})		Frequency Assigned to group
Control	Treated	
3586-3024	3606-3033	N-H stretching
2893-2826	2881-2817	C-H stretching
1702-1660	1714-1680	C=O asymmetrical stretching
1563-1533	1556-1506	N-H bending
1404	1422	C=O symmetrical stretching
1342-1281	1340-1292	C-H deformation
1050-1016	1043-1006	C-O stretching (ester)

Table 1: FT-IR vibrational peaks observed in ammonium acetate.

Wave number (cm^{-1})		Frequency Assigned to group
Control	Treated	
3156	3124	v_3 single degenerate N-H oscillation
3010	3029	v_1 doubly degenerate N-H oscillation
1724	1741	v_2 Triply degenerate N-H oscillation
1402	1401	v_4 Triply degenerate N-H oscillation
710	665	N-Cl stretching

Table 2: FT-IR vibrational peaks observed in ammonium chloride.

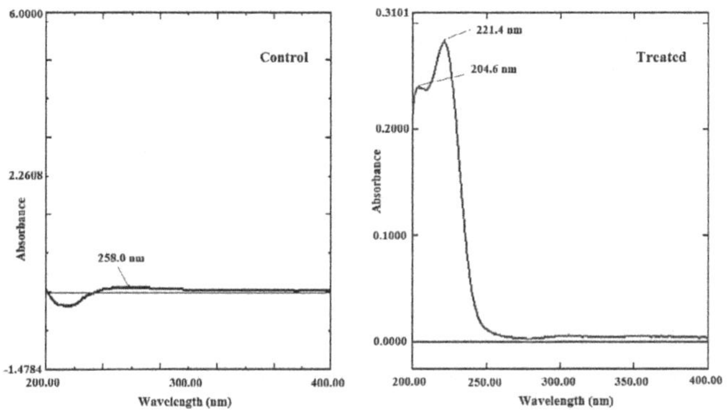

Figure 3: UV spectra of control and treated ammonium acetate.

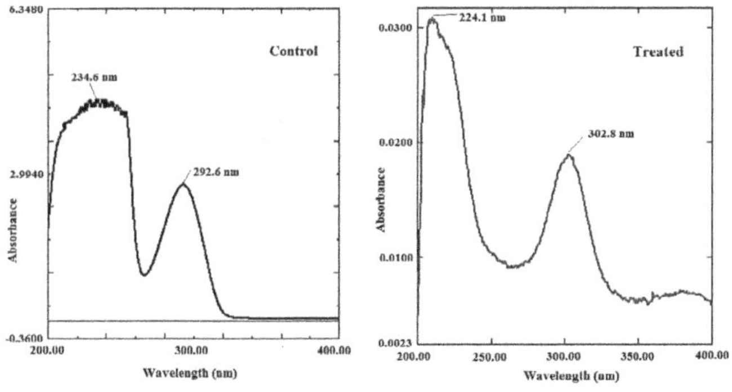

Figure 4: UV spectra of control and treated ammonium chloride.

ammonium chloride showed the alteration in wavenumber of IR peaks assigned to three (v_1, v_2, and v_3) out of four distinct internal oscillations of NH_4 group as well as N-Cl stretching with respect of control. UV spectra of ammonium acetate and ammonium chloride showed the alteration in absorption maxima (λ_{max}) after biofield treatment as compared to respective control.

Altogether, the FT-IR results suggest an impact of biofield treatment on atomic level like dipole moment, force constant, bond strength, and flexibility of treated compounds with respect to control. Likely, the UV result suggests the impact of biofield treatment on bonding and nonbonding electron transition of treated compounds with respect to control.

Acknowledgement

The authors would like thank Trivedi Science™, Trivedi Master Wellness™ and Trivedi Testimonials for their consistent support during the work. Authors also like to acknowledge the whole team of MGV Pharmacy College, Nashik for providing the instrumental facility.

References

1. Budavari S (1989) The Merck Index: An encyclopedia of chemicals, drugs and biologicals. Merck and Co. Inc., Rahway, NJ, USA.
2. Ding L, Tan W, Zhang Y, Shen J, Zhang Z (2008) Sensitive HPLC-ESI-MS method for the determination of tiotropium in human plasma. J Chromatogr Sci 46: 445-449.
3. Rossoff IS (1974) Handbook of Veterinary Drugs: A Compendium for Research and Clinical Use. Springer Publishing Company, New York, USA.
4. Thellend A, Battioni P, Mansuy D (1994) Ammonium acetate as a very simple and efficient cocatalyst for manganese porphyrin-catalysed oxygenation of hydrocarbons by hydrogen peroxide. J Chem Soc Chem Commun 1994: 1035-1036.
5. Bothara KG (2007) Inorganic Pharmaceutical Chemistry. (9th edn) Nirali prakashan, Pragati Books Pvt. Ltd. Pune, India.
6. Mathew JT, Bio LL (2012) Injectable ammonium chloride used enterally for the treatment of persistent metabolic alkalosis in three pediatric patients. J Pediatr Pharmacol Ther 17: 98-103.
7. Oetzel GR, Olson JD, Curtis CR, Fettman MJ (1988) Ammonium chloride and ammonium sulfate for prevention of parturient paresis in dairy cows. J Dairy Sci 71: 3302-3309.
8. Costa E, Teixidó N, Usall J, Atarés E, Viñas I (2002) The effect of nitrogen and carbon sources on growth of the biocontrol agent Pantoea agglomerans strain CPA-2. Lett Appl Microbiol 35: 117-120.
9. Chesworth W (2008) Encyclopedia of soil science. Springer science & Business Media, New York, USA.
10. Blessy M, Patel RD, Prajapati PN, Agrawal YK (2014) Development of forced degradation and stability indicating studies of drugs-A review. J Pharm Anal 4: 159-165.
11. Trivedi MK, Patil S, Shettigar H, Bairwa K, Jana S (2015) Phenotypic and biotypic characterization of Klebsiella oxytoca: An impact of biofield treatment. J Microb Biocshem Technol 7: 203-206.
12. Trivedi MK, Patil S, Tallapragada RMR (2015) Effect of biofield treatment on the physical and thermal characteristics of aluminium powders. Ind Eng Manage 4: 151.
13. Einstein A (1905) Does the inertia of a body depend upon its energy-content? Ann Phys 18: 639-641.
14. Planck M (1903) Treatise on thermodynamics. (3rd edn) English translated by Alexander OGG, Longmans, Green, London (UK).
15. Maxwell JC (1865) A dynamical theory of the electromagnetic field. Phil Trans R Soc Lond 155: 459-512.
16. Rivera-Ruiz M, Cajavilca C, Varon J (2008) Einthoven's string galvanometer: the first electrocardiograph. Tex Heart Inst J 35: 174-178.
17. Rubik B (2002) The biofield hypothesis: its biophysical basis and role in medicine. J Altern Complement Med 8: 703-717.
18. Trivedi MK, Patil S, Shettigar H, Gangwar M, Jana S (2015) Antimicrobial sensitivity pattern of Pseudomonas fluorescens after biofield treatment. J Infect Dis Ther 3: 222.
19. Sances F, Flora E, Patil S, Spence A, Shinde V (2013) Impact of biofield treatment on ginseng and organic blueberry yield. Agrivita J Agric Sci 35.
20. Lenssen AW (2013) Biofield and fungicide seed treatment influences on soybean productivity, seed quality and weed community. Agricultural Journal 8: 138-143.

21. Patil SA, Nayak GB, Barve SS, Tembe RP, Khan RR (2012) Impact of biofield treatment on growth and anatomical characteristics of Pogostemon cablin (Benth.). Biotechnology 11: 154-162.

22. Altekar N, Nayak G (2015) Effect of biofield treatment on plant growth and adaptation. J Environ Health Sci 1: 1-9.

23. Trivedi MK, Tallapragada RR (2008) A transcendental to changing metal powder characteristics. Met Powder Rep 63: 22-28.

24. Dabhade VV, Tallapragada RR, Trivedi MK (2009) Effect of external energy on atomic, crystalline and powder characteristics of antimony and bismuth powders. Bull Mater Sci 32: 471-479.

25. Pavia DL (2001) Introduction to Spectroscopy. (3rd edn) Thomson Learning, Singapore.

26. Stuart BH (2004) Infrared spectroscopy: Fundamentals and applications analytical techniques in the sciences (AnTs). John Wiley & Sons Ltd., Chichester, UK.

27. Alias SS, Chee SM, Mohamad AA (2014) Chitosan-ammonium acetate-ethylene carbonate membrane for proton batteries. Arab J Chem http://dx.doi.org/10.1016/j.arabjc.2014.05.001.

28. Krishnan RS (1947) Raman spectrum of ammonium chloride and its variation with temperature. Proc Ind Acad Sci A 26: 432-449.

29. Max JJ, Chapados C (2013) Aqueous ammonia and ammonium chloride hydrates: Principal infrared spectra. J Mol Struct 1046: 124-135.

Study of Loading SO_4^{2-} on Sb-SnO_2 Nanocrystal and its Calcination Temperature to Make Solid Superacid $SO_4^{\oplus}$/Sb-SnO_2

Xuejun Zhang[1,2*], Xiao-Ning Zhang[2], Qin-Qin Ran[1], Han-Mei Ouyang[1], Hui Zhong[2] and Han Tao[1,2]

[1] *Guizhou Province Key Laboratory of Fermentation Engineering and Biological Pharmacy, Guizhou University, Guiyang, China*
[2] *School of Brewing and Food Engineering, Guizhou University, Guiyang 550025, China*

Abstract

SO_4^{2-}/SnO_2 were reported to be a solid superacid with an acid strength equal to that of SO_4^{2-}/ZrO_2. But papers concerning the SO_4^{2-}/SnO_2 catalyst have been quite few, because of difficulty in preparation of the oxide gels from its salts $SnCl_4$. A highly dispersed light yellow powder, Sb-SnO_2 nanocrystal, was obtained by the synthesis method of "P-CNAIE" and the drying method of "AD-IAA". The Sb doping made the energy gap of nano-crystalline SnO_2 narrower. A saturated solution of ammonium sulfate was dropped into organic solutions containing a fixed amount of Sb-SnO_2 nano-powders in different ratio in order to load Sb-SnO_2 powder with ammonium sulfate. This method has an outstanding advantage that is the loading ratio of $(NH_4)_2SO_4$ to Sb-SnO_2 can come to very high and no free water causes the aggregation of Sb-SnO_2 nano powder. The methods of Differential Scanning Calorimetry (DSC) and Thermogravimetric analysis (TG) demonstrated that the working ratio of Sb-SnO_2 to $(NH_4)_2SO_4$ was 1:1.4 to 1:1.6 wt% and the most favorable calcination temperature for the generation of superficially sulfated groups of Sb-SnO_2 particles should fall between 380°C and 400°C. The adsorption reaction of indicator reveals that the solid acid, calcined Sb-SnO_2 with a bluish color had a $H_0 \leq -14.5$ at least.

Keywords: Solid superacid; Stannic oxide; Nanocrystal; Impregnation; Ammonium sulfate; Calcination temperature

Introduction

Acid catalysts, especially superacid catalysts, play a vital role in the chemical industry of our time. Many organic reactions such as esterification, condensation, cracking, alkylation, saturated hydrocarbon isomerization, can be economically and effectively accomplished with the presence of acid catalysts.

In 1979, Hino et al. [1] indicated, for the first time, that the acid strength of the SO_4^{2-}/ZrO_2 catalyst is estimated to be H_0 (Hammett indicator) ≥ -14.52, one of the strongest solid superacids. Sulfated zirconia (SO_4^{2}/ZrO_2) is a typical solid superacid and exhibits a high catalytic activity for the skeletal isomerization of saturated hydrocarbons and other reactions [2-6]. Sulfated tin oxide (SO_4^{2-}/SnO_2) was later reported by Matsuhashi et al. [4] to be one of the candidates with the strongest acidity, acid strength of which is almost equal to that of SO_4^{2-}/ZrO_2 at least [7-9]. And SnO_2 is more readily available and cheaper than ZrO_2 [10].

Matsuhashi et al. [11] concluded in 2001 that the preparation of many solid superacids of sulfated metal oxides commonly underwent three steps: (i) preparation of amorphous metal oxide gels as precursors; (ii) treatment of the gels with sulfate ion by exposure to a H_2SO_4 solution or by impregnation with $(NH_4)_2SO_4$; and (iii) calcination of the sulfated materials at a high temperature in air. For the synthesis of solid superacid SO_4^{2-}/SnO_2, however, it is difficult to prepare the tin oxide gel precursors from the $SnCl_4$ salts. [11] Hence, the synthesis and application of SO_4^{2-}/SnO_2 catalysts are seldom reported.

Herein we propose a novel three-step method for the preparation of solid superacid SO_4^{2-}/SnO_2. In contrast to the three-step process proposed by Matsuhashi et al., the present method uses metal oxide crystals, instead of metal oxide gels, as precursors. Specifically, this method includes: (i) preparation of high purity nanometer metal oxide crystal with a lot of superficial hydroxyls; (ii) treatment of the gels with sulfate ions, where the as-prepared Sb-SnO_2 nanoparticles were dispersed in organic solvent and then impregnated with saturated ammonium sulfate solution to associate with $(NH_4)_2SO_4$ by water molecule adsorbed on $(NH_4)_2SO_4$; (iii) calcination [12] of the impregnated nano-powders. A coupling reaction of superficial hydroxyls with $(NH_4)_2SO_4$ by losing NH_3 and H_2O undergoes at a proper temperature.

After calcination, the obtained solid powder has been firmly bonded with a group $=SO_4$ on its surface, which means that SO_4^{2-} is by no means a sulphate radical attached to nano particle any more. In virtue of Bronsted's and Lewis' acid-base theory, the attached SO_4^{2-} should be a base but an acid since its negative charge. Our experiments, however, demonstrated such calcined nano particles were a superacid. So we believed that the molecular structure of solid superacid should be noted as $SO_4^{\oplus}/SnO_2$ but SO_4^{2-}/SnO_2, the latter written form of solid superacid, including SO_4^{2-}/ZrO_2, is being widely and incorrectly adopted.

Matsuhashi et al. [11] further indicated that papers concerning the SO_4^{2-}/SnO_2 catalyst have been quite few, because of difficulty in preparation, compared with the relative ease of preparation of the SO_4^{2-}/ZrO_2 material, in particular owing to the difficulty in preparation of the oxide gels from its salts $SnCl_4$.

Experiments

The synthesis of the precursor, antimony doped stannic oxide

The Sb-SnO_2 nanocrystals were synthesized using the method "precipitation–condensation with non-aqueous ion exchange (P-CNAIE)" and dried with the assistance of the Iso-Amyl Acetate (AD-IAA). These two methods were developed in our lab and reported in the published literatures [13-15]. A typical procedure includes the following steps. In an airtight flask containing 200 mL anion-exchange

***Corresponding author:** Xuejun Zhang, Guizhou Province Key Laboratory of Fermentation Engineering and Biological Pharmacy, Guizhou University, Guiyang, 550003, China, E-mail: xzhang203@yahoo.com.cn

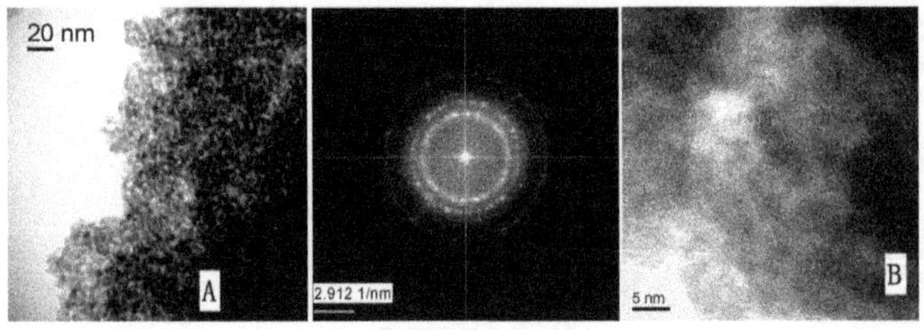

Figure 1: The TEM photos of Sb-doped SnO_2 nano-material synthesized by method of P-CNAIE associate with AD-IAA and the EDP picture (JEM-2010FEF, JEOL, Japan). A and B are TEM images with different amplification scale, and the middle image is an electro diffraction pattern.

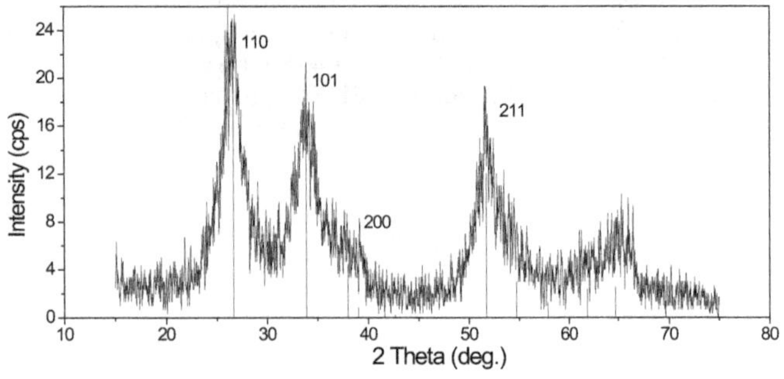

Figure 2: X-ray diffraction of Sb-SnO_2 nano-material synthesized by method of P-CNAIE associate with AD-IAA.

resin, 100 mL alcohol, and 10 mL of ammonia water, 200 mL ethanol solution containing $SnCl_4$·$5H_2O$ (18.0%, w/v) and $SbCl_3$ (0.665%, w/v) were added dropwise with fast stirring. At the same time, NH_3 gas was aerated in the reaction solution. The reaction solution was held close to neutral pH by adjusting the speed of addition. After the addition was complete, the reaction solution was separated from the ion-exchange resin particle through a glass-sand funnel and reacted repeatedly with fresh anion-exchange resin on a shaker. The final chlorine-free colloid solution was held idle on a bench to allow the stratification of the turbid liquid. The upper lightly turbid solution was removed and kept aside for final recovery of all solid content, and the lower dense precipitated slurry was added ~80 mL of iso-amyl acetated to make a co-boiling system. The pale-yellow dispersive fine powders were obtained by co-distilling off water absorbed on the colloid and solvents.

All the exchanged ion-exchange resins were collected and repeatedly washed with fresh solvent to collect any residual precipitate on the surface of the resins. The washed solvent were applied to a short column of ion exchange to remove any remaining chlorine, and were combined with the upper lightly turbid solution, in which the resulting dried powders were added and dispersed on a shaker. In succession, the combined solution was distilled and left behind a fine light-yellow powder of Sb-SnO_2. In this way, all of the metal hydrolysate can be recovered and an exact doping as experimenter desires was achieved.

The impregnation with $(NH_4)_2SO_4$

The sulfated Sb-doped SnO_2 crystals were prepared in our study as follows. 2 g of Sb-doped SnO_2 powder obtained in the synthesis of the precursor, antimony doped stannic oxide was placed in a 50 mL plastic centrifuge tube containing 45 mL of methanol. After the powders were dispersed on a shaker, 3.0 mL of saturated ammonium sulfate, equal to ~2 g of $(NH_4)_2SO_4$, was added in methanol solution, and then the tube continued to be shaken on a shaker violently as the saturated solution was dropped in methanol, when very tiny $(NH_4)_2SO_4$ precipitate was separated in solution. The Sb-doped SnO_2 powders loaded with $(NH_4)_2SO_4$ were separated by centrifugation and further washed in anhydrous alcohol. The process was repeated for three times and finally centrifuged at 4000 r/min. The final sediment was dried under an infrared ray lamp and a dispersed powder was obtained.

The coupling reaction of superficial hydroxyls with $(NH_4)_2SO_4$

The mixed powders obtained in the impregnation with $(NH_4)_2SO_4$ were transferred on a corundum plate, and then calcinated in a muffle furnace. The calcination to couple "SO_4" on superficial hydroxyls of Sb-SnO_2 nanocrystals was carried out at 380°C for 2~3 h.

Results and Discussion

Through the synthesis method of "P-CNAIE" and the drying method of "AD-IAA", highly dispersed pale yellow powders were obtained. Based on our observation, without doping of the antimony, the colloidal solution of stannic chloride and finally dried powders always presented white colour, which implicates that the yellow color of as-prepared powders is caused by doping antimony or, more exactly, by Sb doping into crystal lattice of stannic oxides, because yellow is caused by the formation of crystal with variation of band gap, instead by cluster or hydrolysate that has a forbidden band.

Figure 1 shows TEM images and electron diffraction pattern of nano-meter sized material synthesized in the experiment section. The electro diffraction pattern, the middle image, indicates that the obtained nano material has a determinate crystal structure, which is also confirmed by the TEM image B, from which a layer lattice structure can be distinctly identified. The TEM image A shows the size of as-prepared powders is significantly less than 20 nm. In addition, XRD pattern in Figure 2 illustrates the degree of crystallization and the size of nano particle. Diffraction peaks and their position in the pattern indicate the nano material is stannic oxide crystal, and broad and weak peaks suggest that crystals are nano-meter sized. The positions of peaks are consistent with the standard one that showed in the X-Ray Powder Diffraction Standards of SnO_2, PDF No. 41-1445 from Jade 5.0, see the red bar in Figure 2.

Crystal structure is of course important because the structure endows the material with some special properties, such as optical, semiconductor and electrical properties. On the other hand, superficial hydroxyl is, however, critical for the surface modification of nano-materials, and for hybrid nano-composites to mix with polymers.

In the calcination, it was found that superficial hydroxyl on Sb-SnO_2 nanoparticles had significant effect on the sulfating and roasting of Sb-SnO_2 nanocrystal, which had been demonstrated by Differential Scanning Calorimetry and Thermogravimetric analysis (DSC-TG). The fewer the number of superficial hydroxyl exist, the fewer the sulfated groups exist on the surface of Sb-SnO_2 nanocrystals. The thermogravimetric analysis (Figure 3) and differential scanning calorimetry (Figure 4) on the as-prepared powders support this view of point. Compared with curves 1 (Sb-SnO_2) and 8 ($(NH_4)_2SO_4$), curves 2 to 7 (Sb-SnO_2 + $(NH_4)_2SO_4$) have an additional segment from a to b in Figure 3. It is easy to understand that this segment probably implies the generation of superficially sulfated groups. Curves 2 to 7 are thermogravimetric curves of Sb-SnO_2 nanocrystals that were pretreated at different temperatures, from 25°C to 550°C, for 3 h and then impregnated with a given amount of $(NH_4)_2SO_4$ before thermo-gravimetric analysis. It can be seen that the higher the preprocessing temperatures is, the shorter the line segments from a to b, thus the fewer the amount of sulfated group is. As the preprocessing temperature increases, especially at/over 320°C (Figure 3), the weight losses of Sb-

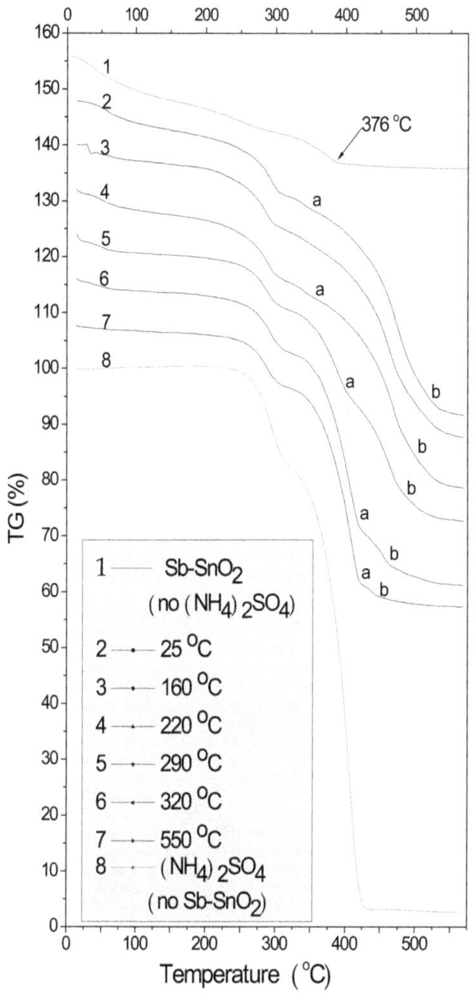

Figure 3: Thermogravimetric curves of Sb-SnO_2 (red one), $(NH_4)_2SO_4$ (blue one) and Sb-SnO_2 pretreated at different temperatures with given amount of $(NH_4)_2SO_4$.

SnO_2 nanocrystals are heavier, resulting from the dehydration between hydroxyls and leading to the decrease of the quantity of the superficial hydroxyls. The decrease in amount of superficial hydroxyls brought about the decline of the quantity of superficially sulfated groups, and the decrease of acid strength or catalytic activities of nanoparticles.

Figure 4 is Differential Scanning Calorimetry (DSC) curves, which more clearly showed the variation of and the difference between samples 1 to 8 due to the distinct images of endothermic peaks and exothermic peaks. The red line has a clear exothermic peak that was caused by the crystallization of superficial hydroxyls of Sb-SnO_2 nanoparticles at ~ 376°C. And the blue one is the differential thermal curve of $(NH_4)_2SO_4$ with two glaring endothermic peaks. The two endothermic peaks are associated with the decomposition of $(NH_4)_2SO_4$ into NH_3, H_2O and SO_3, corresponding to the chemical reaction on following equations:

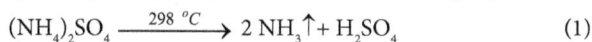

$$(NH_4)_2SO_4 \xrightarrow{298\ ^oC} 2\,NH_3\uparrow + H_2SO_4 \quad (1)$$

$$H_2SO_4 \xrightarrow{410\ ^oC} H_2O\uparrow + SO_3\uparrow \quad (2)$$

It should be pointed out that as the reaction of $(NH_4)_2SO_4$ with superficial hydroxyls of Sb-SnO_2 nanoparticles progressed, the amount of free $(NH_4)_2SO_4$ decreased and the decomposition temperature of H_2SO_4 decreased as well, see the peak B on curve 5 in Figure 4. Nevertheless, it is noted that a third endothermic peak appeared in differential thermal curves of samples 7 to 2. The third endothermic peak only appeared in the curves of Sb-SnO_2 plus $(NH_4)_2SO_4$ and become more obvious as the pretreatment temperatures of Sb-SnO_2 decreased. The appearance of the third peak suggests the cleavage of a chemical bond. As compared with differential thermal curves of Sb-SnO_2 and $(NH_4)_2SO_4$, the third peaks on curves 6 to 2 suggests a bonding reaction took place between superficial hydroxyl and $(NH_4)_2SO_4$, or more exactly, between superficial hydroxyl and H_2SO_4. Therefore, the breaking of bonds represented by the third peak should belong to superficially sulfated groups, which were newly generated groups in the calcination process. We speculate the breaking of bonds contributing to the absorption of heat might follow the cracking reaction as equations (3) and (4) show.

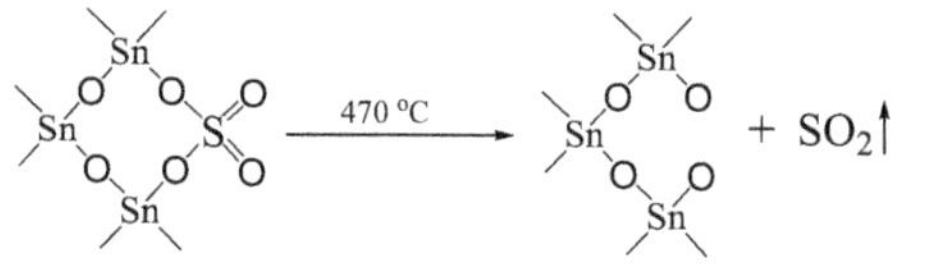

(3)

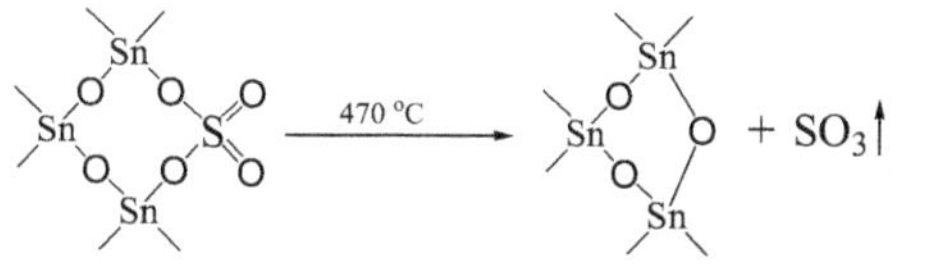

(4)

Based on the above discussed, a summary is drawn in Figure 5, which simply and clearly illustrates the three endothermic peaks. It can be easily understood that the third endothermic peak on the blue curve in Figure 5 should belong to the splitting action of a new group. This

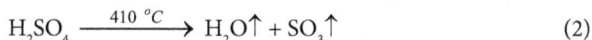

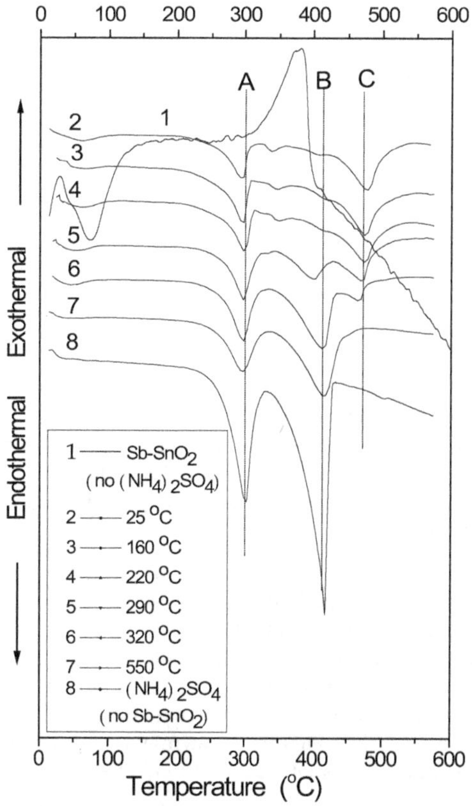

Figure 4: Differential thermal analysis profiles of Sb-SnO_2 (red one), $(NH_4)_2SO_4$ and Sb-SnO_2 pretreated at different temperatures with given amount of $(NH_4)_2SO_4$.

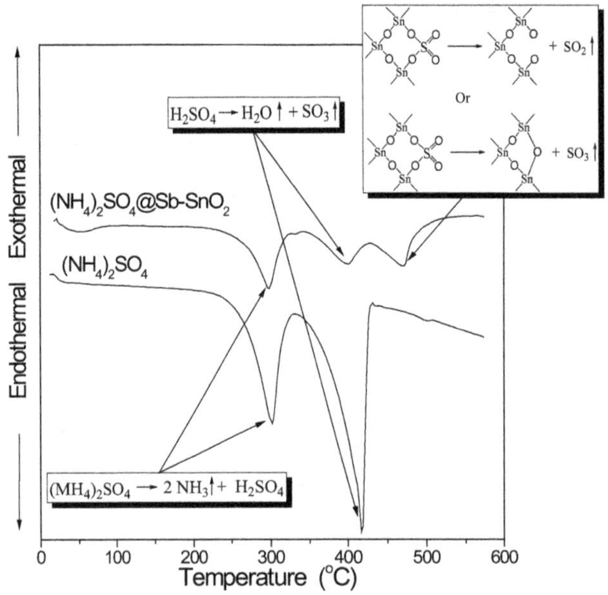

Figure 5: In the calcination, the decomposition of $(NH_4)_2SO_4$ and H_2SO_4 and the generation and the thermal cracking of superficially sulfated groups.

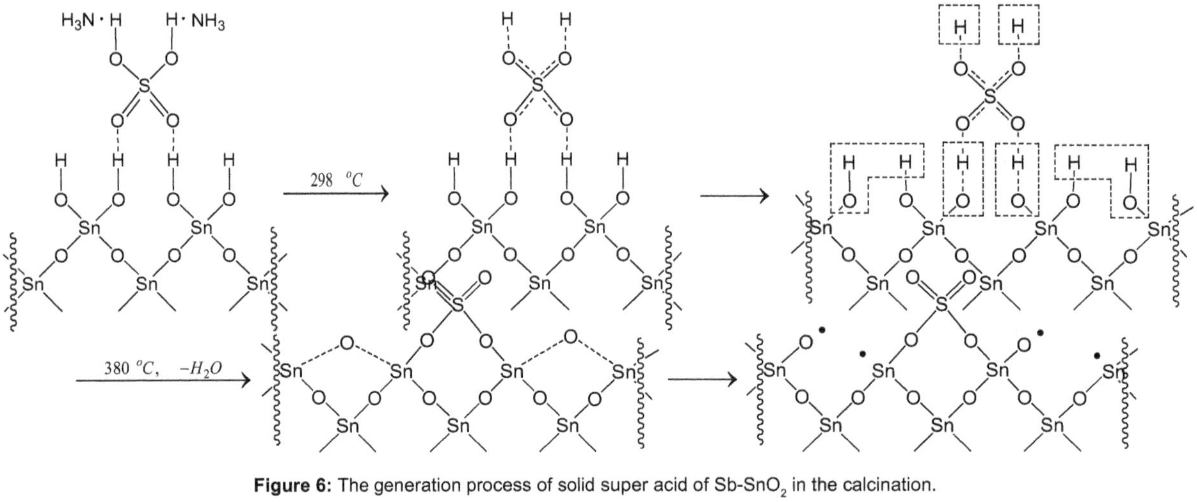

Figure 6: The generation process of solid super acid of Sb-SnO$_2$ in the calcination.

new group generated in the calcination process by "=SO$_4$" bonding to Sb-SnO$_2$ nano-particle against the endothermic peaks on black curve of $(NH_4)_2SO_4$. In other words, the calcination did make "=SO$_4$" group loaded on the Sb-SnO$_2$ nanoparticles forming a solid superacid with a stable "=SO$_4$" group.

According to the data shown in Figures 3-5, we proposed that the most favorable temperature for the generation of superficially sulfated groups of Sb-SnO$_2$ particles should fall between 380°C and 400°C, before the decomposition temperature of H_2SO_4 and after the crystallization temperature of Sb-SnO$_2$. To illustrate the generation of solid superacid of Sb-SnO$_2$, the authors here proposed that a series of chemical reaction such as Figure 6 shows might occur on the surface of Sb-SnO$_2$ as the preparation of solid superacid of Sb-SnO$_2$ underwent.

According to the proposed, group "=SO$_4$" is absolutely impossible to attach to Sb-SnO$_2$ nano-particle in the form of SO_4^{2-}. It should be a group bonded on Sb-SnO$_2$ particle since the dissociation temperature of bonded "=SO$_4$" is up to 470°C.

The relative acid strength of the calcined Sb-SnO$_2$ powders was measured by the adsorption reaction of indicator. The powders (ca. 0.5 g) were calcined at 380°C ~ 390°C in air for 3 h and then placed in a glass vacuum desiccator as the powder was hot. After the sample was pretreated in a vacuum for 2 h and cooled down to room temperature, some cyclohexane solution containing 5% of Hammett indicator was sucked into the vacuum desiccator. The desiccator was heated to 60°C by placing it in a constant water bath, which resulted in the exposure of powder to the indicator vapor. The present powder sample was gradually colored by indicator and changed distinctly the colorless basic form of p-nitrotoluene (pKa or H_0=-11.4), m-nitrotoluene (-12.0), m-nitrochlorobenzene (-13.2), 2,4-dinitrotoluene (-13.8) and 2,4-dinitrofluorobenzene (-14.5) to the yellow conjugate acid form, that is to say, the acid strength of the solid acid is estimated at least to be Ho < -14.5. All of the measurements convincingly demonstrated the calcined Sb-SnO$_2$ was a solid acid, more exactly solid superacid (Figure 7).

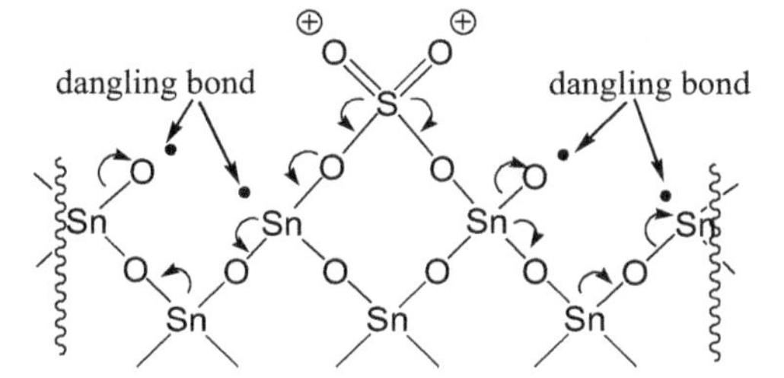

Figure 7: A proposed structure of Sb-SnO_2 solid super acid.

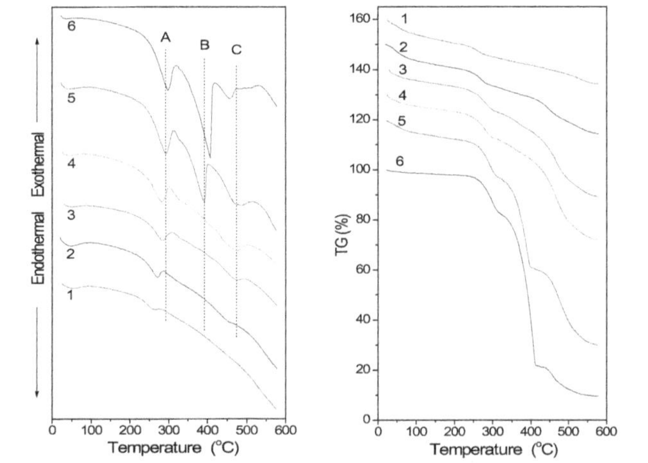

Figure 8: Thermogravimetric (TG) and Differential Scanning Calorimetry (DSC) profiles of 1.0 gram of Sb-SnO_2 exposed to 0.2 g (1), 0.4 g (2), 0.8 g (3), 1.2 g (4), 2.4 g (5), and 4.8 g (6). The optimum exposure ratio is 1:1.2 to 2.4 in weight.

As an acid, the sulfated Sb-SnO_2 should be able to release hydrogen proton or have electron pair acceptors to accept molecules bearing electron pair or negative ions in term of Bronsted's proton theory or in the light of Lewis theory of acids and bases. According to the molecular structural forms put forward by Hino et al [1]. however, SO_4^{2-}/ZrO_2 and SO_4^{2-}/SnO_2 are absolutely impossible to show any acidity because group SO_4^{2-} is a conjugate base of H_2SO_4. Based on the derivation of a series of chemical reaction in calcination and through analysis of the possible structures of Sb-SnO_2 solid acid, a more reasonable structure is proposed in Figure 7.

Because the calcination had the group "SO_4" bonded on Sb-SnO_2 nanoparticles and become a stable group "=SO_4" of Sb-SnO_2 but an attached acid radical "SO_4^{2-}", authors believed that the great enhancement of acidity of sulfated Sb-SnO_2 resulted from a number of dangling bonds around group=SO_4. The both oxygen and tin with dangling bond are electron deficient groups and have electron-withdrawing effects, which leads to the transfer of negative charge from=SO_4 to dangling bonds and make =SO_4 a positive group with acidity. So, we proposed the solid superacid of Sb-SnO_2 should be noted as $SO_4^{\oplus}/SnO_2$ that, as a Lewis' acid, owns a great affinity for molecules bearing electron pair or negative ions.

Some further experiments concerning the impregnation of Sb-SnO_2 with ammonium sulfate were conducted using methods of Differential Scanning Calorimetry (DSC) and Thermogravimetry (TG) to study an optimal impregnation ratio of ammonium sulfate to Sb-SnO_2.

The nano-crystalline Sb-SnO_2 powders had to be dispersed in organic solvent since it could not be recovered if it scattered in water. The ammonium sulfate, however, had to be dissolved in water for its solubility in organic solvent is very low. In the impregnation of Sb-SnO_2 powders with ammonium sulfate, a saturated solution of ammonium sulfate was dropped into organic solutions containing a fixed amount of Sb-SnO_2 nano-powder in different ratio. The ammonium sulfate precipitated as it dropped into organic solvent and Sb-SnO_2 nanoparticles coupled the precipitate via water molecules that adsorbed on $(NH_4)_2SO_4$ fine particles. Without free water, for all water molecules were adsorbed on $(NH_4)_2SO_4$ fine particles. The dried powder was a uniform dispersion of powder of $(NH_4)_2SO_4$ fine particles and Sb-SnO_2 nano-particles. The outstanding advantage of the method presented here, that is the impregnation of Sb-SnO_2 powder with saturated ammonium sulfate, is that the impregnation ratio of $(NH_4)_2SO_4$ to Sb-SnO_2 can come to very high and no free water that will cause the aggregation of Sb-SnO_2 nano powder.

To obtain an optimal impregnation ratio of ammonium sulfate to Sb-SnO_2, a series of Sb-SnO_2 nano-powder impregnated with $(NH_4)_2SO_4$ in different ratio were studied on the Simultaneous TG-DSC Apparatus, STA 409PC, NETZSCH, Germany. The resulted analysis diagrams are showed in Figure 8. Here the Sb-SnO_2 nano-powder did not undergo any heat treatment and just impregnated with $(NH_4)_2SO_4$ directly in organic solvent. It can be simply and clearly identified the endothermic peaks and their height from the Thermo gravimetric (TG) and Differential Scanning Calorimetry (DSC) profiles of Sb-SnO_2 impregnated with different amount of $(NH_4)_2SO_4$. The height of peaks told us if there were excessive or deficient $(NH_4)_2SO_4$, by which an optimal impregnation ratio of ammonium sulfate to Sb-SnO_2 was easily discovered. We have got the knowledge of what the peaks implied based on foregoing discussion, that is, the Peak A was an endothermic peak that caused by $(NH_4)_2SO_4$ being resolved into NH_3 and N_2SO_4, the Peak B an endothermic one which resulted from the decomposition of H_2SO_4, and the Peak C, without a doubt, was brought about by the absorption of heat contributed by the dissociation of a newly generated group=SO_4. The Peak A was always presented in curves of all samples since the decompositions of $(NH_4)_2SO_4$ occurred for all samples but were different in their peak height due to different impregnation ratio, whereas, the Peak B only appeared as the amount of $(NH_4)_2SO_4$, or exactly H_2SO_4, was excessive against superficial hydroxyl of Sb-SnO_2 because only the H_2SO_4 that did not associate with superficial hydroxyl would decomposed.

Obviously, the optimal impregnation ratio should be located between curves 4 and 5 in Figure 8 because curve 5 has a large endothermic peak but curve 4 does not. To diagnose a more accurate optimum ratio of ammonium sulfate to Sb-SnO_2, the authors carried out a series precise experiments in the small range of ratio of Sb-SnO_2 to $(NH_4)_2SO_4$ from 1:1.2 to 1:2.4 wt%. The quantitative analyses of the ratio were conducted by the methods of differentia scanning calorimetry and themogravimetry. The thermal analysis curves, especially DSC curves, in Figure 9, showed the ratio at 1:1.2 wt% did not have endothermic peak B, suggesting impregnated $(NH_4)_2SO_4$ was not enough against the superficial hydroxyl, and the ratios at 1:1.6, 1:2.0 and 1:2.4 wt% all had projecting endothermic peaks at peak B, meaning impregnated $(NH_4)_2SO_4$ were excessive.

In the preparation of solid superacid of Sb-SnO_2 nanocrystal, the working ratio was selected at 1:1.4 to 1:1.6 wt%, a little excessive, in order to make full use of the superficial hydroxyl of Sb-SnO_2 and get more superacid group=SO_4.

Figure 10 is a picture of solid superacid of Sb-SnO_2. The picture shows the deficiency (1:0.8 wt%) or excess (1:2.4 wt%) of impregnated $(NH_4)_2SO_4$ would result in the blue solid superacid of Sb-SnO_2 with

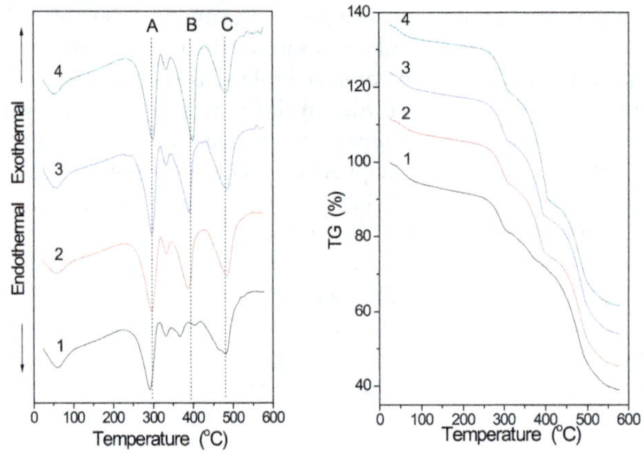

Figure 9: Thermogravimetric (TG) and Differential Scanning Calorimetry (DSC) profiles of 1.0 gram of Sb-SnO_2 exposed to 1.2 g (1), 1.6 g(2), 2.0 g (3) and 2.4 g (4). The optimum exposure ratio is 1:1.2 to 1.6 in weight.

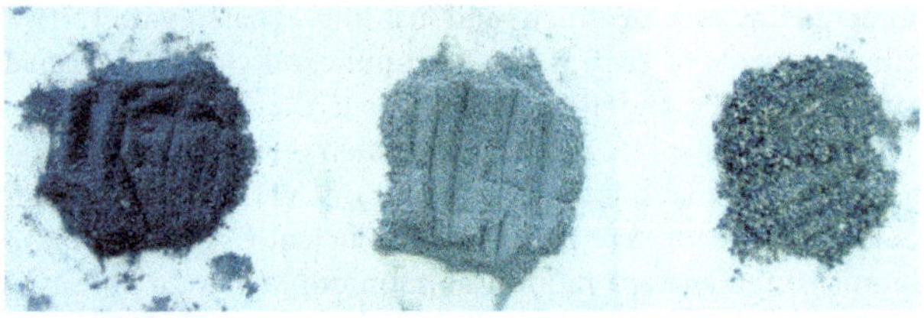

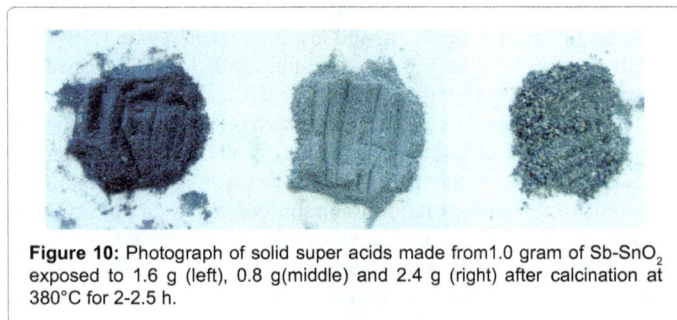

Figure 10: Photograph of solid super acids made from1.0 gram of Sb-SnO_2 exposed to 1.6 g (left), 0.8 g(middle) and 2.4 g (right) after calcination at 380°C for 2-2.5 h.

little light yellow, while, a suitable or little excessive amount (1:1.4 and 1:1.6 wt%) of $(NH_4)_2SO_4$ could get a solid superacid of Sb-SnO_2 with an even blue after the calcination at 380°C. See the left one in Figure 10.

Conclusion

The nano-crystalline SnO_2 doped with Sb (III) that was synthesized by method of "P-CNAIE" and the drying method of "AD-IAA" could be calcined to give a solid superacid after sulfated. It must be noted that a suitable amount of impregnated ammonium sulfate and a calcination temperature adapting to SnO_2 are crucial to prepare the solid superacid of Sb-SnO_2, which is written as $SO_4^{\oplus}/SnO_2$.

This paper proposed the impregnated ratio of Sb-SnO_2 to $(NH_4)_2SO_4$ should be between 1 g to 1.4 g and 1.6 g and the calcination temperature be 380°C to 400°C, and hence an even blue solid superacid powder was obtained. About the catalytic properties of Sb-SnO_2 solid superacid will be discussed in another paper.

Acknowledgement

The authors would like to extremely thank the "Chun Hui" Project from the Ministry of Education of China for funding the researches.

References

1. Hino M, Kobayashi S, Arata K (1979) Solid catalyst treated with anion. 2. Reactions of butane and isobutane catalyzed by zirconium oxide treated with sulfate ion. Solid superacid catalyst. J Am Chem Soc 101: 6439-6440.
2. Arata K (1990) Solid Superacids.Advances in Catalysis 37: 165-211.
3. Arata K (1996) Preparation of superacids by metal oxides for reactions of butanes and pentanes.Appl Catal A: General 146: 3-32.
4. Matsuhashi H, Arata K (2006) Adsorption and desorption of small molecules for the characterization of solid acids. Catalysis Surveys from Asia 10: 1-7.
5. Liu F, Martin-Mingot A, Jouannetaud MP, Bachmann C, Frapper G, et al. (2011) Selective synthesis of gem-chlorofluorinated nitrogen-containing derivatives after superelectrophilic activation in superacid HF/SbF5.J Org Chem 76: 1460-1463.
6. Liu F, Martin-Mingot A, Jouannetaud MP, Zunino F, Thibaudeau S (2010) Superelectrophilic activation in superacid HF/SbF(5) and synthesis of benzofusedsultams. Org Lett 12: 868-871.
7. Matsuhashi H, Hino M, Arata K, Tanabe K, Hattori H, et al.(1989) Acid-Base Catalysis. Kodansha Press, Tokyo, Japan, pages 357.
8. Matsuhashi H, Hino M, Arata K (1990) Solid catalyst treated with anion: XIX. Synthesis of the solid superacid catalyst of tin oxide treated with sulfate ion. Appl Catal 59: 205-212.
9. Yamato T, Hideshima C, Surya Prakash GK, Olah GA (1991) Solid superacid-catalyzed organic synthesis. 4. Perfluorinatedres in sulfonic acid (Nafion-H) catalyzed Friedel-Crafts benzylation of benzene and substituted benzenes. J Org Chem 56: 2089-2091.
10. Au C, Wang Z, Ji W, Liu KC, Dai H (2006) Activation and conversion of propane over nano-sized MnOx-Moy(M=Cu, V) promoted sulfated yttria-stabilized zirconia-titania catalysts.React Kinet Catal Lett 87: 59-66.
11. Matsuhashi H, Miyazaki H, Kawamura Y, Nakamura H,Arata K (2001) Preparation of a Solid Superacid of Sulfated Tin Oxide with Acidity Higher Than That of Sulfated Zirconia and Its Applications to Aldol Condensation and Benzoylation. Chem Mater 13: 3038-3042.
12. Xiang B, Xu H, Li W (2008) Effect of calcination temperature on the performance of nano-size iron oxide catalysts for ethylbenzene dehydrogenation.React Kinet Catal Lett94: 175-182.
13. Zhang X, Liang H, Gan F (2006) Novel Anion Exchange Method for Exact Antimony Doping Control of Stannic Oxide Nanocrystal Powder. J Am Ceram Soc 89: 792-798.
14. Yang F, Zhang X, Mao X, Gan F (2007) Synthesis and Characterization of Highly Dispersed Antimony-Doped Stannic Hydroxide Nanoparticles: Effects of the Azeotropic Solvents to Remove Water on the Properties and Microstructures of the Nanoparticles.J Am Ceram Soc 90: 1019-1028.
15. Sun Q, Gao ZX, Wen B, Sachtler WMH (2002) Spectroscopic Evidence for a Nitrite Intermediate in the Catalytic Reduction of NOx with Ammonia on Fe/MFI. Catal Lett 78: 1-4.

Permissions

All chapters in this book were first published in MCA, by OMICS International; hereby published with permission under the Creative Commons Attribution License or equivalent. Every chapter published in this book has been scrutinized by our experts. Their significance has been extensively debated. The topics covered herein carry significant findings which will fuel the growth of the discipline. They may even be implemented as practical applications or may be referred to as a beginning point for another development.

The contributors of this book come from diverse backgrounds, making this book a truly international effort. This book will bring forth new frontiers with its revolutionizing research information and detailed analysis of the nascent developments around the world.

We would like to thank all the contributing authors for lending their expertise to make the book truly unique. They have played a crucial role in the development of this book. Without their invaluable contributions this book wouldn't have been possible. They have made vital efforts to compile up to date information on the varied aspects of this subject to make this book a valuable addition to the collection of many professionals and students.

This book was conceptualized with the vision of imparting up-to-date information and advanced data in this field. To ensure the same, a matchless editorial board was set up. Every individual on the board went through rigorous rounds of assessment to prove their worth. After which they invested a large part of their time researching and compiling the most relevant data for our readers.

The editorial board has been involved in producing this book since its inception. They have spent rigorous hours researching and exploring the diverse topics which have resulted in the successful publishing of this book. They have passed on their knowledge of decades through this book. To expedite this challenging task, the publisher supported the team at every step. A small team of assistant editors was also appointed to further simplify the editing procedure and attain best results for the readers.

Apart from the editorial board, the designing team has also invested a significant amount of their time in understanding the subject and creating the most relevant covers. They scrutinized every image to scout for the most suitable representation of the subject and create an appropriate cover for the book.

The publishing team has been an ardent support to the editorial, designing and production team. Their endless efforts to recruit the best for this project, has resulted in the accomplishment of this book. They are a veteran in the field of academics and their pool of knowledge is as vast as their experience in printing. Their expertise and guidance has proved useful at every step. Their uncompromising quality standards have made this book an exceptional effort. Their encouragement from time to time has been an inspiration for everyone.

The publisher and the editorial board hope that this book will prove to be a valuable piece of knowledge for researchers, students, practitioners and scholars across the globe.

List of Contributors

Snezana Agatonovic-Kustrin, Anindita Chakrabarti and David W Morton
School of Pharmacy and Applied Science, La Trobe Institute of Molecular Sciences, La Trobe University

Pauzi A Yusof
Physiology Department, Medical School, Universiti Teknologi Mara, Selangor, Malaysia

Mohamed I Aref
Clinical pathology Department, Al Azhar University, Cairo, Egypt

Hamdy Ahmed
Biochemistry of National Research Center (NRC), Al Azhar University, Egypt

Salem E Samra
Chemistry Department, Faculty of Science, Mansoura University, Mansoura, Egypt

Bakir Jeragh
Chemistry Department, Faculty of Science, Kuwait University, Kuwait

Ahmed M EL-Nokrashy
Central Laboratory of drinking water, Dakahliya Comp. for Water, Mansoura, Egypt

Ahmed A El-Asmy
Chemistry Department, Faculty of Science, Mansoura University, Mansoura, Egypt Chemistry Department, Faculty of Science, Kuwait University, Kuwait

Prabal Giri
Department of Chemistry, Guskara Mahavidyalaya, Burdwan 713128, West Bengal, India

Churala Pal
Department of Chemistry, Basanti Devi College, Kolkata 700029, West Bengal, India

Million Mulugeta and Belisti Lelisa
Department of Chemistry, Arba Minch University, Arba Minch, Ethiopia

Urmi Roy, Alisa G Woods, Izabela Sokolowska and Costel C Darie
Department of Chemistry and Biomolecular Science, Clarkson University, USA

Marcin Konior and Edward Iller
National Centre for Nuclear Research, Radioisotope Centre Polatom, 05-400 Otwock, Andrzej Sołtan 7, Poland

Miroslava Kačániová, Jana Petrová, Simona Kunová, Peter Haščík and Ľubomír Lopašovský
Faculty of Biotechnology and Food Sciences, Slovak University of Agriculture in Nitra, Nitra, Slovak Republic

Maciej Kluz
Department of Biotechnology and Microbiology, University of Rzeszow, Rzeszow, Poland

Martin Mellen
Hydina Slovakia, s.r.o., Nová Ľubovňa 505, Nová Ľubovňa 065 11, Slovakia

Saeed Arayne M
Department of Chemistry, University of Karachi, Pakistan

Najma Sultana, Sana Shamim and Asia Naz
Research Institute of Pharmaceutical Sciences, Faculty of Pharmacy, University of Karachi, Pakistan

Nadana Shanmugam, Balan Saravanan, Rajaram Reagan, Natesan Kannadasan, Kannadasan Sathishkumar and Shanmugam Cholan
Department of Physics, Annamalai University, Annamalai Nagar, Chidambaram 608 002, Tamilnadu, India

Mahendra Kumar Trivedi, Shrikant Patil and Harish Shettigar
Trivedi Global Inc., 10624 S Eastern Avenue Suite A-969, Henderson, NV 89052, USA

Ragini Singh and Snehasis Jana
Trivedi Science Research Laboratory Pvt. Ltd., Hall-A, Chinar Mega Mall, Chinar Fortune City, Hoshangabad Rd., Bhopal- 462026, Madhya Pradesh, India

Brian Gulson
Graduate School of the Environment, Macquarie University, Sydney, Australia
Commonwealth Scienti ic and Industrial Research Organisation (CSIRO), Earth Science and Resource Engineering, Sydney, Australia

Alan Taylor
Department of Psychology, Macquarie University, Sydney, Australia

Apoorva V Rudraraju, Mohammad F Hossain, Anjuli Shrestha, Prince NA Amoyaw and Faruk Khan MO
College of Pharmacy, Southwestern Oklahoma State University, 100 Campus Drive, Weatherford, Ok 73096, USA

Babu L Tekwani
National Center for Natural Products Research, University of Mississippi, University, MS 38677, USA

Nagwa ABO EL-Maali and Asmaa Yehia Wahman
Department of Chemistry, Faculty of Science, Assiut University, Assiut, Egypt

Mahmoud Mohamed Alou-El-Makarem, Moussa Madany Moustafa, Mohamed Abdel-Aziz Fahmy and Aamer Mohamed Abdel-Hamed
Department of Medical Biochemistry, Al Azhar University, Cairo, Egypt

Medhat Mohamed Abdel-Salam Darwish
Faculty of Medicine, Al Azhar University, Damietta, Cairo, Egypt

Khaled Nagy El-fayomy
Department of Internal Medicine, Al Azhar University, Damietta, Cairo, Egypt

Snezana Agatonovic-Kustrin
Faculty of Pharmacy, Universiti Teknologi MARA (UiTM), Bandar Puncak Alam, Selangor 42300, Malaysia

David Babazadeh Ortakand and David W Morton
School of Pharmacy and Applied Science, La Trobe Institute of Molecular Sciences, La Trobe University, Bendigo 3550, Australia

Lutsyk VI
Institute of Physical Materials Science, Siberian Branch of Russian Academy of Sciences, Russia
Buryat State University Ulan-Ude, Russia

Vorob'eva
Institute of Physical Materials Science, Siberian Branch of Russian Academy of Sciences, Russia

Abdul Rafiq Khan, Ali Al-Othaim, Shazia Mrtaza, Sara Altraif, Waleed Tamimi and Ibrahim Altraif
National Guard Health Affairs, Department of Pathology and Laboratory Medicine, Biochemical Metabolic Laboratory, Riyadh, Kingdom of Saudi Arabia

Khalid Muhammed Khan
Haji Ibrahim Jamal Research Institute of Chemistry, University of Karachi, Karachi, Pakistan

Waqas Jamil
University of Jamshoro, Jamshoro, Sindh, Pakistan

Hekmat F and Rahmanifar MS
Faculty of Basic Science, Shahed University, Tehran, Iran

Sohrabi B
Department of Chemistry, Surface Chemistry Research Laboratory, Iran University of Science and Technology, Iran

Samina K Tadavi, Jamatsing D Rajput, Suresh D Bagul and Ratnamala S Bendre
School of Chemical Sciences, North Maharashtra Jalgaon, Maharashtra, India

Amar A Hosamani
Solid State and Structural Chemistry Unit, Indian Institute of Science, Bangalore, Karnataka, India

Jaiprakash N Sangshetti
YB Chavan College of Pharmacy, Dr. Rafiq Zakaria Campus, Aurangabad, Maharashtra, India

Fakhra Jabeen and M Sarfaraz Nawaz
Department of Chemistry, Jazan University, Jazan, Saudi Arabia

Alexander S
NRC Kurchatov Institute, Moscow, Russia

Nekhoroshkov PS and Frontasyeva MV
Frank Laboratory of Neutron Physics, Joint Institute for Nuclear Research, Russia

Kravtsova AV and Yermakov IP
Faculty of Biology, Lomonosov Moscow State University, Moscow, Russia

Kamnev AN and Bunkova OM
Faculty of Soil Science, Lomonosov Moscow State University, Moscow, Russia

Duliu O
Department of the Structure of Matter, Faculty of Physics, University of Bucharest, Romania

Mahdi H
College of Engineering, University of Tehran, Tehran, Iran

Davood M
Faculty of Engineering, University of Zanjan, Zanjan, Iran

Mohsen V and Behzad S
Research and Engineering Company for Non-Ferrous Metals, Zanjan, Iran

Ahmed Fawzy and Saleh A. Ahmed
Chemistry Department, Faculty of Applied Science, Umm Al-Qura University, 21955 Makkah, Saudi Arabia
Chemistry Department, Faculty of Science, Assiut University, 71516 Assiut, Egypt

Ishaq A. Zaafarany, Rabab S. Jassas and Rami J. Obaid
Chemistry Department, Faculty of Applied Science, Umm Al-Qura University, 21955 Makkah, Saudi Arabia

Yahya M Hodeeb, Hassan M Hassan and Dina A Samy
Department of Dermatology, Veneriology and Andrology, Faculty of Medicine, Al Azhar University, Cairo, Egypt

Amal M Al Dinary
Department of Community Medicine, Faculty of Medicine, Al Azhar University, Cairo, Egypt

Mohammad Faisal Hossain, Cassandra Obi, Anjuli Shrestha and MO Faruk Khan
SCRiPS, College of Pharmacy, Southwestern Oklahoma State University, USA

Moosavifar M and Jafarbahmani M
Department of Chemistry, Faculty of Science, University of Maragheh, Iran

Haile Hassena
Department of Chemistry, Hawassa College of Teacher Education, Hawassa, Ethiopia

Hasan A
Department of Chemistry, Faculty of Science, Gazi University, Ankara, Turkey

VT Vader
Walchand College of Arts and Science, Solapur, Maharashtra, India

Hamdy A Salem, Shadia E Abdel-Aziz and Noeman B Aref
Department of pests and plant protection, Division of Agriculture and Biological Research, National Research Center, Dokki, Cairo, Egypt

Mahendra Kumar Trivedi, Alice Branton, Dahryn Trivedi and Gopal Nayak
Trivedi Global Inc., 10624 S Eastern Avenue Suite A-969, Henderson, NV 89052, USA

Khemraj Bairwa and Snehasis Jana
Trivedi Science Research Laboratory Pvt. Ltd., Hall-A, Chinar Mega Mall, Chinar Fortune City, Hoshangabad Rd., Bhopal, Madhya Pradesh, India

Qin-Qin Ran and Han-Mei Ouyang
Guizhou Province Key Laboratory of Fermentation Engineering and Biological Pharmacy, Guizhou University, Guiyang, China

Xuejun Zhang and Han Tao
Guizhou Province Key Laboratory of Fermentation Engineering and Biological Pharmacy, Guizhou University, Guiyang, China
School of Brewing and Food Engineering, Guizhou University, Guiyang 550025, China

Xiao-Ning Zhang and Hui Zhong
School of Brewing and Food Engineering, Guizhou University, Guiyang 550025, China

Index

www.ingramcontent.com/pod-product-compliance
Lightning Source LLC
Chambersburg PA
CBHW080629200326
41458CB00013B/4559